Electron Beam X-Ray Microanalysis

Electron Beam X-Ray Microanalysis

Kurt F. J. Heinrich

VNR VAN NOSTRAND REINHOLD COMPANY
NEW YORK CINCINNATI ATLANTA DALLAS SAN FRANCISCO
LONDON TORONTO MELBOURNE

Van Nostrand Reinhold Company Regional Offices:
New York Cincinnati Atlanta Dallas San Francisco

Van Nostrand Reinhold Company International Offices:
London Toronto Melbourne

Library of Congress Catalog Card Number: 80-607032
ISBN: 0-442-23286-1

Manufactured in the United States of America

Published by Van Nostrand Reinhold Company
135 West 50th Street, New York, N.Y. 10020

Published simultaneously in Canada by Van Nostrand Reinhold Ltd.

15 14 13 12 11 10 9 8 7 6 5 4 3 2 1

Library of Congress Cataloging in Publication Data

Heinrich, Kurt. F. J.
Electron beam x-ray microanalysis.

Includes index.
1. Electron probe microanalysis. 2. X-ray spectroscopy. I. Title.
QD98.E4H44 543'.0812 80-607032
ISBN 0-442-23286-1

Da sich hieraus ergiebt, dass die chemische Analyse nicht nur ein Wissen, sondern auch ein Können erfordert, so liegt der Schluss nahe, dass eine bloss geistige Beschäftigung damit, eben so wenig als ein rein empirisches Betreiben derselben, zum Ziele führen kann, und dass dahin nur die vereinigten Wege der Theorie und Praxis gelangen lassen.

(It follows that chemical analysis requires not only knowledge but also skill; one is led to the conclusion that neither pure speculation nor mere empirical action can lead to the goal, which can be reached only by combining theory and practice.)

C. Remigius Fresenius
Anleitung zur Qualitativen Chemischen Analyse
Braunschweig, 1856.

Preface

The exciting discussions of the theory of electron probe analysis in the sixties and early seventies have now abated, and the controversies about "empiricism" versus "theory," and concerning the best value of the "X-ray takeoff angle" are a thing of the past. On the other hand, the application of the lithium-drifted silicon detector to scanning and transmission electron microscopes has greatly broadened the scope of electron probe microanalysis. Old wine is thus served in new bottles, and at times the flavor has suffered in the transfer. This seems therefore a good time for presenting to the users and students of X-ray microanalysis what I hope is a reasonably complete, consistent, and critical description of the theory and practice of electron probe microanalysis and of its physical and instrumental foundations. Concerning the fields of application, I do not claim complete coverage. Rather, my aim was to refer the reader to the relevant literature.

To render the treatment self-consistent, I have found it necessary to somewhat modify the current usage of symbols. For this change, my apologies go to those who will miss familiar landmarks such as $f(\chi)$ and $\phi(\rho z)$.

In many instances, the order of presentation of the material was determined by its practical use. For this reason, electron stopping power and backscattering, as well as the ionization cross section, are discussed in Chapter 9 which deals with the primary X-ray generation calculation, while fluorescent yields and relative line intensities are found in Chapter 4, and the continuous X-ray excitation is described in Chapter 6 in which the background correction is discussed. I hope that the reader, aided by the index where required, will find this arrangement acceptable.

The title of this book has been changed from "Electron Probe Microanalysis" to its present form in order to indicate that it covers the analysis of X-rays produced in electron microscopes, while analysis by means of Auger electrons is at present excluded.

I wish to express my thanks to the many friends and colleagues who have aided and encouraged me in this task, including foremost J. Henoc, H. Yakowitz, D. E. Newbury, C. E. Fiori, and R. L. Myklebust. The shortcomings of the book are my own. I also thank my family and others who have tolerated my short temper during the period in which this book was written.

K. F. J. HEINRICH
National Bureau of Standards
Washington, D.C.

Contents

List of Symbols

(Equations or figures in which Symbols are defined or first used are shown in in parenthesis)

A	ampere (3.2.18)
Å	angstrom (2.1), (see also p. 68)
A_i	atomic weight of element i (4.3.7)
a	area (9.2.8)
$a, b, c \ldots i$	symbols denoting elements (4.3.10)
a, a_1, a_2	constaints in absorption calculation (10.3.32)
a_p	particle cross-section
B	parameter in Hutchins' expression for analysis of thin film (13.3.8)
b	brightness (3.2.19), constant in expression for fluorescence yield (4.2.5)
b_q	constant in Bethe's expression for ionization cross-section (9.2.10)
C	capacitance
C_i	mass fraction of element i (4.3.8)
C_s	spherical aberration constant (3.2.8)
C_t	thermal conductivity (16.3.1)
c	speed of light ($c = 2.99793 \times 10^8$ m/sec) (4.2.1) unspecified constant
D	diameter of lens bore (3.2.6)
d	beam diameter (3.2.9), spacing between diffracting planes (4.2.4)
E	electrostatic field (5.3.6), energy of a photon, kinetic energy of an electron (2.1)

$\bar{E}$	average electron energy
E_o	initial electron energy (Fig. 2.5, 6.1.1)
E_b	energy of backscatter electron (p. 249)
E_q	critical ionization energy (4.2.3)
E_{qm}	energy of a photon of shell q, line m (6.1.1)
e, e^-	charge of electron (5.3.4, p. 67)
e	basis of natural logarithms (5.3.1)
(e)	equivalent (13.5.2)
erf	error function (10.3.57)
exp	exponential (3.2.13)
eV	electron volt (4.2.2)
F	focal point, focal plane (figs. 3.9, 3.15)
F	unspecified function (e.g., 12.1.1), Fano factor (5.3.8)
F_A	absorption correction factor (9.1.8)
F_F	fluorescence correction factor (9.1.7)
F_Z	atomic number correction factor (9.1.9)
$F(a)$, $F(p)$, $F(V)$, $F(\mu)$, $F(\psi)$	factors entering the fluorescence equation (11.1.41)
$F(\chi)$	absorption correction of Castaing & Descamps (10.1.8)
FWHM	full width at half maximum (5.3.12)
f	focal length (3.2.4), unspecified function (3.g., 10.1.17) (X-ray) absorption factor, X-ray attenuation factor (8.1.2)
f_{12}, f_{13}, f_{23}	Coster-Kronig transition probabilities (4.2.7)
f_{ex}	experimental X-ray absorption factor (10.3.36)
f_f	absorption factor for fluorescent radiation (8.2.8)
f_p	absorption factor for primary radiation (8.2.8)
f_y	correction factor for X-ray loss due to electron sidescatter (13.5.7)
f_z	correction factor for X-ray loss due to electron transmission (13.5.7)
$f(t)$	absorption factor for thin film (13.3.26)
$F'_a{}^*(k_a)$	analytical function relating C_a to k_a in a multielement specimen (12.1.3)
G	parameter in fluorescence calculation (11.1.23)
g	parameter in fluorescence calculation (11.1.27)

H	magnetic field strength (3.2.5)
h	Planck's constant: $h = 6.626 \times 10^{-34}$ J · sec (4.2.1), parameter for composition dependence in Philibert's equation (10.3.9)
I	current in electron lenses (3.2.6), quantum efficiency of X-ray production (6.1.1), beam intensity in Beer's law (4.3.2), signal input (14.4.1)
I_c	quantum efficiency of generation of secondary X-ray photons excited by continuous radiation (9.1.2)
I_e	average energy of X-ray photons produced by one electron (6.1.1)
I_f	quantum efficiency of generation of secondary X-ray photons excited by characteristic radiation (9.1.2)
I_g	gun emission current (3.2.12)
I_p	quantum efficiency of generation of primary X-ray photons (9.1.2)
$I(\lambda), I(E), I(\nu)$	quantum efficiency of continuum generation per unit interval of λ, E, ν · (6.3.5-6.3.7)
$I_e(\lambda), I_e(E), I_e(\nu)$	average energy of continuum generation by one electron, per unit interval of λ, E, ν · (6.3.1-6.3.3)
I_u	quantum efficiency of generation from unsupported layer (p. 259)
$I(z)dz$	quantum efficiency of primary X-ray generation within layer of thickness dz at depth z (10.1.1)
I'	intensity after partial absorption (4.3.2), probability of emergence of photon within the solid angle subtended by the spectrometer from a target, per electron (9.1.1)
$I' \uparrow_{fa}$	intensity of secondary radiation of element a produced above the depth of primary excitation (11.1.26)
$I' \downarrow_{fa}$	intensity of secondary radiation of element a produced below the depth of primary excitation (11.1.35)
i	angle of incidence, in Snell's law (3.2.1), electron beam flux (no. of electrons) (9.2.8)
i_b	beam current (9.3.1)
i_r	backscattered current (9.3.1)
J	joule (4.2.2), mean excitation energy
j	current density of emitted electrons (3.2.13)
$j(t)$	observed ratemeter signal at time t
K	K-level of X-ray emission (Table 4.2)

K	constant in Richardson's equation (3.2.13)
k	unspecified constant (e.g., 3.2.6), Boltzmann's constant (3.2.13), mobility of electron (5.3.6), Kramers' constant (6.3.1), X-ray intensity ratio (7.3.2), exponent in Philibert's equation (10.3.6)
k^s	X-ray intensity ratio for standard (p. 347)
L	L-level of X-ray emission (Table 4.2)
L	auxiliary variable for generation calculation: $L = \Sigma\, m_i Z_i$ (9.2.18)
$\mathcal{L}$	Laplace transform (10.1.9)
l	azimuthal quantum number (Table 4.1)
$li(x)$	logarithmic integral of x (9.2.21)
M	M-level of X-ray emission (Table 4.2)
M	magnification (3.2.8), auxiliary variable (9.2.18)
M_a	mass per area of element a (13.4.1)
m	meter (p. 61)
m	tally for iterations (Figs. 12.1-12.3), mass of electron (3.2.5)
m_i	atomic fraction of element i (8.2.5)
N	number of photons counted in an experiment (6.1.2), number of atoms per unit of area (9.2.8)
N_{av}	Avogadro constant (no. of atoms/mole) (4.3.7)
n	principal quantum number (p. 62, Table 4.1), order of diffraction (4.2.4), exponent in the calculation of μ (4.3.3), number of ionized pairs per electron, in detector (5.3.2)
n_a	number of ionizations of element a per electron (9.2.16)
n_1, n_2	indices of refraction (3.2.1, 3.2.14)
$n_{\text{I}}, n_{\text{II}}, n_{\text{III}}$	ionization probabilities of shells I, II, III. (4.2.7)
NI	magnetic excitation (ampere-turns) (3.2.6)
nm	nanometer (1 nm = 10^{-9} m) (p. 78)
O	output (14.4.2)
P	efficiency of detection (5.3.1), peak diffraction coefficient (p. 102)
P_{ij}	factor in Reed's fluorescence correction (11.1.43)
p	pressure (5.3.6), fraction of photons produced below depth z_r (13.1.3)

p'	fraction of photons produced below depth z'_r (13.1.6)
p_{qm}	relative transition probability (weight of line) of line m, shell q (p. 76)
Q_q	atomic ionization cross-section (9.2.8)
q	general symbol for shell (p. 67), charge collected from detector avalanche (5.3.4)
R, r	radius (3.2.5, 5.1.1, 5.3.3)
R	resistor (3.2.12), integral X-ray reflection coefficient (p. 102), backscatter correction factor (9.2.24), range of a sample (A.13)
R_i	region of interest for a line of element i, (Fig. 12.26)
$R(z)dz$	mean electron path length within the layer dz, at depth z (10.3.3)
r	angle of refraction (Snell's law) (3.2.1), electron reflection coefficient (3.2.13), back diffusion parameter in Philibert's equation (10.3.2)
r_e	path length of electrons (Bethe range) (13.1.1)
r_v	virtual radius (3.2.8)
r_q	absorption jump ratio per edge q (4.3.11)
r^S	line-to-background ratio (8.2.11)
S	lens gap (3.2.6), stopping power (9.2.1), tally (9.2.22)
S_{at}	atomic stopping power (9.2.5)
s	linear thickness of absorber (4.3.2.), estimate of standard deviation (A.11)
s_1, s_2	distances from lens center to object and image (3.2.4)
T	absolute temperature (3.2.13), time constant of ratemeter (6.5.4)
t	time (6.1.2), auxiliary variables for integration, (9.2.19)
U	$(= E/E_q)$ overvoltage (9.2.10)
$\mathrm{u_o}$	$(= E_o/E_q)$ initial overvoltage (9.2.20)
u	absorption parameter in fluorescence correction (11.1.27), normalized variable for Gaussian distribution (A.6)
$\mathrm{u_t}$	circumference of particle (13.3.5)
V	volt
V	accelerating potential of an electron (p. 5), electrostatic potential (3.2.2), coefficient of variation (p. 5.5.9)

$\overline{V}$	mean pulse height of detector distribution
V_o	operating potential (Fig. 3.14, 3.2.12)
V_c	relativistically corrected potential (p. 23)
V_e	height of escape peak (5.3.15)
V_q	critical ionization potential (p. 67)
ν	velocity (5.3.6), parameter in fluorescence correction (11.1.33), (Fig. 5-3, Table 5-3)
W	with of rocking curve at half-maximum intensity
W_o	power of the beam
w	$= E_b/E_o$ (9.3.6)
w_q	$= E_q/E_o = \mathrm{I}/\mathrm{U}_\mathrm{o}$ (9.3.6)
x	path length of electron (g/cm^2), (9.2.1), specimen coordinate parallel to surface (Fig. 3.31)
xu	x-unit (p. 68)
Y	true photon count rate (5.3.17)
Y'	observed photon count rate (5.3.17)
y	specimen coordinate parallel to surface (Fig. 3.31)
y_o, y_i	optical images (3.2.14)
Z	atomic number (4.2.3)
ZAF	a correction scheme involving multiplication factors (p. 223)
z	specimen coordinate normal to surface (Fig. 3.31), mass thickness of absorber (g/cm^2) (4.3.5)
$\overline{z}$	mean depth of x-ray production (10.2.2)
$\widetilde{z}$	effective depth of x-ray production (13.1.9)
$\overline{z}^1$	mean depth of generation of emergent x-rays (13.1.8)
z_d	depth of complete diffusion (10.3.11)
z_q	number of orbital electrons in filled shell or sub-shell (9.2.10)
z_r	depth range of electrons (10.3.58)
α	aperture angle (3.2.7), internal amplification factor of detector (5.3.4), efficiency factor in hyperbolic approximation (8.2.2), screening constant in Rutherford's equation (13.6.4), third central moment of a frequency distribution (A.3)
α_i	coefficient for element i in Castaing's third approximation (9.3.4)

β angle in fluorescence calculation (11.1.2), parameter in Wells' beam diameter calculation (13.2.7), angle of deflexion in electron scattering (13.6.4), fourth central moment of a frequency distribution (A.4)

Γ spread of energy of detector pulses (5.3.14)

γ energy dependence parameter in the absorption calculation (10.3.31), azimuthal electron scattering angle (p. 459, fig. 460), gamma transform parameter (14.4.1).

δ secondary electron emission coefficient (14.3.1)

ϵ energy required for pair formation (5.3.2), error (5.3.25), internal efficiency factor of detector (5.3.7), energy efficiency of x-ray production (6.1.1), probability of escape peak photon formation (5.3.16), $=\sqrt{e/2}$, ($\epsilon = 1.166$) in Bethe's equation (9.2.2), angle of electron beam with specimen surface (9.5.1)

ζ angle of change of electron trajectory (10.3.1)

η electron backscatter coefficient (9.3.1)

θ Bragg angle (4.2.4, 4.4.3)

θ_m temperature rise of irradiated specimen (16.3.1)

Λ mean free path (13.6.1)

λ wavelength (2.1)

μ x-ray mass absorption coefficient (4.3.6), arithmetic mean, (A.1)

$\mu(i, \lambda)$ mass absorption coefficient of absorber i for radiation of wavelength λ (14.3.7)

μ_{at} atomic x-ray absorption coefficient (4.3.4)

μ_l linear x-ray absorption coefficient (4.3.7)

μm micrometer ($1\mu m = 10^{-6}$ m) (Fig. 2.3)

ν frequency (4.2.1)

π $= 3.141529 \ldots$ (3.2.19)

ρ density (4.3.5)

Σ	specimen plane (Figs. 5.9 and 5.11)
σ	screening constant in Moseley's law (4.2.3), standard deviation (5.3.8, A.2), Lenard coefficient of electron beam attenuation (10.3.5)
σ_c	coherent x-ray scattering cross-section (Fig. 4.13)
σ_i	incoherent x-ray scattering cross-section (Fig. 4.13)
σ_r	index of refraction (4.4.4)
σ_E	electron scattering cross-section (13.64)
σ_{el}	electron cross-section for elastic collisions (13.61)
σ_{in}	electron cross-section for inelastic collisions (13.63)
σ_β	angular electron cross-section (13.6.5)
τ	photo electric x-ray cross-section (Fig. 4.13, p. 80), dead-time (p. 144, 5.3.17)
Φ	plane of focal circle (Figs. 5.9 and 5.11)
$\Phi(0)$	ratio of intensities from supported and unsupported films (p. 263, 12.4.4)
$\Phi(z)$	depth distribution function, scaled to $\Phi(0)$. (10.1.8)
ϕ	electronic work function (3.2.13)
$\phi(z)$	depth distribution of primary x-ray generation ($\int_0^\infty \phi(z)dz = 1$) (10.1.11)
χ	$= \mu \csc \psi$ (10.1.6)
ψ	x-ray emergence angle (Fig. 10.1, 10.1.2)
Ω	solid angle covered by the detector (8.2.8), ohm (Fig. 6.13)
ω	angle in diffraction from gratings (4.4.2)
$\omega_{\mathrm{K}}, \omega_{\mathrm{L}} \ldots \omega_{\mathrm{q}}$	fluorescent yield, shell K, L, . . . q (4.2.5)

Subscripts:

A	atomic number correction (9.1.10)
$a, b, c \ldots i, j$	elements (4.3.10)
b	background (6.4.1), beam (9.3.1)
c	continuum (9.1.2)
d	detector (5.3.7)
e	energy (6.1.1)
F	fluorescence correction (9.1.10)
f	fluorescent (9.1.2)
g	Gaussian (13.2.2)

K, L, M . . . Q	shell K, L. M . . . Q (4.3.12)
l	linear (4.3.7)
p	primary
q	shell q (p. 67)
r	range
s	spherical aberration (13.2.), standard (12.3.4)
t	total (6.4.1), film thickness (13.3.1)
u	unsupported (13.3.5)
w	window (5.3.7)
x, y, z	coordinates
Z	atomic number correction (9.1.10)
0	initial
ab	aberration (13.2.1)
at	atomic (4.3.3)
av	Avogadro (4.3.7)
min	minimum (13.2.7)
opt	optimum (p. 428)
pu	pulse pile-up (p. 142)

Superscripts:

*	unknown specimen; multielement (p. 194)
s	standard (7.3.2)
′	attenuated (9.1.1)

Electron Beam X-Ray Microanalysis

1. Introduction

In 1948, Raymond Castaing, then a student of A. Guinier, presented to the University of Paris a doctoral thesis entitled: "Application of Electron Beams to a Method of Local Chemical and Crystallographic Analysis." The instrument he described had been produced by modifying an electron microscope. An electron beam focused to a diameter of less than a micrometer ($1 \mu m = 10^{-6}$ m) was used to excite X-rays within a microscopic region on the specimen surface. Spectral analysis of these X-rays provided information concerning the composition of the excited region. The crystalline structure of the area of impact could also be investigated by means of the Kossel-line X-ray diffraction technique [1.1].

Castaing's *electron probe microanalyzer*[1] has become the most important tool for elemental microanalysis. The significant characteristics of Castaing's technique are its high spatial resolution, its nondestructive nature, the possibility of quantitative application, the wide range of elements that can be determined, and the variety of specimens that can be analyzed.

Before Castaing, Hillier [1.2] had obtained a patent in the United States for a practically identical instrument using, however, an electron energy loss spectrometer, rather than an X-ray spectrometer, in the instrument he built in 1944-1945. An electron probe similar to Castaing's electron probe was developed independently by I. B. Borovsky in the Soviet Union [1.3].

The physical foundations of electron probe microanalysis were known for years before the work of Castaing. Electron optics of high perfection had been developed for the electron microscope; the bases of X-ray spectrochemical analysis were described in great detail, e.g., in the books of Siegbahn [1.4] and von

[1] Castaing's instrument is commonly called the electron probe microanalyzer. In the United States, the abbreviation "electron probe"–a translation of the French "sonde éléctronique"–or "microprobe" is commonly used, while in England, the name "microanalyser" is preferred. The term "X-ray probe" should only be used for instruments employing secondary, or X-ray excited (fluorescence) X-ray emission; the name "microbeam probe," sometimes found in the literature, is redundant, since both "beam" and "probe" are translations of the French word "sonde."

Hevesy [1.5]. Curved crystal X-ray spectrometers had been designed by Johann [1.6] and Johansson [1.7], and a scanning electron microscope was built by von Ardenne in 1938 [1.8]. Castaing's brilliant contribution was the skillful combination of known principles in a device which fulfilled an entirely new function. His instrument contained almost all of the important features found in the electron probe analyzers of more recent construction. He also established principles for quantitative analysis which are still basically valid and used today; he described in detail the application of the Kossel-line technique for X-ray microdiffraction analysis, and he indicated correctly the main potential applications. While many persons have since made valuable contributions to the art of electron probe microanalysis, Castaing is the originator and the most prominent investigator of this technique.

Before delving into technical details, we may ask ourselves if the benefits derived from elemental microanalysis really warrant the construction of so complicated and costly an instrument, not to mention the training and research effort of many investigators. This question is answered by the observation that important properties of many materials depend on their microscopic structure, more than on their overall composition. The dimensions of the electron probe enable us to study grain boundaries and precipitates in alloys, and intracellular structures in biological tissue. The microscopic distribution of certain elements is a key to the past history of lunar rocks and meteorites, and the behavior of materials under adverse conditions—such as submersion in seawater or extreme heat within a gas turbine—can be predicted from corrosion studies in which the microprobe plays a prominent part. Thousands of electron beam instruments are used in research, industrial laboratories, and academic institutions, and, as will be described in this book, indispensable information has been obtained by them from a wide variety of specimens. Thus, it can be said without fear of exaggeration that the electron probe has added a new dimension to the characterization of materials.

1. REFERENCES

1.1 Castaing, R., Thesis, Univ. of Paris, Paris, France, 1951.
1.2 Hillier, J., U.S. Patent 2 418 029 (1947).
1.3 Borovsky, I. B. and Il'in, N. P., *Dokl. Akad. Nauk SSSR* **106**, 655 (1956).
1.4 Siegbahn, M., *The Spectroscopy of X-Rays*, Oxford University Press, London, 1925.
1.5 von Hevesy, G., *Chemical Analysis by X-Rays and Its Application*, McGraw-Hill, New York, 1932.
1.6 Johann, H. H., *Z. Phys.* **69**, 185 (1931).
1.7 Johansson, T., *Z. Phys.* **82**, 507 (1933).
1.8 von Ardenne, M., *Z. Phys.* **109**, 553 (1938).

PART I. THE INSTRUMENT

2. A Summary Description of the Events in an Electron Probe Microanalyzer

Electron probe microanalysis is based on the identification of characteristic X-ray lines emitted by the specimen under electron bombardment—and, hence, of the emitting elements—(qualitative analysis), and on the comparison of the intensity of characteristic X-ray emission from the specimen with that of a standard (quantitative analysis).

The interaction of the electrons with the target atom is a complicated process; several phenomena besides characteristic X-ray emission must be taken into account by the analyst. In order to become acquainted with the foundations of electron probe microanalysis, we will start with a short general and qualitative description of the events which take place in the instrument (Fig. 2.1).

In the electron gun, a V-shaped wire (the filament) is heated by passing a current (filament current) through it to a temperature at which it emits electrons. These electrons are accelerated by applying an electric potential—typically 10 to 20 kV—between the filament and the anode plate which faces the gun.[1] A fraction of the electrons passes through a hole in the center of the anode plate; these electrons are focused into a narrow beam (primary electron beam) by means of two or more electromagnets of special shape called electromagnetic lenses.

The electron beam, focused to 0.1-1 μm in diameter, strikes the surface of the specimen. The electrons, due to their kinetic energy, penetrate the target, colliding along their trajectory with the target atoms. As a consequence of these collisions, both the speed and the direction of the electrons change: they are *decelerated* and *scattered*.

[1] The energy of an accelerated electron, E [measured in electron volts (eV) or kiloelectron-volts (keV)], is numerically equal to the accelerating potential, V [measured in volts (V) or kilovolts (kV)], with minor corrections for the energy of electron emission from the cathode and electron-electron interactions in the beam.

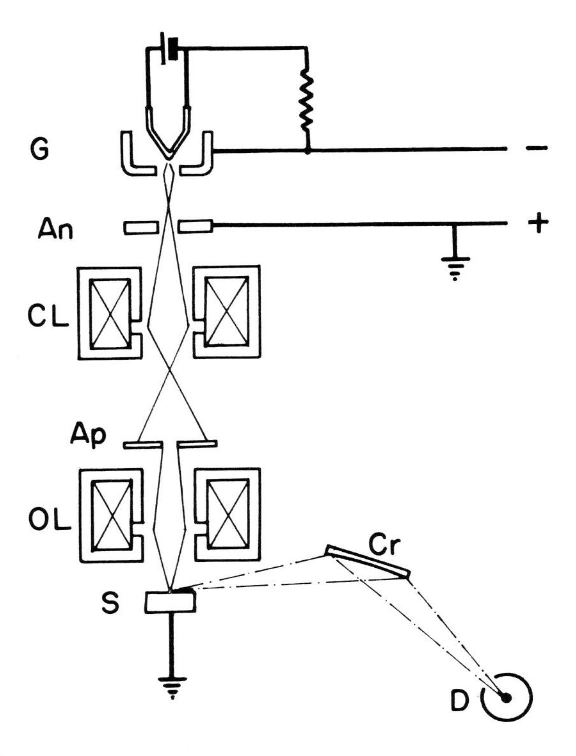

Fig. 2.1. Schematic of an electron probe microanalyzer. G:gun, An:anode plate, CL: condenser lens, Ap:aperture, OL:objective lens, S:specimen, Cr:analyzer crystal, D:Detector. In typical operating conditions, most electrons emitted by the filament are intercepted by the limiting aperture (Ap in Fig. 2.1).

The fate of an individual electron entering the target cannot be predicted since randomness appears in the sequence of electron-atom interactions. Although all impinging electrons have virtually the same direction and energy, after penetration these parameters vary over ranges that can be described by probability (or frequency) distributions. To obtain a quantitative description of the target events–including X-ray emission–we must use simplifying models of action. We frequently give average values to parameters–such as the energy of the electron at a certain depth–which are really spread over a significant range, and we ignore complicating factors such as chemical binding effects and the orientation of the atomic lattice of the target.

In spite of the randomness of the individual events, the range of penetration of monoenergetic electrons in a target is fairly well defined [2.1]. This range can be visually observed in the discharge of electron beams into gases [2.2] since the volume penetrated by the primary electrons (excited volume) emits visible light (Fig. 2.2).

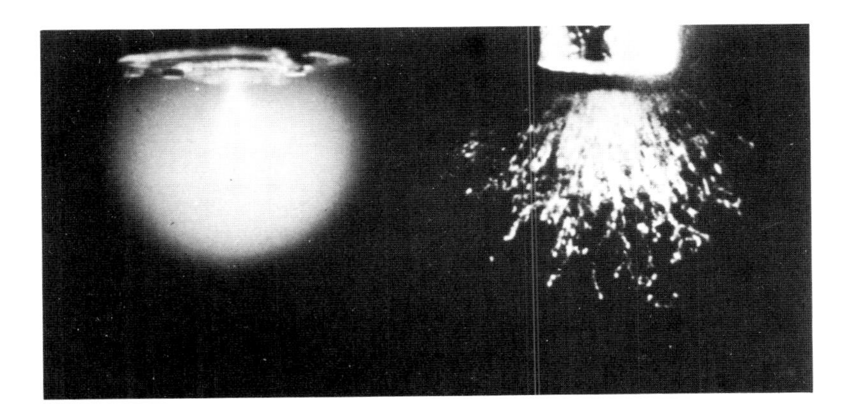

Fig. 2.2. Cathodoluminescence (electron-induced optical emission) produced by an electron beam of 60 keV, in air (left), and individual traces of electrons in a cloud chamber (right). The region of electron-air interaction is sharply defined, and nearly spherical (Schumacher [2.2]).

The depth of penetration of the electrons into a traget depends on the energy of the primary electrons and on the target material. In solids and for the energies used in electron probe analysis, this depth typically ranges from 1 to 10 μm. The depth distribution of the generation of X-rays, which is related to the penetration of electrons, will be discussed in detail in Chapter 10.

The light emission of excited gases shows that the primary electrons diffuse laterally as they penetrate into the specimen. This lateral diffusion is caused by scattering; the combination of penetration and lateral scattering determines the shape of the excited region, which is close to that of a sphere truncated by the specimen surface [2.3]-[2.6] (Fig. 2.3). In elements of high atomic number, the effects of scattering are stronger; the center of the sphere is close to the surface, and a large fraction of the primary electrons is reemitted before their kinetic energy is spent. Such reemerging primary electrons are said to be *backscattered*. In targets of low atomic number, the center of the sphere of electron diffusion is lower, and the fraction of primary electrons which is backscattered (backscatter coefficient, η) is also smaller (Fig. 2.4).

The dependence of the backscatter coefficient on the atomic number of the specimen can be used to quantitatively analyze binary materials (see Section 9.3). It also plays a role in the formation of scanning images by means of the target current (specimen current) or of backscattered electrons (see Chapter 14). But this dependence also affects the intensity of X-ray emission, and must therefore be taken into account in the theory of quantitative electron probe analysis (see Subsection 9.3.1).

Unless otherwise specified, we will assume that the specimen surface is flat and oriented normally to the electron beam. Deviations from either flatness or

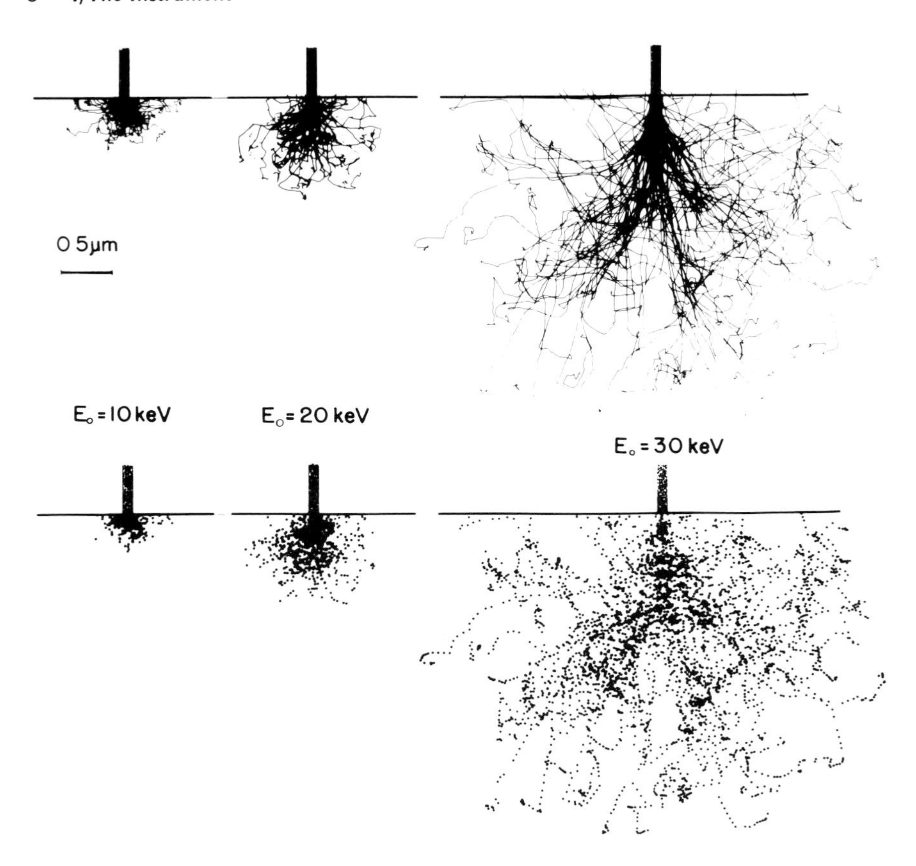

Fig. 2.3. Paths of electrons (above) and sites of primary production of Kα X-ray photons (below) in copper, simulated in a Monte Carlo procedure. The excited volume is roughly spherical and truncated by the specimen surface. The depth of the center of the sphere decreases with increasing atomic number of the target. Note the emergence of backscattered electrons. For the procedure, see Section 13.6.

normal beam incidence cause a change in the backscatter coefficient and in phenomena related to it. Therefore, backscattered electrons (see p. 241) can reveal topographic features as well as differences in atomic number. Quantitative analysis is adversely affected by deviation from flatness, and becomes particularly difficult when the analyzed specimen is smaller—or shallower—than the range of the penetrating electrons.

Besides the backscattered electrons—which have kinetic energies comparable to that of the primary electrons—the specimen surface also emits secondary electrons, with much smaller kinetic energies (see Fig. 2.5), which are produced by quite different mechanisms. At the voltages commonly used in microprobe

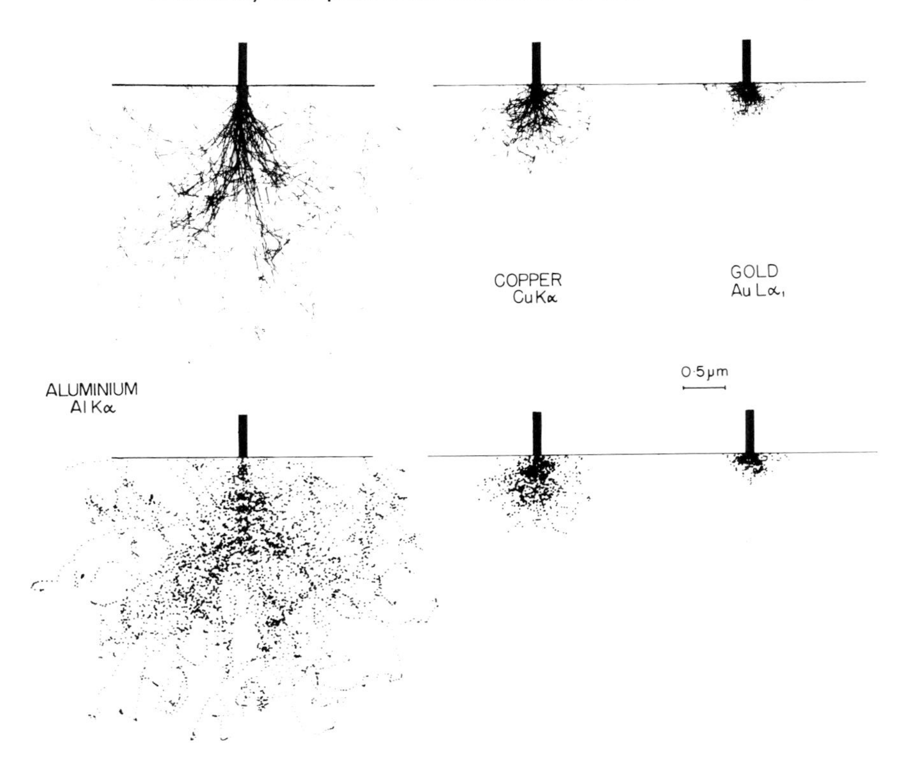

Fig. 2.4. Comparison of electron diffusion and X-ray excitation in targets of aluminum, copper, and gold at 20 kV. If the scale of dimensions were in g/cm^2, rather than μm, the volumes would be of the same order of magnitude for the three elements.

analysis, the sum of the numbers of backscattered and secondary electrons is smaller than the number of primary electrons entering the specimen. Therefore, the irradiated specimen is charged negatively by the electron beam. If it is connected to the positive (ground) side of the power supply used to accelerate the primary electrons, a negative current will flow from the specimen to the ground connection. This current is called *specimen current* or *target current*. Due to backscattering and secondary emission, it is smaller than the beam current. If, however, the beam is trapped in a cavity in the specimen, so that no electrons are reemitted (Faraday cage), then the measured specimen current becomes equal to the *beam current*. The use of a Faraday cage is a simple and accurate way to measure the intensity of the primary beam.

If the specimen is not electrically conductive, local charges build up quickly, disturbing the normal operation, unless a path for the target current has been created by coating the surface of the specimen with a metal or with carbon.

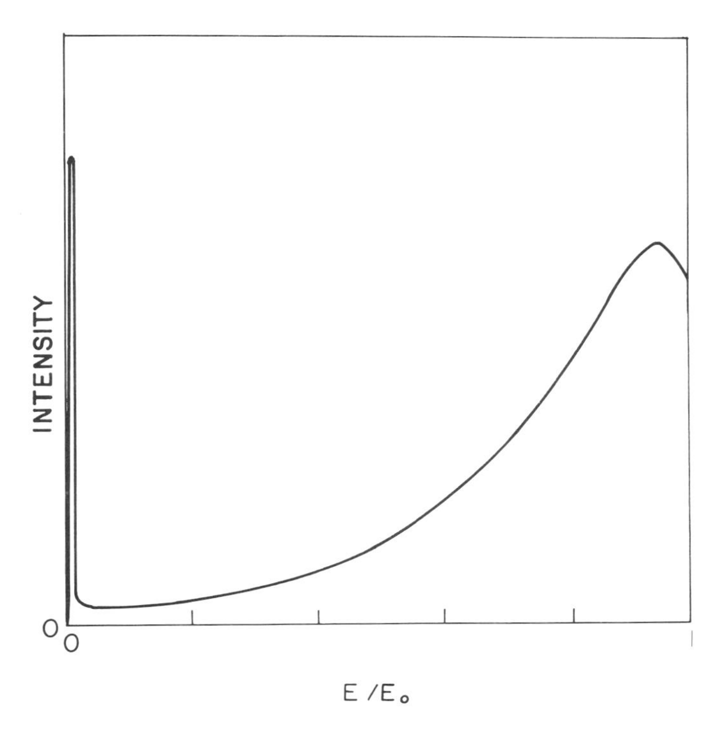

Fig. 2.5. Energy distribution of reemitted electrons. The vertical scale indicates intensity; the horizontal scale indicates energy, E, as a fraction of the energy of the primary electrons, E_0. The narrow peak on the left side is formed by secondary electrons. The broader peak on the right side is formed by backscattered electrons. (Schematic; areas under peaks are not at the same scale.)

Such a coating also protects specimens of poor heat conductivity from overheating on exposure to the beam.

Although most of the energy of the electron beam is converted to heat, other processes, while less frequent, are of more interest to us. The most important process is the ionization of inner shells of target atoms which causes the emission of characteristic X-rays (and of Auger electrons).

X-rays, as all electromagnetic radiation, are emitted as photons, the energy of which, E, is related to the wavelength, λ, of the radiation by the equation

$$\lambda \cdot E = 12\,397 \text{ Å} \cdot \text{eV}. \tag{2.1}$$

If the intensity of photon emission from a source of X-rays is plotted as a function of wavelength, an X-ray spectrum is obtained. The X-ray emission caused by the impact of electrons contains peaks of high intensity within narrow regions of the spectrum. These intense and virtually monochromatic emissions are called *characteristic X-ray lines*; they are superimposed upon a background (continuous radiation) of much lower intensity (Fig. 2.6). The wavelengths of the emitted lines are characteristic of the atoms which emit them, and their relative intensities depend on the mass fractions ("concentrations"[2]) of the emitting elements. Hence, the X-ray emission spectrum provides a basis for qualitative and quantitative elemental microanalysis.

The X-ray lines can be separated and isolated in *X-ray spectrometers*, and their intensities measured by means of *X-ray detectors*. Diverse electronic de-

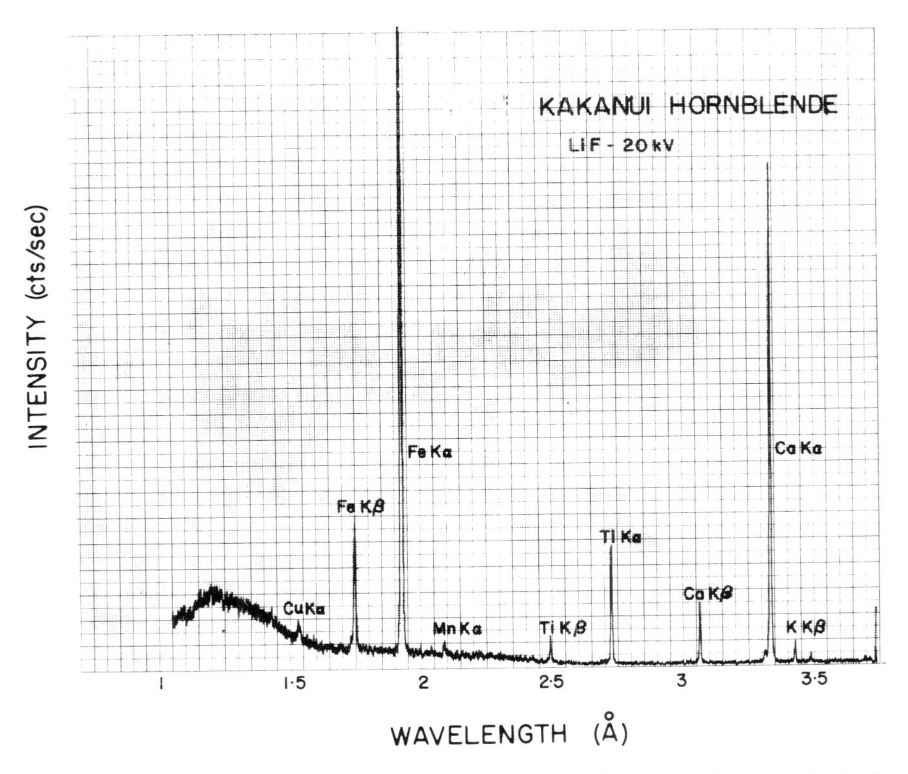

Fig. 2.6. X-ray wavelength spectrum of a mineral, obtained with a lithium fluoride (LiF) (200) crystal spectrometer. The Kα and the Kβ peaks of several elements are identified by the respective symbols. The background is due to continuous radiation. The weight fractions of the elements detected in the spectrum are: potassium (K): 0.017; calcium (Ca): 0.074; titanium (Ti): 0.026; manganese (Mn): 0.0008; iron (Fe): 0.085. The copper peak is an artifact. Present in the specimen, but not emitting lines within the range of wavelength shown are the elements oxygen, sodium, magnesium, aluminum, and silicon.

---

[2] In electron probe microanalysis, composition is customarily expressed in terms of mass fractions ("weight fractions"); these are usually referred to as "concentrations," although in chemistry the term "concentration" more frequently denotes a measure of mass per unit volume.

vices are used to amplify and manipulate in various ways the signals emitted by the detectors (Fig. 2.7).

Many materials–including some minerals–emit brightly colored light when excited by the electron beam. This phenomenon–*cathodoluminescence*–can be an auxiliary source of information on specimen composition.

Electron beams and soft X-rays[3] are strongly absorbed by air. Therefore, their paths must be contained in vacuum enclosures. In some instruments, the same enclosure contains both the electron and the X-ray optics; in other instruments, a window permeable to X-rays separates the spectrometers from the electron optics.

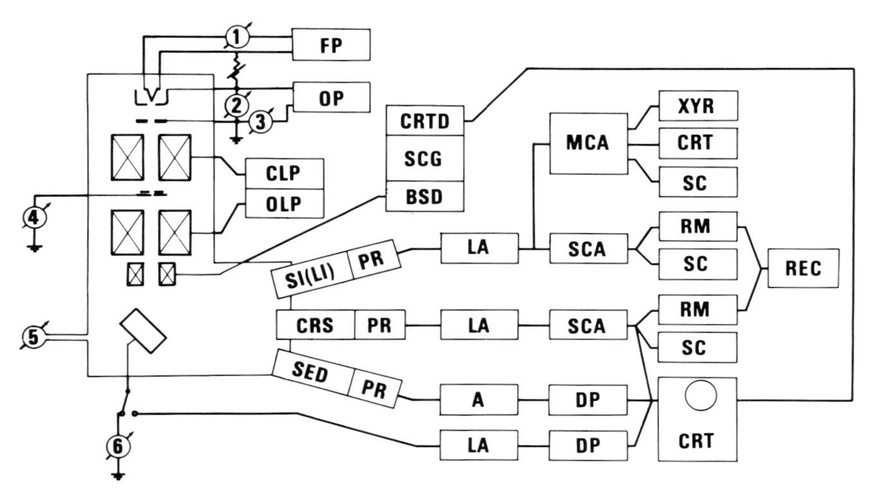

Fig. 2.7. Schematics of electron probe microanalyzer and associated circuitry. Meters: 1, filament current; 2, operating potential; 3, gun current; 4, monitor current; 5, vacuum; 6, specimen current.

A	amplifier
BSD	beam scan driver
CLP	condenser lens supply
CRS	crystal spectrometer
CRT	cathode-ray tube
CRTD	cathode-ray tube driver
DP	data processor
FP	filament power supply
LA	linear amplifier
MCA	multichannel pulse height analyzer
OLP	objective lens power supply
OP	operating potential
PR	preamplifier
REC	recorder
RM	ratemeter
SC	scaler
SCA	single-channel pulse height analyzer
SCG	scan generator
SED	secondary electron detector
SI(LI)	lithium-drifted silicon detector
XYR	*x-y* recorder

For accurate measurement, the operating potential read on meter 2 must be corrected for the drop of potential through the gun bias resistor (above meter 2).

[3] Low- and high-energy X-ray photons are frequently called "soft" and "hard," respectively.

A *mechanical stage* is used to move the specimen with respect to the electron beam. This enables the operator to place the areas of interest of the specimen under the beam. Most instruments also have provisions for electrostatic or electromagnetic deflection of the beam (*electron beam scanning*).

To find the regions on the specimen which one wishes to analyze, one must observe the point of impact of the electron beam on the specimen and its surroundings. This can be done by means of an *optical microscope* which is built into most electron probe microanalyzers, or with images obtained by beam scanning.

The components of the instrument will be discussed in more detail in Chapters 3 and 5. The measurement and the use of the X-ray emission are described in Chapters 6 through 10, and Chapter 12.

2. REFERENCES

2.1 Cosslett, V. E. and Thomas, R. N., in *The Electron Microprobe*, McKinley, T. D., Heinrich, K. F. J., and Wittry, D. B., Eds., John Wiley & Sons, New York, 1966, p. 248.

2.2 Schumacher, B. W., *Proc. First Int. Conf. on Electron and Ion Beam Science and Technology*, Bakish, R., Ed., John Wiley & Sons, New York, 1965, p.5.

2.3 Archard, G. D. and Mulvey, T., *Brit. J. Appl. Phys.* **14**, 626 (1963).

2.4 Bishop, H. E., *Proc. Phys. Soc.* **85**, 855 (1965).

2.5 Curgenven, L. and Duncumb, P., TI Res. Rep. 303, Tube Investment Ltd., Hinxton, Saffron Walden, Essex, England, July 1971.

2.6 Duncumb, P., *Proc. 25th Anniv. Meeting of EMAG*, Institute of Physics, London, 1971, p. 132.

3. Probe-Forming Optics

3.1 THE VACUUM

Because electrons are decelerated and scattered by matter, a well-focused electron beam cannot be formed and maintained in air; therefore, the electron-optical section of the electron probe, including the specimen, must be contained in an evacuated enclosure. A vacuum of 10^{-4} to 10^{-6} torr, necessary for the operation of the microprobe, can be achieved by means of an oil-diffusion pump backed up by a mechanical forepump. However, the diffusion of gases at low pressure is slow, and instrument sections distant from the diffusion pump may remain at a much higher pressure than the pressure measured close to the pump connection. Leaks in mechanical feedthroughs in the wall of the enclosure and outgassing specimens may contribute to raising the pressure within the instrument. For these reasons, the capacity of the vacuum pumps should be chosen generously; in some instruments separate exhaust connections are provided for various sections of the electron optics (Fig. 3.1).

The components of residual gases and vapors can be identified with a small mass spectrometer (residual gas analyzer) (Fig. 3.2); such an analysis may indicate the source of vacuum problems. Several types of failure can be associated with specific contaminants. Air leaks can be recognized by the presence, at abnormal levels, of oxygen and nitrogen. The location of the leak can be found by blowing helium at suspected points; the helium leaking into the enclosure can be detected with high sensitivity with relatively simple leak detectors. The presence of water vapor may be due to leaks in cooling lines, to adsorption of moisture on instrument components, or to outgassing of specimens or other objects recently introduced in the vacuum enclosure. Both oxygen and water may cause premature filament failure (3.1).

Organic materials in the residual gas frequently produce conspicuous contamination deposits on the specimen surface upon impact of the electron beam (Fig. 3.3 [3.2]). The mechanism of formation of the contamination is not fully understood. One important source of organic contaminants is the diffusion-

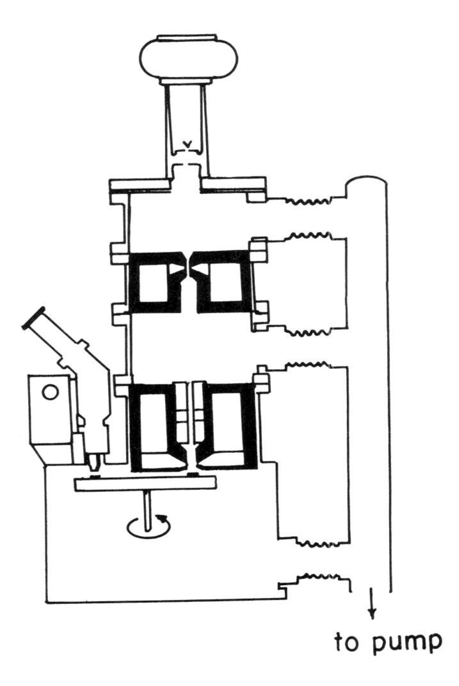

Fig. 3.1. If the lenses and the specimen restrict the passage of residual gases, separate vacuum connections may be provided for various sectors of the vacuum enclosure. In the type of instrument shown here, the specimen cannot be seen during analysis; a turntable is provided which transports it from the position under the electron beam to the optical light microscope.

pump oil. Breakdown is particularly severe when the hot oil makes contact with air. Catastrophic air leaks (for instance during introduction of specimens in the enclosure) can produce cracking and discoloration of the oil, carbonaceous deposits within the diffusion pump, and severe deterioration of the vacuum. If the electron beam is maintained, serious contamination by condensation products of the gun, of apertures, and of the specimen may occur. There are now silicone oils available which are much more resistant to oxidation. Experimentation with novel types of oil may, however, be risky; for instance, Adler et al. reported that the use of sulfur-containing oil caused the deposition of contamination spots which emitted the lines of the sulfur spectrum [3.3].

Contaminants may also emanate from rubber gaskets and other organic materials besides the pump oil. A common contaminant is the argon-methane mixture frequently used in flow-proportional detectors which may seep through the detector window.

The contamination rate depends significantly upon the target material. Substances extracted from the target under electron bombardment can play a role

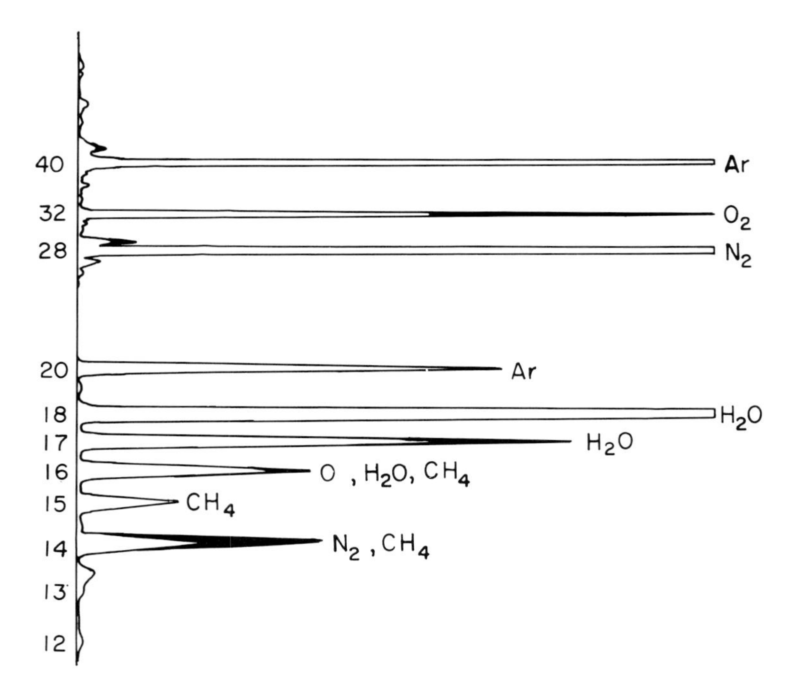

Fig. 3.2. Residual gas mass spectra from the main vacuum enclosure of an electron probe analyzer. Numbers to the left are mass numbers and symbols indicate species producing the peak. White peaks: spectrum from column and spectrometers. Valves connecting to gun and to specimen chamber are closed. Measured pressure: 3×10^{-5} torr. Methane (CH_4) and argon (Ar) are seepage from flow-proportional detectors. Full peaks: same, but valves to gun and specimen chamber are opened. The three strongest peaks are off scale.

in the formation of contamination, particularly if volatile compounds used in the surface preparation have not been removed prior to analysis.

The contamination deposits are frequently used to make the point of impact of the electron beam visible, and to judge the shape and size of the electron beam cross section. Their usefulness is, however, limited to fairly coarse beams, since the contamination spot is larger than the true beam cross section, and increases with exposure time, due to surface migration.

The contamination of the specimen does not cause excessive local electrostatic charges, but even a tenuous deposit can affect the emission of secondary electrons. For this reason, the site of previous exposures is often visible in secondary electron or target current scans (Fig. 3.4). Small amounts of contamination on a metal surface which are not microscopically visible can sometimes be observed after etching, since they protect the underlaying metal from attack by the etchant.

The contamination is particularly undesirable when soft X-rays are observed

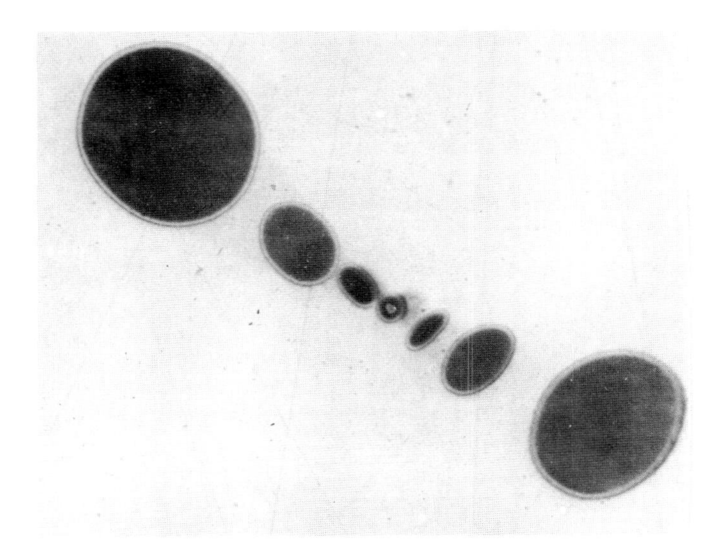

Fig. 3.3. Contamination spots produced by an astigmatic electron beam. Magnification: 680× (see p. 29).

or measured, because it absorbs such radiation. A stream of air or even of inert gases directed upon the point of beam impact can be used to inhibit its formation. Another approach is to eliminate the contaminants from the vicinity of the specimen by condensing them on cooled areas in appropriate locations [3.4]. The best solution to the problem is an improvement in the quality of the vacuum. One of the difficulties in such an effort is that the contamination may be introduced with the specimens; another difficulty is that the seals of the feedthroughs needed for operating spectrometers and specimen stages may emit contaminating substances.

Once good vacuum has been obtained, it should not be disturbed when filaments are changed or specimens introduced. Therefore, it is advisable that the manufacturer provide gates by means of which the gun section and the specimen chamber can be isolated from the rest of the enclosure. Other useful measures are to provide specimen stages for a large number of specimens which can be switched in position without opening the sample chamber, and to keep the vacuum enclosure of the electron-optical column as small as possible (Fig. 3.5).

The quality of the vacuum becomes particularly important when soft X-rays or low-energy electrons are observed. The use of such signals requires lower residual pressures than those routinely obtained in most present-day instruments, and alternatives to the conventional oil-diffusion pump system such as ion or turbopumps may have to be employed.

Fig. 3.4. Traces of area scans on an electropolished iron single crystal. The image was obtained with an electron beam of normal incidence using secondary electrons as image-forming signals. The larger square is 256 μm wide. (Courtesy of H. Yakowitz, NBS.)

In the mechanical design of electron probes and scanning electron microscopes, great care must be taken to avoid vibrations of the instrument caused by the action of the mechanical forepumps. It is preferable that these not be mounted on the same mechanical structure as the electron-optical column.

In the design of the electrical circuits, one must consider the possibility of corona discharges in the electron gun and in the detector assembly as a consequence of vacuum failures. The corresponding circuits should therefore be protected against accidental overloads.

3.2 ELECTRON OPTICS

Because of their negative charge, the movement of free electrons is affected by electrostatic and electromagnetic fields [3.5], [3.6]. As shown by Busch in 1926 [3.7], it is possible to construct devices, based on such interactions, which focus the flow of electrons emanating from a small source toward a point. This effect is similar to the focusing of light by optical lenses. With appropriate field configurations, it is also possible to obtain electron prisms and mirrors. For these reasons, the technology of electron beam manipulation by such devices is called *electron optics*.

The theory and application of electron optics has been studied extensively

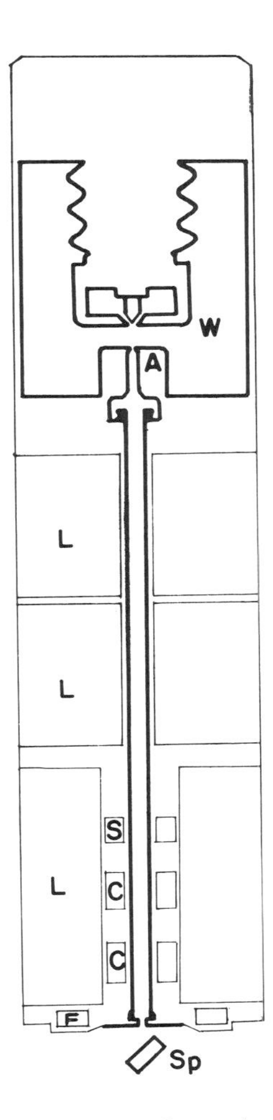

Fig. 3.5. Schematic cross section of the ETEC scanning electron microscope column. W stands for the Wehnelt cylinder. Between the anode plate (A) and the specimen (Sp), the electron beam is confined within a narrow tube. The lenses (L), stigmator coils (S), scanning coils (C), and coils for the compensation for change of working distance during scans (F) are outside the vacuum enclosure.

by the developers of transmission electron microscopy [3.6]. A full treatment of this subject is beyond the scope of this book [3.5]. Because most electron probe users do not build their instruments, we will limit ourselves to the basic knowledge needed for the proper use and maintenance of the electron probe microanalyzer.

The essential properties of optical lenses are related to Snell's law, which describes the change of direction of a beam of light when passing from one medium into another:

$$\frac{\sin\ i}{\sin\ r} = \frac{n_2}{n_1}. \tag{3.2.1}$$

Here, i is the angle of incidence, r is the angle of refraction, and n_1, n_2 are the refractive indices of the two media (Fig. 3.6).

Similarly, we can describe the change of direction of an electron beam passing from a region of electrostatic potential V_1 into another of potential V_2, by

$$\frac{\sin\ i}{\sin\ r} = \left(\frac{V_2}{V_1}\right)^{1/2}. \tag{3.2.2}$$

The analogy to light optics is obvious. There is, however, one significant difference. In light optics, the change of refractive index is usually abrupt (on a macroscopic scale), because the interface of the two media is well defined. On the contrary, changes in electrostatic potential within the field through which the electron moves are gradual. Therefore, the changes of path of the electrons are also gradual.

A simple electrostatic lens is formed by two adjacent cylindrical sections differing in electrostatic potential (Fig. 3.7). The *electrostatic field* can be described in terms of equipotential surfaces. These surfaces are perpendicular to the *lines of force*—the lines along which a positive charge is accelerated by the field.

A device frequently used in electron optics is the equipotential lens (Fig. 3.8). This lens consists of three circular apertures, of which the outer two are at

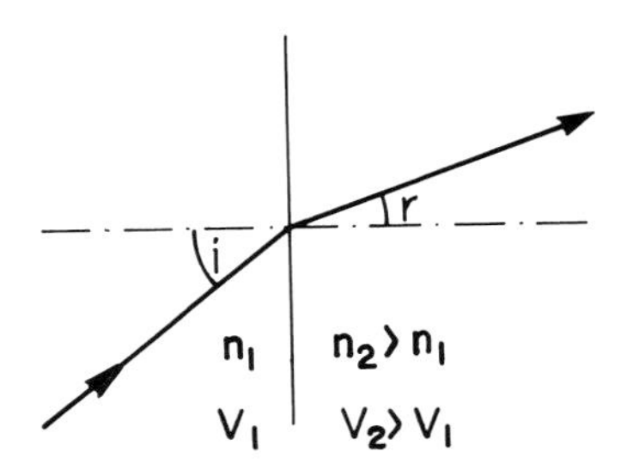

Fig. 3.6. Snell's law for the change of direction of a light beam at the interface of two media of optical indices n_1 and n_2 [Eq. (3.2.1)] is also applicable to the change of direction of an electron beam at the interface of two electrostatic potentials (V_1, V_2), if in the equation n_1 and n_2 are replaced by $\sqrt{V_1}$ and $\sqrt{V_2}$. A practical difference arises from the fact that refractive indices in optical devices usually change abruptly at interfaces, while electrostatic potentials in an electric field change continuously.

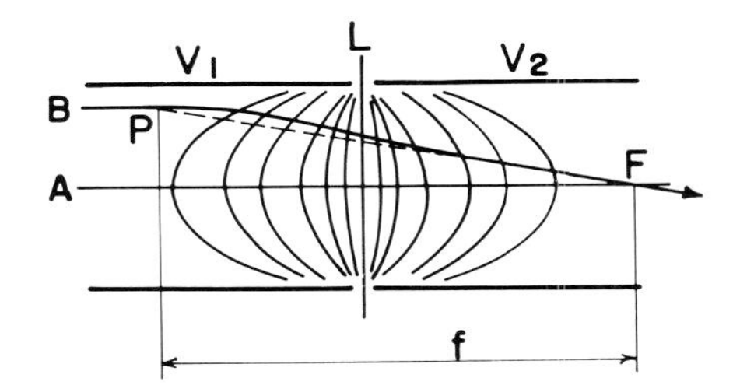

Fig. 3.7. Electrostatic lens formed by adjacent coaxial cylindrical sections at different potentials ($V_2 > V_1$). The device is symmetrical with respect to the plane L, which is normal to the optical axis A. A beam B, parallel to the optical axis, crosses the equipotential lines, and joins the axis at the focus F. Point P is the projected intersection of the entrant and exiting rays, and it determines the focal length, f. This length is larger than the distance from L to F. Hence this lens—as are all electron lenses—is a real lens of negative thickness.

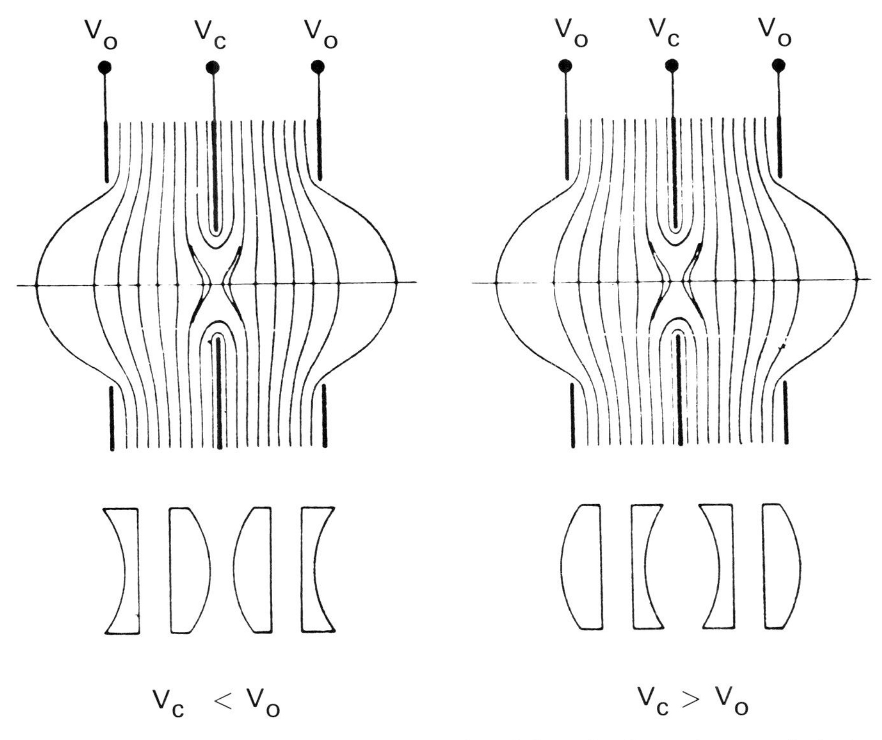

Fig. 3.8. The equipotential lens. This lens consists of three circular apertures, or electrodes. The two outer electrodes are at ground potential. The lens is always convergent, regardless of the potential of the central electrode. (From Hall [3.6].)

ground potential. The lens is always convergent, regardless of the potential of the central element.

The electrons emanating from a small source, after passing through a lens, are focused into an image of the source. If the dimensions of the object (source) and the image are y and y', we define the *magnification* by

$$M = y'/y. \tag{3.2.3}$$

The value of M is below unity in optical systems such as the electron probe microanalyzer, in which the image of the filament tip is *demagnified.* If we denote by s_1 and s_2 the distances from the center of the lens to the object and the image, respectively, we define the focal length of the lens, f (Fig. 3.9) by

$$\frac{1}{f} = \frac{1}{s_1} + \frac{1}{s_2}. \tag{3.2.4}$$

As in light optics, the above equation is an approximation which is inadequate for strong and thick lenses. However, the deviations from the law are not in the same direction as in light optics; electron lenses act analogously to a hypothetical optical lens of negative thickness.

The distance between the focus and the nearest lens component is called the *working distance*. This parameter is of importance in the construction of electron-optical devices.

In some electron microscopes, and in the first electron probe microanalyzer

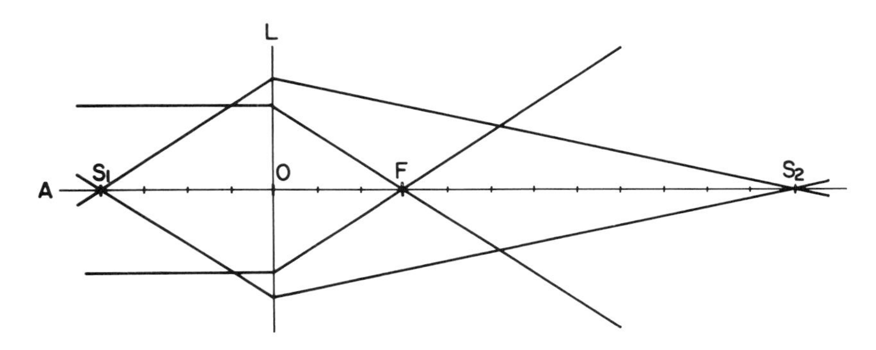

Fig. 3.9. The Gaussian lens. Rays which enter the lens L in direction parallel to the optical axis A cross this axis at the focal point, F. If the rays passing through a point at the optical axis S_1 are focused by the lens at the point S_2, then the distances from the center of the lens, $\overline{OS_2} = s_1$, $\overline{OS_2} = s_2$, and $\overline{OF} = f$, are related by the equation

$$\frac{1}{f} = \frac{1}{s_1} + \frac{1}{s_2}.$$

built by Castaing, electrostatic lenses were used to demagnify the beam diameter. However, the electrostatic lens has several disadvantages. It requires high voltages for its operation, produces serious distortions (aberrations), and the quality of the lens suffers severely if small dust particles are deposited in the lens gap. For these reasons, electromagnetic lenses are used in the design of electron probe microanalyzers.

3.2.1 The Magnetic Lens

An electron traveling through a magnetic field is accelerated at a right angle to its direction and to that of the field (Fig. 3.10). If the electron enters a uniform field normally, its path forms an arc of a circle of radius

$$r = \frac{1}{H} \frac{2mV_C}{e} \tag{3.2.5}$$

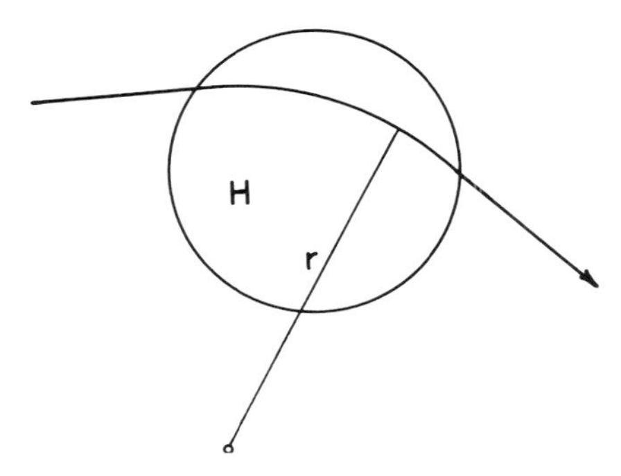

Fig. 3.10. Change of direction of an electron traversing the magnetic field H [Eq. (3.2.5)]. The field vector is perpendicular to the plane of the drawing.

where H denotes the magnetic field strength (in gauss), V_c denotes the (relativistically corrected) acceleration voltage, and m and e denote the mass and the charge of the electron, respectively. The relativistic voltage correction can be estimated by the equation

$$V_C = V(1 + 10^{-6}\ \mathrm{V})$$

with V_c and V given in volts. Thus, at 20 kV it amounts to approximately 2%. The vector component of the electron movement which is parallel to the field is unaffected. Hence, electrons which enter a magnetic field obliquely follow a helical trajectory.

Any axially symmetrical magnetic field acts as a lens for electrons which move parallel to the field axis. Such a field can be schematically represented by an equivalent uniform field, bounded by two surfaces. Between these boundary surfaces, the movement of the electrons is helical, but the changes of angle with respect to the field axis, which account for the focusing of the lens, occur in the nonuniform regions of the field, which are represented by the boundary surfaces (Fig. 3.11).

In practice, such a field can be produced by an electrical current flowing through a circular conductor (solenoid), or by a magnet–permanent or electric–having hollow cylindrical poles (Fig. 3.12). The most common design consists of a coil enclosed by a soft iron shroud with a narrow gap (Fig. 3.13). The field of such a lens is limited to the region between and close to the gap. A variety of characteristics can be obtained, depending upon the dimensions of the gap and of the bore; it is also possible to use asymmetrical lenses, in which the bore of one pole piece differs from that of the other.

The properties of such magnetic lenses were studied by Liebmann [3.8], who showed that the focal length can be calculated with good approximation by

$$\frac{f}{S + D} = k \frac{V}{(NI)^2} \tag{3.2.6}$$

where S is the gap in the shroud, D is the diameter of the lens bore (or the mean diameter in the case of asymmetrical lenses), and NI is the magnetic excitation (in ampere-turns). The proportionality factor k varies slowly–from 25 to 35–within the range of dimensions $0.5 \leqslant S/D \leqslant 2$ (Fig. 3.14).

Quite a different type of lens–a shroudless, elongated solenoid of small dimensions, called the minilens–was used in the construction of an electron probe by LePoole [3.9] (Fig. 3.26). The compact shape of this lens provides easy accessibility to the specimen area if the minilens is used as an objective.

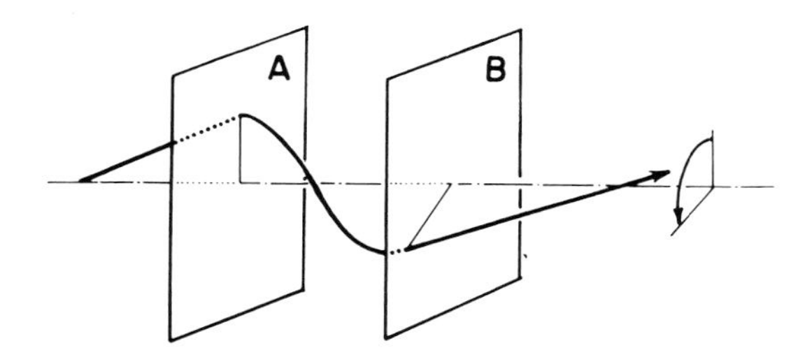

Fig. 3.11. Focusing action of a longitudinally uniform field between planes A and B on an electron trajectory. The sense of rotation changes when the direction of the electron path is reversed. Thus, the principle of reversibility of paths is not strictly applicable in electron optics.

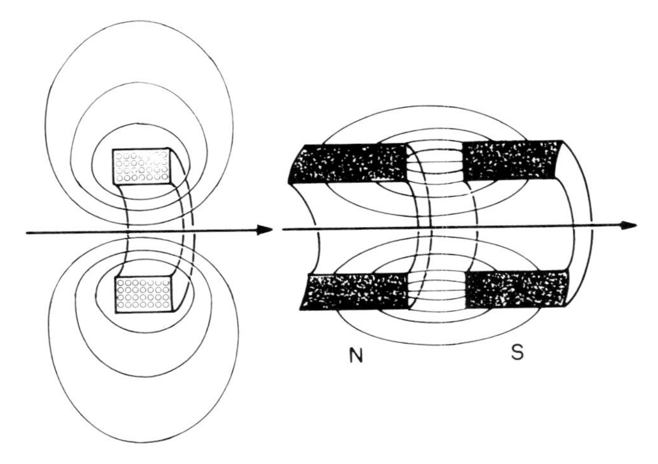

Fig. 3.12. An axially symmetrical magnetic field can be produced by a solenoid or by a magnet having hollow cylindrical poles.

Such a lens was used for this purpose in a commercially built electron microscope-microanalyzer (EMMA), and in the CAMEBAX electron probe-scanning electron microscope.

Electron lenses dissipate heat, and must be cooled, usually by contact with a coil through which water is circulated. The cooling is particularly critical for the minilens, due to the strong heat production within a relatively small volume.

Lens Aberrations. Like their optical counterparts, electron lenses follow

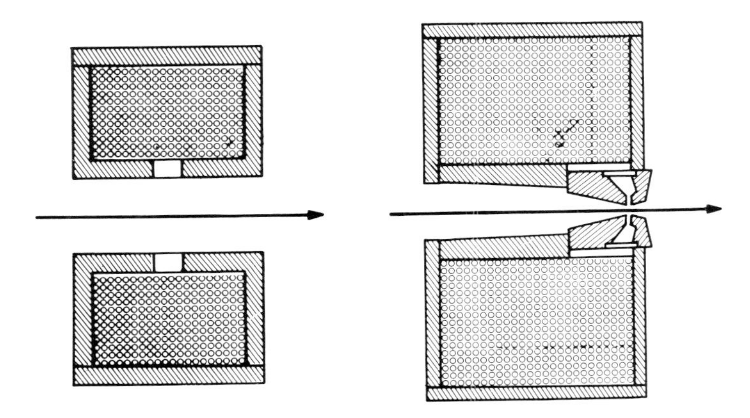

Fig. 3.13. Electromagnetic lenses. Left: symmetrical lens for long focal length; right: asymmetrical lens for short focal length. In this lens, the pole pieces are interchangeable and adjustable in position.

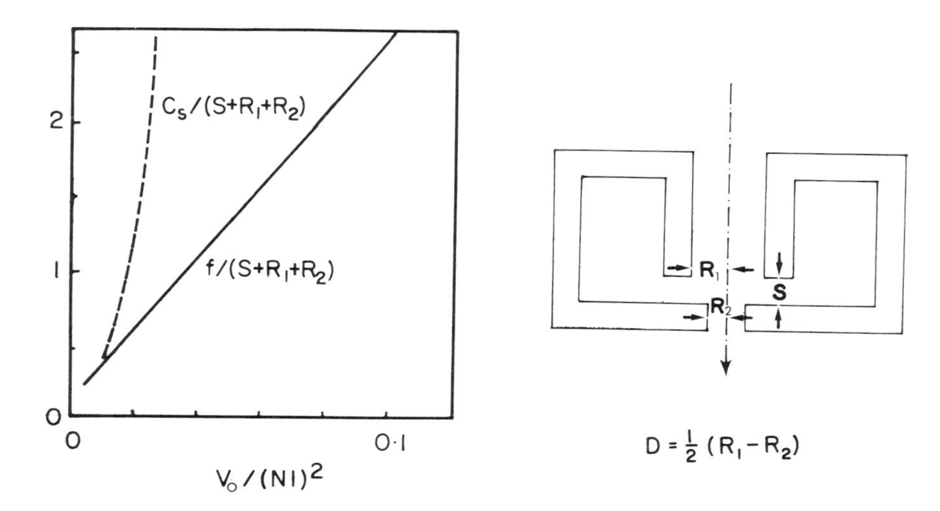

Fig. 3.14. Spherical aberration, C_s, and focal length, f, as a function of the electron acceleration potential V_0, the magnetic excitation NI (ampere-turns), and the dimensions of lens gap and bore. (Simplified from Liebmann [3.8].)

imperfectly the laws of geometrical (Gaussian) optics. The lack of precise correspondence with the theoretical model gives rise to defects in the Gaussian images produced by real electron lenses; these defects are called *aberrations*. Our interest centers on the effects of aberrations on the focused electron beam which we use to excite X-rays, or to produce scanning images. For a full theoretical discussion of aberrations see [3.5].

While some aberrations are due to imperfections in the electron lens, others are inherent to its properties, and therefore unavoidable. Nevertheless, the degree to which the lens approaches the theoretical Gaussian lens, and, hence, the magnitude of the inherent aberration, depends on the design and operation parameters of the lens. In particular, the aberrations are related to the aperture angle α, which contains the rays refracted by the lens. It can be shown that a physical lens approximates a perfect lens to the extent to which $\sin \alpha$ approximates α. (α is typically of the order of 5×10^{-3} rad.) Since $\sin \alpha$ can be represented by a series

$$\sin \alpha = \alpha - \frac{\alpha^3}{3!} + \frac{\alpha^5}{5!} - \cdots \tag{3.2.7}$$

the real lens thus follows the laws of the ideal lens to the extent that the third- and higher-order terms in the above series can be neglected. The effects of the third-order terms, which rapidly increase with increasing α, are called *third-order aberrations*. Several third-order aberrations (spherical aberration, distortion, curvature of field, field astigmatism, and coma) exist. All of these aberrations, except the spherical aberration, vanish when the source of the image lies on the optical axis of the lens. (Therefore, they are called *field aberrations*.)

Due to the *spherical aberration*, the electron lens focuses more strongly the electrons which enter the periphery of its field than those which arrive close to the optical axis (see Fig. 3.15). The effect of the spherical aberration on the image of a point source can be discussed with the aid of Fig. 3.16 which describes a demagnifying lens such as the one used in the electron probe microanalyzer. Consider a point source, S, located on the optical axis of the lens, L, at a distance from the lens much larger than the focal length f. A Gaussian lens would produce a point image S', at the intercept of the axis with the Gaussian plane, G, which is at a distance from L slightly larger than f. But as a consequence of the third-order terms, the power of the lens increases with increasing distance from the axis at the lens, r_α. Hence, the electrons emitted from S cross the axis at a point which is closer to the lens than the Gaussian plane, G, intercepting this plane at the point S'', at a distance r_i from S'. The arrival of electrons within an aperture angle α produces on the Gaussian plane a disk of radius r_i; this radius is proportional to the cube of r_α. If M is the magnification of the lens, then the point S will be seen by an observer in the Gaussian plane to have an apparent (virtual) radius of $r_v = r_i/M$. If α is small, r_α is approximately proportional to α, so that r_v and r_i are both proportional to α^3:

$$r_i = C_s\alpha^3 \quad \text{and} \quad r_v = C_s\alpha^3 M^{-1}. \tag{3.2.8}$$

The constant C_s is called the *spherical aberration constant*. The focused beam is narrower than the intersection with G, at a point on the axis close to G; the

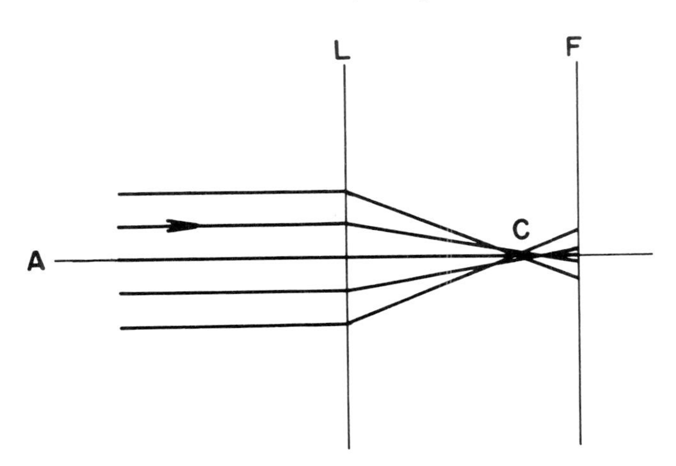

Fig. 3.15. Spherical aberration. L:plane of the lens, A:optical axis, F:focal plane, C:circle of minimum confusion.

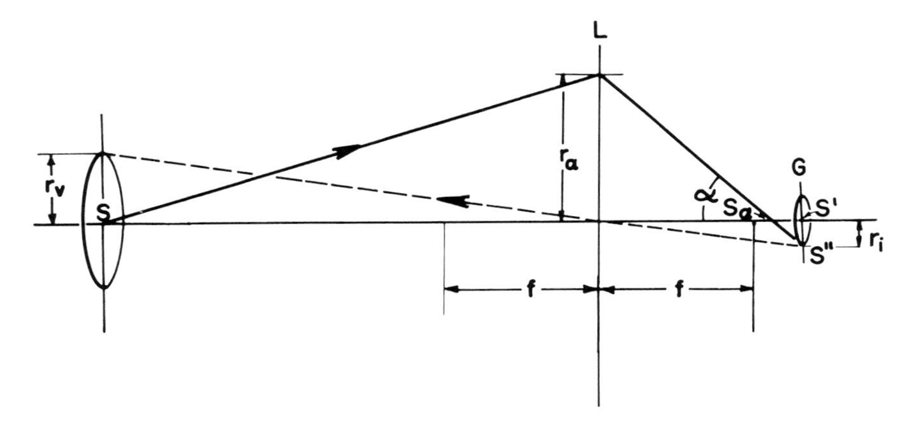

Fig. 3.16. The effects of spherical aberration on the image of a real source. See explanation in the text.

minimum beam diameter (minimum circle of confusion) is one-fourth of the intercept on G:

$$d_{\min} = \frac{1}{2} C_s \alpha^3. \tag{3.2.9}$$

If we assume that the source, instead of being a point, is a circle of diameter d_0, we obtain for the minimum circle of confusion

$$d_{\min} = M \sqrt{d_0^2 + \frac{1}{4} C_s^2 \alpha^6}\,. \tag{3.2.10}$$

The spherical aberration constant has the dimension of length. Its numerical value can be related to the lens parameters by the equation

$$C_s = kfV/(NI)^2 \tag{3.2.11}$$

in which f is the focal length of the lens, V is the electron acceleration potential in volts, NI is the magnetic excitation in ampere-turns, and k is a constant of the order of 100. The paper by Liebmann [3.8] also relates the value of the spherical aberration constant to the dimensions of the pole gaps (Fig. 3.14). In the objective lens of the electron probe microanalyzer, C_s is typically 1.5 - 4 cm.

Equations (3.2.8) through (3.2.10) show that to obtain a small beam diameter, the spherical aberration, and hence the focal length, must be made as small as possible. The beam diameter can also be reduced by using a smaller aperture angle, a smaller source, or a higher demagnification. However, all these variables reduce the intensity of the beam as the beam diameter is decreased. The aperture angle is usually defined by an aperture located in the field of the last (objective) lens. By this arrangement, the effects of aberration of previous lenses are minimized.

While the spherical aberration seriously limits the resolution of electron probes and must be carefully considered in the design of the instrument, the following third-order field aberrations are of negligible importance:

1. *distortion* (change of magnification as a function of distance from the axis);
2. *curvature of field* (curvature of the image obtained from a flat source);
3. *coma* (a comet-like image of point sources removed from the axis);
4. *field astigmatism*, which causes the rays emitted from a point source to focus into two mutually perpendicular lines (not to be confused with the *elliptical* astigmatism which will be discussed later).

As already mentioned, these aberrations vanish where the sources and their images are on the optical axis. This condition is obtained by aligning the optical column (see Subsection 3.2.3). Once the column is aligned, the field aberrations are present only when the instrument is used in the beam-scanning mode; even there, they only appear on the edges of large rasters, and are usually unimportant.

We will now consider several aberrations which are not third-order effects.

Chromatic aberration is a focusing defect caused by the energy spread of the electrons forming the beam. This effect is normally negligible (except for high-resolution microscopy), as are the *space-charge effects* which are due to mutual repulsion of the electrons in a dense beam. The increase in size of the focal spot at high currents in conventional instruments is not caused by space-charge effects, but rather by the increase in the aperture required to obtain a more intense beam, and hence of the spherical aberration. The effects of *diffraction* of an electron beam passing through a small aperture are not significant for beams in the diameter range useful for x-ray analysis.

Mechanical Defects of the Optical System–Elliptical Astigmatism. In spite of care in manufacture and maintenance, instrumental imperfections, such as deformation of the bore, magnetic hysteresis in the core, and dirt particles or burrs in apertures, may cause significant distortion of the lens field. The resulting defect is called *elliptical astigmatism*. As in field astigmatism, the resulting beam cross section is elliptic at positions close to the focus. In the exact focusing position, an enlarged circular cross section is obtained. Further advance along the axis–or adjustment of the lens field–produces another ellipse with axes at a right angle to those of the first ellipse. (See Figs. 3.3 and 3.17.)

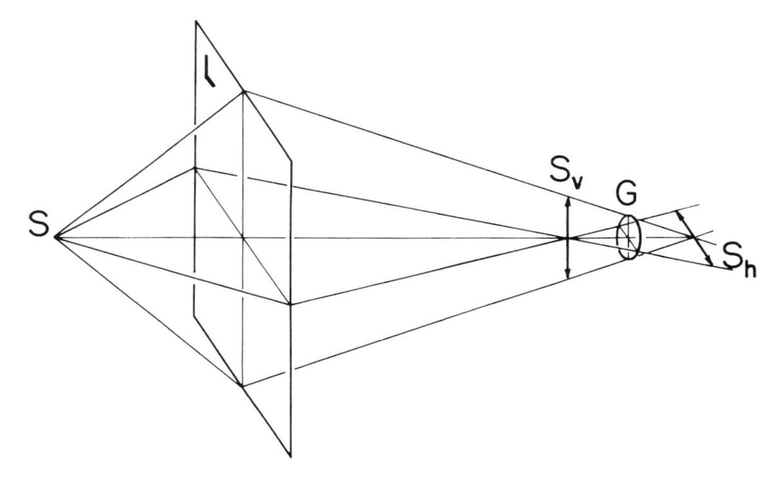

Fig. 3.17. Elliptical astigmatism. L is the plane of the lens. The focal position of the vertical component S_v differs from that of the horizontal component S_h. Hence, in the intermediate plane G, the point object S produces an image in the shape of a disk. (See also Fig. 3.3.)

The misalignment, and the resulting field astigmatism, are usually less important than the above-mentioned mechanical or magnetic defects. While some accidental causes of astigmatism can be removed by cleaning bores and cleaning or changing apertures, elliptical astigmatism due to defects of symmetry of the lens may persist. Most instruments have coils or other stigmator devices to correct such deficiencies. Stigmators will not, however, completely remove the strong astigmatism caused by dust particles in the apertures.

Effects of External Magnetic Fields. Magnetic fields generated outside the instrument sometimes cause observable aberrations, particularly in finely focused beams (scanning electron microscopes). Such fields may be caused by transformers, such as the ones used in power supplies. The resulting defect resembles elliptical astigmatism. It can be distinguished from the latter by the fact that the orientation of the elliptical beam cross section does not change when passing through the focal position.

3.2.2 The Electron Source

To accurately measure the relative X-ray intensities required in quantitative microprobe analysis, we need a stable source of electrons. This requirement can be satisfied by a gun of simple design—the self-biased thermal emission gun—which is used in most electron probe microanalyzers and scanning electron microscopes. The gun is shown schematically in Fig. 3.18 [3.10], [3.11].

The electrons are emitted from an electrically heated hairpin-shaped filament. Their initial kinetic energy is small, but they are accelerated towards the anode

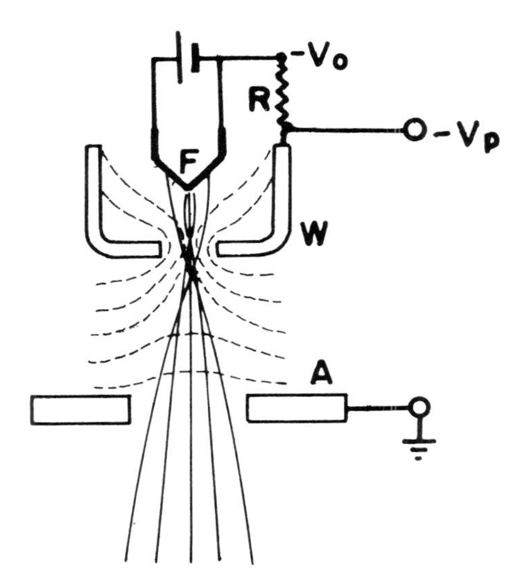

Fig. 3.18. The self-biased electron gun. The distance from filament tip to Wehnelt cylinder has been exaggerated in this schematic drawing.

plate, A, by an electrostatic field between the filament and the anode plate. The field is produced when a negative voltage (usually 5 to 20 kV) is applied to the filament with respect to the anode plate which is at external ground potential. A perforated shield–the Wehnelt cylinder, or grid cap–W, is placed between the filament and the anode. The Wehnelt cylinder is directly connected to the negative side of the accelerating voltage source, V_p, while the filament is connected to this point through a variable resistor, R, called the bias resistor. As long as no heating current is applied, the filament, F, does not emit electrons, and no current flows through the resistor R. Therefore, F and W are at the same potential. When the filament is heated by the battery included in Fig. 3.18, or by an equivalent circuitry, the filament emits electrons, and a current flows through the resistor R, so that, according to Ohm's law, the filament becomes more positive than the Wehnelt cylinder. The emitted electrons are collected by instrument components–e.g., the specimen–which are connected to the positive side of the accelerating power supply.

The equipotential lines drawn in Fig. 3.18 show the configuration of the field within the gun. The electron emission is limited to a small region near the filament tip. Since the filament is more positive than the Wehnelt cylinder, the electrons are forced to travel within a channel in the hole in the Wehnelt cylinder, proceeding in a narrow bundle towards the anode plate. As the emission from the filament increases, so does the potential difference between filament and Wehnelt cylinder (called gun bias). This bias tends to counteract the increase in gun emission as the filament heating is increased. In consequence, a plot of gun emission current as a function of heating current (Fig. 3.19) shows a region in which further increase of the heating current does not result in a significant increase of gun emission. A gun working in this region is called saturated. Emission from a saturated gun is characterized by high stability. Further increase of the filament current would shorten the filament life and should therefore be avoided.

The operating characteristics of the self-biased gun vary with the distance

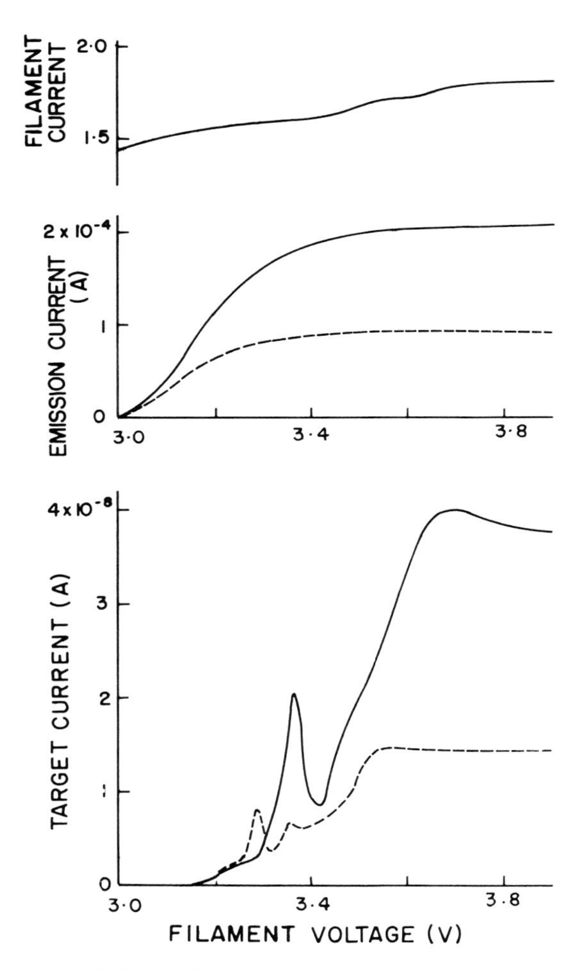

Fig. 3.19. Filament, emission, and target currents as a function of filament voltage in a self-biased gun. Bias resistor was 2 MΩ (upper curves) and 5 MΩ. The operating voltage was 20 kV. Saturation voltages are 3.7 and 3.55 V, respectively.

from the filament tip to the Wehnelt cylinder, with the accelerating voltage, and with the value of the bias resistor. Typically, the bias voltage is adjusted to about 1% of the accelerating voltage. If the bias is too high, the gun emission may be completely suppressed; if it is too low, the regulation may be insufficient.

The electrons emitted from the filament tip converge at a point called crossover. At this point, a real image of the emitting region of the filament tip is formed. It is the purpose of the electron-optical system to form a real demagnified image of this crossover on the surface of the specimen. Since we wish this image to be as small as possible, it is also desirable to restrict the emitting area on the filament tip. This area increases with decreasing gun bias [3.10]; therefore, a low gun bias limits the sharpness of focus of the electron beam on the specimen (see Fig. 3.20).

The need for a small final image of the electron-emitting part of the filament tip is particularly important in instruments such as the electron probe microanalyzer or the scanning electron microscope, in which the size of the final beam

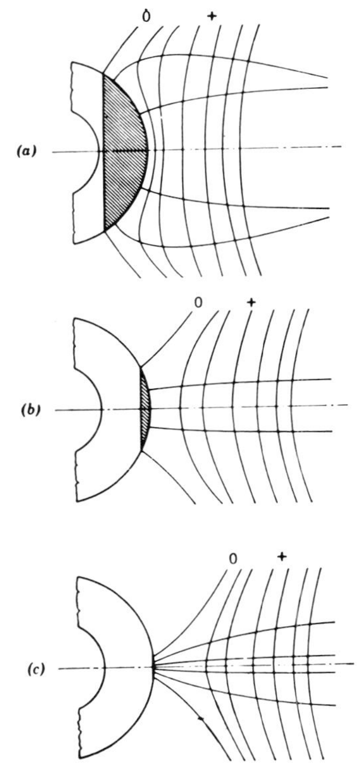

Fig. 3.20. Variation of the area of electron emission at the filament tip as a function of gun bias. The bias increases from (a) to (c). (See Ogilvie [3.12].)

cross section determines to a large extent the spatial resolution of the technique. In the conventional transmission electron microscope, the beam is not focused on the specimen, and the purpose of the condenser lens is to increase the current density on the specimen. The resolution of the transmission microscope depends of the quality of the effective magnification of the illuminated specimen.

For maximum gun current, the position of the filament must be maintained well centered with respect to the opening of the grid cap. As the distance from the filament tip to the grid cap surface must also be adjustable, it is necessary to provide for mechanical translation of the filament position in three orthogonal directions. But, the filament may warp due to heating, particularly during the first hours of filament life. Hence, the position of the filament must be controllable externally during operation. It is equally important to adjust the position of the anode plate in order to maximize the fraction of the beam emission which passes through the hole in the anode plate. The gun can thus be aligned by alternately changing the positions of the filament and the anode plate until the emission through the anode plate is maximized. This alignment is made with the gun in saturation. The total gun current can be varied by changing either the distance from filament tip to grid cap, or the bias resistor.

In some instruments, the position of the anode plate is secured by the action of two screws against the tension of a spring. The anode plate is heated by the gun during operation, and may thus expand and be displaced laterally. Therefore, it may be necessary to periodically readjust the anode plate position during prolonged operation in order to restore maximum beam current.

For quantitative analysis, it is necessary to know the energy of the electrons impinging upon the specimen. For this purpose, most power supplies carry a meter which indicates the voltage across the output. It should be noted, however, that the energy of the electrons depends on the potential of the filament with respect to the specimen. In the arrangement shown in Fig. 3.18, this potential differs from the voltage across the power supply output by the voltage drop through the bias resistor. If the operating potential is measured from ground to the Wehnelt cylinder, the corresponding correction, which is frequently neglected, can be calculated according to Ohm's law if the gun emission current and the value of the biasing resistor are known:

$$V_0 = V_p - I_g R . \tag{3.2.12}$$

In this equation, V_0 is the true operating potential, V_p is the potential across the power supply, I_g is the gun emission current, and R is the biasing resistance.

The energy of the impinging electrons can be measured by determining experimentally the Duane-Hunt limit for continuous radiation (see Section 6.3).

The factors affecting the life of the filament have been investigated by Bloomer [3.13]. The filament life is limited by thermal evaporation of the filament material, and by corrosion due to oxygen or water vapor. The evaporation increases with the filament temperature, while corrosion is caused by poor vacuum conditions. The cause of filament failure can be determined by microscopic observation of the failed filament. Chemical attack produces a uniform thinning of the heated area, while filaments burned due to evaporation are thinned noticeably only near the point of failure [3.14]. Under appropriate conditions, filament life should exceed 30 hours.

The quantity of main interest in the evaluation of electron emitters is the electron emission per unit area, or current density of emitted electrons, j, which, according to Richardson's equation [3.15], is

$$j = K(1 - r)T^2 \exp(-e\phi/kT). \qquad (3.2.13)$$

In this equation, K is, for metals, a combination of fundamental constants ($K = 120$ A/cm^{2}deg^2), r is an electron reflection coefficient of the order of 0.05, T is the absolute temperature, ϕ is the electronic work function given in volts, so that $e\phi$ is the work required for extracting an electron through the surface of the emitter, and k is Boltzmann's constant. The value of the work function varies somewhat with the temperature. In metals it is typically 4-5.5 eV, with about half this value for the elements calcium, strontium, and barium.

The beam intensity (A/cm^2) at any image plane can be related to the intensity at another plane by Abbé's sine condition (Fig. 3.21). Assume an optical system in which the object of height (normal to the axis of symmetry of the system) y_o produces an image y_i. If a ray from y_o enters the optical system (at a) at an angle α_o, and leaves it (at b) at the angle α_i with respect to the axis of the system, then the following relation holds:

$$n_o y_o \sin \alpha_o = n_i y_i \sin \alpha_i . \qquad (3.2.14)$$

The ratio y_i/y_o is the magnification of the system. The factors n_o and n_i are the

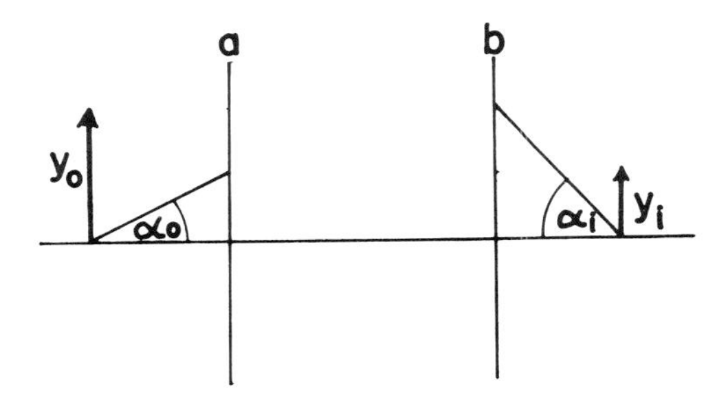

Fig. 3.21. The Abbé sine condition.

respective refractive indices. This principle has an equivalent in electron optics, where the refractive index is replaced by the square root of the potential, measured from a reference point at which the electrons would be at rest. (For small angles, $\sin \alpha$ can be replaced by α.) We obtain

$$y_o \sin \alpha_o V_o^{1/2} = y_i \sin \alpha_i V_i^{1/2}. \tag{3.2.15}$$

If the object y_o is the heated filament tip of the gun, then $V_o = kT/e$. Let us now make an approximation in which the object and the image are disks within which the current densities for two disks crossed by the same flux of electrons will be inversely proportional to the squares of the radii of the disks:

$$j_i = j_o \frac{eV_i \sin^2 \alpha_i}{kT \sin^2 \alpha_o}. \tag{3.2.16}$$

If the entire cathode emission is gathered, the half-angle of emission is: $\alpha_o = \pi/2$. Hence, $\sin \alpha_o = 1$, and we obtain Langmuir's equation which gives the maximum intensity obtainable in a system with thermionic emitter:

$$j_i = j_o \frac{e}{kT} V_i \sin^2 \alpha_i = j_o \frac{11\,600}{T} V_i \sin^2 \alpha_i \tag{3.2.17}$$

in which V_o is the operating potential given in volts [same as V_i in Eq. (3.2.12)]. For small values of α, we can approximate:

$$j_i = j_o \frac{11\,600}{T} V_i \alpha_i^2 \text{ A/cm}^2. \tag{3.2.18}$$

Langmuir's equation can be restated by saying that in an ideal system the current density per unit of solid angle, or *brightness*, is the same at any image plane:

$$b = \frac{j_i}{\alpha_i^2 \pi} = j_o \frac{11\,600}{\pi T} V_i. \tag{3.2.19}$$

The brightness of a gun at the crossover point is a measure of its efficiency. It is often lower than that predicted by Eq. (3.2.19), because the collection of electrons may be over a smaller angle than $\pi/2$.

The brightness of the gun is important when small beam cross sections are required since the achievable beam current at the specimen level drops quickly with diminishing beam diameter [3.16] (see Section 13.2). High current densities are also necessary for scanning transmission microscopy [3.17] and for the analysis of thin films. Therefore, there is a demand, particularly in the scanning electron microscope, for electron sources having higher emission current densities than the conventional tungsten filament. Oxydic electron sources are widely used in the electron tube industry, and the use of platinum filaments covered with oxides has been reported [3.18], as well as that of dispenser electrodes containing a core of oxides [3.19]. However, electron sources of this type are not stable enough for use in quantitative X-ray intensity measurements.

The Lanthanum Hexaboride Gun. Broers [3.20] described a gun based on electron emission from a heated rod of lanthanum hexaboride with a lifetime of 8000 hours, when operated at the current densities equivalent to those of a tungsten filament which would last about 50 hours. The main problem in the construction of such a gun is the high reactivity of lanthanum hexaboride at the temperatures required for electron emission. Broers resolved this difficulty by heating indirectly the emitting region of the hexaboride rod with a coil, while the region of the rod which is in contact with other materials is cooled with a heat sink (Fig. 3.22). A gun of these characteristics can be operated at a brightness several times higher than that of the conventional tungsten filament gun, with high emission stability, and with the aforementioned gain in the life of the assembly. Lanthanum hexaboride guns are now available for several commercial instruments.

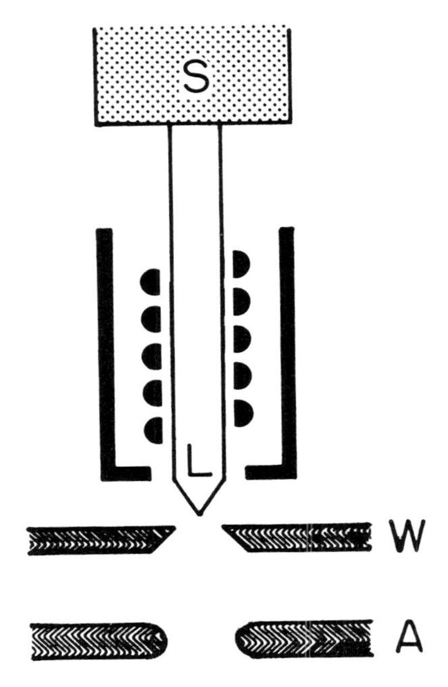

Fig. 3.22. Lanthanum hexaboride gun [3.20]. *L*:lanthanum hexaboride rod, *S*:heat sink, *W*:Wehnelt cylinder, *A*:anode plate. The lanthanum boride rod is not in direct contact with the heating coil. The assembly is surrounded by a heat shield.

The Field Emission Gun is a source of electrons of very high emitted current density. By etching a whisker of tungsten to a tip of about 500-Å radius, and subjecting this tip to an electrostatic field of appropriate strength, an electron source is obtained which emits electrons at room temperature, from a very small virtual source (smaller than the actual tip), and at a very high current density. Crewe, who adapted the field emission gun to scanning electron microscopes, reported an observed emission density of 10^6 A/cm^2 [3.21]. This compares very favorably with the typical densities at a hot filament tip (approximately 10 A/cm^2). Given the small source size and high emission density, one obtains beam diameters of less than 100 Å with the aid of a single auxiliary lens, and with a beam intensity which exceeds considerably that obtainable with a hot source and a conventional three-lens system. Crewe's gun is shown in Fig. 3.23.

These characteristics render the field emission gun very attractive for the construction of a simple scanning electron microscope; with additional lenses, very small spot sizes are attainable. However, the small size of the emitting tip makes it difficult to obtain stable electron emission, even if the gun is operated at a vacuum of 10^{-9} torr. Furthermore, the advantages in brightness of the field emission gun compared with the hot filament are significant only for beam diameters below 10^{-7} m (0.1 μm). Therefore, this gun is of more interest to the electron microscopist than to the analyst.

3.2.3 Electron-Optical Column and Alignment

Figure 3.24 shows schematically the electron-optical system of the electron probe. The gun crossover, f_1, is demagnified by the action of the upper, or condenser lens which forms the focus f_2. Below this point, an aperture (ap) limits

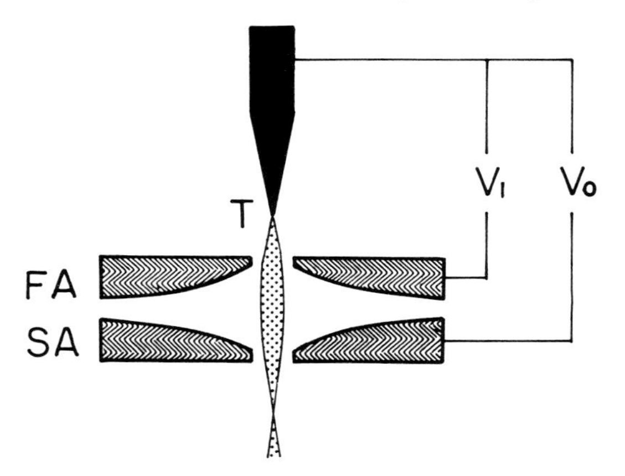

Fig. 3.23. Field emission electron gun [3.21]. *T* is the tip of the emitter, FA is the first anode, SA is the second anode, $V_1 \sim$ -3000 V, and V_o is the accelerating (operating) potential (-20 to -100 kV).

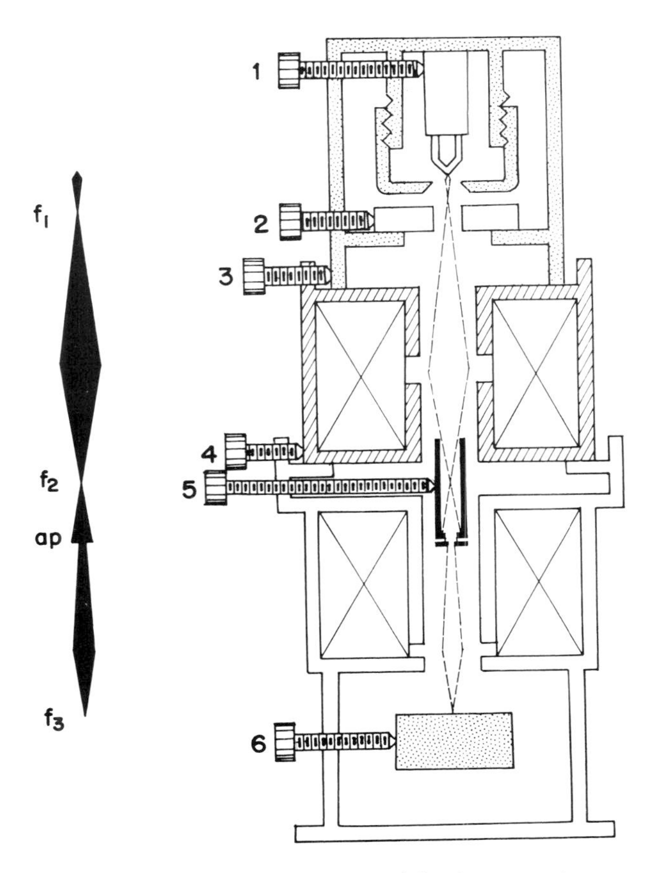

Fig. 3.24. Mechanical alignment of the electron optics.

the width of the expanding beam. A second lens, called the objective lens, further demagnifies the image of the filament tip at f_2, forming the final focus, f_3, at the specimen surface.

In most instruments, some or all of these components can be moved laterally for alignment (or electronic alignment is possible), while tilting adjustments can rarely be performed. Therefore, the adjustable parameters are the lens currents and the lateral positions of the components. We must now discuss the effects of changes in these parameters.

A change of intensity of the lens current varies the strength of the lens and, hence, the position of the corresponding focus along the optical axis. The current of the objective lens must be adjusted so as to make the position of f_3 coincide with the specimen surface. As will be discussed later, the correct position of the specimen surface is defined by the focusing of the X-ray spectrometers and controlled with the aid of the optical microscope. We can assume that the specimen position is fixed, and that focusing of the electron beam must be achieved by varying the strength of the objective lens.

The variation of the field strength of the condenser lens shifts the position of f_2 along the optical axis. Such a shift changes the diameter of the cone of electrons impinging on the aperture (ap); this, in turn, varies the fraction of the beam which passes through the aperture, and hence, the intensity of the beam impinging on the specimen. Let us assume that the current density within the cone at the level of the aperture (ap) is uniform. If the diameter of the aperture is denoted by d_{ap}, and that of the cone of electrons impinging upon the aperture by d_{co}, then, after passage through the aperture, the beam intensity is reduced by the factor $(d_{ap}/d_{co})^2$, provided that d_{ap} is smaller than d_{co}. If the width of the electron beam at the aperture level is smaller than the diameter of the aperture, and if the beam is centered, the beam intensity will not be diminished by its passage through the aperture. It follows that the intensity of the electron beam impinging upon the specimen can be regulated by varying the current flowing through the condenser lens, and that high beam intensities are obtainable only when the lateral position of the aperture is centered around that of the focus f_2.

We recall that astigmatism due to field aberrations causes deterioration of focusing as the distance of the focus from the optical axis increases. To minimize this defect in the objective lens, we must laterally shift all components above the objective lens, until f_3 is located on the optical axis of the objective lens. It is also important that f_2 be located at the optical axis of the condenser lens. Failure to achieve this condition not only increases the aberration due to the condenser lens, but also interferes with the regulation of the beam intensity, as will be demonstrated later.

We will now discuss some simple tests of alignment. We have already mentioned the observation of the gun current as a criterion of gun saturation. In a similar manner, the observation of the beam current passing through the anode plate is useful in the alignment of the filament tip and of the anode plate. For this purpose, we may measure the beam current impinging upon the specimen (by means of a Faraday cage), or the specimen current. It is easy to prepare a Faraday cage by drilling a hole of about 2-mm diameter and several millimeters deep in a metal surface, which can be put in the specimen position, and to cover it with an aperture of 100-μm diameter. For this measurement, it is also possible to use a collector which intercepts the beam below the anode plate. (The use of a monitor current device by means of a double aperture at the position of ap will be discussed later in more detail.)

The production of visible light by impact of electrons on appropriate materials can be used to visually observe beam cross sections at various levels of the electron-optical column. However, some materials–such as zinc sulphide–do not withstand the effects of a concentrated electron beam. The use of particulate phosphors in the upper sections of the column is dangerous, since particles may be displaced and deposited at the rim of apertures, producing severe beam aberrations. The most useful position for a cathodoluminescent screen is at the specimen level. A luminescent specimen can be observed there through the optical microscope, and it is valuable in the alignment of both lenses and the aperture. Luminescent materials suited for this purpose include the oxides of aluminum and magnesium, calcium tungstate, calcium molybdate, various fluorides, zirconia, and the mineral benitoite. Since these materials have low electrical conductivity, they must be coated with a thin transparent layer of carbon. Although the cross section of a well-focused beam is close to or below the resolution of most of the optical microscopes used in microprobes, an experienced operator can observe surprisingly well, by means of the luminescent image, defects such as elliptical astigmatism. In some luminescent materials, the image of the beam intercept is more diffuse than in others; alumina seems to give particularly sharp images. The definition of the point of impact is thus an indication of the proper setting of the objective lens current. Other tests for focusing the objective lens will be discussed later when line and area scanning is described.

Let us now reduce the strength of the objective lens so that f_3 falls below the surface of the luminescent screen. We see on the screen a circular luminescent spot which is the cross section of the beam. The edge of the spot, which is sharply defined, images the rim of the limiting aperture, and imperfections of this aperture—such as dirt particles and burrs—can be clearly observed if the objective lens current is regulated to obtain a spot of 100-200-μm diameter.

If such features are distinguishable at the rim of the aperture, then it can be observed that the image of the aperture rim rotates as the setting of the objective lens is changed. This is due to the helicoidal movement of the electrons obliquely crossing a magnetic field (see p. 24). The center of rotation of this image coincides with the intersection of the optical axis of the lens with the screen surface. Consequently, the manner in which the circular field collapses into the image of the focus f_3, as the objective lens is focused, reveals the position of this focus with regard to the lens axis (Fig. 3.25). If f_3 lies on the lens axis, then the image of the aperture collapses symmetrically into the image of f_3. If it is off axis, the center of the aperture image describes a spiral path, which may remove it from the field of observation of the light microscope. It is then necessary to move the components above the objective lens (or the lens itself if it can be moved) so as to obtain symmetrial collapse. After this alignment, the optical microscope should be adjusted to bring the image f_3 into the center of the field of vision. If, after complete alignment of this instrument, this shift has been significant (i.e., more than a few micrometers), then X-ray spectrometers will also have to be realigned, as described later.

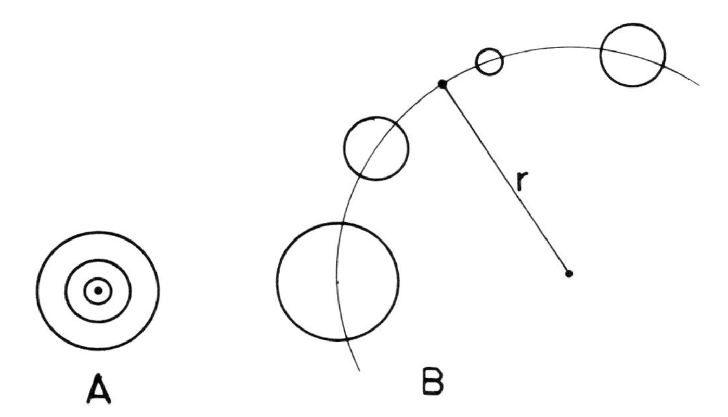

Fig. 3.25. (A) Symmetrical and (B) asymmetrical collapse of the electron beam cross section observed on a cathodoluminescent specimen, while adjusting the objective lens. r is the distance from focused beam impact point to optical axis of the objective lens. In case (A), $r = 0$.

The above adjustment should be made with a beam current intensity such as normally used in analysis. If the instrument has provisions for beam scanning, the scan generators should be turned off. We will now describe further observations on the screen in the specimen position as a function of changes in the field strength of the condenser lens. To interpret these observations, we must consider the events at the level of the limiting aperture.

We recall that the cone of electrons emerging from the condenser lens is intercepted by the aperture, and that the width of this cone depends upon the position of the image f_2, and hence, upon the field strength of the condenser lens. If f_2 is lowered to the level of the aperture, the entire beam emerging from the condenser lens will pass through the limiting aperture. This, however, is only possible if the components are appropriately aligned.

Let us assume that this alignment is not perfect. At a moderate beam intensity, the entire aperture will be illuminated by the electrons emerging from the condenser lens. We first adjust the objective lens in such a way that the image of the aperture on the screen almost fills the field of vision of the light microscope. Then we lower the condenser lens current in order to bring f_2 closer to the limiting aperture. As the cone of electrons which illuminates the aperture becomes narrower, the current density at the aperture level increases. Therefore, the image on the screen becomes brighter, and the target current increases. But, as the cone intercept at the aperture level decreases in diameter, it eventually fails to cover the entire aperture. The edge of the cone intercept then becomes visible on the screen which previously was illuminated uniformly. As this "clipping" of the aperture image progresses, the target current starts to drop. We can now alternately adjust the aperture position and further reduce the condenser lens current until the target current cannot be further increased. As now the entire beam cross section is contained within the aperture, the outlines of the aperture on the screen become indistinct, and a very bright diffuse spot on the screen marks the position of f_2. The electron density at this spot is frequently sufficient to bring the screen to thermal glow or to fuse or alter it.

Once the image of f_2 is obtained on the screen, we can also test if f_2 is located on the optical axis of the condenser lens. This test is analogous to that of the objective lens. If, on altering the strength of the condenser lens, the image of f_2 diffuses symmetrically, f_2 is on the optical axis of this lens. If it drifts laterally, the lens position should be shifted to bring f_2 into its axis. If the section above the condenser lens moves with the lens (as in Fig. 3.24), it must also be readjusted to restore the position of f_2 with respect to the limiting aperture.

In summary, the observation of a luminescent screen at the specimen position provides the following information:

1. sharpness of focus on the specimen, beam stigmatism;
2. condition of the aperture edge;
3. position of the focus f_3 with regard to the objective lens axis;
4. position of the focus f_2 with respect to the aperture;
5. position of the focus f_2 with respect to the axis of the condenser lens.

The measurement of the maximum obtainable beam current with a Faraday cage in the specimen position provides an additional indication of alignment. The procedure is complicated, however, by the fact that the alignment of one component may interact with that of other components. These interactions may vary from one type of instrument to another, depending on the mechanical relations of the components. For an instrument such as that shown in Fig. 3.24 we suggest the following alignment sequence, which should be adaptable to other instruments with minor changes.

We first visually align all components, including the filament and the anode plate, to symmetry. This should enable us to obtain a beam at the specimen level. The gun is saturated, and the filament and anode plate are centered for maximum beam current. Now a luminescent screen is inserted at the specimen level. The specimen height position is carefully adjusted to maximum sharpness of the microscope image. Next, we observe the sharpness of focus on the screen, and the conditions of the aperture rim. If the aperture rim is not perfectly circular, or if excessive stigmatism is observed, the aperture must be changed and, if necessary, the pole pieces and aperture holders, as well as the Wehnelt cylinder and anode plate, must be cleaned. A mildly abrasive paste can be used for this purpose. Particles clinging to apertures can frequently be removed in an ultrasonic cleaner. With scanning devices and stigmators disconnected, the focus f_3 is now brought into the center of the microscope field, by adjusting the position of the microscope objective, and symmetrical collapse of the aperture image is obtained by adjusting the aperture position. It may be necessary to readjust the microscope field, the filament, and the anode plate position. Next, the positions of the condenser lens and the components above it are adjusted to bring f_2 into the center of the aperture as well as into the axis of the condenser lens. After this step, the whole procedure is repeated to assure optimum alignment of all components.

In some instruments, a luminous screen can be inserted below the anode plate. This permits aligning the gun section by observing the patterns which appear with the gun below the saturation point, as described by Hall [3.6, p. 147].

Some of the freedoms of movement required for the alignment procedure discussed above may not be available in certain intruments. Presumably, pre-alignment by the manufacturer of certain elements will simplify the alignment task of the operator. If this alignment has been imperfect, or if alignment conditions change in time, the user may wind up with a premisaligned optical column. The following list is a synthesis of the beam alignment procedure just discussed.

1. Visually center all components, particularly the filament and the anode plate.
2. Heat the filament until the gun current meter indicates gun saturation.
3. Adjust anode plate to obtain maximum target current.
4. Adjust the aperture position to obtain concentric collapse of the beam cross section on the luminescent specimen.
5. Adjust the condenser lens position, and readjust the gun and anode plate positions, to obtain maximum target current, and symmetrical illumination of the cross section at specimen level at lower beam intensities.
6. If drift is noted during operation, periodically adjust the anode plate position (if it is movable) or else adjust the filament position.

Note that adjustment 5 can usually be omitted, except for complete realignment.

3.3 THE TARGET ASSEMBLY

The electron probe microanalyzer is a special X-ray tube, the target of which is a region of the specimen. This target must be physically manipulated by means of the specimen stage, optically observed through a light microscope, bombarded by an electron beam issuing from an objective lens of short working distance, and positioned in such a way as to permit the free exit of the X-rays to be observed by the spectrometer or spectrometers. We will now consider some of the problems arising from the concurrence in a narrow space of these instrument components, and some of the constructional solutions to these problems.

The early instruments had objective electron lenses of conventional asymmetrical design, and the specimen was positioned normally to the impinging electron beam. The advantage of having more than one X-ray spectrometer available soon became apparent. As a result, most instruments had two or more X-ray spectrometers which viewed the specimen at a small X-ray emergence angle ($\psi = 10° - 20°$), through the space between the pole piece of the objective lens and the specimen plane (Fig. 3.26). When electron probe microanalysis was extended to X-rays of long wavelength (> 4 Å), the shallow emergence resulted in excessive absorption losses of the X-rays within the specimen (see Chapter 10 and Section 12.3). Although a controversy concerning the appropriate value of the emergence angle ensued and persisted for several years, it is now clear that, at least when soft X-rays are observed, emergence angles below 30° are to be avoided.

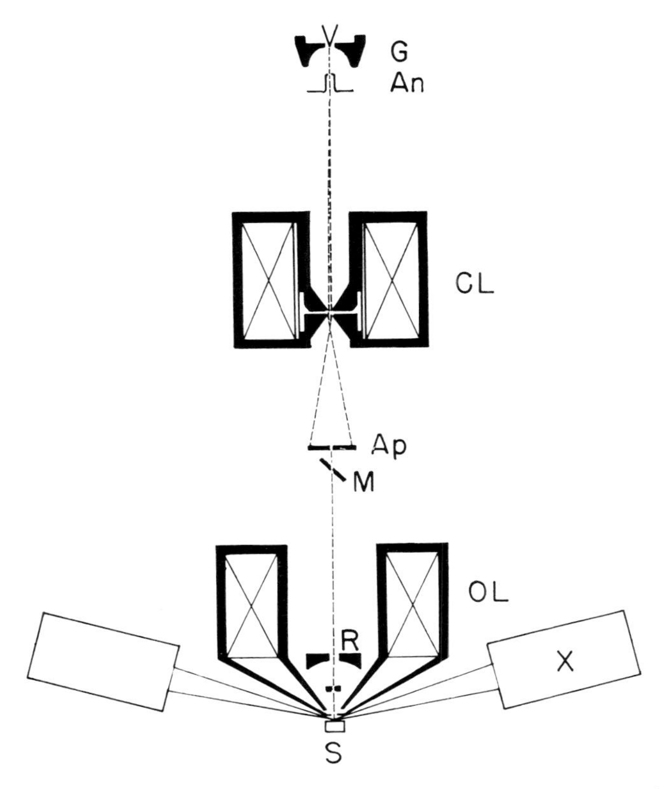

Fig. 3.26. Schematic diagram of a conventional microanalyzer (see Castaing [3.22]). G:gun, An:anode plate, CL:condenser lens, Ap:aperture, M:mirror for light microscope, R:reflective objective, OL:objective lens, S:specimen, X:X-ray spectrometer.

Configurations which result in a higher X-ray emergence angle can be obtained by modifying the design of the objective lens (Fig. 3.27). The possible variations of design include windows in the pole pieces, through which the X-rays pass towards the spectrometer, inverted or flat objective lenses, and minilenses of the LePoole design. Another alternative is to incline the specimen with respect to the electron beam (oblique beam incidence).

The design of the assembly of components in the specimen region also involves the objective of the light microscope and the specimen stage. These parts of the electron probe microanalyzer will be discussed next.

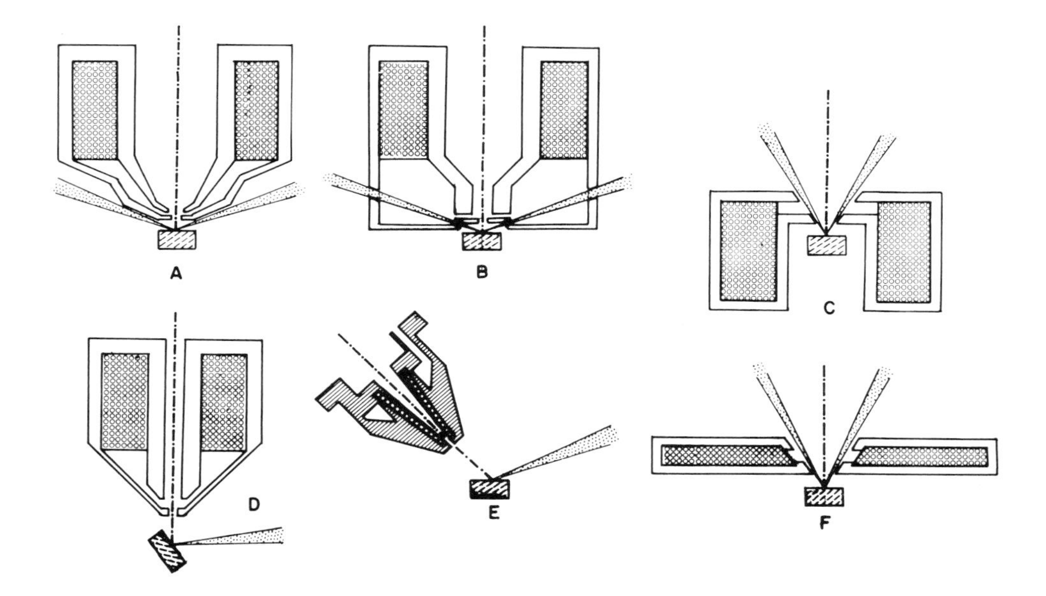

Fig. 3.27. Diverse objective lens configurations.

(A) Traditional design, resulting in a low X-ray emergence angle. The light optics objective can be located within the lens bore.

(B) Windows within the shroud of the lens permit the emergence of X-rays at a higher angle.

(C) The "inverted-lens design" permits the use of a large solid angle for detectors at a high X-ray emergence angle. However, this design limits the size and movement of the specimen.

(D) With oblique electron beam incidence, a high X-ray emergence angle, and ample space for good diffracting light optics are obtained. However, the calculation of the backscatter effects is more uncertain than with normal electron beam incidence.

(E) Even more space is obtained with oblique beam incidence and the use of a minilens, a shroudless solenoid enclosed in a nonferromagnetic cooling mantle.

(F) A "pancake lens" provides ample room for both X-ray emergence and specimen movement.

3.3.1 The Optical Light Microscope

The diameter of a typical specimen is about ten thousand times that of the region analyzed with a focused static electron beam. It is therefore important to know which part of the specimen is excited by the beam. The position of the specimen with respect to the beam is usually observed with a light microscope which is focused on the point of electron beam impact and its surroundings. Such a light microscope is incorporated in all modern electron probe microanalyzers, although it is typically absent in scanning electron microscopes.

The development of beam-scanning techniques, and the use of target current or secondary electrons for image formation, have provided an alternative to the use of a light microscope for imaging the specimen. The resolution of the scanning electron image is usually higher than that of the microscope; more importantly, the use of the same electron beam which analyzes the specimen to produce a topographic image assures that the imaging device is in alignment with respect to the analyzing assembly. However, certain features, such as staining patterns in biological preparations and colors of minerals with normal or polarized light, cannot be observed by electron beam scanning. Therefore, light microscopes, with provision for transmitted illumination and, for mineralogical studies, polarizers, are required. This is well illustrated by the instrument shown in Fig. 3.28 which was built around a high-quality transmission light microscope [3.23], with certain compromises as to the electron beam resolution and X-ray measurement capabilities.

The light microscope has still other important functions. Its use for the alignment of the electron beam has been discussed in the previous section. The observation of cathodoluminescence of specimens as an auxiliary analytical tool will be discussed in Chapter 7. The most important function of the microscope, however, is related to the focusing of the X-ray spectrometers. The cylindrically curved crystal spectrometers used in the electron probe microanalyzer require for proper operation that the X-ray source be located on a line on the specimen surface [3.24], [3.25]. The position of this line can be changed by mechanical alignment of the spectrometer. When all spectrometers are aligned, their lines of focus intersect at the point of impact of the electron beam on the specimen. But the position of each line of focus changes as the specimen surface elevation is changed. Therefore, it is important to maintain the position of the specimen surface rigorously constant, and the optical microscope serves this purpose. One of the stage movements—the elevation—is used to adjust the image of the specimen surface to maximum sharpness; the alignment calibration of the spectrometers is performed with the specimen in the elevation position corresponding to the best focusing of the microscope image. For this reason, the microscope must have a shallow depth of focus, and the sharpness of the optical image should be checked every time an X-ray measurement is made. The scanning electron microscope capabilities of the electron probe microanalyzer are

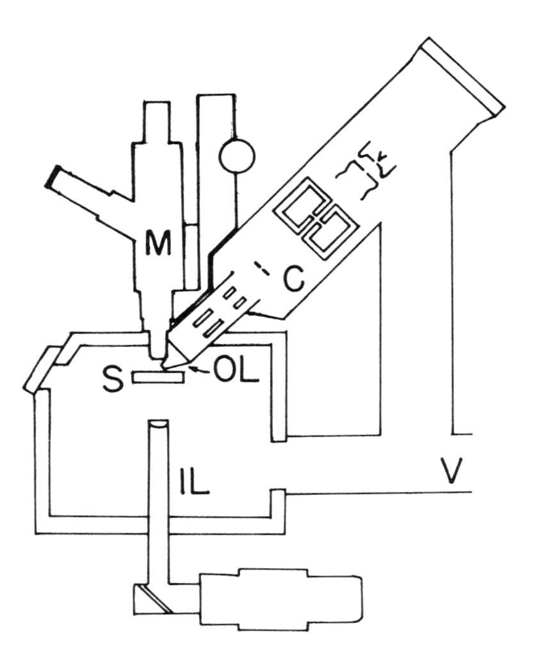

Fig. 3.28. Electron probe microanalyzer for the study of minerals [3.23], incorporating a high-quality microscope, M. C:electron optical column, with gun, anode plate, condenser lens, limiting aperture deflection plates, and a small minilens used as objective lens (OL), S:specimen, IL:illuminator for transparent specimens, V:connection to vacuum pumps.

not useful in this respect. One of the remarkable characteristics of the scanning microscope is its large depth of focus which renders the scanning image insensitive to small changes in the elevation that might, nevertheless, affect the focusing of the X-ray spectrometers.

For the observation of the features of interest for electron probe analysis, the light microscope should have an optical resolution of at least 1 μm. This requirement, as well as the needed shallowness of field, demand an objective lens of short focal length. In order to maintain the entire field of view in focus, it is necessary that the axis of the light-optical microscope be normal to the specimen surface. If the electron beam impinges normally on the specimen surface as well, the two optical axes coincide. But since the working distance of the electron objective lens is also short, there is very little space available for both objective lenses above the specimen, particularly if space for the emerging X-rays must also be provided.

To solve this problem, the instrument builder may choose one of the following solutions.

1. He may use a refractive or reflective light objective, the center of which is perforated to allow the passage of the electron beam. The light objective may be located inside the electron objective lens—as shown in Fig. 3.26, or an inverted or pancake electron objective lens (Fig. 3.27) may be used. As it is difficult to build refracting light objective lenses perforated at the center, the Cassegrainian spherical mirror objective is frequently used in this context (Fig. 3.29).
2. He may incline the specimen with respect to the electron beam (Figs. 3.27, 3.28, and 3.30).
3. He may provide a specimen stage with two separate positions for the specimen, one for light-optical observation, and the other for X-ray analysis. The specimen is transferred from one position to the other by rotation of an arm or turntable (Fig. 3.1).

The first approach adversely affects the quality of the light microscope, since the important central part of the light-optical system is eliminated. The objective must remain within the vacuum enclosure, and the system is basically limited to

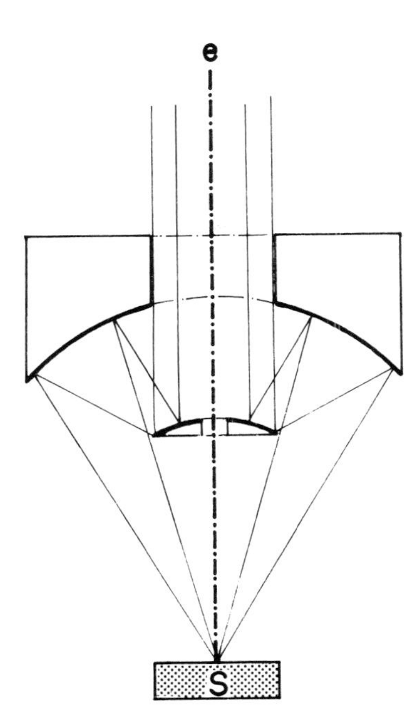

Fig. 3.29. The Cassegrainian spherical mirror objective. *e* is the electron beam and *S* is the specimen.

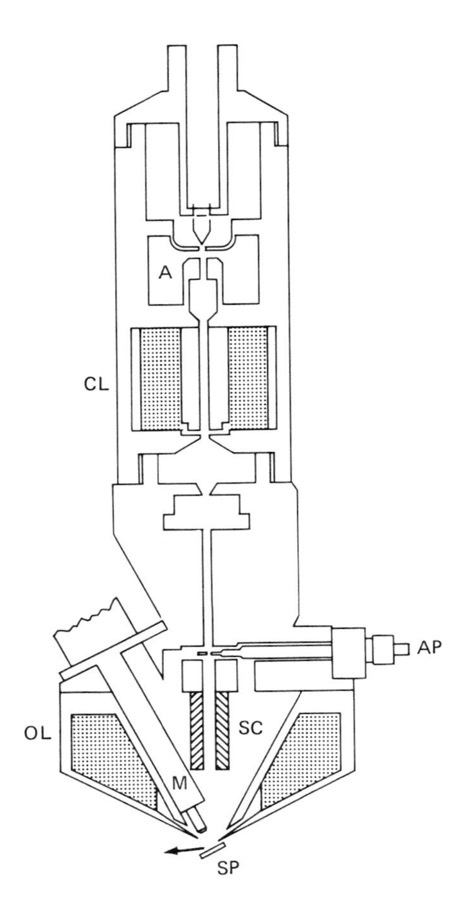

Fig. 3.30. Electron beam microanalyzer with inclined specimen [3.26]. A:anode, AP:limiting aperture, M:microscope, OL:objective lens, CL:condensor lens, SC:scanning coils, SP:specimen.

one fixed magnification and field of view. There are limitations in the use of oblique illumination and polarized light, and contamination or damage to the lens due to the effects of the beam are possible. On the other side, this solution is very satisfactory from the point of view of the electron-optical design, and, with an inverted or pancake objective lens, there is ample space for the escape of X-rays at a high emergence angle.

The second approach is very satisfactory from the point of view of visual observation. A microscope with exchangeable objectives can be used, as well as accessories for oblique and polarized illumination. The tilted-specimen configuration is thus particularly useful in the analysis of biological and mineralogical specimens. The oblique electron beam incidence is, however, less desirable for quantitative X-ray analysis. As will be discussed later in detail, it is at present more difficult to apply theoretical correction calculations for this configuration than for normal beam incidence. This difficulty is not fundamental; rather, it is related to our present limitations in the knowledge of the phenomena of beam-target interaction.

The third, or turntable, approach, is the least advantageous, since it precludes the viewing of the specimen during analysis. Therefore, cathodoluminescence cannot be directly observed. This renders the judgment of the quality of the electron beam, and the control of spectrometer focusing, more difficult. The orientation of the beam with respect to the specimen cannot be directly verified during analysis, and depends on a mechanism of high precision. Although this approach would permit the use of microscopes of high optical quality, it can be considered obsolete.

In scanning electron microscopes, X-ray emission is usually detected and measured with energy-dispersive detector systems. These spectrometers are much less sensitive to changes in specimen elevation than crystal spectrometers. For this reason, an optical microscope is not necessary in such instruments. If crystal spectrometers are used, they should be mounted in an inclined focusing circle position, as discussed in Section 5.1.

3.3.2 The Specimen Stage

The purpose of the specimen stage is to provide mobility to the specimen so that it can be oriented in any desired way with respect to the focused electron beam. In order to discuss the requirements for motions of the specimen, we must consider the geometric relations of the elements concurring in the specimen area (Fig. 3.31).

We first define three mutually orthogonal linear displacements along the directions x, y, and z. The specimen surface which will be analyzed is parallel to x and y. If the electron beam, e, intersects the plane formed by x and y, then any point of the specimen surface can be translated towards the beam intercept by moving the specimen along the directions x and y. If an arbitrary point is established as the origin of coordinates along x and y, any point on the specimen surface can be defined by indices along x and y. Movements along x and y (particularly electron beam movements) are frequently referred to as "horizontal and vertical movements." The direction of z, normal to the specimen surface, is frequently called "elevation." The elevation movement enables us to bring the specimen surface into the focal plane of the light microscope. As mentioned before, the specimen surface must be normal to the axis of the microscope, o; hence, o coincides with z.

Consider now the intersection of the electron beam with the specimen sur-

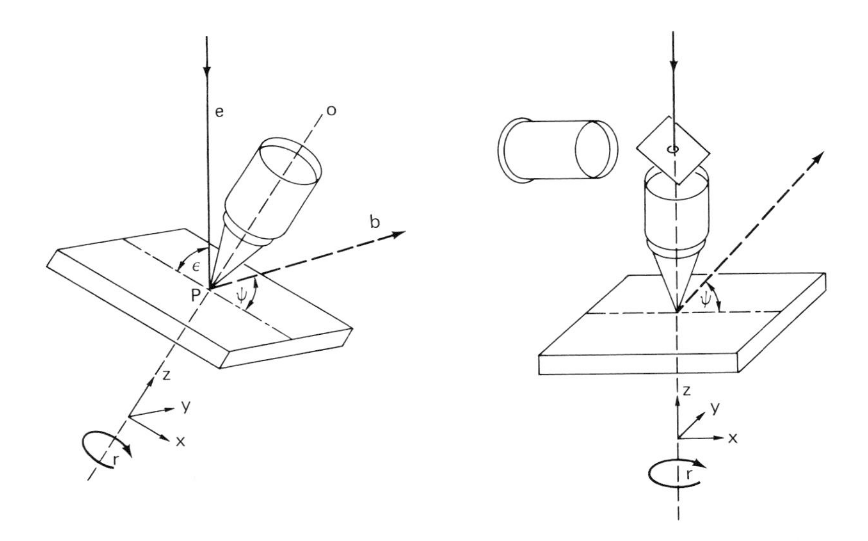

Fig. 3.31. Specimen stage movement: general case (left) and normal electron beam incidence (right). x, y, and z are the coordinates of the stage movement, ϵ is the electron beam incidence angle, e is the electron beam, b is the X-ray path, ψ is the X-ray emergence angle, o is the axis of the microscope objective, and r indicates the sense of the optional stage rotation.

face, at point P (we neglect the depth of penetration of the electrons). In order to obtain the best spatial resolution, we will focus the beam by adjustment of the current through the objective lens, so that its narrowest cross section coincides with this point. The angle of the electron beam with the specimen surface, ϵ, is the complement of the angle between the vector z and the electron beam, e. If the direction of the electron beam is normal with respect to the specimen surface, e and o coincide.

The X-ray spectrometers must focus on the point P. Since this position is, however, determined by the optical axis of the magnetic objective lens, it follows that the light microscope objective should be adjustable during analysis, so that the point P can be positioned into the center of its field of vision. For the calculation of the X-ray absorption losses within the specimen, the mean X-ray emergence angle towards each X-ray spectrometer, ψ, must be accurately known.

The requirements for the stage, including all devices related to the orientation of the specimen (sample holders), can be recapitulated as follows.

We need two rectilinear movements at right angle, parallel to the specimen surface, both indexed, and free of backlash and wobble (to 1 μm or less), and one motion normal to the surface, also resettable to 1 μm, with the help of the optical microscope. In any position, the X-ray emergence angle must be known to one degree or better. The movements which are parallel to the specimen surface should enable us to direct the beam over most of the specimen surface. As specimen mounts of sizes up to 25-mm diameter are frequently used in metallography and in the mounting of minerals, the stage and specimen holders should accommodate at least this size.

In quantitative determinations, the X-ray intensities emitted by the specimen are compared with those obtained from standards. Therefore, we should have at least two positions for standard and specimen mounts within the vacuum enclosure, from which these specimens can be brought under the beam without breaking the vacuum or shutting off the gun emission. Some instruments have sets of elemental standards available in special positions. However, one often uses additional special standards in practice, which must be introduced in the instrument with their specimen holders simultaneously with the specimens, unless standards and specimens were mounted in the same receptacle.

For the study of concentration changes along a line on the specimen, it is useful to have automatic stage motions along x and y, which can be continuous or stepwise. The speeds or step lengths should be precisely known, and several speeds should be available. To facilitate the orientation of such lines of scan, a rotation around the axis o should also be incorporated in the stage.

It is sometimes postulated that for accurate analysis of steep concentration gradients the direction of the X-ray emergence should be in an isoconcentration plane of the specimen (i.e., normal to the concentration gradient). In such a case the emerging X-rays would travel through a region of composition equal to that directly excited by the electrons. With the X-ray emergence angles used in modern instruments, this argument is specious. However, in an instrument of low X-ray emergence angle, this alignment would also require rotation of the specimen.

Rotation may be provided, either around the center of the view field of the optical microscope (i.e., around the point P), or around the geometrical center of the specimen. The latter mode of operation, although simpler to achieve, renders the manipulation of the specimen more cumbersome.

For many types of analysis, including automatic programming or analysis of selected small particles in a field, it is necessary to index positions on the specimen. The additional freedom provided by the specimen rotation complicates the use of such orthogonal indices to orient the electron beam upon a predetermined point. It is useful for such operations to have a locking arrangement which resets the rotation movement at a reproducible angle.

Rotation of the specimen around an axis *normal* to the electron beam is useful in positioning of nonflat objects for scanning electron microscopy, and for producing stereoscopic image pairs. However, since the position around this axis, which determines ϵ and ψ, must be precisely defined for quantitative analysis with X-rays, some instruments provide a different type of stage for each of these modes of operation.

The electrical current entering the specimen through the electron beam must be returned to the ground connection of the accelerating power supply. Since this current can be used to form target current images, or to perform quantitative analyses, the specimen holder must be insulated and shielded, so that a noise-free target current signal, accessible from the outside of the vacuum enclosure, can be obtained.

The above list of requirements will not necessarily satisfy the needs of all analysts. Techniques such as the functional testing of semiconductor devices, optical microscopy with transmitted light, electron diffraction, and X-ray diffraction by the Kossel technique, require special provisions. Sometimes, unusually large specimens must be analyzed, and these require special stages. For such reasons, it is advantageous that the specimen stage be removable and that it can be replaced by stages adapted to such functions. The space requirements for special stages must be taken into account in the design of the objective lens and other components close to the specimen position.

3. REFERENCES

3.1 Bloomer, R. N., *Brit. J. Appl. Phys.* **8**, 83 (1957).
3.2 Castaing, R. and Descamps, J., *C. R. Acad. Sci.* **238**, 1506 (1954).
3.3 Adler, I., Dwornik, E. J., and Rose, Jr., H. J., *Brit. J. Appl. Phys.* **13**, 245 (1962).
3.4 Campbell, A. J. and Gibbons, R., in *The Electron Microprobe*, McKinley, T. D., Heinrich, K. F. J., and Wittry, D. B., Eds., John Wiley & Sons, New York, 1966 p. 75.
3.5 Grivet, P., *Electron Optics*, rev. ed., Pergamon Press, London, 1965.
3.6 Hall, C. E., *Introduction to Electron Microscopy*, McGraw-Hill, New York, 1966.
3.7 Busch, H., *Ann. Phys.* **81**, 974 (1926).
3.8 Liebmann, G., *Proc. Phys. Soc.* **B68**, 679, 737 (1955).
3.9 LePoole, J. B., *Proc. 3rd European Conf. on Electron Microscope*, Czechoslovak Academy of Science, Prague, 1964, p. 439.
3.10 Haine, M. E. and Einstein, P. A., *Brit. J. Appl. Phys.* **3**, 40 (1952).
3.11 Haine, M. E., Einstein, P. A., and Borchards, P. H., *Brit. J. Appl. Phys.* **9**, 482 (1958).
3.12 Ogilvie, R. E., *Norelco Reporter* **11**, 75 (1964).
3.13 Bloomer, R. N., *Proc. Inst. Elec. Eng.* **104**, 153 (1957).
3.14 Fisher, R. M., in *X-Ray and Electron Probe Analysis*, ASTM Special Tech. Publ. 349, American Society for Testing and Materials, Philadelphia, Pa., 1963, p. 88.
3.15 Smith, Ll. P., in *Handbook of Physics*, Condon, E. U. and Odishaw, H., Eds., McGraw-Hill, New York, 1958, pp. 8-74.
3.16 Wittry, D. B., *J. Appl. Phys.* **29**, 1543 (1958).
3.17 Kuypers, W., Thompson, M. N., and Andersen, W. H. J., in *Scanning Electron Microscopy / 1973*, Johari, O. and Corvin, I., Eds., ITT Research Institute, Chicago, Ill., 1963, p. 10.

3.18 Bishop, F. W., *Rev. Sci. Instr.* **31**, 124 (1960).
3.19 Kamigato, O., *Japan. J. Appl. Phys.* **4**, 604 (1960).
3.20 Broers, A. N., *J. Appl. Phys.* **38**, 1991, 3040 (1967).
3.21 Crewe, A. V., et al., *Rev. Sci. Instr.* **39**, 576 (1968).
3.22 Castaing, R., *Adv. Electr. Elec. Phys.* **13**, 317 (1960).
3.23 Fontijn, L. A., Bok, A. B., and Kornet, J. G., *Proc. 5th Int. Congr. on X-Ray Optics and Microanalysis*, Möllenstedt, G. and Gaukler, K. H., Eds., Springer, Berlin, 1969, p. 261.
3.24 Malissa, H., *Elektronenstrahl-Mikroanalyse*, Springer, Vienna, 1966, p. 98.
3.25 Smith, J. P. and Pedigo, J. E., *Anal. Chem.* **40**, 2028 (1968).
3.26 Macres, V. G., Preston, O., Yew, N. C., and Buchanan, R., *Proc. 5th Int. Congr. on X-Ray Optics and Microanalysis*, Möllenstedt, G. and Gaukler, K. H., Eds., Tübingen, 1968, Springer, Berlin, 1969, p. 343.

PART II. X-RAY PHYSICS

4. The Physics of X-Rays

4.1 THE DISCOVERY AND NATURE OF X-RAYS

X-rays were discovered in 1895 by Wilhelm Conrad Roentgen, Professor of physics at Würzburg, Germany [4.1], [4.2]. While studying the properties of cathode rays–i.e., electron beams–produced by electrical discharges through gases at low pressures, he observed fluorescence effects due to a new type of radiation outside the glass apparatus in which the discharge took place. He soon established that this radiation could traverse all materials, to varying degrees, that it produced visible fluorescence with many substances, and that it darkened photographic plates. He also observed that the novel rays, which he called X-rays, travel in straight lines and are not deflected by magnetic fields, and that they discharge electrified bodies by what is now called ionization.

Soon, the "Roentgen rays" attracted much attention among laymen as well as among scientists. Some hoped that with the aid of special goggles they could look through walls and curtains. While these exciting perspectives were not realized, the general public was compensated for this failure by the use of X-rays for medical diagnosis and treatment.

Meanwhile, the scientific exploration of X-rays advanced rapidly. In 1896, Winkelmann and Straubel discovered the production of fluorescent X-rays; in 1899, Haga and Wind demonstrated that X-rays can be diffracted [4.3]. They concluded that the wavelength of X-rays must be of the order of 10^{-10} m. A more detailed study of the X-ray spectrum became possible when a practical method of wavelength dispersion was proposed by Laue, who pointed out that the wavelength of X-rays was of the order of magnitude of the spacing of atoms in solid matter. Therefore, the ordered array of atoms in a crystal should act toward an X-ray beam like a diffractive grating. This hypothesis was experimentally confirmed in 1912 by Friedrich and Knipping [4.4].

W. H. Bragg and W. L. Bragg formulated a theory of the diffraction of X-rays by crystals [4.5], and by diffracting the radiation emitted from a platinum target with a sodium chloride crystal, they obtained the first X-ray spectrum (Fig.

4.1). This spectrum contained, besides a slowly varying continuous background, several emission lines. It was soon found that the position of these lines was characteristic of the elements in the emitting X-ray target.

The X-ray emission of various elements was studied in more detail by Moseley, who observed great similarity in the spectra [4.6], [4.7]. He noticed series, or sequences, of lines (Fig. 4.2) the wavelengths of which shifted systematically from one element to the next. Before the publication of Moseley's work, Bohr had developed the concept of the atomic number [4.8]-[4.10]. Moseley's experiments supported Bohr's theories, and thus gave an experimental foundation to the periodic system of elements which had been postulated in 1869 by Mendele'ev.

On the basis of the regularity of X-ray spectra shown by Moseley, Coster and Hevesy discovered the element hafnium in 1923 [4.11], and Noddack et al. found rhodium in 1925 [4.12]. The further development of X-ray spectrometry until 1930 is described by Siegbahn [4.13]. An excellent account of the knowledge of the physics of X-rays before the Second World War is presented by Compton and Allison [4.14]. Complementary information is contained in the treatise by Blokhin [4.15]. For a modern treatment of atomic inner shell processes see Crasemann [4.17], and the review articles on X-ray physics in [4.16].

4.2 X-RAY EMISSION SPECTRA

X-rays, the same as visible light or gamma rays, are a form of electromagnetic radiation. They differ from other types of such radiation by their wavelength range, which—in the context of electron probe microanalysis—goes from 10^{-8} m to 10^{-11} m. In practice, the angstrom unit is often used to define wavelengths

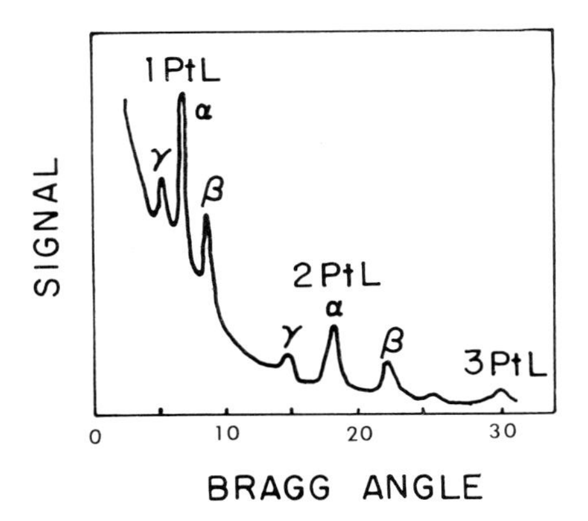

Fig. 4.1. The first X-ray spectrum, obtained by W. H. Bragg (redrawn from [4.5, p. 29]). The radiation, produced by a platinum (Pt) target, was diffracted by a sodium chloride crystal. From left to right, we observe the Pt $L\gamma$, Pt $L\alpha$, and Pt $L\beta$ lines in the first and second order of diffraction, as well as the Pt $L\gamma$ and Pt $L\alpha$ lines in the third order of diffraction.

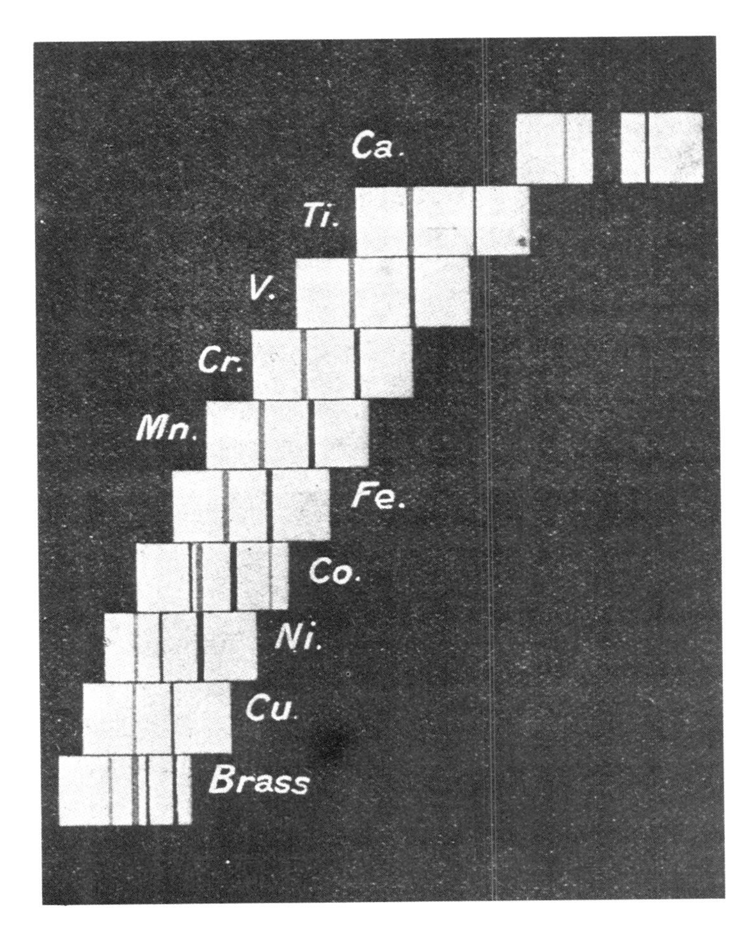

Fig. 4.2. X-ray spectra showing the change in position of Kα and Kβ-lines of various elements, in the order of the atomic number, obtained by Moseley. The cobalt (Co) target also contains iron (Fe) and nickel (Ni). Brass shows the lines of both zinc (Zn) and copper (Cu).

(1 Å = 10^{-10} m). The range of wavelengths of interest thus is from one-tenth to one hundred angstroms.[1]

Like all electromagnetic radiation, X-rays exist as indivisible units called photons, each of which has a well-defined energy, E. The X-ray counts observed in conventional counting systems are the amplified detector pulses produced in the detection of such individual X-ray photons.

[1]The Ångström unit, or angstrom (1 Å = 0.1 nm) is not recommended as part of the International System of Units (SI) [4.18], but in view of existing practice its use is temporarily accepted.

The wavelength of a photon, λ, its frequency, ν, and its energy, E, are related by

$$E = h \cdot \nu = h \cdot c/\lambda. \tag{4.2.1}$$

The values of the constants are as follows:

$$h = 6.626 \times 10^{-34} \text{ J} \cdot \text{sec} = 6.625 \times 10^{-27} \text{ erg} \cdot \text{sec (Planck's constant)}$$

$$c = 2.99793 \times 10^{8} \text{ m/sec (speed of light in vacuum).}$$

(With these constants, E is obtained in joules.) In X-ray physics the electron volt is frequently used as a measure of energy:

$$1 \text{ eV} = 1.6021 \times 10^{-19} \text{ J},$$

so that

$$h = 4.135 \times 10^{-15} \text{ eV} \cdot \text{sec}.$$

Using the angstrom unit for λ, we also obtain:

$$E \cdot \lambda = h \cdot c = 12\,397 \text{ eV} \cdot \text{Å} \tag{4.2.2}$$

The order of magnitude of the energy of X-ray photons ($\sim$0.1 – 100 keV) suggests that they are produced by processes within individual atoms, and Moseley's law (see Section 4.2.1) indicates a simple relation between the energies involved in producing characteristic X-ray photons and the atomic number of the emitting element. A model for the production of X-ray photons can be derived from the theories concerning the nature of the atom.

Rutherford's experiments on electron and α-particle scattering showed the existence within the atom of a discrete nucleus of positive charge, surrounded by layers, or shells, of orbiting electrons, which are charged negatively. The potential energy of the orbital electrons increases with the distance from the nucleus. Quantum physics imposes the restriction that only a limited number of energy states can be occupied by the orbital electrons. It is more useful, in fact, to describe these electrons by their energy levels rather than by their location or trajectories. The permitted energy levels can be defined by three numbers called, respectively, the principal quantum number, n, the azimuthal quantum number, l, and the inner quantum number, j. Electrons having the same principal quantum number are said to occupy the same shell, and electrons having all three quantum numbers in common share the same subshell (Table 4.1)

Table 4.1. Energy Levels of Inner Orbital Electrons.

Atomic Levels																
n	1	2	2	2	3	3	3	3	3	4	4	4	4	4	4	4
l	0	0	1	1	0	1	1	2	2	0	1	1	2	2	3	3
j	1/2	1/2	1/2	3/2	1/2	1/2	3/2	3/2	5/2	1/2	1/2	3/2	3/2	5/2	5/2	7/2
X-Ray nomenclature	K	L_I	L_{II}	L_{III}	M_I	M_{II}	M_{III}	M_{IV}	M_V	N_I	N_{II}	N_{III}	N_{IV}	N_V	N_{VI}	N_{VII}

The number of orbital electrons in a neutral atom is equal to its atomic number, Z. The number of electrons which can occupy each shell is limited to $2n^2$; each subshell can contain up to $2j + 1$ electrons. The first shell is therefore filled when it contains two electrons, the second shell is filled with eight electrons, and the third shell is filled with eighteen electrons. The filling of successive shells with increasing atomic number gives rise to the periods of Mendele'ev's table; those properties of the atom which are related to the number of electrons in orbit (e.g., chemical properties and X-ray emission energies) are determined by the atomic number.

Ionization is the removal of an orbital electron (or more than one) from an atom. The energy required for this process increases from the outer to the inner shells; within one shell or subshell, it increases with increasing atomic number.

Ionization of an inner shell produces a vacancy which is immediately filled by the descent of an electron from a more peripheral shell. This produces another vacancy in the shell which the descending electron had originally occupied. In the process an amount of energy is freed which is equal to the difference of the potential energies of the two orbital levels. This energy can manifest itself in two ways. Either another electron is ejected from an outer orbit (Auger electron), and the energy excess over the ionization potential for this electron is used to impart kinetic energy to the Auger electron, or the entire energy is emitted as an X-ray photon (Fig. 4.3). The probability that this second event, rather than an Auger electron emission occurs as a consequence of the ionization, is called the fluorescent yield (see Subsection 4.2.4).

In the electron probe specimen, an ionization of the inner level of an atom which may produce the emission of an X-ray photon may be caused by the impact of one of the electrons of the beam, or by interaction with an X-ray photon. In the first case we speak of *direct* or *primary excitation*, and in the second case of *secondary* or *fluorescent X-ray excitation*. Other ionization mechanisms are of no interest for electron probe microanalysis.

If an electron causes ionization, its energy is diminished by that required to remove the orbital electron, plus the kinetic energy imparted to that electron. The minimum energy required is equal to the product of the *critical excitation potential* and the electric charge of the electron. In fluorescent X-ray excitation,

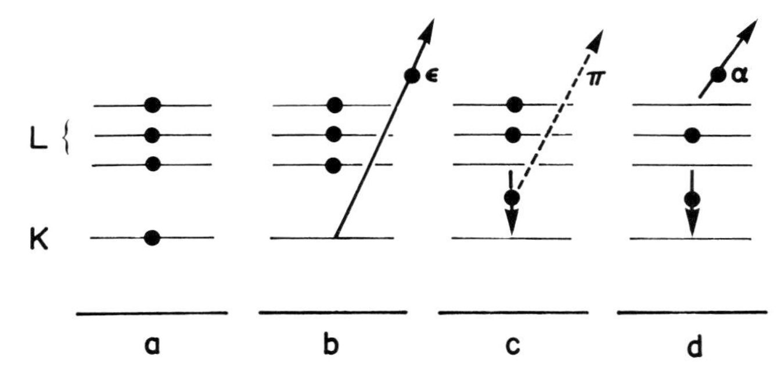

Fig. 4.3. Schematic representation of phenomena related to inner-shell ionization.
(a) Neutral state. All levels are occupied by electrons.
(b) Ionization. An electron from an inner shell is ejected. If the ionization was caused by a photon, the electron ϵ is called a photoelectron. A vacancy is created in the K-shell.
(c) X-ray photon emission. An electron from a higher level descends into the vacant level; the energy difference is invested in the photon π.
(d) Auger electron emission. Same as (c), but the energy excess is used to remove the Auger electron α from its orbit, and to impart to it kinetic energy.

the ionizing photon is absorbed, and the difference in energy between the primary and the secondary X-ray photon is imparted to a photoelectron. In either case, the emission of an Auger electron instead of a photon may occur. The energy spectra of the electrons produced in these processes can be used to obtain analytical information (Auger spectroscopy [4.19], electron spectroscopy for chemical analysis, or ESCA [4.20]). A discussion of electron spectroscopy is, however, outside the scope of this book.

In shells other than the innermost, or K-shell, there exist also nonradiative transitions (Coster-Kronig transitions) which are based on the passage within one shell of electrons from one subshell to another.

In what follows, we will discuss the wavelengths and relative intensities of the X-ray lines emitted from atoms ionized in the inner shells. A quantitative treatment of the probability of primary ionization by electrons will be given in Section 9.2.

The traditional nomenclature of the levels used in X-ray spectroscopy is somewhat arbitrary. The shells and subshells are listed in Table 4.1, where the corresponding values of the quantum numbers are also given. The most important transitions which produce characteristic X-ray lines are shown, for gold, in Fig. 4.4. The heavily marked transitions are highly probable, and thus produce intense lines. A more complete listing of lines is shown in Table 4.2. The lines most commonly used are underlined in this table, and the initial and final energy levels are indicated.

The X-ray lines are classified into series, according to the level at which ionization took place initially. The ionization of the K-shell produces K-lines, the

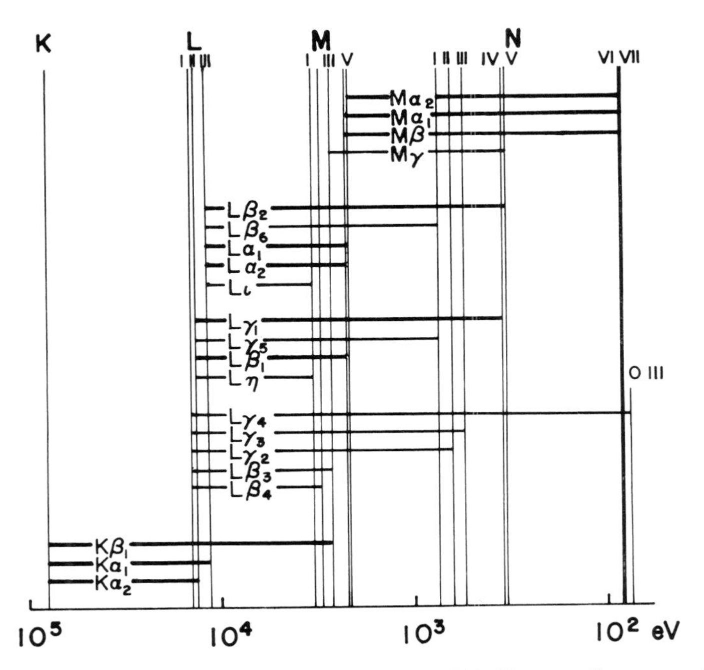

Fig. 4.4. The atomic energy level diagram of gold (Z = 79). Lines usually measured in electron probe microanalysis are drawn heavily. Lines not commonly observed are omitted. See Table 4.1 for definition of energy levels by quantum numbers.

ionization of the L-shell produces L-lines, etc. Within each series the lines are traditionally labeled by Greek letters which follow the capital letter designating the shell, and which are usually followed by an arabic number as subscript (e.g., $L\alpha_1$). The subscripts classify the lines of a series in the order of intensities. The Greek letters themselves are not subscripts.

Some of the transitions which, according to the diagram of levels (Fig. 4.4), would appear possible, do not occur at all or are extremely infrequent. Such transitions are said to be forbidden. The allowed transitions are those which obey the following transitions rules:

$$\Delta n < 0, \quad \Delta l = \pm 1, \quad \Delta j = -1, 0, +1.$$

The emission of a characteristic X-ray photon requires the passage of an electron from one orbital level to another. Therefore, a characteristic X-ray photon cannot be emitted if the initial position for this electron is empty in the neutral atom. For this reason, L-lines are not normally emitted from elements of atomic number below eleven, and hydrogen and helium do not emit any characteristic X-rays at all. However, in solid materials, levels which are empty in the neutral

Table 4.2. X-Ray Lines and Transitions.[a]

	INITIAL LEVEL OF IONIZATION								
FINAL LEVEL	K	L_I	L_{II}	L_{III}	M_I	M_{II}	M_{III}	M_{IV}	M_V
L_I	-								
L_{II}	$\underline{K\alpha_2}$								
L_{III}	$\underline{K\alpha_1}$								
M_I	-		$L\eta$	L_1					
M_{II}	$K\beta_3$	$L\beta_4$							
M_{III}	$\underline{K\beta_1}$	$L\beta_3$							
M_{IV}	$K\beta_{10}$		$L\beta_1$	$\underline{L\alpha_2}$					
M_V	$K\beta_9$			$\underline{L\alpha_1}$					
N_I			$L\gamma_5$	$L\beta_6$					
N_{II}	$K\beta_2$	$L\gamma_2$							
N_{III}	$K\beta_2$								
N_{IV}			$L\gamma_1$	$L\beta_{15}$			$M\gamma_2$		
N_V				$L\beta_2$			$M\gamma_1$		
N_{VI}			$L\nu$					$M\beta_1$	$\underline{M\alpha_2}$
N_{VII}			$L\nu$						$\underline{M\alpha_1}$
O_I			$L\gamma_8$	$L\beta_7$					
O_{II}			$L\gamma_4$						
O_{III}									
O_{IV}			$L\gamma_6$						
O_V				$L\beta_5$					

[a]Strong lines which are frequently used in electron probe microanalysis are underlined.

atom may be occupied. This explains, for instance, the emission of $L\alpha_{1,2}$ radiation from lithium and beryllium.

The X-ray spectrometers used in electron probes frequently lack the resolving power to separate doublets (lines at close distance in the spectrum). The unresolved doublets are measured in such a case as if they were a single line, and this is indicated by dropping in the notation of subscripts. Thus, the notation $K\alpha$ refers to the measurement of the unresolved doublet $K\alpha_1 + K\alpha_2$. Sometimes, the notation $K\alpha_{1,2}$ is used for the same purpose.

4.2.1 Moseley's Law

The minimum potential V_q required to remove an orbital electron from a shell of an atom is called the critical ionization potential of the respective atom and shell. It is related to the energy required to dislodge the electron from its orbit, E_q, by

$$V_q = E_q/e$$

where e is the charge of the electron. The subscript q defines the shell or subshell (K, L_I, . . .) and is appended to symbols of variables which depend on the shell or subshell.

The critical ionization potential is, in first approximation, proportional to the square of the electrical charge of the nucleus (i.e., the atomic number, Z). Because of the repulsion of the dislodged electron by the negative charge of the other shell electrons, the ionization energy is diminished by a screening constant σ:

$$E_q \propto (Z - \sigma)^2. \tag{4.2.3}$$

If an electron moves from one shell level to another, the energy emitted or absorbed in the process is the difference in the energies of the two levels. It follows that the energies, and hence the frequencies, of homologous lines vary smoothly with the atomic number of the emitting element, and are approximately proportional to the square of the atomic number. This regularity in the X-ray emission spectra was first described by Moseley [4.6], [4.7], and is known as Moseley's law. While Moseley made reference only to the emitted lines, similar relationships between the atomic number and the absorption edges–which represent the critical excitation potentials on the wavelength or frequency scale–are also referred to as applications of Moseley's law (Fig. 4.5).

Units of Wavelength. The X-ray diffraction technique proposed by Laue and used by Bragg and Moseley can yield very precise wavelength measurements [4.17, Vol. 2, ch. 3]; it is the most common approach to X-ray spectroscopy (see p. 99). Diffraction takes place when the wavelength, λ, the distance between planes of diffracting atoms within the crystal, d, and the angle of diffraction, θ (see Fig. 4.26), between the X-ray beam and the diffracting planes, are related by Bragg's law:

$$n\lambda = 2d \sin \theta. \tag{4.2.4}$$

In this equation n is an integer called the order of diffraction. For very accurate measurements, a small correction is applied for the refraction of X-rays entering and leaving the crystal. After corrections for refraction and for thermal expansion of the crystal are made, the accuracy of a careful wavelength measurement is limited by our knowledge of the lattice constant of the crystal. This limitation led Siegbahn to propose in 1919 that wavelengths should be calibrated in x-units,

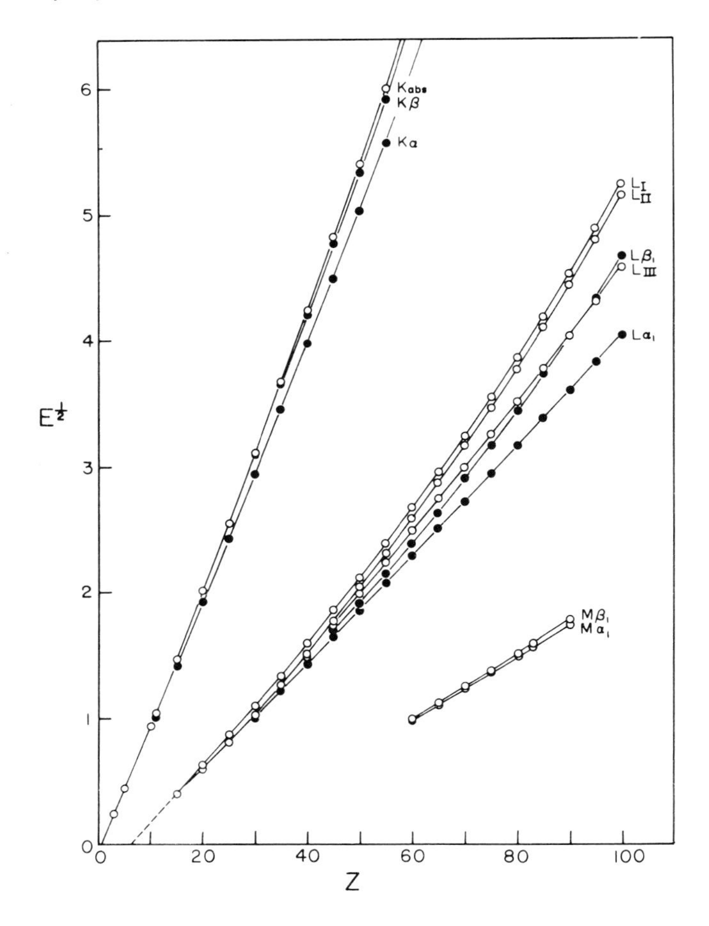

Fig. 4.5. Diagram showing the validity of Moseley's law for several lines and absorption edges.

defined by assigning the value of 3029.04 xu to the first-order grating constant of calcite at 18°C. This value was chosen to obtain as closely as possible the relation

$$1000 \text{ xu} = 1 \text{ Å} \equiv 10^{-10} \text{ m}.$$

It was then assumed that all perfect calcite crystals have the same lattice constant. When the limitations of this assumption became known, it was proposed that the wavelengths of certain X-ray lines be used as primary standards [4.21], [4.22]. Although the angstrom unit (Å) is metric, while the x-unit is not, this distinction is of no practical concern to electron probe microanalysis, in view of the limited precision of the X-ray spectrometers used.

4.2.2 Chemical effects on X-ray lines

The electrons contained in the outer shells are involved in chemical binding (valence electrons), and their energy states vary accordingly with the nature and strength of chemical bonds. On the contrary, the inner-shell electrons are not involved in chemical processes, and therefore, the X-ray generation due to interaction of these electrons is virtually independent of the chemical or physical state of the emitting atoms. This is the basis upon which elemental analysis by X-ray spectrometry is founded. Chemical effects on X-ray emission can only be observed when radiation of low energy is generated in transitions from the outer levels involved in chemical bonds. In X-rays of low energy, the relative energy distribution of the emitted photons is broader than for harder X-ray lines, and can be affected by the chemical states, Instead of narrow lines, we observe emission bands, and their shape and position may change with the chemical state (e.g., with the valence) of the target atom. Moreover, the intensity of emission depends on the degree to which the level from which the electron descends is filled. Hence, the integral intensity of band emission may also be affected. Such chemical effects can be observed in the K-line emission of elements of atomic number below 20 (particularly in the $K\beta$-lines), and in the L-spectra of elements of atomic number below 30.

Chemical effects on X-ray lines can be used to obtain information on binding, although it is much more difficult to interpret them than to observe them [4.23], [4.24]. However, they may also cause errors in quantitative elemental analysis. When K-lines of elements of atomic number below twenty are used for this purpose, the element of reference should be similarly bound in the standard as in the specimen, in order to exclude the possibility of such an error [4.25].

Satellite Lines. Lines which cannot be explained on the basis of the orbital energy levels of singly ionized atoms can frequently be observed. They are relatively weak and appear at wavelengths close to those of strong diagram lines; therefore, they are called satellite lines. The energy of the satellite lines suggests that they were produced by doubly ionized atoms.

For the electron probe analyst, these lines are usually too weak to be useful. They can be used, however, to demonstrate remarkable chemical effects; an interesting example is the change, from element to oxide, of the intensity ratio and wavelengths of the $K\alpha_3$ and $K\alpha_4$ satellite lines of magnesium and aluminum (Fig. 4.6) [4.26].

4.2.3 X-Ray Line Intensities

We will now consider the factors which govern the probability of the emission of an X-ray photon of a given line from an atom previously ionized in the corresponding shell. The *fluorescence yield* indicates the probability of a radiative transition (i.e., of the emission of an X-ray photon). The *relative transition probabilities* (weights of the lines) give the proportions in which the photons of various lines

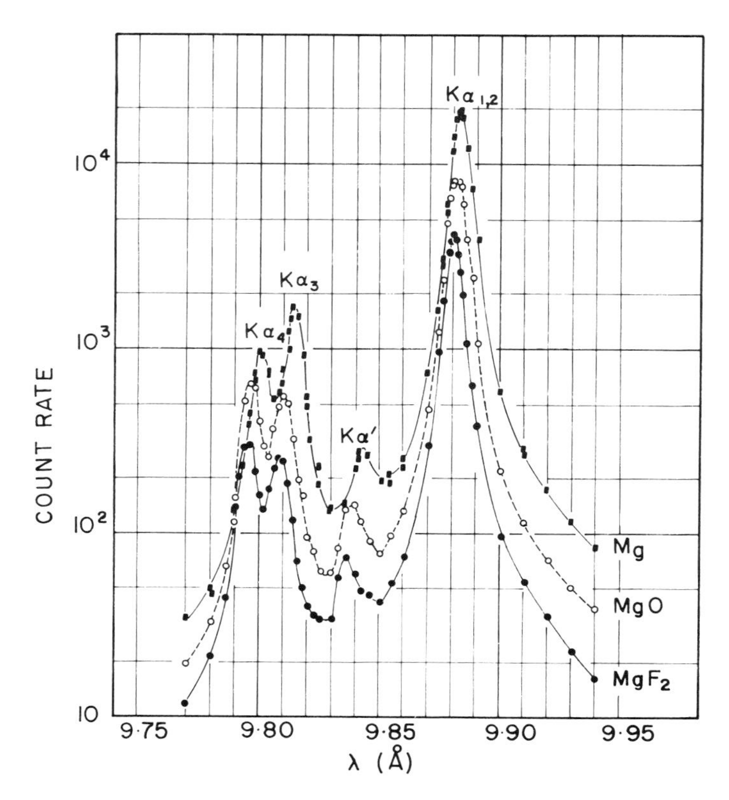

Fig. 4.6. Chemical effects on the emission spectra of magnesium (Mg), magnesium oxide (MgO), and magnesium fluoride (MgF_2). Count rates are logarithmically plotted, and vertically shifted for easier comparison. Note the differences in intensity between Kα_3 and Kα_4, and the perceptible peak position shifts in all lines.

of the same level are produced. Finally, the *Coster-Kronig coefficients* refer to nonradiative transitions of vacancies from one subshell to another, in shells other than the K-shell. The knowledge of these parameters, and hence of the relative intensities of lines emitted by each element, is useful in the qualitative interpretation of X-ray spectra. But a more important role is assigned to them in the data reduction procedures for quantitative electron probe microanalysis. It is therefore useful to describe this role before we discuss the emission parameters in more detail.

Quantitative analysis is based on the measurement of the ratio of X-ray intensities emitted from the specimen and from a standard of known composition, respectively. The values of parameters which equally affect both emissions have no effect on the result of the analysis. In general, the fluorescence yields, weight of lines, and Coster-Kronig coefficients for a given X-ray line have the same value for the specimen and the standard. The situation differs, however, if the specimen also emits secondary radiation. In this case, the emission parameters for the exciting radiation affect the signal intensity of the specimen, but not necessarily of the standard. For instance, in the measurement of chromium in steel, we must take into account the indirect excitation of the chromium K-radiation by the K-radiation of iron. If the standard is pure chromium, an analogous secondary emission does not take place in it. Hence, the fluorescent yield for Fe K-radiation, and the weight of each exciting iron line, must be known. The same parameters for the chromium lines apply, however, to both the specimen and the standard, and thus cancel in the calculation of the relative intensity of Cr K-line emission.

Fortunately for the analyst, the fluorescent excitation by lines from the K-level is usually the most important: the emission parameters for higher levels are less accurately known, but their effects are usually less significant.

4.2.4 The Fluorescence Yield

The probability of X-ray emission after ionization is called the *fluorescence yield* because in early experiments the X-ray excitation was by fluorescence. However, the mechanism of ionization has no effect on the value of the fluorescence yields; they apply equally to the X-ray emission from atoms excited by electrons. The mechanism of ionization, and the conditions under which ionization occurs, do, however, determine the probability of ionization. The cross sections for ionization by electrons will be discussed in Section 9.2. For excitation by means of X-rays, see the photoelectric absorption coefficients in Section 4.3.

In the K-shell, the only possible outcomes of an ionization are the emission of a K-line X-ray photon or of an Auger electron. Hence, the sum of the probability of X-ray emission (fluorescence yield) and that of Auger-electron emission (Auger yield) is unity. The K-fluorescence yield, ω_K, increases monotonically with the atomic number of the emitting element. Many algebraic equations expressing this relation have been proposed. Most are of the form

$$\omega_K = \frac{Z^4}{a + Z^4} \quad \text{or} \quad \omega_K = \frac{(Z-1)^4}{b + (Z-1)^4}. \tag{4.2.5}$$

Burhop [4.27] proposed an expression which permits a closer approximation to experimental data:

$$\left(\frac{\omega_K}{1 - \omega_K}\right)^{1/4} = a' + b'Z + c'Z^3 \tag{4.2.6}$$

so that

$$\omega_K = \frac{d^4}{1 + d^4}$$

with

$$d = a' + b'Z + c'Z^3 .$$

The two simpler expressions in Eq. (4.2.5) are special cases of Eq. (4.2.6), with ($a' = 0$, $b' = a^{-1/4}$, $c' = 0$) for the first formula, and ($a' = -b^{-1/4}$, $b' = b^{-1/4}$, $c' = 0$) for the second formula. For the fit of experimental values [4.2] to such an approximation, see Fig. 4.7.

The mechanisms of X-ray production from the L- and higher levels are more complex, and less accurately known. Although mean fluorescent yields, which are weighted averages of the yields from the subshells, are found in the literature, for the calculation of emission of lines each sublevel should be considered separately. For the L-shell, we have the three yields, $\omega_{L\text{-}I}$, $\omega_{L\text{-}II}$, and $\omega_{L\text{-}III}$; for the M-shell, there will be five yields, and so on. Because of the difficulties in measuring these yields, and of the large number of parameters needed, our knowledge of the fluorescent yields in levels above the K-level is quite incomplete. Since the strongest L-line, the $L\alpha_1$ line, is generated from a vacancy in the L_{III} level, the yield for the L_{III} level is the most important in practice. The variation with atomic number in the higher levels is not as smooth as in the K-level, so that interpolation procedures are not as accurate for the L- and M-yields (Figs. 4.8 and 4.9).

A further complication arises in the shells above the K-level from the transitions from one subshell to another. These *Coster-Kronig transitions* consist of the descent of an electron from an outer (i.e., higher in energy) to an inner subshell, while accordingly the vacancy migrates in the opposite direction. Since transitions leading to a state of ionization of higher potential energy do not occur, the allowed transitions of the vacancy in the L-shell are $L_I \rightarrow L_{II}$, $L_I \rightarrow L_{III}$, and $L_{II} \rightarrow L_{III}$. The number of ionizations of the subshell L_I which lead to X-ray production is not affected by these transitions, but additional ionizations are created in the L_{II} and L_{III} levels which in turn increase the X-ray production from these subshells. Let us call the respective Coster-Kronig transition probabilities f_{12}, f_{13}, and f_{23} (Fig. 4.10). If the probabilities of direct ionization of the three levels are called n_I, n_{II}, and n_{III}, the total probabilities of X-ray emission from the subshells, including the effects of the Coster-Kronig transition, are:

$$\begin{aligned}
I_{L\text{-}I} &= n_I \omega_{L\text{-}I} \\
I_{L\text{-}II} &= (n_{II} + n_I \cdot f_{12}) \cdot \omega_{L\text{-}II} \qquad (4.2.7) \\
I_{L\text{-}III} &= [n_{III} + n_{II} \cdot f_{12} + n_I (f_{13} + f_{12} \cdot f_{23})]\, \omega_{L\text{-}III} .
\end{aligned}$$

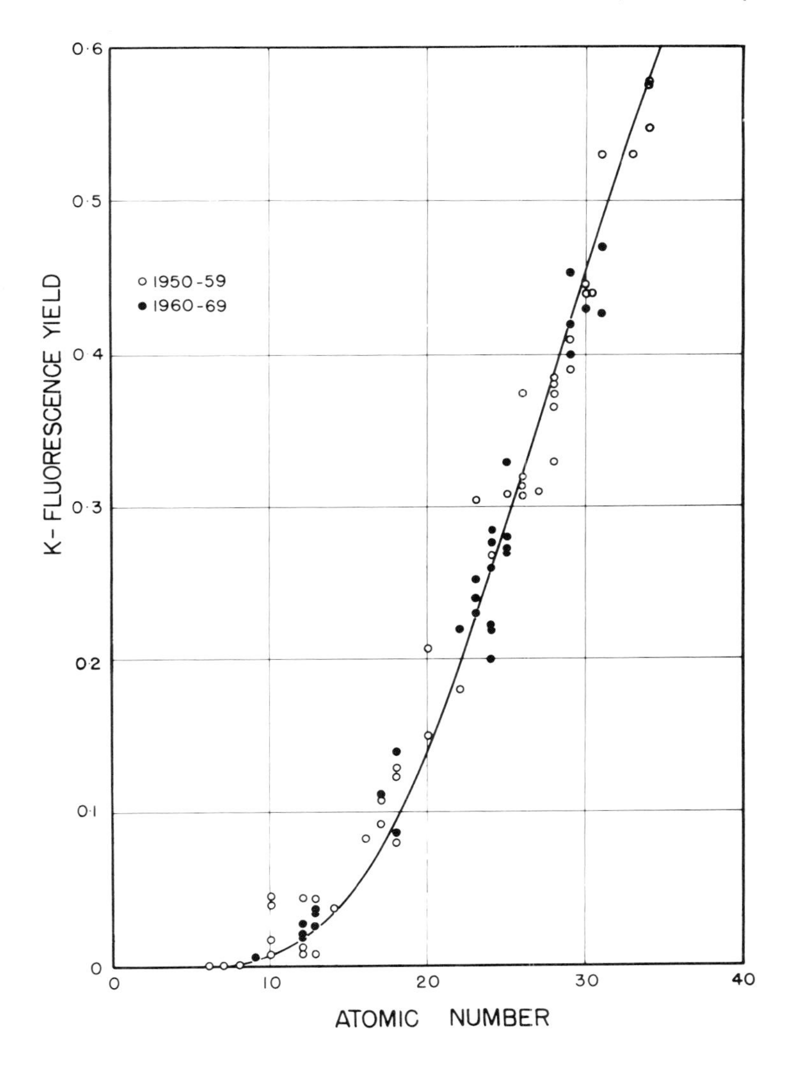

Fig. 4.7. K-fluorescence yield as a function of atomic number. The experimental data are extracted from [4.28]. The curve is an approximation due to Burhop [4.27].

The excitation probabilities n_{I}, n_{II}, and n_{III} depend somewhat on the excitation conditions. However, the calculation of the intensities according to the equations given above cannot be performed with a high degree of accuracy since, in the first place, the transition probabilities, f_{ij}, are not well known. It is therefore justifiable for approximate calculations to neglect the differences in excitation potentials (and, in the case of fluorescence, the differences of absorption co-

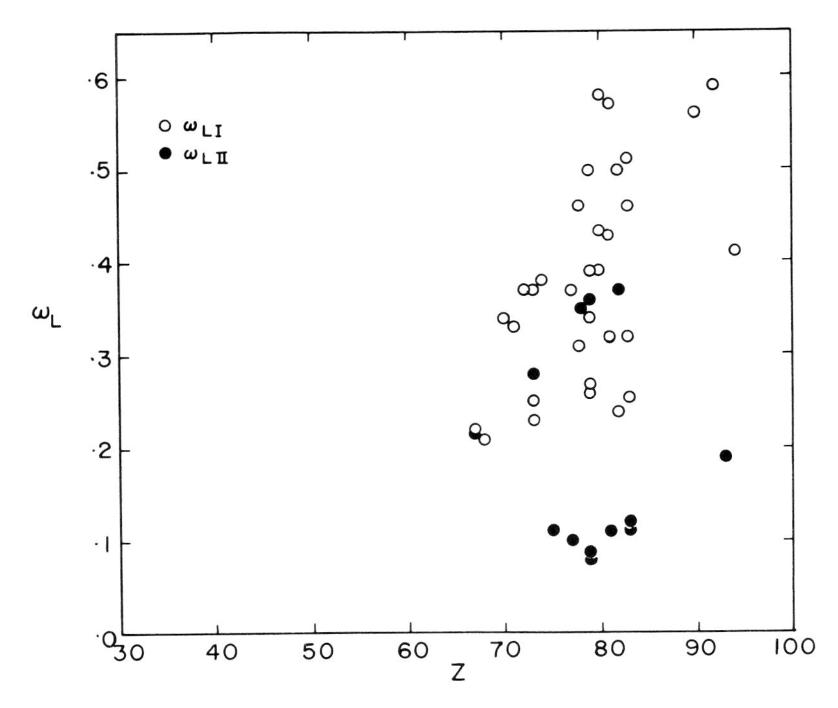

Fig. 4.8. L_I- and L_{II}-fluorescence yields. Data from [4.28].

efficients), and to postulate that in accordance with the number of electrons in each subshell

$$n_I : n_{II} : n_{III} \simeq 1{:}1{:}2$$

so that

$$\begin{aligned} I_{L\text{-}I} &= n_I \omega_{L\text{-}I} \\ I_{L\text{-}II} &\simeq n_{II}\,(1 + f_{12})\omega_{L\text{-}II} \\ I_{L\text{-}III} &\simeq n_{III}\,[1 + \tfrac{1}{2}f_{23}\,(1 + f_{12}) + \tfrac{1}{2}f_{13}]\;\omega_{L\text{-}III}. \end{aligned} \tag{4.2.8}$$

These relations are sometimes expressed in terms of "effective yields":

$$\begin{aligned} \omega_{LII}\,(\mathrm{eff}) &= (1 + f_{12})\omega_{LII} \\ \omega_{LIII}\,(\mathrm{eff}) &= [1 + \tfrac{1}{2}f_{13} + \tfrac{1}{2}(1 + f_{12})f_{23}]\;\omega_{LIII}. \end{aligned} \tag{4.2.9}$$

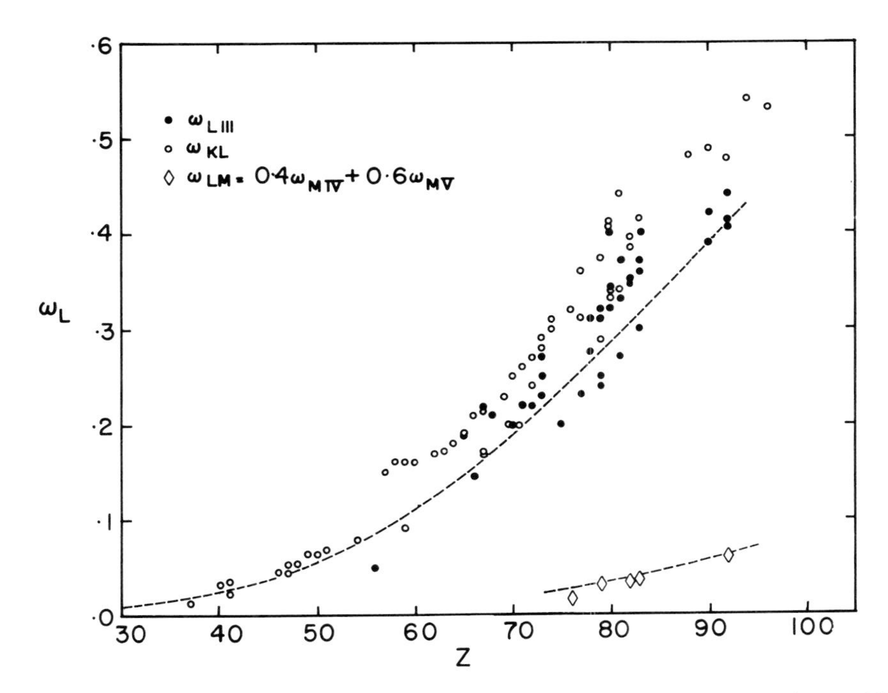

Fig. 4.9. Fluorescence yields from the L_{III} level, mean L-fluorescent yields, and mean M-fluorescent yields. Data from [4.28].

The large uncertainties in the effective yields from shells other than the K-shell are usually not of grave consequence for the accuracy obtainable in electron probe microanalysis, since, except for indirect excitation, the yields are the same for specimen and standard. For further information on atomic fluorescence yields see [4.28], [4.29].

4.2.5 Relative Transition Probability

Let us assume that a spectrum has been measured and that the number of photons collected for every line is large so that statistical fluctuations can be neglected. In such a case, the relative transition probability, or "weight" of a line is equal to the number of generated photons of this line, divided by the number of photons generated in all lines produced by ionization of the same shell or subshell. For a $L\alpha_1$ line, for instance, the relative transition probability would be the number of photons of the $L\alpha_1$ line generated within the specimen, divided by the number of photons in all lines which are produced after ionization of the L_{III} shell. The weights of K- and L-lines are shown in Figs. 4.11 and 4.12, respectively. Experimental data can be obtained from [4.15], [4.29], as well as from [4.17, Vol. 1, Ch. 6]. As will be seen later, these parameters are used in the

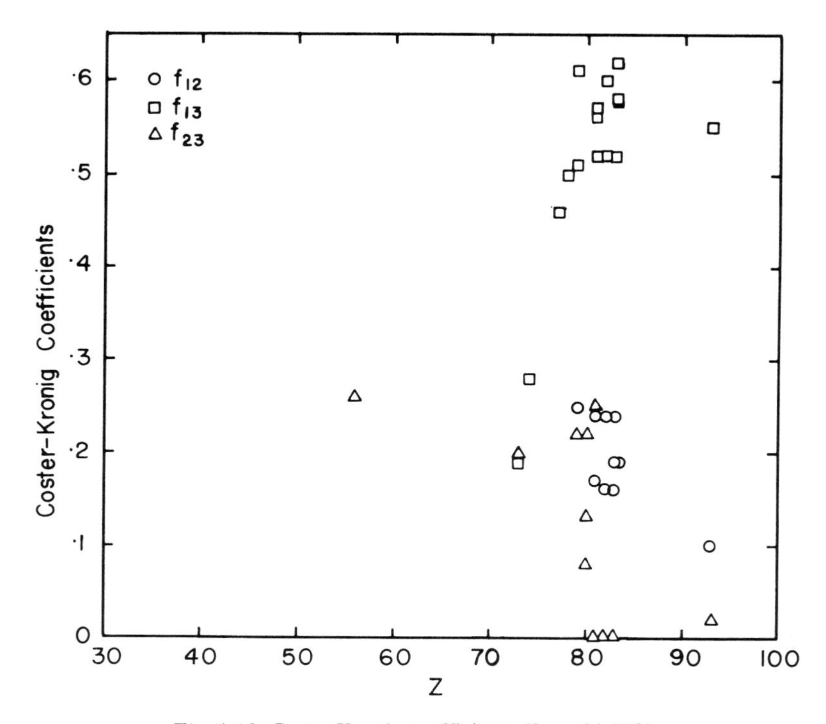

Fig. 4.10. Coster-Kronig coefficients (from [4.28]).

theoretical calculation of X-ray intensities. We calculate a combined relative transition probability for doublets which are not separated in the measurement, such as the Kα and Mα doublets. The weights of line can be considered as virtually invariable with respect to the mechanism of atomic ionization, or to the specimen composition. Significant exceptions, particularly with satellite lines, have been mentioned in the discussion of chemical effects (see Subsection 4.2.2). Ratios of intensities of lines originating from different shells or subshells, in contrast, depend on the conditions of ionization, and should not be considered in the same context. It should be noted that the weights of lines are those of the generated intensities rather than those which are observed after emergence from the specimen. Hence, in the experimental determination the absorption of X-rays within the target, and the variation of spectrometer efficiency with wavelength must be taken into account, and integrated peak intensities rather than observations at the spectrometer position of maximum intensity must be used.

The relative transition probability for K-lines is fairly well known. The sum of the weights of the K$\alpha_{1,2}$ and K$\beta_{1,2,3}$ lines is equal to one. Calling the weights of lines p, with subscripts indicating the respective line, we can write:

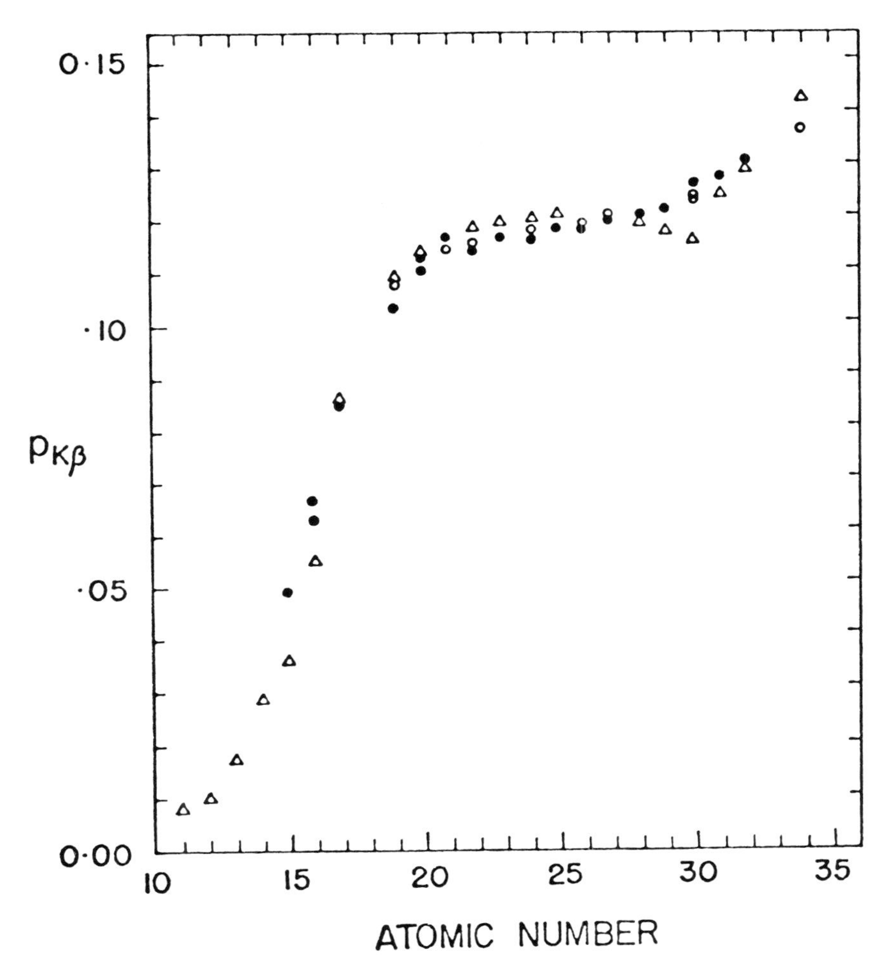

Fig. 4.11. Relative transition probability of the Kβ-line. Full circles: measurements with Si(Li) detector at NBS. Empty circles: measurements with Si(Li) detector by other authors. Triangles: measurements with crystal detector, by Salem and Falconer (1972).

$$p_{K\alpha} + p_{K\beta} = 1.$$

Similar relations for the L-lines can be deduced from Fig. 4.4 or Table 4.2. It can also be appreciated that the weights of $M\alpha_{1,2}$, of $M\beta$ and of $M\gamma$ are all close to one since each of these emissions comes from another subshell. The greatest need in improved knowledge of weights of X-ray lines is thus in the L-spectrum.

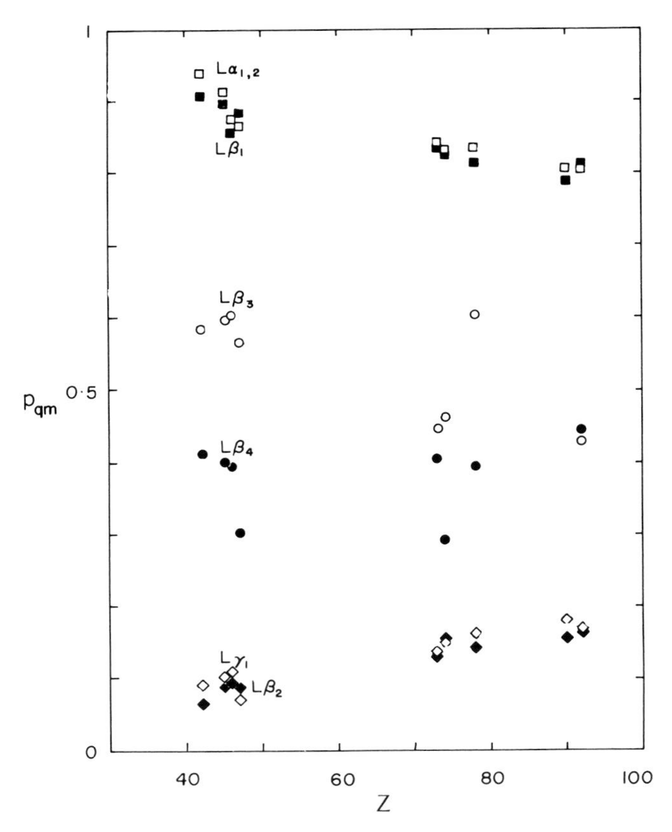

Fig. 4.12. Relative transition probabilities of the most important L-lines, as a function of the atomic number of the emitter. For the transitions see Table 4.2.

4.3 The Absorption of X-Rays

When a beam of X-rays traverses matter, its intensity diminishes. This is due both to X-ray scattering–which produces X-ray photons moving in directions other than that of the primary beam–and to photoelectric absorption. At wavelengths shorter then 0.1 nm (1 Å), especially for absorbers of low atomic number, the contribution of scattering to the total attenuation is large (Fig. 4.13). But in the wavelength range of chief interest in electron probe microanalysis ($\lambda > 0.1$ nm), scattering is of relatively little importance, and photoelectric absorption, i.e., the interaction of the X-ray photons with the various orbital electrons, is the main mechanism for X-ray attenuation [4.17, Vol. 1, ch. 3]. The result of this interaction is the emission of a photoelectron, and ionization of the absorbing atom [see Fig. 4.3(b)] which will, in turn, emit a secondary X-ray photon or an Auger electron. The absorption, with subsequent ionization, can take place in several orbital levels of any atom of the absorber, although the

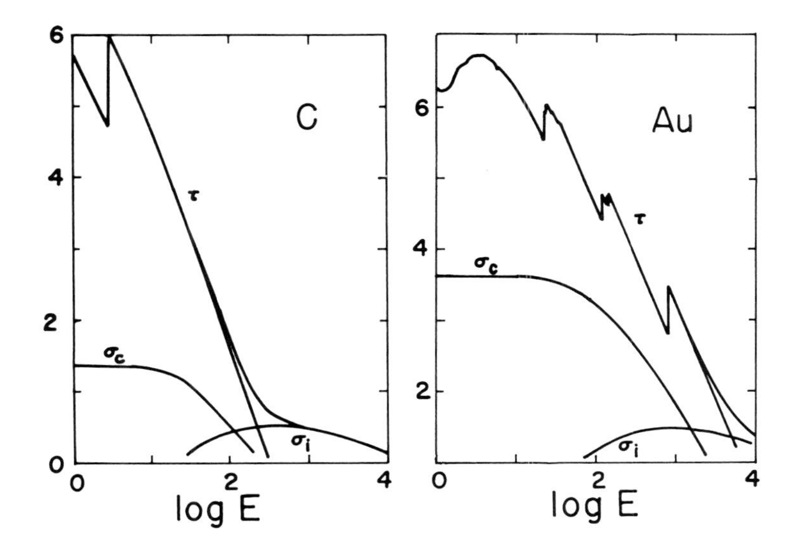

Fig. 4.13. Logarithms base 10 of the atomic cross sections (barns/atom) for the absorbers carbon (left) and gold (right), as a function of photon energy, E (keV) = 2.397/λ (Å). τ:photoelectric cross section, σ_c:coherent scattering cross section, σ_i:incoherent scattering cross section.

probabilities of absorption vary from level to level, and are strongly dependent on the atomic number of the absorber and the energy of the X-ray photon. Roentgen had observed that heavy (i.e., high atomic number) materials are stronger absorbers than light materials. In view of the effects of X-ray absorption on the measured X-ray intensities we will discuss this dependence more quantitatively.

Because of the small role played by scattering of X-rays within the wavelength range of interest, we seldom find in electron probe literature a distinction drawn between the photoelectric absorption and the total attenuation of X-rays. Scattering will be discussed in Section 4.4 in relation with X-ray diffraction.

Photoelectric absorption does not occur unless the energy of the ionizing photon is large enough for the removal of the orbital electron of interest to infinity (i.e., out of the field of the nucleus). Ionization thus requires that the photon to be absorbed should have, for each orbital level, an energy at least equal to the critical ionization energy mentioned in Section 4.2. If the energy exceeds this level, the difference is imparted to the photoelectron which is ejected, as kinetic energy.

In a thin layer of an absorber, where only a small fraction of the incident radiation is absorbed, this fraction is proportional to the layer thickness, ds:

$$dI = -\mu I \cdot ds. \tag{4.3.1}$$

The proportionality factor μ is called the X-ray *absorption coefficient*. The negative sign in Eq. (4.3.1) keeps this coefficient positive. For thick layers, integration over s yields:

$$I' = I \exp(-\mu s) \quad \text{(Beer's law)}. \tag{4.3.2}$$

In this equation, I is the incident intensity and I' is the emergent intensity. Since both refer to radiation of the same wavelength, the scale of measurement of intensities is irrelevant; the same coefficient applies to energy measurements as to count-rate measurements of X-ray fluxes.

The layer thickness s of an absorber can be defined in various ways. If it is expressed in units of length, the corresponding absorption coefficient, μ_l [cm^{-1}], is called the *linear X-ray absorption coefficient*. The photoelectric attenuation of X-rays is proportional to the number of atoms encountered by a photon along its path through the absorber. Therefore, the value of the linear absorption coefficient depends on the density of the absorber, and hence of temperature, pressure, state of aggregation, and chemical combination. This dependence on density can be avoided by defining an *atomic absorption cross section*, which is the attenuation caused by a layer containing one atom of absorber per unit of area. If 1 cm^2 is chosen as such a unit of area, the numerical values of the coefficients so defined are inconveniently small. It is therefore common to express atomic cross sections in terms of 1 atom/barn (1 barn = 10^{-24} cm^2) [4.30], [4.31].

Because X-ray absorption is an atomic process which occurs mainly in the inner orbital electron levels, the atomic absorption coefficients, μ_{at}, are, for all elements and wavelengths, virtually independent of chemical and allotropic changes, and change smoothly as a function of atomic number. Exceptions to these rules will be discussed later.

The atomic absorption coefficients of an element of atomic number Z can be calculated with good approximation by an equation of the form

$$\mu_{at}(Z, \lambda) = c_{at} Z^4 \lambda^n \tag{4.3.3}$$

in which the power n lies between 2.5 and 3, varying slightly with element and wavelength range. The parameter c_{at} changes slowly with the atomic number of the absorber.

If the absorber contains more than one chemical element, the following addition rule is rigorously valid:

$$\mu_{at}(*, \lambda) = \Sigma\, m_i \mu_{at}(i, \lambda). \tag{4.3.4}$$

In this equation, m_i is the atomic fraction of element i (i.e., the ratio of the number of atoms of i to the number of all atoms). The symbol $\mu_{at}(i,\lambda)$ denotes the atomic cross section of the element i for the radiation of wavelength λ. The asterisk (*) in the left-hand term of Eq. (4.3.4) denotes an absorber containing more than one element. This symbol will be reserved in this book to mark parameters and functions of multielement specimens.

A tabulation of atomic cross sections is found in [4.32]. It is more common in literature to find X-ray absorption defined by means of the X-ray *mass absorption coefficient*. The thickness of the absorber is expressed as mass per area (g/cm^2). As the density of the absorber, ρ, is equal to mass per volume (g/cm^3) the mass thickness, z, can be expressed as the product of linear thickness and density:

$$\rho s = z. \tag{4.3.5}$$

The equation which defines the mass absorption coefficient is thus

$$dI = -\mu I \cdot d\,(\rho s) = -\mu I \cdot dz. \tag{4.3.6}$$

Hence, for a thick absorber $I' = I \exp(-\mu z)$, and the unit of μ is cm^2/g.

In view of the common use of mass absorption coefficients in literature, absorption is always described in this book by means of mass absorption coefficients. The symbol μ without subscript will always denote the mass absorption coefficient.

In the literature, the mass absorption coefficient is frequently represented by the symbol μ/ρ, in which μ denotes the linear absorption coefficient. This notation is cumbersome in practice, and unnecessary since we will not use linear absorption coefficients.

The relations among the three absorption coefficients are:

$$\mu(i, \lambda) = \mu_l\,(i, \lambda)/\rho = 10^{-24} \times \mu_{\mathrm{at}}(i, \lambda) \cdot (N_{\mathrm{av}}/A_i) \tag{4.3.7}$$

where N_{av} is the number of atoms per mole (Avogadro constant, N_{av} = 6.022 169 $\times$ 10^{-23} mol^{-1}) and A_i is the atomic weight of element i.

We can now derive the summation rule for mass absorption coefficients of multielement absorbers. Calling C_a the mass fraction of element a, we observe that a layer of unit area and thickness $d(\rho s)$ contains $N_{\mathrm{av}} \cdot (C_a/A_a) \cdot d\,(\rho s)$ atoms of element a. From the definition of μ_a, and Eq. (4.3.7), we obtain:

$$\mu(*\lambda) = \frac{-dI}{I \cdot d(\rho s)} = \sum_i \frac{N_{av}}{10^{24}} \cdot \frac{C_i}{A_i} \cdot \mu_{\mathrm{at}}(i, \lambda) = \sum_i C_i\,\mu(i, \lambda). \tag{4.3.8}$$

The mass absorption coefficient of a multielement absorber is thus the sum of the mass absorption coefficients of the components, weighted as mass fractions. This property is very convenient in spectrochemical analysis, where concentrations are usually expressed as mass fractions.

For thick absorbers we obtain

$$I' = I \exp(-z \, \Sigma \, C_i \mu_i), \tag{4.3.9}$$

and Eq. (4.3.3) transforms into

$$\mu(a, \lambda) = \frac{N_{\text{av}}}{A_a} \cdot c_{\text{at}} Z_a{}^4 \lambda^n = c \frac{Z_a{}^4}{A_a} \lambda^n \simeq c' Z_a^3 \lambda^n. \tag{4.3.10}$$

The exponent n is unchanged; c and c' are factors which vary slowly as a function of Z and λ. Figures 4.14 to 4.16 show the fit of experimental data to the wavelength dependence which is expressed in Eq. (4.3.10). The scales on these graphs are logarithmic, so that Eq. (4.3.10) would produce a straight line of slope n. The measured points on Fig. 4.14 are for the elements aluminum and titanium. A slight but significant curvature can be seen for aluminum.

Equation (4.3.10) only applies to photoelectric absorption. Therefore, log-log plots of the total attenuation coefficient as a function of wavelength at short wavelengths (i.e., high photon energy), where scattering is significant, deviate strongly from linearity, particularly if the atomic number of the absorber is low (Fig. 4.13). In such cases, however, the total attenuation is too low to be of great importance in the quantitative procedures for electron probe microanalysis.

Absorption Edges. As can be seen in Fig. 4.14, the mass absorption coefficients of aluminum and titanium change drastically and abruptly at certain wavelengths (absorption edges). The energies corresponding to these edges are those required to ionize the respective K-shells. For elements of higher atomic number, we observe similarly L- and M-absorption edges (Figs. 4.15 and 4.16), and, according to the number of shells and subshells, we obtain one K-edge, three L-edges (L_{I}, L_{II}, L_{III}), and five M-edges. Since the positions of these edges correspond to the critical ionization potentials, they can also be presented in the form of a Moseley diagram (Fig. 4.17).

The ratio of the absorption coefficients at either side of an edge—called the absorption jump ratio, $r_q = \mu_1/\mu_2$—is a measure of the probability that the absorption of a photon at the wavelength just above the edge is caused by the ionization of the corresponding level. This probability is equal to

$$\frac{\mu_1 - \mu_2}{\mu_1} = 1 - 1/r_q. \tag{4.3.11}$$

The K and L_{III} absorption jump ratios diminish with increasing atomic num-

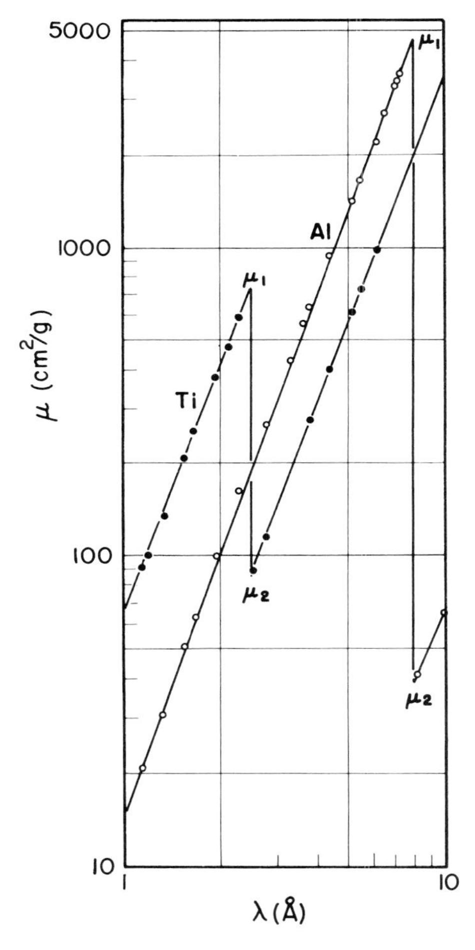

Fig. 4.14. Mass absorption coefficients of titanium (Ti) and aluminum (Al), from 1 to 10 Å. Both elements have a K-absorption edge (between μ_1 and μ_2) in this wavelength region. Circles represent experimental values. The lines are a fit to Eq. (4.3.10).

ber (Figs. 4.18 and 4.19), while the L_I and L_{II} edges change little with atomic number. The K-jump ratio can be obtained by the following empirical expression:

$$r_K = 60 \cdot Z^{-0.6}. \tag{4.3.12}$$

A study of high wavelength resolution of the absorption coefficients of an absorber in the region close to an edge discloses a fine structure which changes with the chemical state of the absorbing element [4.32]. Minor deviations from

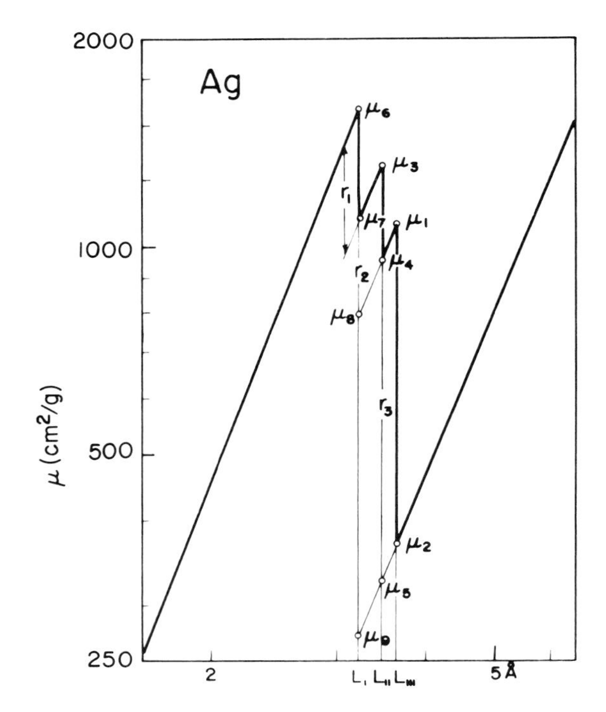

Fig. 4.15. Mass absorption coefficients of silver (Ag) (Z = 47), showing the three L absorption edges.

Eq. (4.3.10) can be observed for some distance above the absorption edges [4.33]. Therefore, it is prudent not to use for quantitative analysis emission lines which are close to an absorption edge of another element present at large concentrations.

Further deviations from Eq. (4.3.10) are found in the long wavelength region [4.34], particularly with absorbers of high atomic number (Fig. 4.13). The mass absorption cross sections in this region are very high, difficult to predict or measure accurately, and subject to the effects of the chemical state. For this reason, the use of lines in the soft X-ray region for quantitative analysis causes a significant uncertainty, as will be discussed further in Chapter 10. Since elements of low atomic number ($Z < 10$) emit X-ray lines or bands only in this region, accurate quantitative analysis for these elements is difficult or impossible (Section 12.4).

Sources and Uncertainties of X-Ray Absorption Coefficients. Mass absorption coefficients or cross sections related to them can either be calculated from theoretical principles [4.35] or determined experimentally. The determination of mass absorption coefficients is performed in experiments based on Eqs.

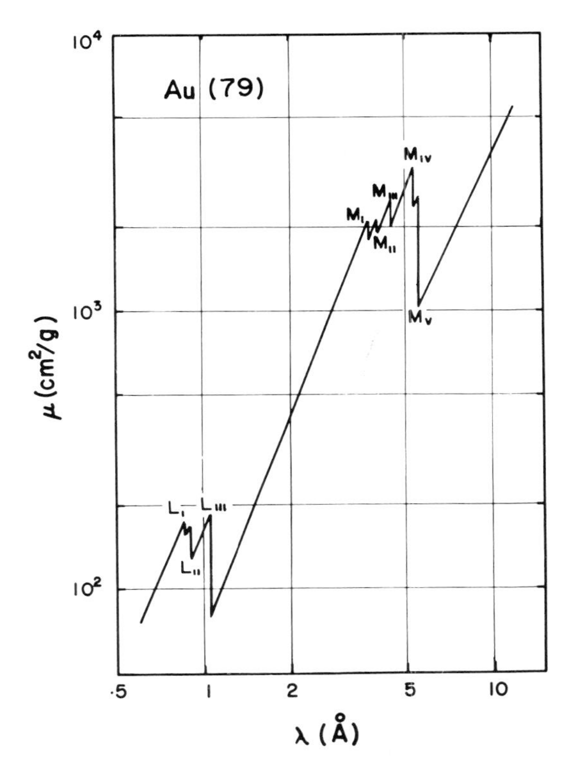

Fig. 4.16. Mass absorption coefficients of gold (Au) (Z = 79), with L- and M-absorption edges.

(4.3.1) and (4.3.2). The absorbers may either be relatively thin foils of solid material, or they may be gaseous [4.36]. Although the experiment is simple in principle, great care must be taken to avoid systematic errors. For many elements the preparation of pure absorbers of uniform thickness is difficult or impossible, especially for long wavelengths which require very thin absorbers. For this reason, our knowledge of mass attenuation coefficients for wavelengths above 10 Å is not very accurate. Between absorption edges of the same series, particularly the M- and N-edges, they are also poorly known [4.37].

For solid absorbers the uniformity of thickness is very important. This fact was frequently overlooked in earlier measurements, in which materials deposited on filter paper were used. When gaseous absorbers are used, the most critical factor in the measurement is the accurate knowledge of the absorber pressure.

The electron probe analyst needs a table of mass absorption coefficients for all elements, and for all lines used in analysis. For data reduction with computers, it is preferable to use functions which can generate the mass absorption

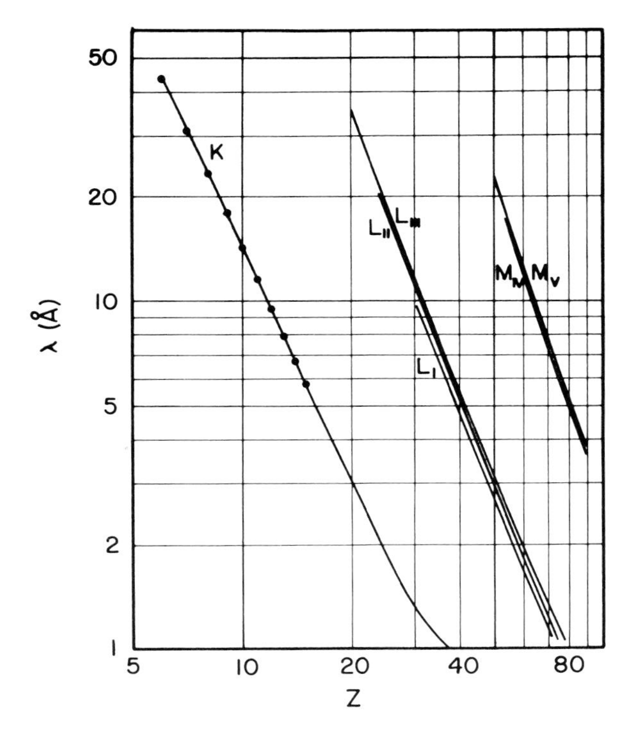

Fig. 4.17. The positions of the absorption edges follow Moseley's law; their plot as a function of atomic number is linear on a log-log scale. The graph shows the position of the K- and L-edges and of two of the five M-edges.

coefficients on demand. The function given in Eq. (4.3.10) can be employed; its use requires a set of values of the coefficients c or c', and of the exponents n, for each element and range between absorption edges. It is also necessary to provide the wavelengths, or energies, of all edges and emission lines of interest. Most proposed sets of coefficients [4.37]-[4.40] operate on the base of the simplification in Eq. (4.3.10), in which the effect of the atomic weight is ignored:

$$\mu(a, \lambda) = c'_{qa} Z_a{}^3 \lambda^{n_{qa}}.$$

The coefficients and the exponents change slowly with atomic number, and the exponents n_{qa} change abruptly when an absorption edge is crossed (Figs. 4.20-4.23). Values for coefficients and exponents are listed in reference [4.37]. The absorption jump ratios at the edges can easily be calculated from such sets of parameters.

The accuracy of tabulated or calculated values of the absorption coefficients has been a matter of investigation and controversy. It should be noted that Eqs.

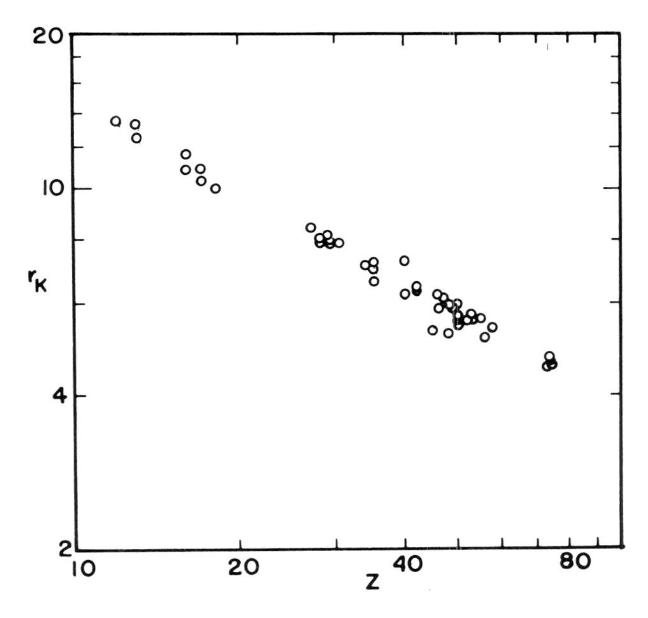

Fig. 4.18. K-jump ratio.

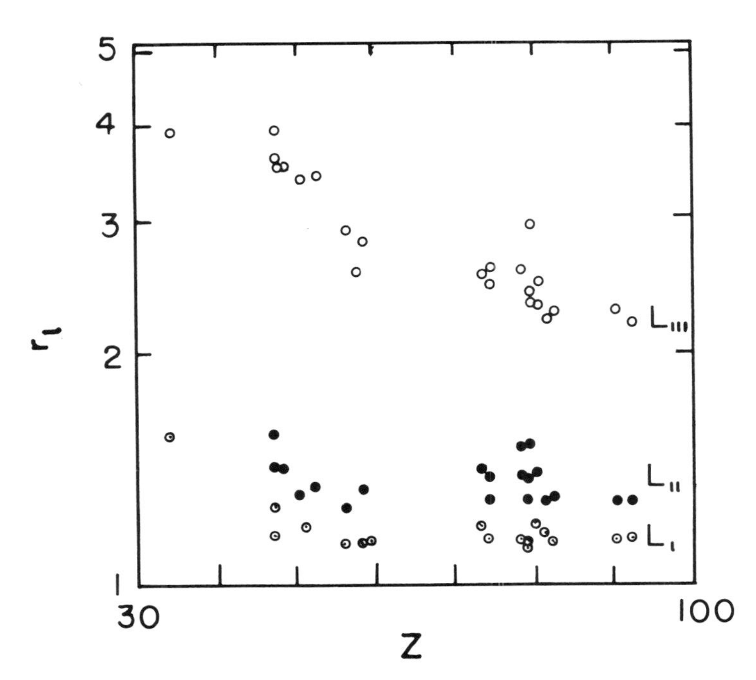

Fig. 4.19. L-jump ratios.

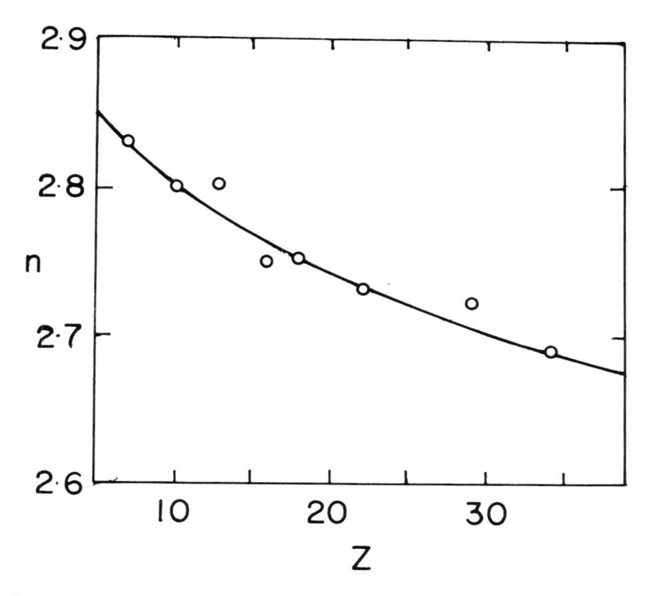

Fig. 4.20. Parameters for the estimation of X-ray mass absorption coefficients. Exponent for the region above K-edges.

(4.3.10) and (4.3.11) are approximations. The main potential source of errors in the generalization of experimental absorption coefficients is the selection and weights attributed to sources of experimental results [4.37], [4.41]. For computer procedures the positions of lines and edges can also be obtained from Moseley's law, rather than being stored in advance [4.42], but great care must be taken to avoid inaccuracies which cause an analytical line to fall on the wrong side of an absorption edge, since such an error may produce significant inaccuracies in the analytical result. On the other hand, the differences between the values for the mass absorption coefficients obtained from [4.37], [4.39], and [4.40] are usually of minor consequences. This is so because the accuracy of the absorption correction depends on the uncertainties in both the mass absorption coefficients and in the model for the absorption correction. The uncertainties in the latter are usually larger, and are therefore in most cases the limiting factor to the accuracy of the corrections. For this reason the analyst should not attempt, in principle, to derive mass absorption coefficients from measurements of X-rays excited by electrons.

The X-ray absorption coefficients for wavelengths longer than 10 Å are very high and difficult to predict, especially for elements of high atomic number (Fig. 12.15). The difficulties and uncertainties which result from these facts will be discussed in Sections 12.3 and 12.4.

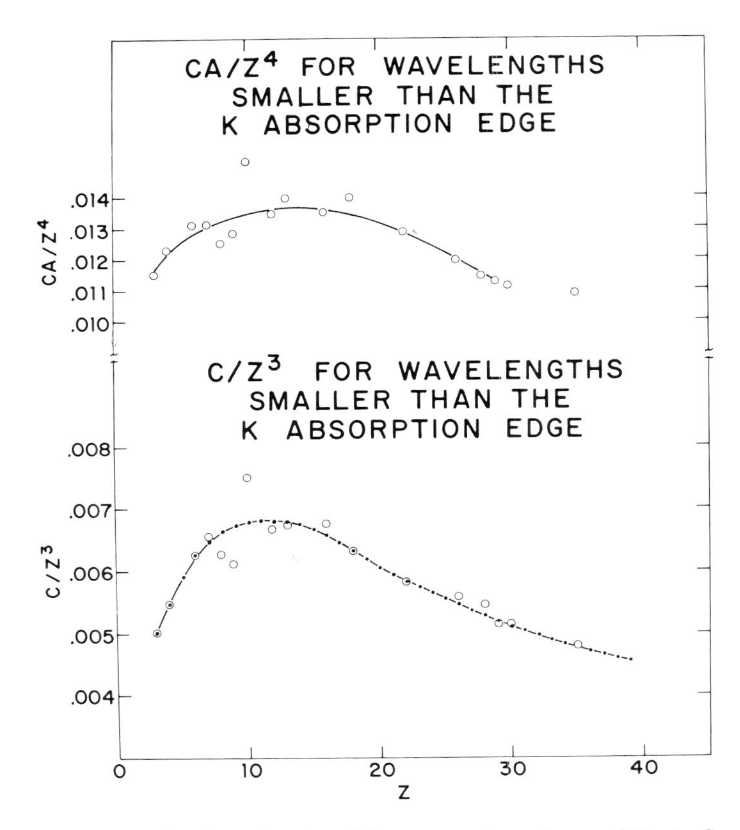

Fig. 4.21. Parameters for the estimation of X-ray mass absorption coefficients. Constant for wavelengths above the absorption edge.

4.4 THE DIFFRACTION OF X-RAYS

We can describe X-ray scattering by assuming that an atom of a material irradiated by X-rays can interact with a photon, absorbing its energy, and immediately reemitting it. If the reemitted photon has the same frequency as the absorbed photon, we call the scattering elastic, or Thomson scattering. If part of the energy of the primary photon is absorbed by the atom so that the emitted photon has a slightly lower frequency, the scattering is inelastic, or Compton scattering. A device such as a crystal or a grating is said to diffract the impinging radiation if elastic scattering at high intensity occurs in one or several well-defined directions. In terms of the wave concept applied to photons, this requires that the scattered photons be in phase and reinforce each other, representing a common wavefront. Consider such a diffracting object, with a parallel

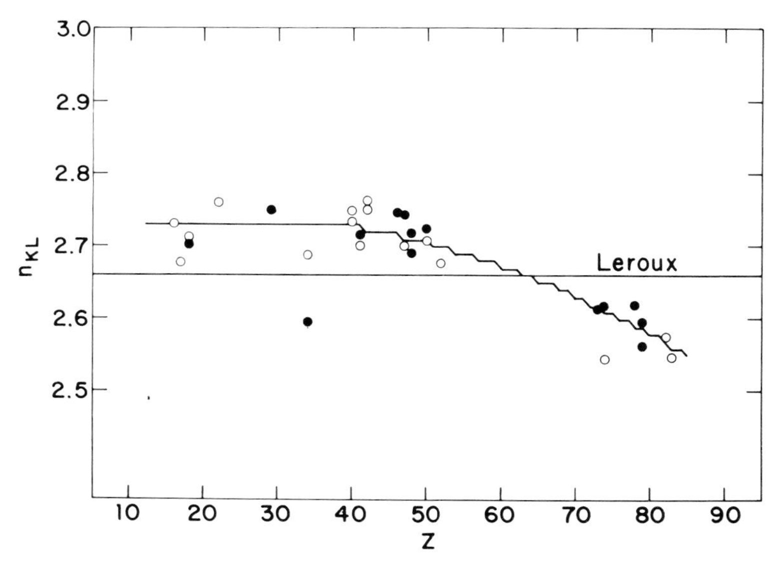

Fig. 4.22. Parameters for the estimation of X-ray mass absorption coefficients. Exponent for the region between K- and L-edges.

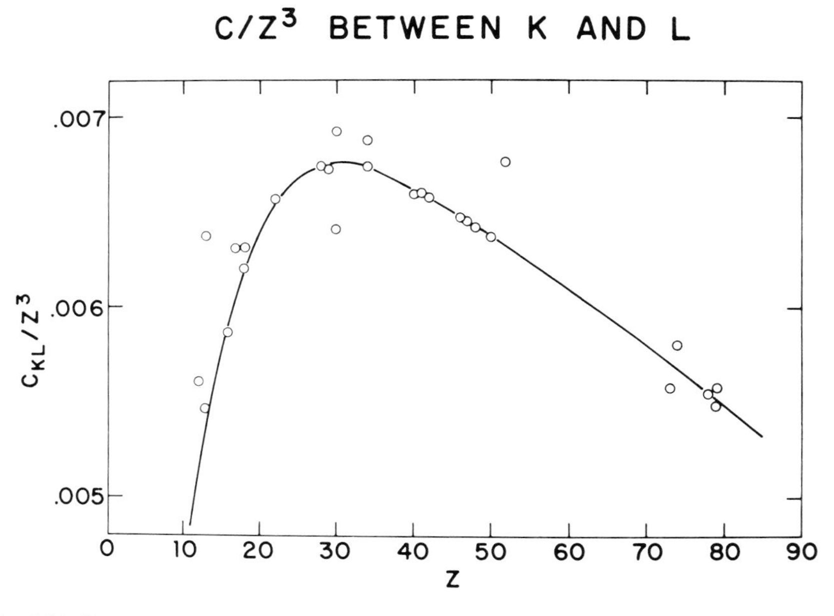

Fig. 4.23. Parameters for the estimation of X-ray mass absorption coefficients. Constant for the region between K- and L-edges.

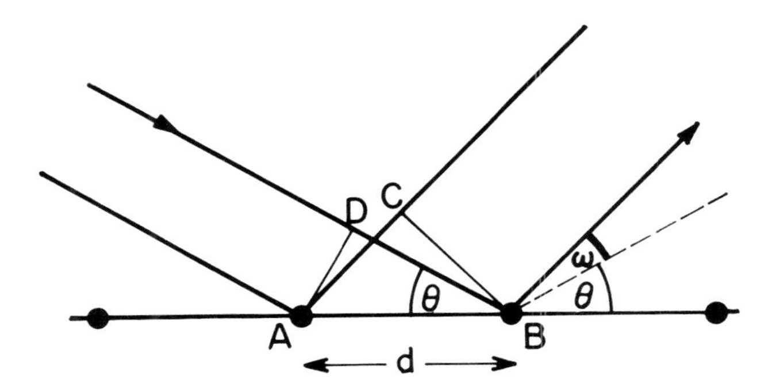

Fig. 4.24. Diffraction from a plane grating.

X-ray beam incident upon it, and another parallel beam of diffracted X-rays emerging from it (Fig. 4.24). If we define a plane normal to the incident beam, $\overline{AD}$, and another normal to the diffracted beam, $\overline{BC}$, then for diffraction to occur, all paths from one plane to the other must be either equal in length, or differ by a whole number of wavelengths, so that the diffracted radiation is in phase. It follows that diffraction by incoherent Compton scattering is impossible; since the scattered Compton photons differ among themselves in wavelength, no common front can be established under any geometric conditions. The diffraction of X-rays is thus due to elastic scattering.

Gratings. X-rays can be diffracted by two-dimensional gratings. A plane grating is a flat surface containing equally spaced parallel lines or grooves (Fig. 4.24). The incident radiation is oriented at a low angle with respect to the grating surface, so that only the ridges or grooves act to produce elastic scattering. We will call d the distance between these scattering centers, θ the incidence angle of the radiation with respect to the surface of the grating, and $\theta + \omega$ the angle of the diffracted beam with respect to the surface. The condition required for diffraction to take place is that the distances $\overline{AC}$ and $\overline{BD}$ differ by a multiple of the wavelength λ:

$$\overline{BD} - \overline{AC} = n\lambda \qquad (4.4.1)$$

where n is the order of diffraction. As

$$\overline{AC} = d \cdot \cos(\theta + \omega) \quad \text{and} \quad \overline{BD} = d \cdot \cos\theta,$$

diffraction will occur when

$$n\lambda = d\,[\cos\theta - \cos(\theta + \omega)]\,. \qquad (4.4.2)$$

At the zeroth order of diffraction ($n = 0$) all wavelengths are diffracted. In this particular case, the grating acts like a plane mirror.

Gratings can be used for diffracting X-rays of large wavelengths [4.43]. Their application is limited by the technical difficulties in preparing gratings having very small distances between scattering centers. If the distance d is much larger than λ, then the term $[\cos \theta - \cos (\theta + \omega)]$ must be very small; hence this condition would require exceedingly small angles of incidence and emission.

Blodgett–Langmuir Devices (Pseudocrystals). The dimensions of large organic molecules, such as the chains of fatty acids, are of the same order as the wavelengths of soft X-rays. Hence, diffracting devices can be built in which the distances between scattering centers are determined by the size of such molecules. The Blodgett-Langmuir pseudocrystals [4.44], [4.45] are constructed by folding monomolecular layers of an insoluble salt of a fatty acid such as barium or lead stearate. These layers are formed when a solution of the fatty acid is made to react with the surface of an aqueous solution of a salt of the respective metal. The hydrophilic metal ions, which are good X-ray scatterers, due to their high atomic number, are at the interface between the layer and the salt solution, and the thickness of the layer is equal to the length of the fatty acid chain. If this layer is carefully collected from the surface of the solution, and folded over many times in the process of collection, we obtain a structure in which the scattering centers are separated by twice the length of the fatty acid chain. There is, however, no order in the distribution of scattering centers within each layer (Fig. 4.25). A device formed by 50 to 200 such layers can be used to diffract soft X-rays.

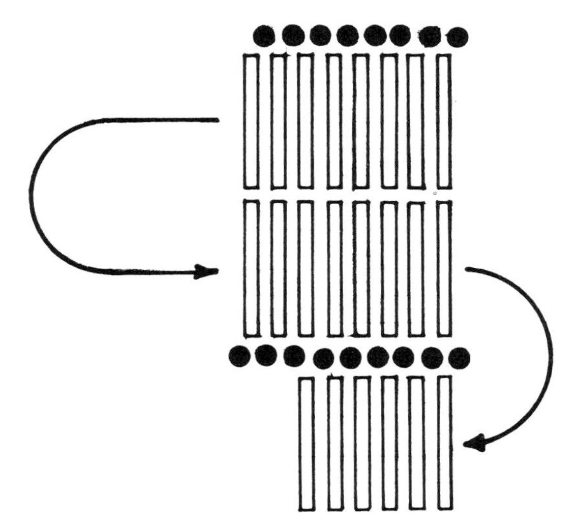

Fig. 4.25. Schematic of the formation of a Blodgett-Langmuir pseudocrystal. The rods represent the fatty acid chains, and the full circles represent the heavy metal atoms. Arrows indicate the folding of the monomolecular layer of which the device is composed.

The conditions necessary for diffraction can be deduced from Fig. 4.26. In order to obtain coherent diffraction from the entire surface, it is necessary that the emergence angle with respect to the layers be equal to the angle of incidence, θ. Only in this condition (specular reflection), the distance between the diffracting atoms within each layer does not play a role in the diffracting conditions, i.e., there is no destructive interference among rays diffracted from the same layer. But a second condition is required to produce diffraction: the difference between paths of X-rays scattered from any two layers must be a multiple of the wavelength, λ. Consider the path of a ray which intercepts the first layer at the point A, and that of another ray impinging on the second layer at the point B, which is offset from the normal to the surface through point A by the distance u. The value of u is not specified, so that we are dealing with the most general case, in which the order of the scattering centers within each plane is immaterial.

The plane going through $\overline{CA}$, normal to the plane of the drawing, is also normal to the incident beam; the corresponding plane through $\overline{AD}$ is normal to the emerging beam. Hence, refraction will occur of the sum of $\overline{CD}$ and $\overline{BD}$ is a multiple of the wavelength, λ. For this sum we obtain the following:

$$\overline{CB} + \overline{BD} = 2\overline{EB} - \overline{EC} - \overline{DF}$$

$$EC = EA \cos\theta = (d \cot\theta - u)\cos\theta$$

$$DF = FA \cos\theta = (d \cot\theta + u)\cos\theta$$

$$CB + BD = 2d/\sin\theta - 2d\cot\theta \cdot \cos\theta = 2d \sin^{-1}\theta\,(1 - \cos^2\theta)$$

$$= 2d \sin\theta.$$

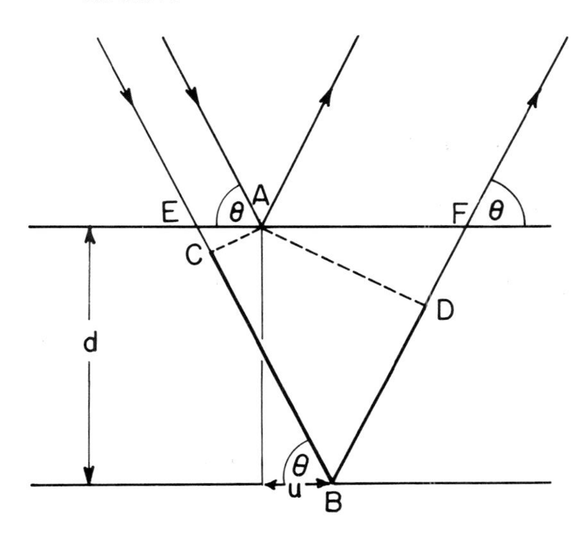

Fig. 4.26. Diffraction from a crystal or from a Blodgett-Langmuir pseudocrystal.

Hence, the condition for diffraction is:

$$n\lambda = 2d \sin \theta . \tag{4.4.3}$$

This condition is called *Bragg's law.*

X-Ray Diffraction by Crystals. The considerations which led us to Bragg's law are equally applicable to crystal lattices. In this case, the distance between layers, d, is replaced by one of the lattice parameters—or distance between one type of lattice planes—of the crystal. This distance is the only property of the crystal which determines the angle of diffraction for a given wavelength. For most crystals, the lattice parameters useful for diffraction are shorter than the lengths of fatty acid chains, and are in the proper range of diffraction of X-rays in the wavelength region of chief interest in electron probe microanalysis (1-15 Å).

For precise measurements of wavelengths the effect of the refractive index of the crystal of the X-ray line, σ_r, must be taken into account. For this purpose, Bragg's law is modified as follows:

$$n\lambda = 2d \sin \theta \left(1 - \frac{1 - \sigma_r}{\sin^2 \theta}\right). \tag{4.4.4}$$

The refractive indices for X-rays are, however, very close to one (Fig. 4.27), so that the practical effect of the above equation is negligible for our purposes.

The diffracting properties of a crystal can be used to determine, with the aid of an X-ray beam of known wavelength, the lattice parameters and orientation of the crystal (X-ray diffraction analysis). Conversely, by means of a crystal of known spacing, the characteristic lines of a spectrum can be separated and measured (X-ray spectrometry). The spectrometric techniques used in electron probe microanalysis will be discussed in the next chapter.

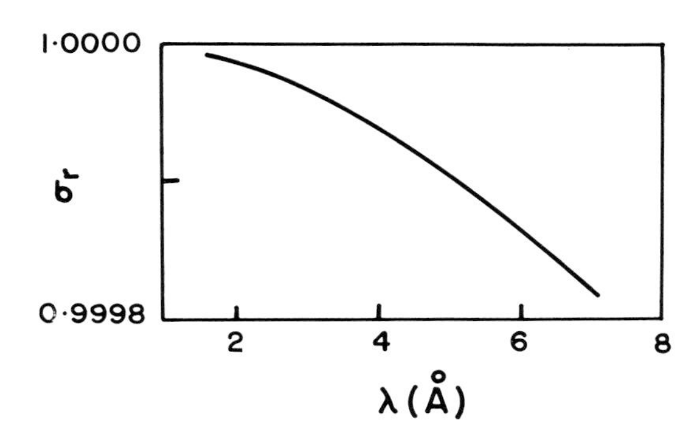

Fig. 4.27. Refractive index for mica, σ_r, as a function of X-ray wavelength.

4. REFERENCES

4.1 Roentgen, W. C., *Am. Phys.* **64,** 1 (1898).
4.2 Roentgen, W. C., *Science* **3,** 227, 726 (1896). (Translated by A. Stanton.)
4.3 Haga, H. and Wind, C. H., *Ann. Phys.* **68,** 884 (1809).
4.4 Friedrich, W., Knipping, P., and Laue, M., *Ber. bayer. Akad. Wiss.,* 303 (1912).
4.5 Bragg, W. H. and Bragg, W. L., *Proc. Roy. Soc. (London), Ser. A* **88,** 428 (1930).
4.6 Moseley, H. G. J., *Phil. Mag.* **26,** 1024 (1913).
4.7 Moseley, H. G. J., *Phil. Mag.* **27,** 703 (1914).
4.8 Bohr, N., *Phil. Mag.* **26,** 1 (1913).
4.9 Bohr. N., *Phil. Mag.* **26,** 476 (1913).
4.10 Bohr. N., *Phil. Mag.* **26,** 857 (1913).
4.11 Coster, D. and Hevesy, G., *Nature* **111,** 79 (1923).
4.12 Noddack, W., Tacke, G., and Berg, O., *Naturwissenschaften* **13,** 567 (1925).
4.13 Siegbahn, M., *Spektroskopie der Röntgenstrahlen,* 2nd ed., Springer, Berlin, 1931.
4.14 Compton, A. H. and Allison, S. K., *X-Rays in Theory and Experiment,* 2nd ed., Van Nostrand, Princeton, N.J., 1935.
4.15 Blokhin, M. A., *The Physics of X-Rays,* 2nd rev. ed., Engl. translation, U.S. Atomic Energy Commission, Office of Technical Services, Dept. of Commerce, Washington, D.C., 1961.
4.16 Flügge, S. (Ed.), *Handbuch der Physik,* vol. 30, Springer, Berlin, 1975.
4.17 Crasemann, B. (Ed.), *Atomic Inner-Shell Processes,* 2 vols., Academic Press, New York, 1975.
4.18 *The International System of Units (SI),* NBS Special Publ. 330, National Bureau of Standards, U.S. Dept. of Commerce, Washington, D.C., 1972.
4.19 Wagner, C. D., *Anal. Chem.* **44,** 967, 1050 (1972).
4.20 Siegbahn, K., et al., *ESCA, Acta Roy. Soc. Sci.* (Uppsala), Ser. IV, **20,** Almqvist and Wiksells, Uppsala, 1967.
4.21 Merrill, J. J. and DuMond, J. W. M., *Ann. Phys.* (New York) **14,** 166 (1961).
4.22 Bearden, J. A., Rep NYO-10586, U.S.A.E.C., Division of Technical Information Extension, Oak Ridge, Tenn., 1964.
4.23 Meisel, A. (Ed.), *Röntgenspektren und chemische Bindung,* Phys. Chem. Institut, Karl Marx Universität, Leipzig, E. Germany, 1966.
4.24 Albee, A. L. and Chodos, A. A. *Amer. Mineralogist* **55,** 491 (1970).
4.25 Sweatman, T. R. and Long, J. V. P., *J. Petrol.* **10,** 332 (1969).
4.26 Sandström, A. E., in *Handbuch der Physik,* vol 30, Flügge, S., Ed., Springer, Berlin, 1957, p. 238.
4.27 Burhop, E. H. S., *J. Phys. Radium* **16,** 625 (1955).
4.28 Fink, R. W., Jopson, R. C., Mark, H., and Swift, C. D., *Rev. Mod. Phys.* **38,** 513 (1966).
4.29 Bambynek, W., et al., *Rev. Mod. Phys.* **44,** 716 (1972).
4.30 Veigele, W. J., *Atomic Data* **5,** 51 (1973).
4.31 Condon, E. U. and Odishaw, H. (Eds), *Handbuch der Physik,* McGraw-Hill, New York, 1958, pp. 9-13.
4.32 Scofield, J. H., Rep. TID-4500, UC-34 Physics (UCRL-51326), Lawrence Livermore Lab., Univ. of California, Livermore, Calif., 1973.
4.33 Nagel, D., in *Quantitative Electron Probe Microanalysis,* Heinrich, K. F. J., Ed., NBS Special Publ. 298, National Bureau of Standards, U.S. Dept. of Commerce, Washington, D.C., 1968, p. 189.
4.34 Henke, B. L. and Ebisu, E. S., *Adv. X-Ray Analysis* **17,** 150 (1974).
4.35 Pratt, R. H., Ron, A., and Tseng, K. H., *Rev. Mod Phys.* **45,** 273 (1973).

4.36 Wuilleumier, F., Ph.D. Thesis, Univ. of Paris, Paris, France, 1969.
4.37 Heinrich, K. F. J., in *The Electron Microprobe,* McKinley, T. D. Heinrich, K. F. J., and Wittry, D. B., Eds., John Wiley & Sons, New York, 1966, p. 296.
4.38 Leroux, J., *Adv. X-Ray Analysis* **5**, 153 (1961).
4.39 Theisen, R., *Quantitative Electron Microprobe Analysis,* Springer, Berlin, 1965, Table D.
4.40 Frazer, J. Z., A computer fit to mass absorption coefficient data. Rep. S.I.O. 67-29, Institute for the Study of Matter, Univ. of California, LaJolla, Calif., 1967.
4.41 Hughes, G. D. and Woodhouse, J. B., *Proc. 4th Int. Conf. on X-Ray Optics and Microanalysis,* Hermann, Paris, 1966, p. 202.
4.42 Yakowitz, H., Myklebust, R. L., and Heinrich, K. F. J., NBS Tech. Note 796, National Bureau of Standards, U.S. Dept. of Commerce, Washington, D.C., 1973.
4.43 Franks, A. and Lindsay, K., in *The Electron Microprobe*, McKinley, T. D., Heinrich, K. F. J., and Wittry, D. B., Eds., John Wiley & Sons, New Yor, 1966, p. 83.
4.44 Langmuir, I., *J. Franklin Inst.* **218,** 153 (1934).
4.45 Blodgett, K. B., *J. Amer. Chem. Soc.* **56,** 459 (1934).

PART III. THE MEASUREMENT OF X-RAYS

5.
X-Ray Spectrometry

In order to measure the intensity of X-ray lines and to study their characteristics, we use an X-ray spectrometer which provides an electric signal for a narrow wavelength fraction of the X-ray spectrum which is being observed. We may either physically separate the X-ray photons which fall within the wavelength range of interest before detection (wavelength dispersion), or we may use electronic means (pulse height analysis) to separate the signals produced by the photons within this wavelength range from other signals. The following list shows the techniques which are commonly used to analyze X-ray spectra. The wavelength dispersion techniques are as follows:

1. diffraction of X-rays by gratings, Blodgett-Langmuir devices, or crystals;
2. specific X-ray filters (Ross filters);
3. total reflection.

The pulse height dispersion techniques are as follows:

4. pulse height selection by means of single-channel analyzers;
5. pulse height analysis with multichannel analyzers.

5.1 CRYSTAL SPECTROMETERS

The relations expressed in Bragg's law can be put to use to construct a simple X-ray spectrometer, based on the use of a flat single crystal as a diffracting device (Fig. 5.1). To obtain parallel incident and emergent beams and to limit the possible paths of X-rays to the proper Bragg angle for the radiation of interest, beam collimators formed by parallel flat plates of a heavy metal such as tantalum are mounted in the paths of the incident and emergent beams. An X-ray detector located at the end of the exit collimator receives the diffracted radiation. Such a flat-crystal spectrometer can be tuned to a particular wavelength by adjusting the angular positions of crystal, collimators, and detector.

Bragg's law requires that the emergence angle be equal to the incident angle, θ. Hence, the angle between the incident beam and the emergent beam is equal to $180° - 2\theta$. It is possible to cover the available wavelength range by scanning through the angular positions described by Bragg's law; in such a case, the angular speed of the exit collimator and detector must be twice that of the crystal.

In practice, a crystal spectrometer has a range of Bragg angles of less than 90°, due to mechanical limitations. The limits of the spectrometer, and the lattice spacing d of the crystal, determine for each crystal the observable wavelength range. Table 5.1 lists some of the crystals which are frequently used in X-ray spectrometers, the orientation of the reflecting planes used in the spectrometer, their d-spacing, and the corresponding wavelength range, for a spectrometer adjustable for values of from 15° to 70°, in the first order of reflection (see Fig. 5.2).

Birks cites the following properties of crystal spectrometers as relevant factors for evaluation: intensity, resolution, line-to-background ratios, and mechanical quality [5.2, p. 53]. The first three factors are interrelated. The emission line is superposed upon a background mainly formed by continuous radiation. The natural widths of line determine a theoretical limit of the achievable line-to-background ratio. Birks calculated for pure elements the values shown in Table 5.2.

The characteristics of a crystal as an analyzer for a given line can be disclosed by an intensity plot obtained when parallel X-rays fall upon a crystal which is rocked about the Bragg angle (rocking curve, Fig. 5.3 [5.2, p. 42]). Such a curve can be obtained with a double-crystal spectrometer [5.3], and can be corrected for the finite width of the X-ray emission line. The parameters of interest are

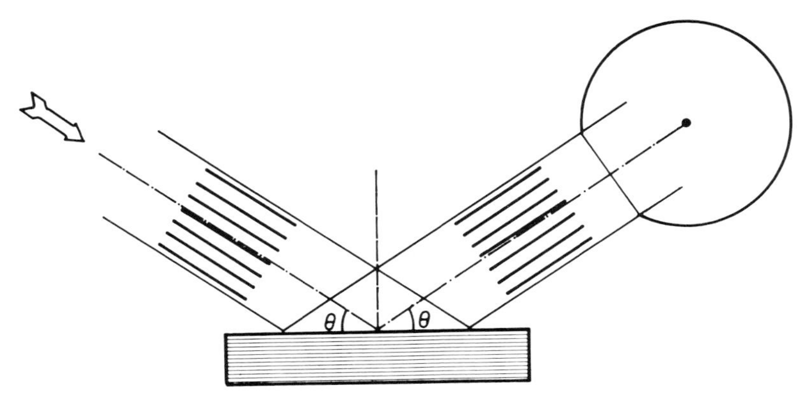

Fig. 5.1. Flat-crystal Bragg spectrometer. The incident (left) and emergent (right) X-rays are constrained to parallel beams by means of plate collimators. The orientation of the crystal and the collimators is such that both the incident and the emergence angles are equal to the angle θ of Eq. (4.4.3).

Table 5.1. Crystals Used in X-Ray Spectrometers [5.1].

CRYSTAL	REFLECTING PLANE	d (Å)	WAVELENGTH (Å) $\theta = 15°$	$\theta = 70°$	RANGE OF ELEMENTS (FIRST ORDER) $K\alpha$	$M\alpha_1$	$M\alpha, M\beta$
Lithium fluoride	200	2.013	1.04	3.78	19-35	49-84	92-94
Sodium chloride	200	2.819	1.45	5.3	17-29	44-74	81-94
Silicon	111	3.136	1.6	5.9	16-28	41-71	78-94
Silica	$10\bar{1}1$	3.343	1.7	6.3	15-27	40-69	76-94
Silica	$10\bar{1}0$	4.255	2.2	8.0	14-24	37-61	70-94
Ethylene diamine *d*-tartrate (EDDT)	020	4.402	2.28	8.27	14-24	37-61	70-94
Pentaerythritol (PET)	002	4.371	2.26	8.21	14-24	37-61	70-94
Ammonium dihydrogen phosphate (ADP)	011	5.320	2.75	10.0	12-21	33-56	65-94
Graphite	0002	6.70	3.45	12.6	11-19	30-50	62-94
Mica	001	9.93	5.14	18.7	9-16	26-42	62-82
Postassium acid phthalate (KAP)	$10\bar{1}0$	13.32	5.3	25	8-16	23-42	62-82
Rubidium acid phthalate (RAP)	$10\bar{1}0$	∿13.05*	5.3	25	8-16	23-42	62-82
Lead stearate	-	∿50	26	94	5-7	20-22	-

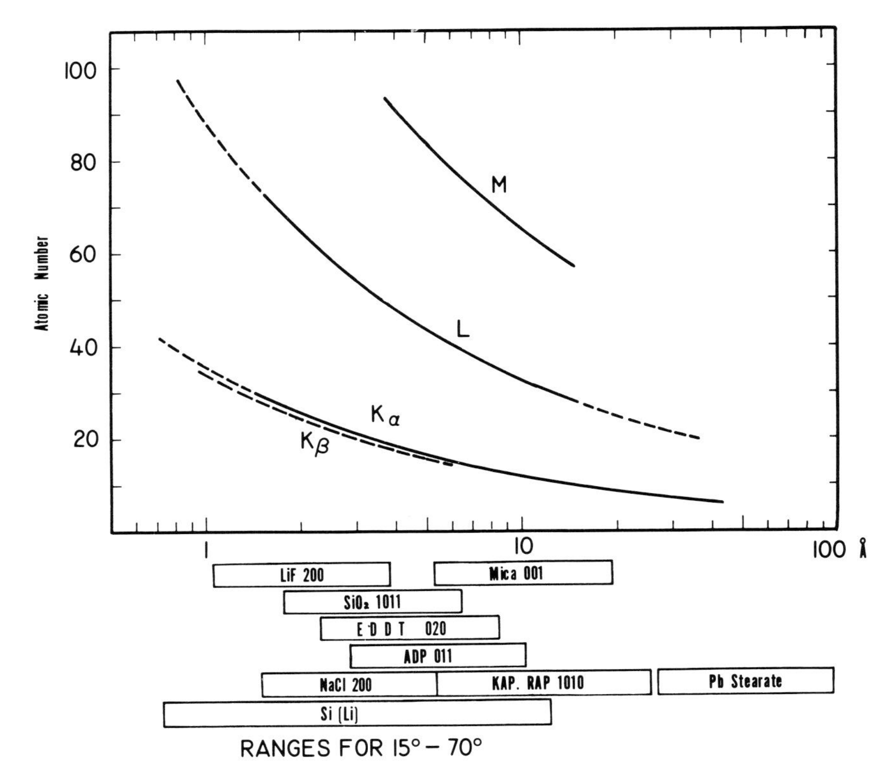

Fig. 5.2. Ranges of wavelength and atomic number covered by spectrometers which are used in electron probe microanalyzers. The range of PET is virtually the same as that of EDDT.

the breadth at half-maximum signal intensity, W, which is an indication of resolution (measured in seconds of arc), the peak diffraction coefficient, P (measured in percent), and the integral reflection coefficient, $R = \int P\, d\theta$ (measured in radians). Table 5.3 (from Birks [5.2, p. 43]), gives the values of these parameters for some of the crystals most frequently used in electron probe microanalysis.

The intensity of diffraction depends on the atomic number of the scattering centers, on the shape of the unit cell, and also, to a large degree, on the degree of

Table 5.2. Line-to-Background Ratios Extrapolated to Natural X-Ray Line Breadth [5.2, p. 53).[a]

Ti $K\alpha$	Cu $K\alpha$	Ge $K\alpha$	Zr $K\alpha$	Ta $L\alpha$	Au $L\alpha$	Au $L\beta_1$
7150	2900	1980	240	52	343	236

[a]Operating voltage (E_0): 30 keV, $\theta = 45°$.

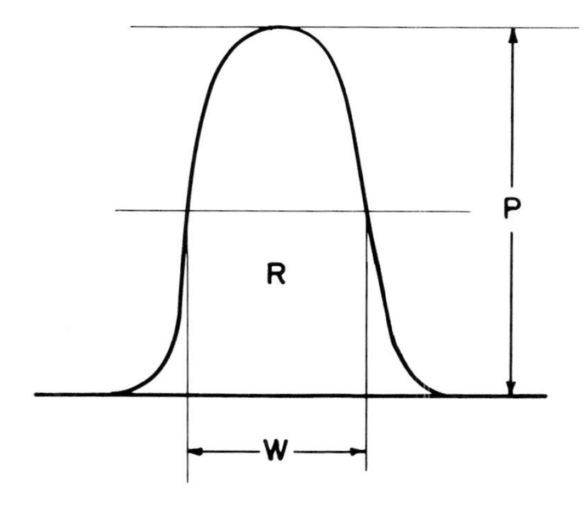

Fig. 5.3. Rocking curve of a crystal. The horizontal scale is calibrated in units of angle and the vertical scale is calibrated in X-ray intensity (or photon count rate). *R*:integral reflection coefficient, *P*:peak diffraction coefficient, *W*:breadth at half-maximum signal intensity.

perfection of the crystal lattice. As illustrated in the figures for lithium fluoride (LiF) in Table 5.3, surface treatments which disturb the perfection of the crystalline lattice, such as abrading and flexing, greatly increase the integrated reflection coefficients of some crystals, due to the formation of a mosaic of small regions of slightly varying orientation. As will be seen, spectrometers useful for point sources must be of the curved-crystal type, and most crystals cannot be curved without affecting the perfection of the lattice. It is therefore difficult to predict or specify exactly the intensities to be obtained from a crystal spectrometer. The choice and preparation of crystals is influenced by empirical observations, and by the availability of large single crystals of appropriate characteristics.

As Fig. 5.1 shows, diffraction of X-rays by a flat crystal, according to Bragg's equation, requires a parallel incident X-ray beam. If a large X-ray source is available, the intensity of the diffracted signal can be increased by increasing the sur-

Table 5.3.

CRYSTAL	*W* (")	*P* (%)	*R*(RAD)
LiF, cleaved	14	40	3×10^{-5}
LIF, abraded	120	50	4×10^{-4}
EDDT, etched	215	20	2×10^{-4}
KAP, cleaved	70	13	3×10^{-5}
graphite hot pressed	1800	30	3×10^{-3}

face of the diffracting crystal. However, the X-ray source in the electron probe microanalyzer is practically a point; for a given crystal and position, only a small solid angle covers the required Bragg condition, and therefore, even with a large flat crystal, only a small part of the crystal will diffract the incoming radiation. Therefore, a flat-crystal spectrometer is inadequate for the analysis of an X-ray point source.

One can, of course, increase the signal by using more than one crystal for the diffraction of the radiation. If we wish the diffracted radiation to converge into one point, at which the detector can be placed, then the crystals must be placed along a circle, called the focal circle (Fig. 5.4, left). In three-dimensional space, such an arrangement could be extended to an array of small crystals arranged on a spherical surface. However, Bragg's law requires that the diffracting planes should not be tangent to the focal circle; rather, they should be perpendicular to a line between the crystal center and a point on the focal circle which is equidistant from the two spectrometer foci, S (specimen) and D (detector).

An approximation to this hypothetical multicrystal spectrometer is the Johann spectrometer (Fig. 5.4, center) [1.6]. The crystal and the two foci (specimen and detector) are located on the focal circle; the Bragg angle, and hence the wavelength settling of the spectrometer, can be changed by sliding the crystal and one of the foci (the detector) along the focal circle; the detector will travel at twice the angular speed of the crystal. To maintain the Bragg angle over the entire crystal surface, the crystal is bent to a radius twice that of the focal circle. Such a spectrometer does not need any collimator. In practice, the position of the detector focus may be occupied by a slit, to exclude stray radiation, with the detector being located behind the slit.

Johann spectrometers are frequently used in electron probe microanalyzers and they are quite satisfactory in practice. It should be noted, however, that the Bragg and focusing conditions are only approximately maintained. Referring to Fig. 5.4 (center), we note that the ray we have drawn from the focus S crosses the focal circle at A and meets the crystal surface at B. However, B has an angular orientation which would correspond for exact focusing conditions to those of point C, rather than A. Hence, the Bragg condition is not precisely met for the ray meeting the crystal wings which are distant from the focal circle. Furthermore, the ray which emerges at B at the proper Bragg angle does not reach the focusing circle at the proper focus, D; rather it arrives at the nearby point E. Consequently, the Johann geometry produces both a broadening of the X-ray line observed by the detector (with respect to perfect focusing), and an imperfect focusing at the detector level.

A perfect focusing condition is obtained in the Johansson spectrometer (Fig. 5.4., right) [1.7]. This is identical to the previous spectrometer, except that the crystal is first bent to twice the radius of the focal circle, and then ground to the radius of the focal circle. Now the crystal surface is the equivalent

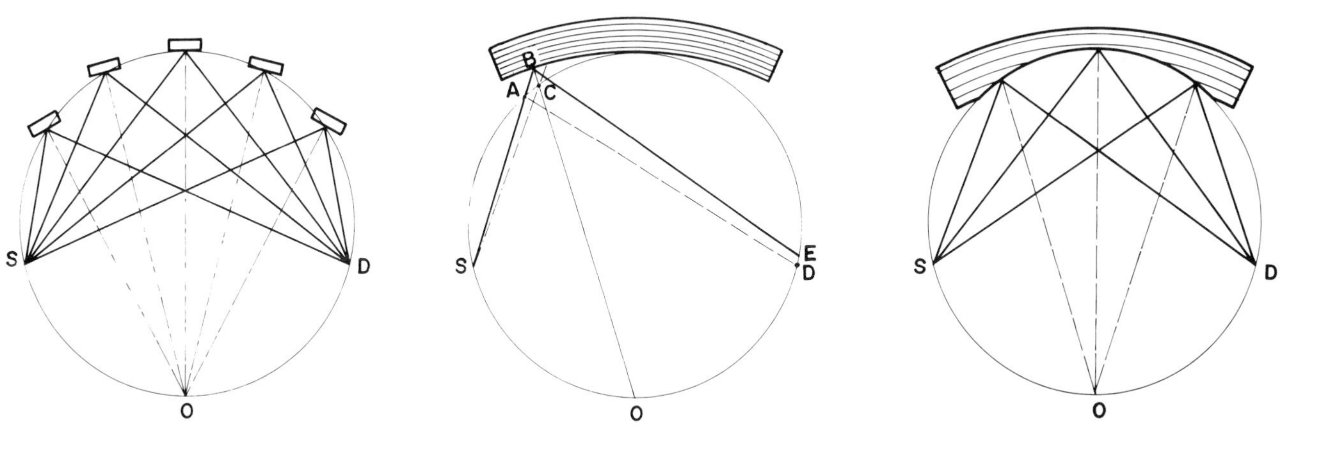

Fig. 5.4. Focusing X-ray spectrometers. Left: Several flat crystals are arranged on a focal circle; Bragg reflections of monochromatic X-rays emitted by the specimen (focus S) are focused at the detector (focus D). The crystals are oriented with their diffracting planes normal to the line connecting the center of each crystal with point O, which is on the focal circle, and equidistant from S and D. Center: An approximation to the preceding principle is given by the *Johann* spectrometer [1.6]. A single flat crystal is bent to twice the focal circle radius. The device is not theoretically perfect. The ray incident on B should have, for correct Bragg reflection, the direction of the ray passing through C, rather than through A. The diffracted ray from point B will pass close to the second focus D, but not exactly through it (point E). If the crystal is small enough, these aberrations can be neglected. Right: A theoretically rigorous solution is the *Johansson* spectrometer, in which the crystal is first bent to twice the focal circle radius, and then ground to the focal circle curvature. Bragg's law is valid over the entire crystal surface, and the emerging rays focus at point D [1.7].

of the set of crystals in Fig. 5.4 (left), and both the Bragg condition and the focusing are improved. It will be appreciated that in real spectrometers the ratio between the crystal size and the focal circle radius is much smaller than that shown in Fig. 5.4, so that the imperfection of the Johann spectrometer is grossly exaggerated in this figure. The Johansson geometry is impossible to achieve with crystals such as mica which cannot be ground; yet, spectrometers of the Johann type with mica crystals may have very good wavelength resolution.

The Blodgett-Langmuir pseudocrystals mentioned previously (Fig. 4.26) are also used in the Johann configuration.

Although crystals can be bent spherically, all crystals used at present in electron probe microanalyzers are cylindrically bent. The effect of the height of the crystal (i.e., the dimension normal to the focal circle) is a line broadening called fanning divergence [5.2, p. 47]. This defect is caused by the change in the incidence angle of the X-rays, which, outside the focal circle, are oblique with respect to the circle (Fig. 5.5). Consequently, for such rays, the effective crystal spacing is increased.

The accuracy of the analysis depends on the precision with which both line and background can be measured. The precision is limited, according to Poisson's law of counting statistics (Section 6.2), by the number of detected pulses, and hence by the intensities achievable with the spectrometer.

For all crystals, the lines in the first order of reflection have both the strongest intensities and the highest line-to-background ratios. Therefore, in most cases the first order is the choice for the observation of a line or the measurement of its intensity. A line of higher order emitted by an element present at high concentration may be used, however, if a weak line of another element is observed simultaneously, and if the count rates of the first-order reflection

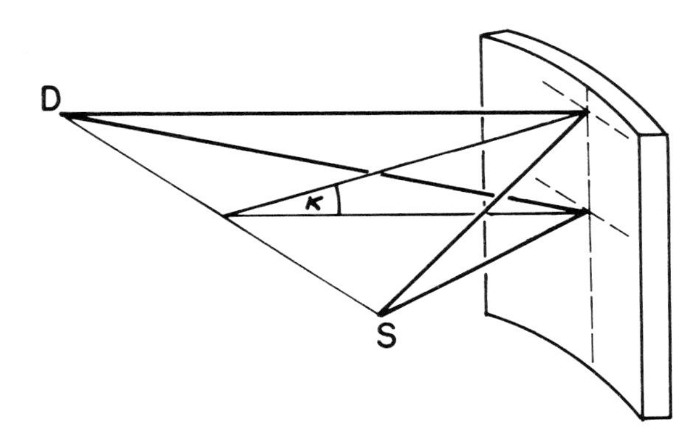

Fig. 5.5. Fanning divergence. At the edge of the crystal, the incident beam is inclined with respect to the focusing circle. Therefore, the effective spacing of the crystal is increased from d to $d/\cos\kappa$.

are too high for accurate measurement at the electron beam intensity required for the measurement of the weak line. (See Section 5.2 for coincidence losses.) Mica has higher-order reflections of very high intensities and is therefore often used for observations of such reflections.

The intensity of the diffracted radiation is proportional to the solid angle covered by the spectrometer crystal.

The angle within the focal circle covered by the crystal does not vary with the crystal position, and hence the Bragg angle (Fig. 5.6). Thus, the solid angle is proportional to the angle covered by the height of the crystal, and therefore to the distance from the specimen to the center of the crystal. If the physical center of the crystal coincides with the point at which the diffracting planes are tangent to the focal circle, this distance is equal, according to Bragg's law, Eq. (4.4.3), to

$$s = 2r \sin \theta = r \frac{n\lambda}{d} \tag{5.1.1}$$

where r is the radius of the focal circle and d, n, and θ have the same meaning as in Eq. (4.4.3). Therefore, if during a wavelength scan this distance increases at constant speed, the observed wavelength also changes at a constant rate (Fig. 5.7).

We have discussed in Section 3.3 the spatial limitations in the region close to the specimen, and the lens configurations which were developed to permit the observation of the emitted X-rays at a high emergence angle. Even with the most

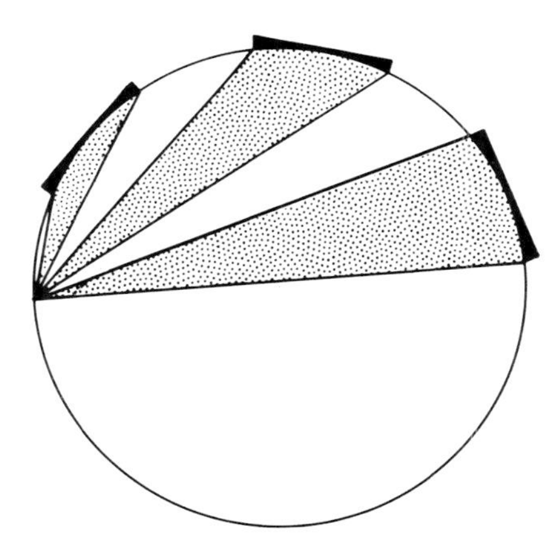

Fig. 5.6. The angle within the focal plane covered by a crystal does not vary with the crystal position and hence with the Bragg angle. However, the angle in the plane normal to the focal plane (width of crystal), and hence the solid angle covered by the crystal, diminishes linearly with the distance from the X-ray source to the center of the crystal.

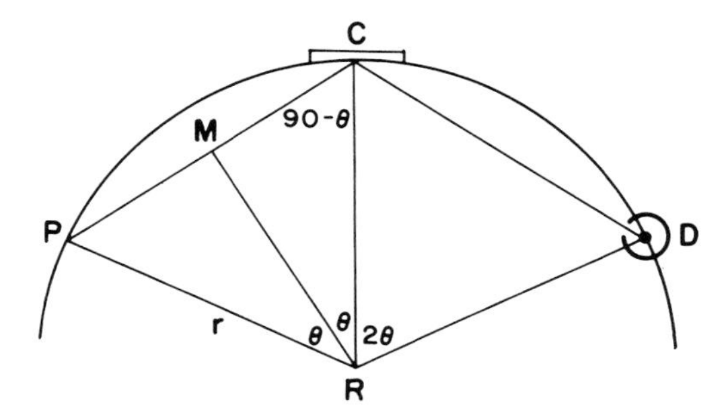

Fig. 5.7. The distance from the specimen (P) to the center of the crystal (C) is equal to $2r \sin \theta$.

efficient design, the solid angle available for the exit of X-ray is limited, particularly if space for three or more X-ray spectrometers must be provided. If the designs of Fig. 5.4 (center and right) were used, and the center of the focal circle were fixed with respect to the rest of the instrument, then a large solid angle would have to be maintained open for the X-rays, since the elevation of the crystal changes as it moves along the circle. If the spectrometer were mounted in a vertical plane, the X-ray emergence angle would also change as a function of the Bragg angle (Fig. 5.6). These complications are avoided by designing the spectrometer in such a way that the crystal is forced to move in a straight line as the Bragg angle is changed. This can be achieved by guiding the crystal along a track which defines a constant X-ray emergence angle for the full wavelength range of the spectrometer. The center of the focal circle will therefore move in a circle, as shown in Fig. 5.8, and the crystal has to be reoriented so as to comply with the Bragg condition at every point of its trajectory along the track. The detector moves along a cloverleaf figure and its window must at all times be oriented towards the crystal.

Crystal spectrometers can be mounted on the electron probe microanalyzer in a vertical or in an inclined configuration. In the first case, the plane containing the focal circle is normal to the specimen surface, while in the second position it is tilted at an angle equal to the X-ray emergence angle, ψ. The vertical configuration occupies a smaller sector of the specimen surface plane, and is preferred in instruments having several X-ray spectrometers. A disadvantage of this arrangement is the sensitivity of the detected X-ray signal to small variations in specimen elevation. The reason for this limitation is illustrated in Fig. 5.9. It can be seen that all X-rays which obey Bragg's law for a given wavelength are contained within a wedge, the edge of which is the focal line from A to B. The spectrometer is properly aligned if the intersection of the electron beam with the specimen surface, Σ, falls upon this line. If the point of electron beam impact is located close to this line, and within the wedge, the

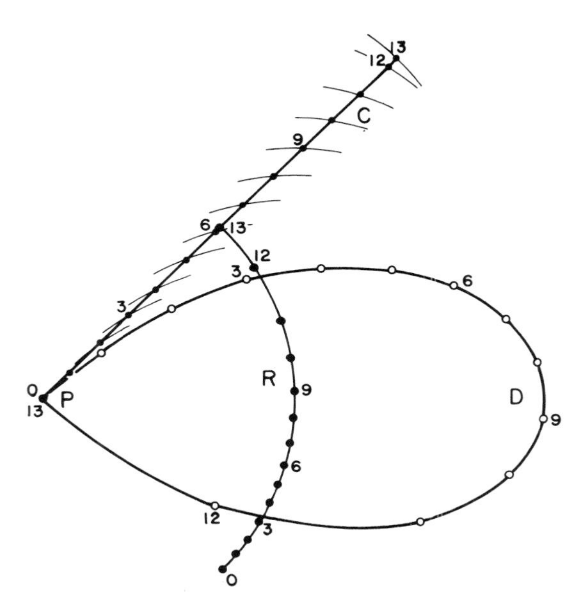

Fig. 5.8. Geometry for a spectrometer of constant X-ray emergence angle. The crystal travels along the straight line from *P* to *C*. The orientation of the crystal is changed along the trajectory, to satisfy Bragg's law. The center of the focusing circle rotates around the specimen position, *P* (track marked *R*), and the detector tracks along the cloverleaf figure part marked *D*. The orientation of the detector assembly must be rotated so as to always look towards the crystal. The limitations in spectrometer range are given by the fact that for a Bragg angle of 0°, the detector would have to be placed at the point *P* (i.e., the specimen), and for a Bragg angle of 90°, both crystal and detector would be at point *P*. Hence, the physical dimensions of these components determine the limiting angles.

X-ray intensity will be only slightly lower than in the proper alignment position. However, the electron beam, *e*, crosses the wedge at a single point; if the specimen is moved to a position above or below this point, the X-ray intensity drops rapidly (Fig. 5.10).

The inclined spectrometer is much less sensitive to the elevation of the spectrometer because the electron beam in an aligned spectrometer passes within the wedge, and, particularly if the X-ray emergence angle is small, close to the line of focusing (Fig. 5.11). Therefore, this spectrometer arrangement is preferred in instruments such as scanning electron microscopes, in which the elevation of the specimen may vary, and particularly if an optical microscope is not incorporated in the instrument.

Spectrometer Alignment. Since the details of construction of focusing X-ray spectrometers differ, it would be futile to attempt to give detailed instructions for spectrometer alignment. It should be useful, however, to discuss the prin-

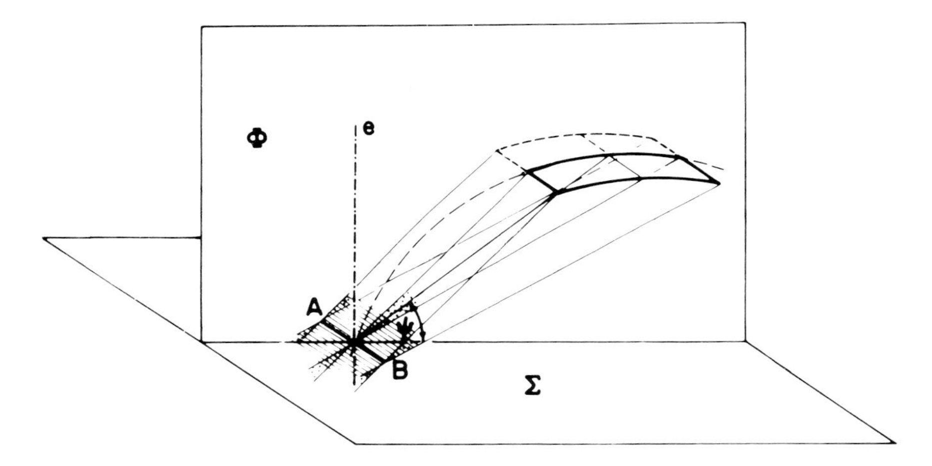

Fig. 5.9. Vertical X-ray spectrometer. The plane containing the focal circle, Φ, is perpendicular to the specimen surface, Σ. The electron beam, *e*, must be aligned to fall upon the focal line, from *A* to *B*. The X-ray emergence angle, ψ, is an average value.

ciples which are involved, with the aid of a spectrometer having a type of freedom for alignment which is available in some commercial electron probe microanalyzers.

As mentioned in Subsection 3.3.1, the optical microscope determines the correct position of the specimen elevation, while the alignment of the electron-optical column determines the point of impact of the beam on the specimen. In reference to Fig. 5.9, we can thus assume that the point of intersection of the beam, *e*, with the specimen plane, Σ, is fixed. The spectrometer alignment requires that the focal circle in the plane Φ be shifted so as to intersect the specimen surface at the same point as the electron beam, and to orient the crystal so as to assume the proper position to satisfy Bragg's law. If the crystal is at the proper wavelength setting (i.e., if the distance $\overline{PC}$ in Fig. 5.7 is properly defined), then it can be aligned by the combination of two degrees of freedom: horizontal displacement of the spectrometer assembly in the plane Φ (focal circle alignment), and tilt of the crystal around an axis perpendicular to Φ, passing through the center of the crystal surface (crystal tilt alignment). There is a unique set of values for these two variables which fulfills the alignment condition, and if the spectrometer is properly constructed, this pair of values should be the same regardless of the wavelength setting. We have remarked, however, that a high signal intensity can be obtained from any point close to the focal line $\overline{AB}$ which lies within the two limiting planes crosshatched in the drawing. For this reason the drop of intensity due to small displacements of the X-ray source from the first focus can be corrected by a small change in the tilt of the

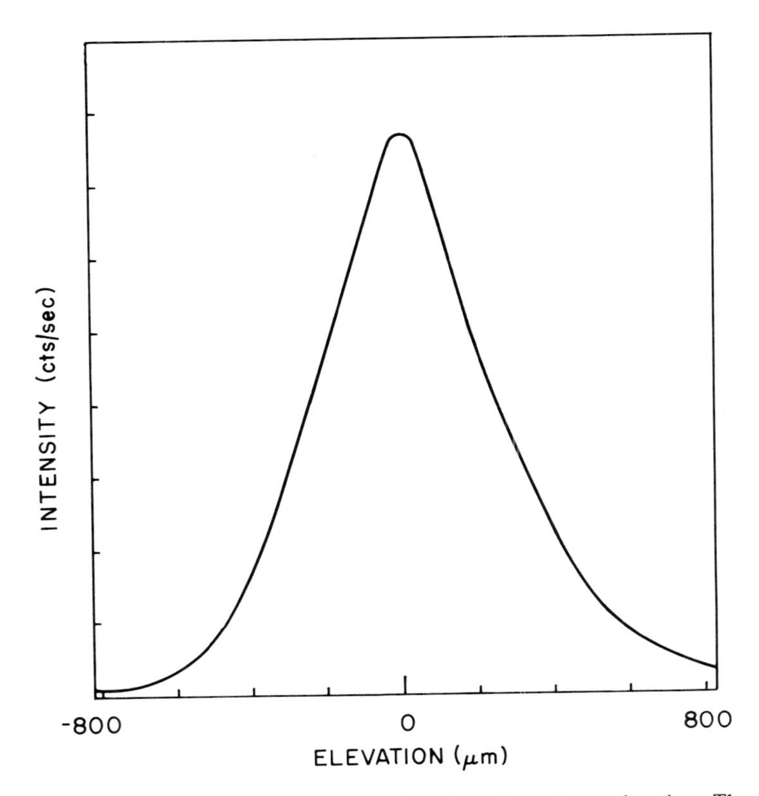

Fig. 5.10. Variation of signal intensity as a function of specimen elevation. The line is Fe Kα, observed with a 4-in. radius LiF crystal, in the first order of reflection.

crystal. If one plots the values for focal circle positions and tilt alignment angles on orthogonal axes (Fig. 5.12), there is a line along which the X-ray intensity is strongest, and along this line the intensity changes very slowly. It is therefore difficult to determine the pair of coordinates corresponding to perfect alignment by measurement of one X-ray emission line only. If the sets of best values for two X-ray lines distant in wavelength are measured and plotted, the point of intersection at which high intensities are obtained for both lines is the spectrometer alignment condition for the entire spectrometer range.

In performing the measurements required to establish the lines shown in Fig. 5.12, it may be necessary to rock the crystal through the Bragg angle for each measurement because of imperfections in the mechanical arrangement of the spectrometer. This precaution also assures that in the final alignment position–which includes the wavelength setting which may have to be readjusted–the detector or the slit in front of it will be accurately positioned.

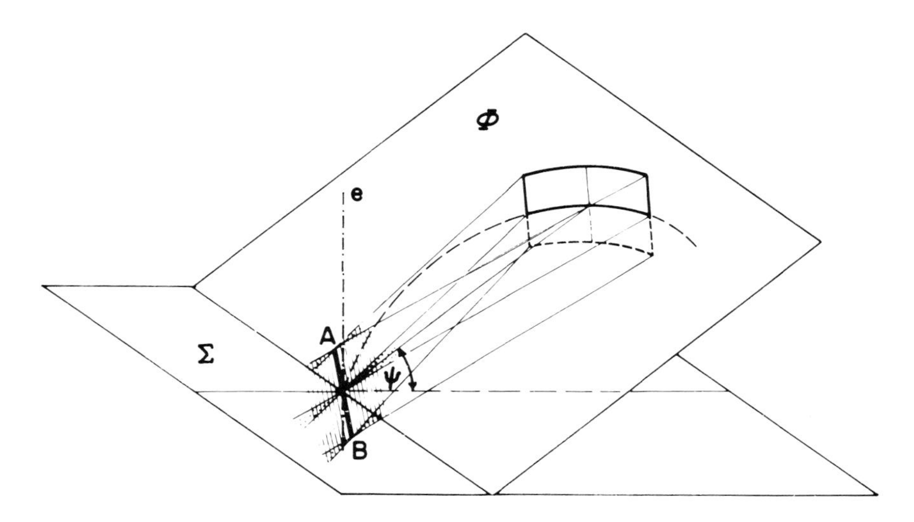

Fig. 5.11. Inclined spectrometer. The symbols have the same meaning as in Fig. 5.9.

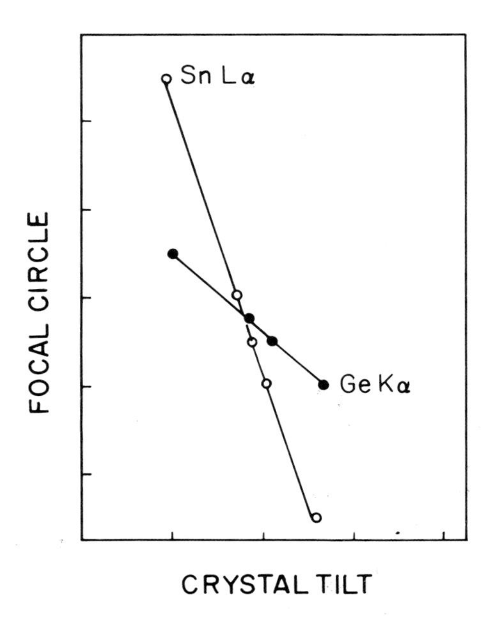

Fig. 5.12. Alignment of a lithium fluoride (LiF) crystal spectrometer. The wavelengths used are 3.5998 Å (Sn $L\alpha_1$), and 1.2553 Å (Ge $K\alpha$).

With crystals of good quality, the curved-crystal spectrometer, in combination with a point source of X-rays, should have a satisfactory resolution and line-to-background ratio, without need for collimation in front of the detector. If, however, the crystals are too mosaic, or there is danger of stray of backscattered electrons entering the detector, then such a slit may improve the performance of the spectrometer.

Because each crystal spectrometer is set on one wavelength at a time, microprobes are typically equipped with two to four spectrometers, and, to overcome the limitation in the spectral range of the crystals, crystal exchangers are installed in the spectrometer. In a good crystal spectrometer it should be possible to exchange crystals without breaking the vacuum of the spectrometer, without need for realignment procedures after the exchange, and the performance of the spectrometer must not be significantly altered by these operations.

5.2 ROSS FILTERS AND TOTAL REFLECTION

Before Blodgett-Langmuir pseudocrystals became available, other methods for monochromatization of soft X-rays were advocated. The method of balanced filters described by Ross [5.4] is based on the large change of mass absorption coefficients at the K-edges. If X-rays pass through two filters of appropriate thickness, made from materials of close atomic numbers, the attentuation for all wavelengths will be the same, except for the region between the respective K-edges. If the transmittances are measured, and one signal is subtracted from the other, the signals effectively cancel, except for X-rays of wavelength between the two edges (Fig. 5.13).

Castaing and Pichoir [5.5], [5.6] used gaseous filters of nitrogen and oxygen for the detection of the K-emission of oxygen, and the pair methane-nitrogen for the measurement of the nitrogen K-band.

In a similar manner, the angle of total X-ray reflection, which is related to the refractive index (Fig. 4.27), is a function of the wavelength of the reflected radiation. A spectrometer for soft X-rays based on this difference in the angle of total X-ray reflection as a function of wavelength was described by Herglotz [5.7]; a similar device was used in electron probe microanalysis by Borovsky [5.8].

The use of differences of X-ray signals in the techniques described above is statistically disadvantageous (see Section 6.2), and for this reason such spectrometers are not widely used at present.

5.3 X-RAY DETECTORS

X-Ray detectors produce a signal by means of which the X-rays can be observed and measured. Such detectors can be mounted at the foci of the spectrometers discussed in the previous section; they can also be used without the aid of diffracting devices, as will be shown later in this section.

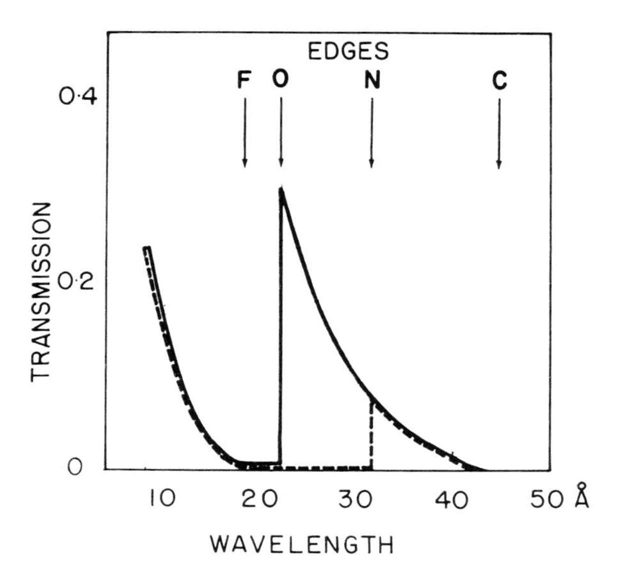

Fig. 5.13. Transmission factors of gaseous absorbers in a Ross filter for the detection of oxygen. Full line: oxygen filter. Broken line: nitrogen-krypton filter [5.6].

In the detectors of interest for electron probe microanalysis, individual electrical pulses are produced by the arrival of X-ray photons. In most detectors, the average height of these pulses in proportional to the energy of the detected photons; such detectors are called *proportional detectors* [5.9, ch. 7]. The energy resolution of these detectors is limited by the spread of the pulse heights around their mean, and by their stability. It is possible to separate and identify X-ray lines by the analysis of the heights of the pulses they generate. The electronic devices used for this purpose are called *pulse height analyzers*. The amplified pulses can be counted in *scalers* or their rate of arrival measured by *ratemeters*. They can also be used to form scanning images, as will be described in Chapter 15.

5.3.1 Gas-Filled Detectors [4.17, Vol. 2, Ch. 5]

The X-ray detectors used in conventional electron probe microanalysis are based on the ionization of a gas by X-ray photons [5.9]. Figure 5.14 schematically shows such a detector. Its basic parts are a metallic enclosure (usually cylindrical), containing the detector gas, with a window for the passage of X-rays into the detector, and an anode wire electrically insulated from the enclosure. A voltage supply maintains the anode at an adjustable stable positive potential, typically between 1200 and 2500 V, with respect to the enclosure.

Most of the photons entering the detector are absorbed in it, ionizing atoms

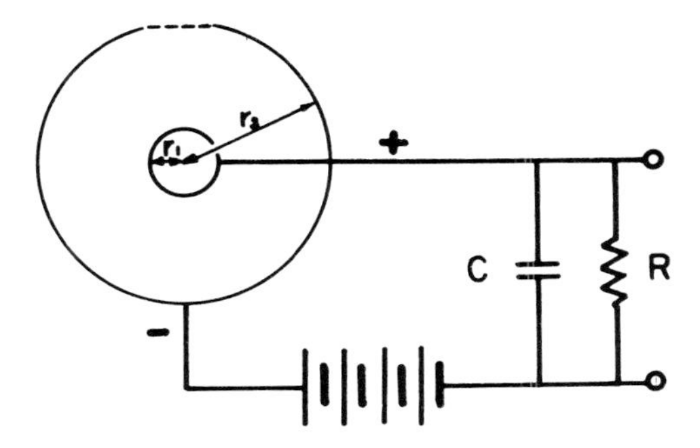

Fig. 5.14. Schematic of a gas-filled detector. W:window, r_1:radius of anode wire, r_2:radius of detector, C:capacitor, R:resistor.

of the detector gas. The probability of ionization, P, depends on the photoelectric mass absorption coefficient of the gas for the radiation, μ, and the mass cross section of the absorber, z:

$$P = \frac{N_0 - N}{N_0} = 1 - e^{-\mu z} \tag{5.3.1}$$

where N_0 is the number of photons entering the absorber, and N is the number of photons exiting the absorber without having been absorbed, and hence, without having produced a signal pulse. The ionization efficiency of the gas detector depends therefore upon the diameter of the detector and the pressure of the gas, as well as upon the mass absorption coefficient of the gas for the radiation.

When ionization occurs, part of the energy of the X-ray photon is spent in overcoming the binding energy of the respective orbital electron, and the rest of the energy is imparted to the electron as kinetic energy. The photoelectron collides with other atoms or molecules and usually causes further ionizations. The vacancy in the ionized level may produce a secondary X-ray photon, which in turn will be absorbed, or a nonradiative transition. The chain of events continues until all energy is spent. The number of ionizations resulting from this chain is, within statistical fluctuations, proportional to the energy of the original X-ray photon.

In the absence of an external electrostatic field, the ions and electrons formed in this chain would rapidly recombine. If a positive potential is applied to the anode, the electrons instead migrate towards it, while the positive ions slowly drift towards the enclosure of the detector. There is competition between migration and recombination; as the potential applied is increased, the fraction of electrons collected at the anode increases until approaching unity.

A detector operating under this condition, in which recombination has become negligible, is an *ionization chamber*. The electrons produced by each photon and collected by the anode of the chamber form a current pulse. If potential is supplied through a resistor, as shown in Fig. 5.14, the anode potential is temporarily lowered, according to Ohm's law. A voltage signal pulse has therefore been generated. However, for practical applications the pulses generated in the ionization chamber are inconveniently small.

Argon is frequenty used as a detector gas. In argon, 26.4 eV are needed, on the average, for the formation of one ion-electron pair. The average number of pairs produced by one photon is therefore

$$n = E/\epsilon; \quad \text{for argon:} \quad \epsilon = 26.4 \text{ eV} \tag{5.3.2}$$

where E is the photon energy, in electron volts.

Once the probability of recombination of the electron-ion pairs has become negligible, a further moderate increase of potential applied to the detector does not change the height of the signal pulse. However, the speed and energy of the electrons approaching the anode increases.

As the electrons are accelerated towards the anode, they are also slowed down by collision with gas molecules. Hence, for a given field, an average drifting velocity is obtained. Because the field strength increases as the anode is approached, the drifting velocity also increases until on impact secondary ionizations with the gas molecules occur. The secondary electrons may, in turn, cause further ionizations, so that an avalanche of electrons (Townsend avalanche) is produced. The ratio of the number of electrons collected by the anode to those produced in the primary ionization chain is called the *internal amplification factor*, α, and a detector operating under the conditions indicated above is called a proportional gas detector (Fig. 5.15).

The electrostatic field produced by the voltage, V, applied at a distance, r, from the anode center can be obtained by the equation

$$E = \frac{V}{r \log (r_2/r_1)} \tag{5.3.3}$$

in which r_1 is the radius of the cylindrical anode wire and r_2 is the radius of the enclosure. Since the field strength increases quickly at short distances from the anode, the secondary ionization in a proportional detector is restricted to a layer of a fraction of a millimeter around the anode.

The charge of the avalanche collected by the anode is

$$q = \alpha \cdot n \cdot e^-. \tag{5.3.4}$$

Here, e^- is the elementary charge of an electron. If the values of the resistor and the capacitance indicated in Fig. 5.15 are known, the speed at which the charge of an avalanche is reduced can be calculated by

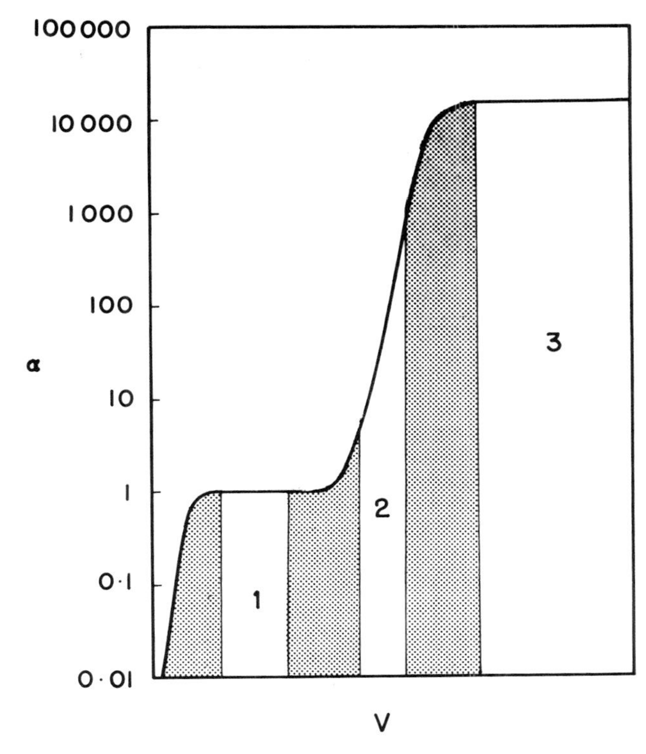

Fig. 5.15. Regions of operation of a gas-filled X-ray detector. 1: ionization chamber, 2: proportional, region, 3: Geiger counter region. Within the proportional region, the amplified pulse is proportional to the photon energy.

$$V = V_0 \exp(-t/RC) \tag{5.3.5}$$

where V is the potential at time t and V_0 is the initial potential. The collection of the charge for further amplification must proceed in a short interval if a significant pulse is to be observed. (This interval is typically on the order of 10 μsec).

The internal amplification factor used in conventional applications of proportional counters is between 10^3 and 10^7. Thus, the signal is considerably increased over that obtained in the ionization chamber. If the detector voltage is raised too high, however, the value of the internal amplification factor becomes dependent upon the number of primary electrons per photon, n. Therefore, the proportionality between photon energy and signal pulse height is affected. This region of operation is called the region of limited proportionality. At even higher voltages (Geiger-Muller region) all signal pulses have identical height regardless of the photon energy. Because the output pulses in this region are even higher than those in the proportional region, the Geiger-Muller counters were widely used at a time when pulse-amplifying electronics had not yet been developed to the present state. However, in this region the proportionality between photon energy and pulse height is lost. More importantly, the duration of pulses of a Geiger counter is in the millisecond range, so that its use is limited to low count rates (up to 500-1000 counts/sec). For these reasons, Geiger counters are not used in electron probe microanalysis.

As mentioned before, as a result of the interaction of acceleration by the field and the collisions with gas molecules, the electron drifting towards the wire anode acquires a constant average drifting velocity, which can be characterized by the mobility:

$$v = kE/p \tag{5.3.6}$$

where v is the velocity, k is the mobility, E is the field, and p is the pressure. Hence, the dimensions of mobility are [$cm^2atm/V \cdot sec$]. The mobility of the ion A^+ at 1 atm is 1.37 $cm^2atm/V \cdot sec$. The mobility of electrons is 10^3 to 10^4 times higher than that of ions, varying considerably over ranges of field and pressure. This difference in mobilities has two consequences of interest.

1. The positive ions, drifting outwards (into regions where the field decreases) do not achieve the speed necessary to produce secondary ionizations.
2. By the time the Townsend avalanche of electrons reaches the anode wire, the ions have moved very little from their points of production. Consequently, they form a field charge which tends to lower the field gradient of the detector. At high counting rates, this lowers the value of α, and hence the pulse height. Since these ions are slow in reaching the detector body, the observed pulse at the detector output consists of a very fast rise of the signal level, caused by the electron avalanche, and a much slower drop of signal which continues until all ions are collected. To eliminate this long pulse tail, the pulse is clipped electronically after the initial fast rise, which is the only part of the initial pulse used in forming the final amplified pulse.

Detector Gases. The mechanism of pulse formation described above can be altered by various artifacts, depending on the composition of the detector gas. If, for instance, pure argon is used, the emission of secondary electrons from the cathode bombarded by the slow ions produces an afterpulse which can be suppressed by the addition of methane. A mixture of 90% argon and 10% methane is widely used as a general-purpose detector gas. For preferential detection of photons of low energy, detector gases have been proposed which absorb less of hard X-rays; among them are neon, methane, carbon dioxide, and a mixture of 75% argon and 25% methane. However, such gases require an operating potential for the detector above 2000 V. Conversely, krypton and xenon are used when the efficient absorption and detection of hard radiation is required. Owing to their higher price, these gases are only applied in hermetically sealed detectors.

A severe anomaly arises from the formation of negative ions due to electron attachment in electronegative gases such as oxygen and chlorine. This process fundamentally alters the avalanche process, even when these gases are present at low concentrations. Therefore, the purity of detector gases must be controlled.

Considerable discussion has arisen over the causes of excessive pulse height shrinkage often observed at high count rates [5.10]-[5.12]. This defect is dependent on the composition of the counter gas, and on the cleanliness of the anode wire. Dirty anode wires also produce broadening of the pulse height distribution at low count rates. It is therefore desirable that the wire be changeable. In some detectors, the wire can also be cleaned by resistive heating.

The Detector Window. The window of the gas-filled detector must permit the entry into the detector of the X-rays, while stopping the passage of gas into the detector or out from it. It must withstand whatever pressure differences may arise in operation or service between the interior of the detector and its surroundings. For shortwave radiation, up to several angstroms, these conditions are easily met, and foils of aluminum or beryllium can be used. For softer radiation vacuum-tight beryllium windows, useful up to a wavelength of 10 Å are now available. For X-rays of even longer wavelength, thin organic films such as nitrocellulose, formvar, or stretched polypropylene are employed. Although such windows have satisfactory transparency for soft X-rays, they are fragile, and must frequently be supported by nylon cords or metal meshes. They are electrically insulating, and can build up charges which distort the field in the detector, unless they are coated at the detector inside with a metallic layer. This precaution also eliminates noise due to the entrance of light into the detector and subsequent emission of photoelectrons.

Thin organic windows are permeable to gases and energetic electrons. If detectors with such windows are used in air, the diffusion of oxygen into the active volume of the detector must be avoided. In an evacuated ambient, the detector will quickly lose pressure and thus become inoperative. For these reasons, the detector must be continuously flushed with the counter gas, and be maintained at a slight overpressure if surrounded by air or other gases. Detectors having provisions for the renewal of the counter gas are called *flow detectors* (Fig. 5.16) [5.13]. When gas bottles are used to store the detector gas, it is good practice to change the bottles before they are completely emptied, since condensable impurities can accumulate in the last fractions released from the bottle.

The efficiency of gas detectors depends on the fraction of the incoming photons absorbed in the gas, and hence on the density of the gas which in turn changes with pressure. If the gas escapes into the atmosphere, the pressure

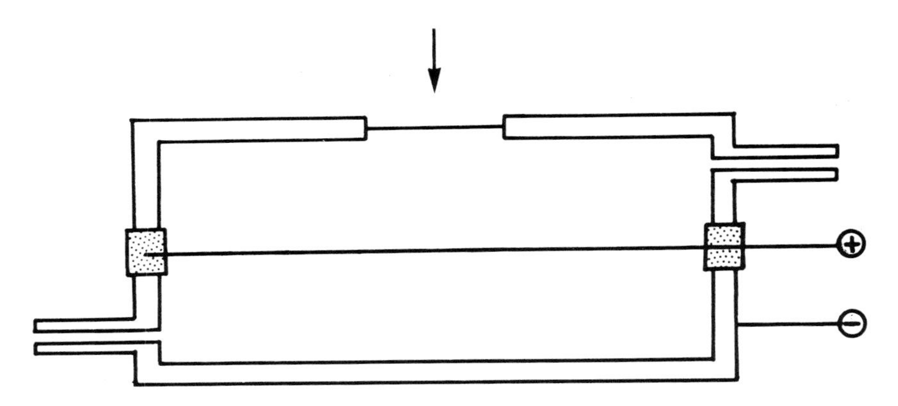

Fig. 5.16. Cross section through a flow-proportional detector. The detector gas is continuously renewed, so that changes in its pressure and composition due to diffusion through the thin detector window are compensated.

inside the detector, and hence its efficiency, may change with the atmospheric pressure. To avoid this effect, it is advisable that the internal detector pressure be regulated to be independent of atmospheric conditions [5.14].

In instruments which must be maintained at a good vacuum to protect field-emission guns, or to avoid the effects of specimen contamination, particularly with Auger-electron analysis, the vacuum contamination from escaping detector gas can be a problem. This is especially the case when the detector gas contains organic compounds such as methane. Some instruments provide protection against the detector gas by means of a second set of windows which separates the spectrometer assemblies from the rest of the vacuum enclosure.

Backscattered or stray primary electrons are able to cross detector windows, particularly at high operating voltages. They produce in the detector charge avalanches which increase the background of the detector output. The effect can be very significant when the detector is in a position close to the specimen. It can be eliminated by placing a strong magnetic field across the detector slit, so that the electrons are deflected from their path into the detector [5.15].

The uniformity of the magnetic field around the anode may be disturbed at the extremes of the anode wire. Such end effects degrade the resolution of the detector. They can be reduced by limiting the entrance of X-rays to the central part of the detector, and by means of guard rings (concentrical shields) located at the extremes of the wire, and maintained at an electric potential at which the pulse height distribution of the detector output is sharpest.

Wavelength-dispersive spectra obtained with curved-crystal spectrometers and gas-proportional detectors are shown in Figs. 2.6, 5.17, and 5.24.

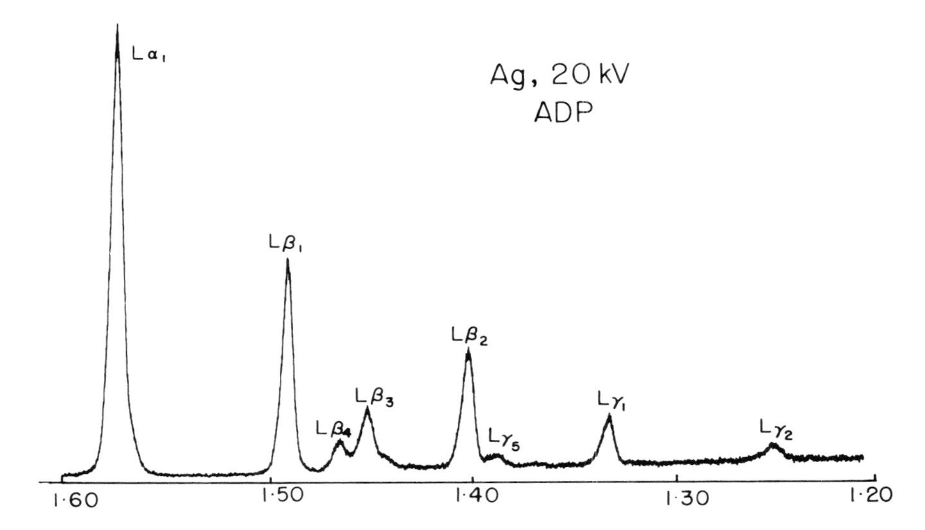

Fig. 5.17. Spectrum of the principal L-lines of silver (Ag), obtained with a curved-crystal spectrometer with an $(NH_4)H_2PO_4$ crystal, of 10-cm radius of curvature, and reflecting on the 101 plane. The detector was an Ar-10% CH_4 mixture gas-proportional detector, E_0 = 20 kV, i_b = 0.45 × 10^{-7} A. The wavelength calibration is in "equivalent LiF-angstroms", i.e., the wavelength in angstroms which would be obtained if the crystal were an LiF crystal. To obtain the wavelength in true angstroms, the numbers must be multiplied with the ratio of the respective lattice spacings, which in this case is 5.32/2.0136 = 2.642 [7.4].

5.3.2 The Lithium-Drifted Silicon Detector (Si(Li) Detector) [5.16]-[5.20]

The lithium-drifted silicon detector operates in a similar way to the gas-filled ionization chamber. It consists of a cylinder of p-type silicon into which lithium has been diffused by simultaneously applying heat and an electrostatic field. The lithium ions pair with atoms of residual impurities, and a region of intrinsic properties (i.e., analogous to a region of extremely high purity) is produced, in which, in the absence of ionization, the intrinsic conductivity is very low. The bases of the cylinder have conductive layers; the Schottky barrier contact on the front side completes a p-i-n type diode, in which most of the volume is depleted of free charge carriers (Fig. 5.18). The conductor at the front through which the X-rays enter the detector is a layer of gold approximately 200 Å thick; beneath this gold layer an inactive layer of silicon extends to approximately 0.1 μm. A negative charge in the order of 1000 V is applied to the front of the detector.

X-ray photons entering the device interact primarily by photoelectric absorp-

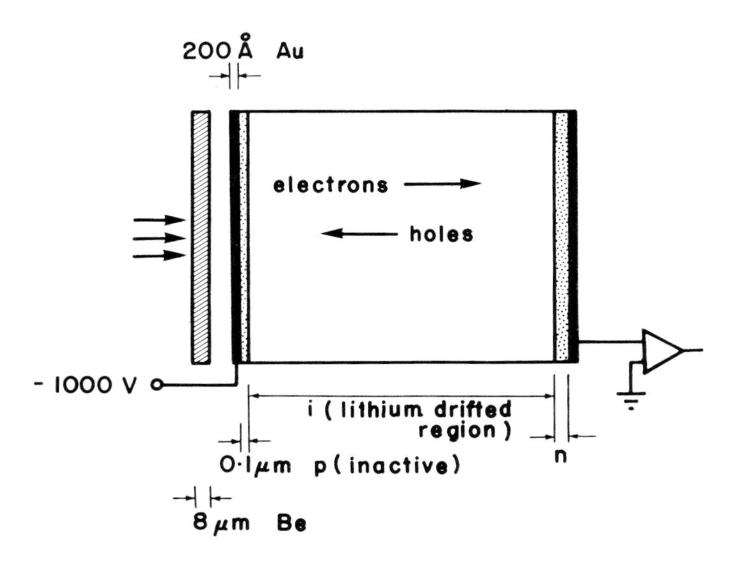

Fig. 5.18 Lithium-drifted silicon X-ray detector.

tion, producing a cloud of electron-hole pairs. As in the case of the ionization chamber, the number of pairs can be obtained by the equation $n = E/\epsilon$. The averrage energy absorbed in the formation of an electron-hole pair is about eight times smaller than that required to ionize one argon atom (ϵ = 3.6-3.8 eV/electron-hole pair). Therefore, the statistics of the event provide a better energy resolution than with the argon detector. The charge is swept from the diode by the bias voltage, and collected for pulse formation. As in the ionization chamber, the probability of recombination is negligible, but there is no internal amplification. Hence, the signal obtained by each photon is very small. The use of this detector for X-ray detection would be impossible without the development of low-noise solid-state (field-effect transistor) amplifiers, and the resolution of the detector is greatly dependent on the noise level of the detector and amplifier. For proper operation, both the detector and the preamplifier must be cooled with liquid nitrogen. This requirement accounts for the large size and weight of the detector assembly, which cannot be mounted on conventional crystal spectrometers (Fig. 5.19). The lithium-drifted silicon detector is therefore always used to observe the full spectrum of emitted radiation, in conjunction with pulse height analysis, which will be discussed later in this chapter.

Even when not in use, silicon detectors should not remain for more than a few hours at room temperature, because further drift of the lithium at room temperature may damage them. Therefore, a constant supply of liquid nitrogen is required where such a detector is used.

Many electron probes operate at residual pressures above 10^{-5} torr. At these

Fig. 5.19. Si(Li) detector mounted in fixed position on an electron probe microanalyzer. (A) is the detector housing. (B) is the preamplifier.

pressures, residual impurities, particularly from oil pumps, tend to deposit on cool surfaces. To protect the front of the cooled silicon detector from the deposition of such impurities, the detector is housed in a special enclosure within which a good vacuum is maintained. A beryllium window provides access of the X-rays to the detector. The absorption of X-rays by this window usually determines the lower limit of photon energy at which X-rays can be measured.

Some detectors have a movable beryllium window which can be retracted once a high vacuum has been established in the instrument. With such detectors, the characteristic X-ray emission of oxygen and carbon can be observed. However, in the region of very low photon energy, the resolution of the silicon detector is not substantially different from that of a proportional gas detector (Fig. 5.20).

The proportionality between the energy of the detected photons and the corresponding mean pulse height (*linearity* of the detector) is excellent (Fig. 5.21), and the stability of the amplification factor, which only relies on external amplification, is also very good, even over long periods. The efficiency of the silicon detector spectrometer is much larger than that of a crystal spectrometer (see Section 12.6), and can be predicted accurately. These properties render the energy-dispersive silicon detector an attractive spectrometer for absolute measurements, in spite of the limited resolution.

5.3.3 Amplification and Pulse Height Analysis

According to Eq. (5.3.2), a pulse produced by a photon of 10 keV in an argon detector having an internal amplification factor of 10^5 consists of approximately 4×10^7 electrons, or 6.4×10^{-12} C. A corresponding pulse on the solid-state detector–which has no internal amplification–is approximately 10^{-4} times smaller. Clearly, these pulses must be carefully amplified before they can be used to provide information. The first stages of amplification are most critical, and the need for cooled field-effect transistors for the solid-state detector has already been mentioned.

In order to take advantage of the proportionality between photon energy and pulse height, it is necessary to use amplifiers which linearly magnify the initial signal. The amplified signal can then provide information concerning the energies of the detected photons (pulse height analysis), and pulses produced by photons of undesirable energy, or by other sources of signals, can be excluded from the signal by electronic devices (pulse height discrimination, pulse height selection).

Figures 5.22 through 5.24 show pulse height spectra obtained by sorting, according to their height, the pulses obtained by amplification of detector signals. In Fig. 5.22, the pulses are shown as seen on an oscilloscope triggered by

Fig. 5.20. Parameters related to spectral resolution. —o—o—o— natural width of $K\alpha_1$ line. ◆—◆—◆— distance between centers of $K\alpha_1$ and $K\alpha_2$ lines. (The resolution of a SiO_2(1011) curved crystal in typical instruments is of the same order of magnitude.) LiF, ADP, RAP, PbSt line widths observed with 10-cm radius curved crystals of typical slit width setting of LiF 200, ammonium dihydrogen phosphate, rubidium acid phthalate, and a lead stearate pseudocrystal, all under conditions of moderate resolution. The points marked ϕ and ϕ are high-order reflections which show the resolution of the spectrometer. Points marked o and ◐, above the respective lines, show the measured width of bands in the first order of reflection. —□—□— K-lines, as resolved with a lithium-drifted silicon detector. ▲, △ same, $L\alpha$ and $M\alpha$ lines. The high values between 1.5 and 5 keV are due to the overlap of L and M lines – – – – – – theoretical resolution of a silicon detector, in absence of noise in the amplification stage [Eq. (5.3.13)]. – . – . – . – theoretical resolution for an argon proportional detector [Eq. (5.3.12)]. ⊟ 90% Ar 10% CH_4 flow-proportional detector. ■ sealed argon-methane proportional detector. All values are full width at half maximum (FWHM).

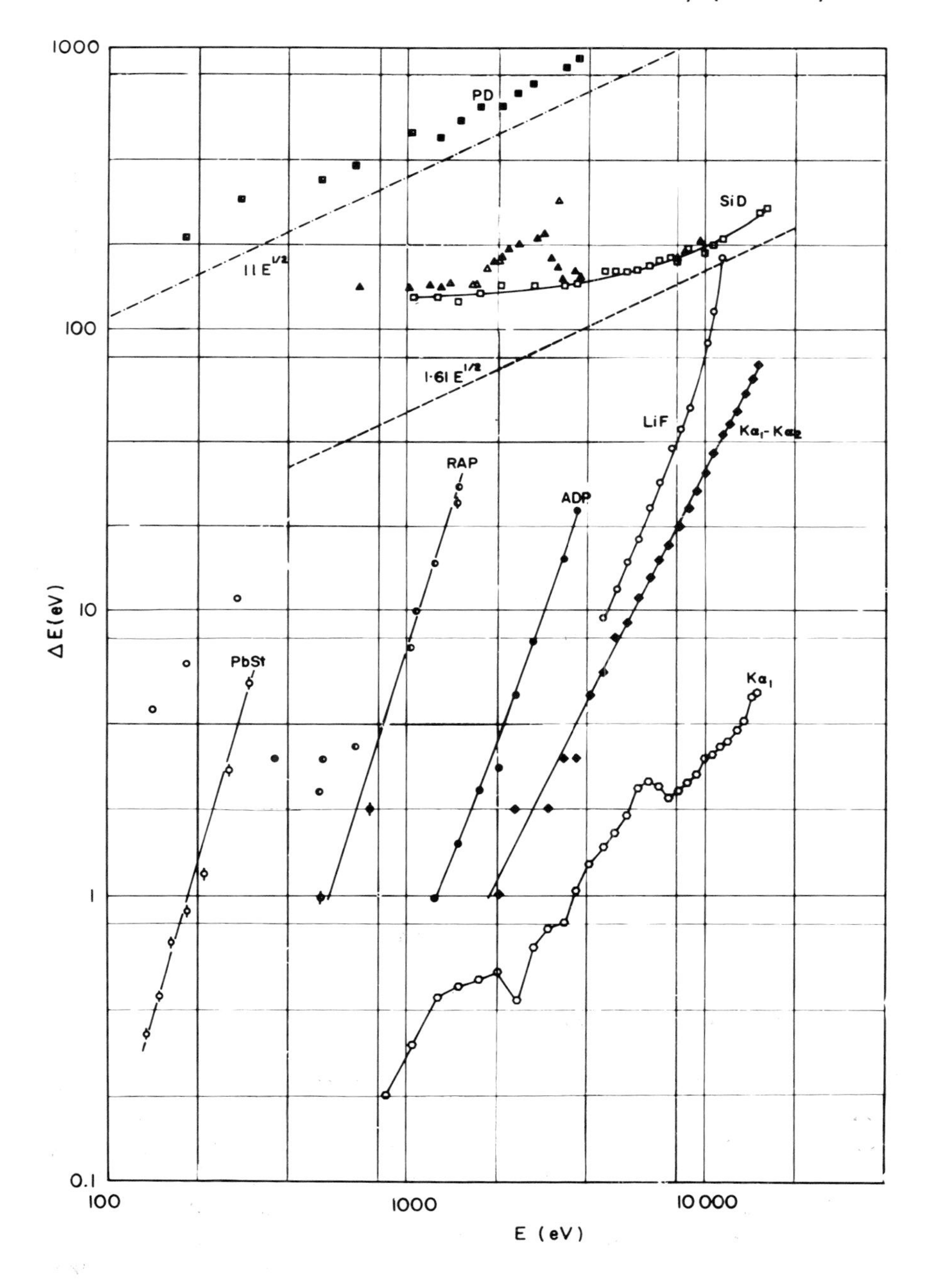

1000
100
10
1
0.1
100
1000
10 000
E (eV)
ΔE (eV)
PD
11 E^1/2
SiD
1·61 E^1/2
LiF
Kα1-Kα2
RAP
ADP
PbSt
Kα1

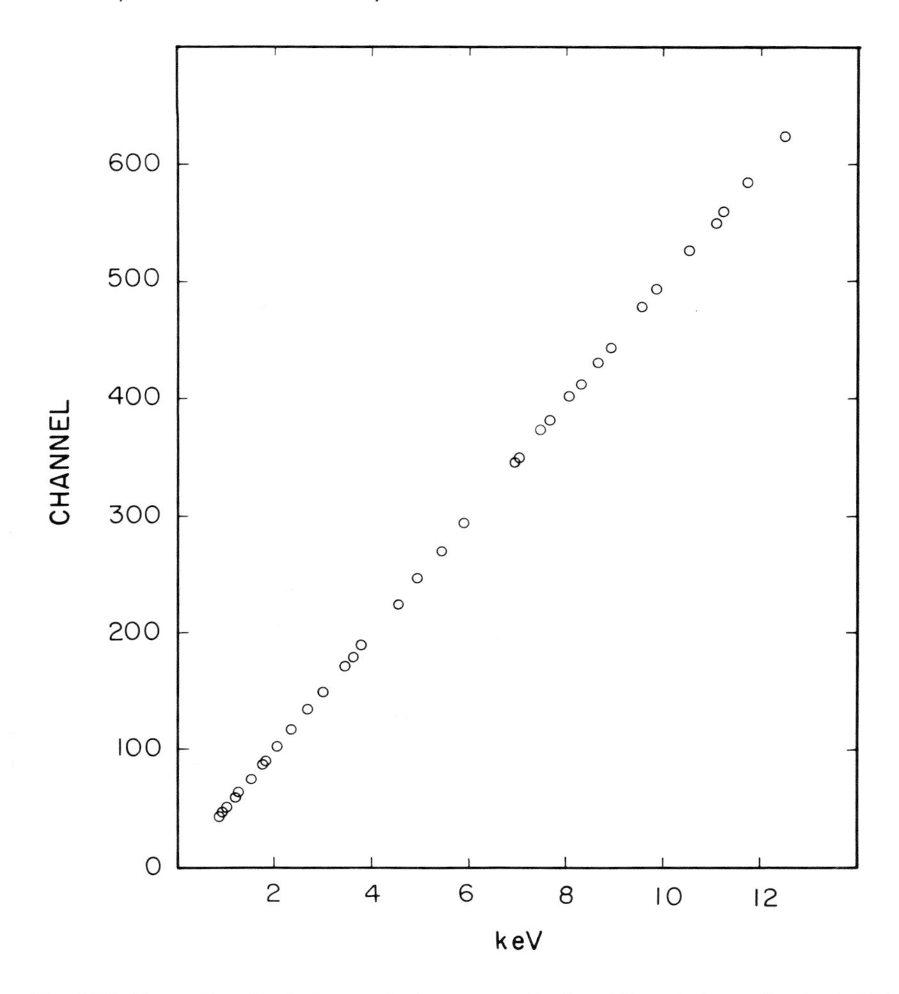

Fig. 5.21. Proportionality between photon energy (horizontal) and observed pulse height (vertical) of a silicon detector signal.

the onset of each pulse. Due to thermal and electronic noise the signal level in the absence of detected pulses is not completely flat. Therefore, the pulse height spectrum shows and intensity of noise pulses which tends to infinity as the pulse height approaches zero. For noise-free signal detection the height distribution of the noise pulses must be completely separable from that of the signal pulses. If this is the case, then an electronic device in the later stages of amplification (discriminator, or threshold) can be used to suppress from further transmission all pulses below an adjustable level, which is set above the noise pulses, and below the distribution of the signal pulses. Such a discriminator is used in all circuits for amplification of pulses from proportional detectors. It was also employed in the oscilloscope with which the images on Fig. 5.22 were made.

The presence of undesirable pulses which are larger than the signal pulses is less frequent. Such pulses can, for instance, be due to higher-order reflections in a crystal spectrometer. To eliminate them, we can determine an upper level of discrimination, and reject all pulses higher than this level. An electronic device operating with two levels of discrimination is called a *single-channel pulse height analyzer*. The lower level is called *baseline*, and the region between the two levels, within which the pulses are considered acceptable, is called the *window*. A single-channel pulse height analyzer can be constructed by means of two gates which allow the passage of pulses above the respective levels of discrimination, and an anticoincidence circuit, which rejects the pulses which have passed both gates (Fig. 5.25). A single-channel analyzer working in this mode is said to operate in the differential mode. In the integral mode, the rejection due to the anticoincidence circuit is inactive so that the device simply acts as a discriminator. The pulses emitted from the pulse height analyzer are normalized in height and shape, to match the requirements of devices such as scalers and ratemeters which receive them.

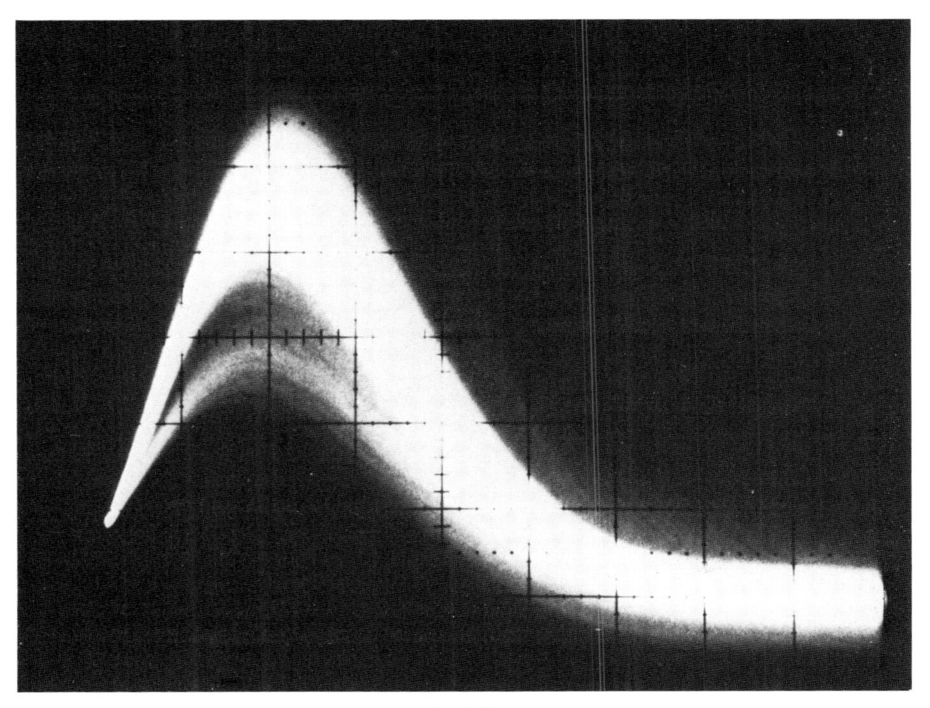

Fig. 5.22a

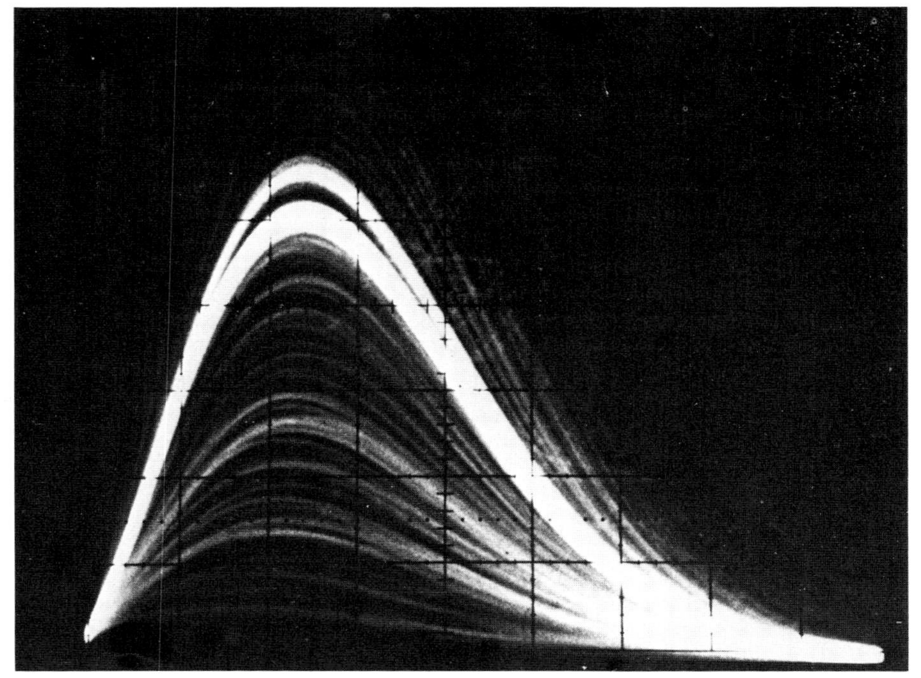

Fig. 5.22b

Fig. 5.22. Amplified detector pulses shown on an oscilloscope screen. Both images show pulses producted by Fe K radiation, and a discriminator setting was used to eliminate the background noise pulses. 5.22a: Fe Kα radiation, separated from the spectrum by an LiF crystal, and detected in an Ar-CH_4 flow-proportional detector. The Fe Kα-Ar Kα escape peak can be seen. Time scale: 0.5 sec/cm. 5.22b: pulses from an iron target, at 20 kV, without separation by crystal, observed by Si(Li) detector. Fe Kα and Fe Kβ are fully separated. The escape peaks are too weak to be seen. Time scale: approximately 5 sec/cm.

In the energy-dispersive configuration in which the detector directly observes the X-rays emitted from the specimen, the differential pulse height analyzer mode permits the observation of one energy peak, with exclusion of peaks which have higher or lower pulse height. This principle was used by Dolby [5.21], [5.22], who combined a flow-proportional detector and three single-channel analyzers to simultaneously detect and observe three soft X-ray lines, such as beryllium, carbon, and oxygen Kα. Each analyzer was set on a "region of interest" for the line of one element. The effects of overlapping peaks were minimized by subtracting part of the ratemeter signals from the regions of interest of the interfering lines. This method has recently attracted renewed interest for the use with Si(Li) detectors, as will be described in Section 12.6. A block diagram for a typical dispersive spectrometer channel is given in Fig. 5.26.

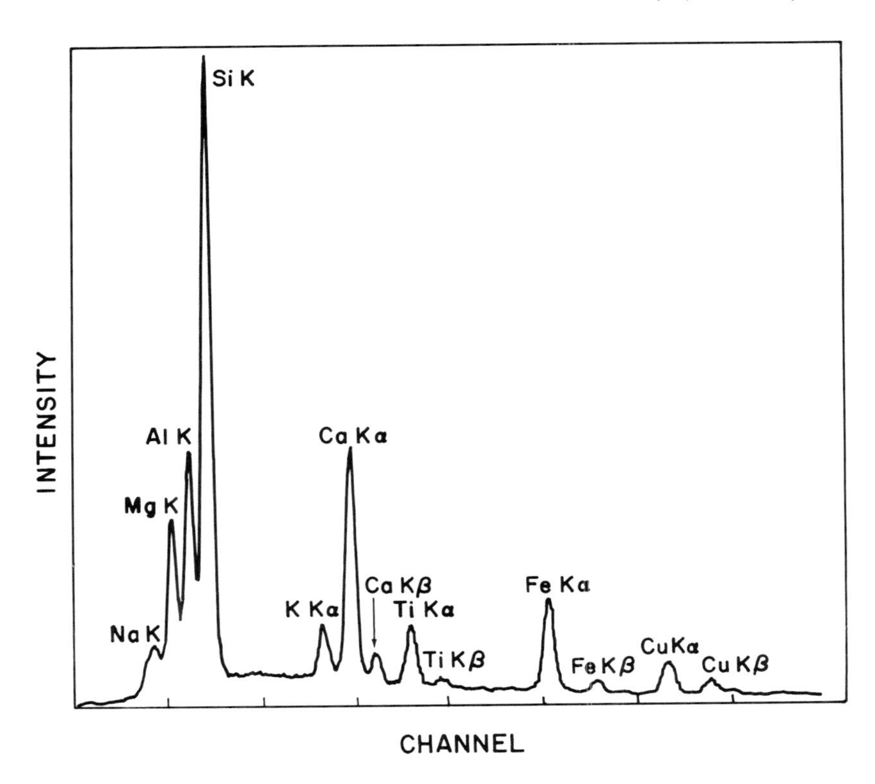

Fig. 5.23. Energy spectrum obtained with a lithium-drifted silicon detector from the same mineral from which the wavelength spectrum shown in Fig. 2.6 was obtained. Operating voltage: 20 kV. Detector resolution (FWHM for MnKα): 170 eV.

When crystal spectrometers are used, interference situations which require the use of a single-channel analyzer are uncommon (such a situation would arise from the interference of a line in a higher order of diffraction). The unnecessary use of the analyzer in the differential mode may enhance drifts due to instability of components; therefore, it is advisable to normally use the analyzer in the integral mode if no interference is present. The lower threshold of the analyzer should be kept halfway between the noise level and the signal level (Fig. 5.27). It should also be noted that the setting of the pulse height analyzer may interfere with the performance of wavelength scans with Bragg spectrometers (Section 7.1).

The relation of the height of signal pulses relative to the window of the pulse height analyzer or the threshold of the discriminator can be modified by adjusting the detector voltage (i.e., the internal amplification factor), the linear amplifier, or the pulse height analyzer settings. In a given experimental setup,

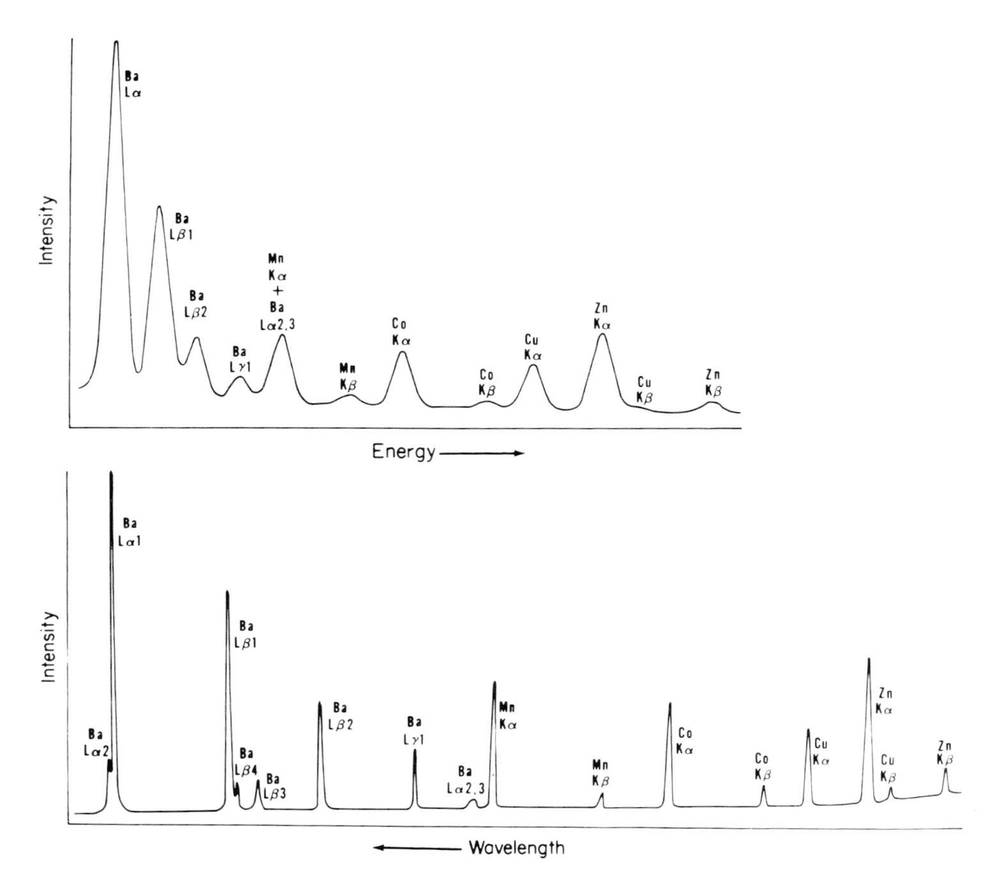

Fig. 5.24. Comparison of spectra of a glass (NBS K252) with a Si(Li) detector (above) and with a 10-cm radius LiF curved-crystal spectrometer (below). Both performed at 20-kV excitation potential. The composition of the glass is as follows: SiO_2: 0.40, BaO: 0.35, MnO: 0.10, MnO_2: 0.05, CuO: 0.05, CoO: 0.05, all mass fractions.

there is an optimum detector voltage at which the detector resolution is sharpest. It is good practice to determine and then maintain this detector voltage, and to maintain always the pulse height analyzer, or discriminator, at the same settings of the baseline, and where applicable, the window. The voltage range within which the pulse height analyzer will accept pulses is noted, and with the help of a monitor oscilloscope, the linear (external) amplification factor is adjusted until the pulses of interest are seen to fall within this preestablished voltage range. After switching to an X-ray line of different wavelength, all the operator has to do is to adjust the linear amplification, while observing the

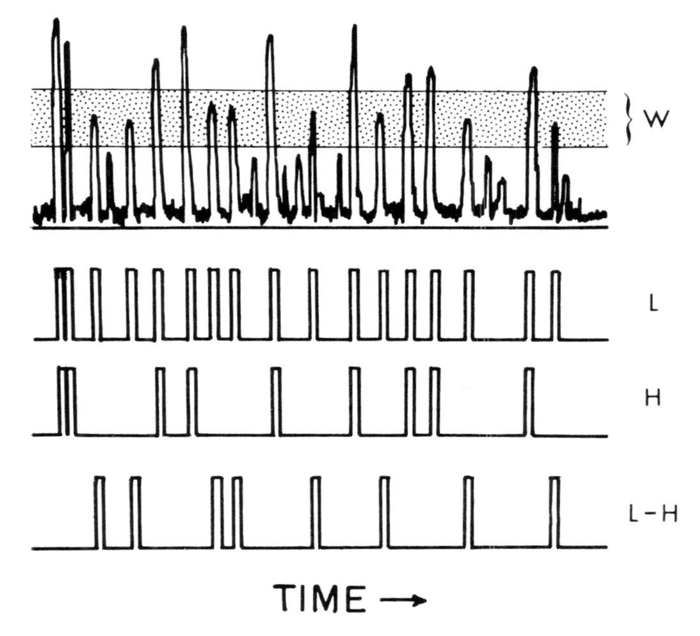

Fig. 5.25. The operation of the single-channel analyzer. Upper row: a set of pulses issued by the linear amplifier. Two discriminator levels determine the window (*W*). When the pulses passing above the upper level (*H*) are subtracted from those passing the lower level (*L*), only the pulses falling within the window (*L* – *H*) remain in the output (differential mode). If the upper discriminator in inactive (integral mode), the output of the pulse height analyzer is that given by (*L*).

amplified pulses on the oscilloscope. The required equipment configuration is shown in Fig. 5.26.

Energy-Dispersive Analysis: The Multichannel Pulse Height Analyzer (MCA). Pulse height discrimination by means of single-channel analyzers is applicable when an analyzer is available for every line to be measured, e.g., in conjunction with a crystal spectrometer. The single-channel pulse height analyzer is, however, impractical when an entire X-ray spectrum is received by a detector which directly observes the X-ray source [5.23]. In this arrangement, which has become increasingly attractive with the development of the lithium-drifted silicon detector, the entire detector output, after linear amplification, can be observed simultaneously by means of a multichannel pulse height analyzer. This instrument consists of the following components.

1. The *analog-to-digital converter* sorts the pulses which arrive within a pre-established counting period, according to their height, into a number of *channels*.

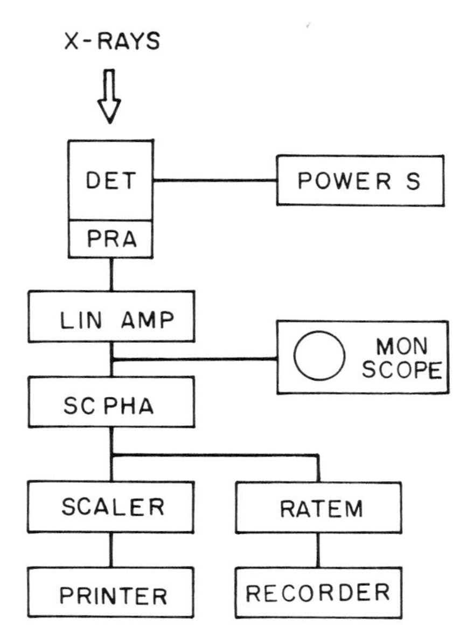

Fig. 5.26. Setup for single-channel pulse height analysis. DET:detector, LIN AMP:linear amplifier, MON SCOPE:monitor oscilloscope, POWER S:power supply, PRA:preamplifier, RATEM:ratemeter, SC PHA:single-channel pulse height analyzer.

2. The *memory* records the number of pulses corresponding to each channel. It is equivalent to a set of scalers, one for each channel.
3. The *readout section* communicates, in several ways, the information collected by the memory, to the experimenter. Typically, the output of a multichannel analyzer can be displayed on an oscilloscope, or on an *x-y* recorder. The number of individual channels or groups of channels, and the counts accumulated in them, can be shown digitally on scalers, or the contents of the memory may be transmitted to a computer for further processing. In systems which rely principally on data handling by means of computers, the direct readout facility may be very simple. It is, however, useful to have a direct output, particularly by means of the cathode-ray tube, for rapid qualitative tests in the course of electron-probe microanalysis, and to have alternative digital readout provisions available for situations in which the computer interface is inoperative.

The *analog-to-digital converter* (ADC) is the heart of the multichannel analyzer (MCA). After the arrival of a pulse an electric signal proportional to the pulse is stored in a capacitor. When the pulse has reached a maximum the capacitor is discharged by a constant-current source. The time from the start

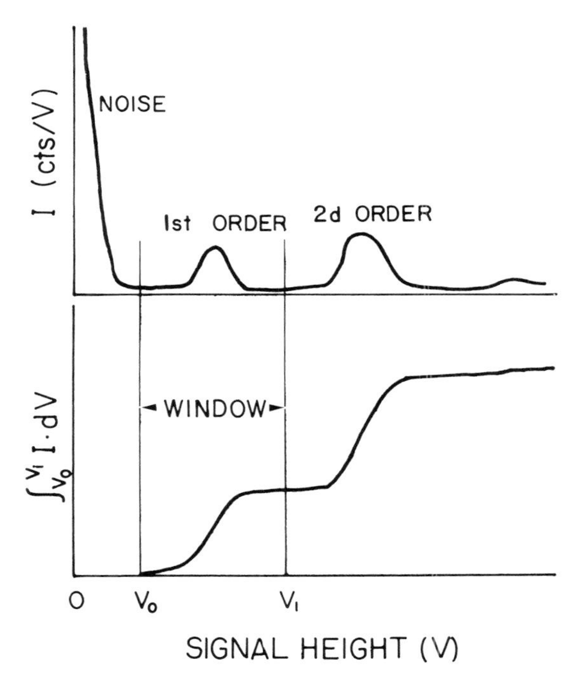

Fig. 5.27. Pulse height spectrum (upper graph) of a crystal reflection, with detector noise and first- and second-order components of characteristic radiation. The threshold (V_0) is set between the noise level and the first-order reflection. The second-order reflection can be eliminated by setting the single-channel analyzer window as shown. The lower graph shows the integrated function obtained when, after setting the lower threshold, the window width is changed from zero to the upper limit. This mode of operation serves to find the appropriate window width.

of the ramp formation to the moment at which the capacitor is fully discharged is taken as a measure of the pulse height. The length of this time period determines the channel to which the pulse is added. For the proper working of the ADC it is necessary that the incoming pulses meet the specifications for the instrument, with regards to range of height, duration, impedance, and shape. In some instruments a preamplifier is incorporated which is used to amplify the detector pulses and which assures that the match of the incoming pulse to the analog-to-digital converter is achieved.

During the interval in which the pulse is compared with the ramp, no other pulses can normally be accepted. The resulting dead-time of the multichannel analyzer is large when compared with that of single-channel analyzers, and it increases with the number of the channel in which the pulse is stored. Hence, the use of an excessive number of channels increases the coincidence losses. On the other hand, pulses which overflow the range of voltage corresponding to the selected number of channels may also disturb the functions of the MCA, and it is therefore desirable to set the range of the ADC so that all arriving pulses fall within it.

Memory. Available multichannel analyzers vary considerably in the size of available memory–i.e., in the number of channels (or scalers) which are available, as well as in the maximum number of counts which can be stored in each channel. The number of channels varies typically from 1024 to 4096, and their counting capacity from 10^5 to 10^6 per channel. The number of channels available in memory can be larger than the maximum number of channels into which the ADC can sort the arriving pulses. It is then possible to use in one counting operation a sector of the memory, while one or more spectra are held for later comparison in other memory locations. Such a procedure can be followed even if the available memory does not exceed the capacity of the channel, since the ADC can be operated over a fraction only of its available range. For a lithium-drifted silicon detector, a minimum of 400 channels is desirable; it is not advisable to extend the pulse spectrum over more than 1000 channels, since the excessive number of channels causes large dead-time losses, and poor counting statistics in the contents of each channel. A minimum of 800 channels of memory permits the storage for later comparison of one spectrum while another is collected. This is a very desirable feature, particularly if the contents of the memory after collection of a spectrum cannot be transferred to a computer.

The size of memory per channel determines the maximum accuracy in the measurement of an intensity of radiation in this channel. According to Poissonian statistics (see the Appendix), the theoretical relative standard deviation of a measurement of 10^5 counts is 0.33, and that of a measurement of 10^6 counts is 0.13. Excessive count rates not only result in high coincidence losses, but they may also affect the operation of the analog-to-digital converter. The sorting of pulses depends on the stability of the baseline above which the pulses rise; if this baseline has not been fully restored to its normal value before the arrival of the next pulse, this pulse may not be sorted into the correct channel. Depending on the characteristics of the device, the baseline may tend to be high or low when the count rate is excessive. Since the distances between arriving pulses vary randomly, the cumulative effect of these shifts is a broadening and displacement of the peaks in the pulse height spectrum. It is therefore important that the count rate of *all* arriving pulses (regardless of their height) not exceed the specifications for the instrument. Most multichannel analyzers have a meter which indicates the percentage of coincidence losses; the investigator can use the readings of this meter to keep the operating conditions within such bounds as not to produce a degradation of the resulting pulse height spectrum.

The coincidence losses can be canceled by means of a "live-time" counting procedure. The duration of the counting period is determined by means of a crystal clock incorporated in the multichannel analyzer. In the live-time mode, this clock is stopped during the processing of a pulse, so that it counts only the time during which the converter can receive pulses. The use of this mode cannot, however, prevent the deterioration of resolution and pulse height shifts which occur at high count rates. The limitations of the multichannel analyzer with respect to count rate are a serious disadvantage in its use, particularly in quantitative analysis. However, the dead-time characteristics of multichannel analyzers have improved rapidly in recent years.

5.3.4 Output

Depending on the intended use, the memory contents of the multichannel analyzer can be transferred into various devices. The oscilloscope output is very fast—spectra can be observed while being collected, and various types of display, such as spectra on a logarithmic scale, or simultaneous display of several spectra, are possible. A permanent record can be obtained by photographing the cathode-ray tube display. A clearer output can be produced with the aid of an *x-y* recorder. The numerical contents of one or several channels can be shown on scalers, or printed on typewriters or line printers. This information can also be recorded on punched paper tape, or on a magnetic tape or disk for storage and later use.

The uncorrected spectra displayed on the oscilloscope or graphically recorded (Figs. 5.23 and 5.24) can be qualitatively interpreted, as described in Chapter 7. However, the limited resolution of energy spectra, and the resulting high background frequently render desirable a further treatment of the data by a computer [5.24]. The information can enter the computer by direct connection, or via a tape or disk. A flow diagram of an instrument assembly which provides several possible lines of procedure is shown in Fig. 5.28. The handling of the spectra, including background correction, deconvolution of lines, application of line-identification schemes, and the corrections for quantitative analysis will be described in Sections 12.6 and 12.7.

5.3.5 Detector Efficiency

An X-ray photon which has reached a detector may fail to produce a signal pulse for several reasons. It may be absorbed by a part of the detector which does not produce a signal, such as the detector window, or it may traverse the active part of the detector without interacting with it (Fig. 5.29). Even if the photon is absorbed in the active zone of the detector, a pulse of appropriate characteristics may not be produced, because of production of fluorescent radiation which escapes the active zone (escape pulses), or because the detector has been rendered temporarily inactive during the processing of a previous pulse (coincidence losses).

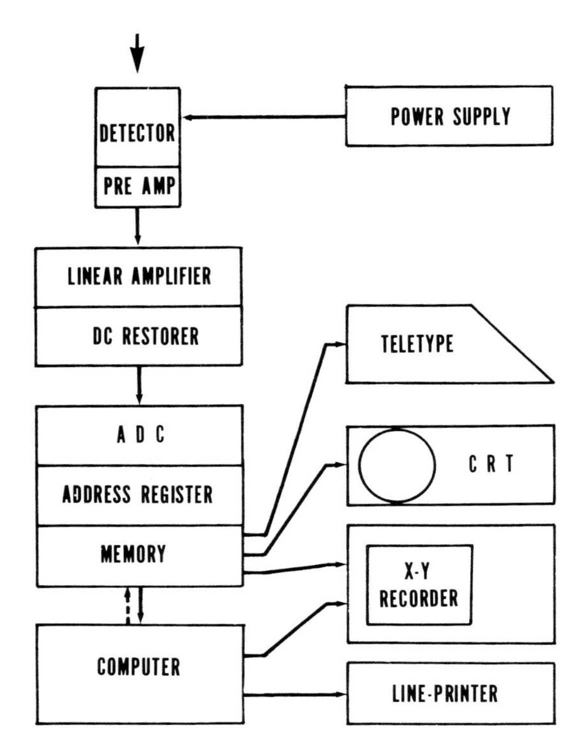

Fig. 5.28. Flow diagram of an energy-dispersive spectrometer assembly with a multichannel analyzer, and various output devices (peripherals). ADC:analog-to-digital converter, CRT: cathode-ray tube, PRE AMP:preamplifier.

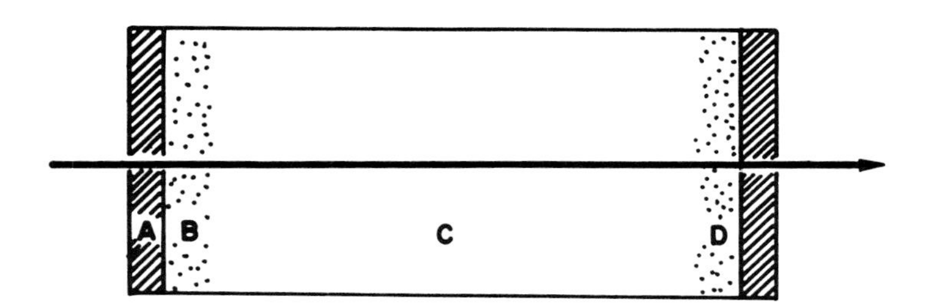

Fig. 5.29. The efficiency of an X-ray detector is determined by the fraction of X-rays entering the detector which are absorbed in the active region, C, and produce a signal. Photons which are absorbed in the enclosure (window, A), or in inactive regions of the detector (B, D), or which emerge after crossing it cannot produce a signal.

The efficiency of a detector is a function of the wavelength of the observed radiation, of the characteristics of the active region of the detector and its envelopes, and of the detector shape and size. Figure 5.30 shows typical efficiency curves for various types of detectors, plotted as a function of the wavelength of the impinging photon. In the case of the solid-state silicon detector, the absorption of X-rays by the protecting beryllium window must be taken into account in calculating the efficiency. Abrupt changes of efficiency occur at the absorption edges of the absorbers. These are due to changes in the probability of the passage without absorption of a photon through the active region of the detector.

The combination of the absorption mechanisms in the detector gives the formula for the detector efficiency, P:

$$P = \epsilon e^{\Sigma - \mu_w z_w} \left(1 - e^{-\mu_d z_d}\right) \tag{5.3.7}$$

where μ and z denote the respective absorption coefficients and thicknesses; the subscripts w and d denote windows and detector, and the summation indicates that more than one window layer (including the inactive silicon layer of the silicon detector) may be present. If the efficiency definition is also to

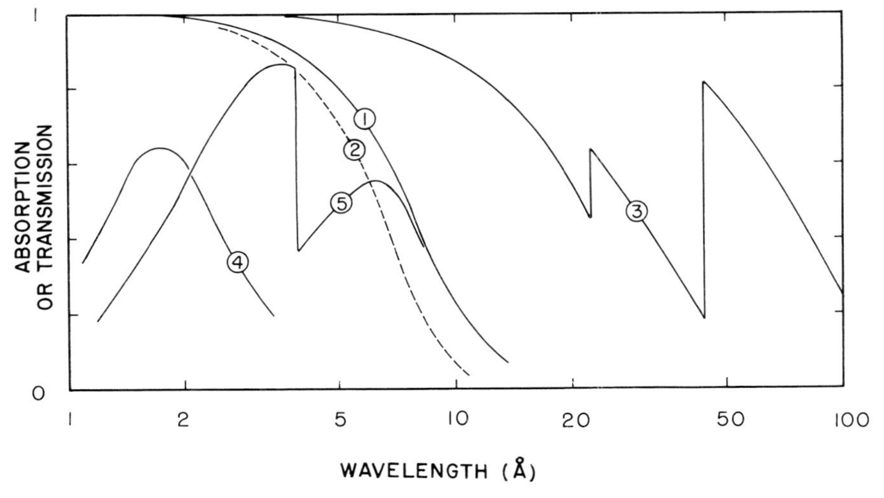

Fig. 5.30. Curves 1, 2, and 3 show the transmittance of X-rays through windows made of 0.025-mm beryllium, 0.05-mm beryllium, and 60 g/cm² formvar. The abrupt variations in transmittance through formvar are due to the absorption edges of oxygen K (23.3 Å) and carbon K (43.8 Å). Curves 4 and 5 are the efficiencies of an argon detector having a thick aluminum window, and of another argon detector with a 0.025-mm beryllium window. The descent of these curves towards short wavelengths is due to incomplete absorption within the detector of the photons which have passed the window. Curve 5 changes abruptly at the wavelength of the argon K-edge, due to the change in absorption within the detector.

include the solid angle Ω covered by the detector receiving radiation from a point source, the above expression is multiplied by $\Omega/4\pi$. The internal efficiency factor, ϵ, close to one, denotes the probability that an absorbed photon will produce a pulse. Its value depends on the dead-time, incomplete charge collection, escape peak formation, and other detector artifacts which will be discussed later in this chapter. The efficiency of a typical silicon detector as a function of photon energy is shown in Fig. 5.31.

5.3.6 Detector Resolution [5.25]

The spread in pulse height of the signal produced in a proportional gas detector by monochromatic radiation is due to a combination of fluctuations in the number of primary ionizations per photon, in the avalanche production, in electronic noise, and due to changes in the electrostatic field with time (e.g., at high count rates) or with respect to the position within the detector. Let us assume that the last two effects are insignificant under proper operating conditions. If the primary ionization were the sum of a number of causally independent events, the spread in the number of ionizations per electron, σ_n, would be Poissonian (see the Appendix). Since, however, chains of causally related events occur in the formation of the primary electron cloud, the spread of n is smaller than that

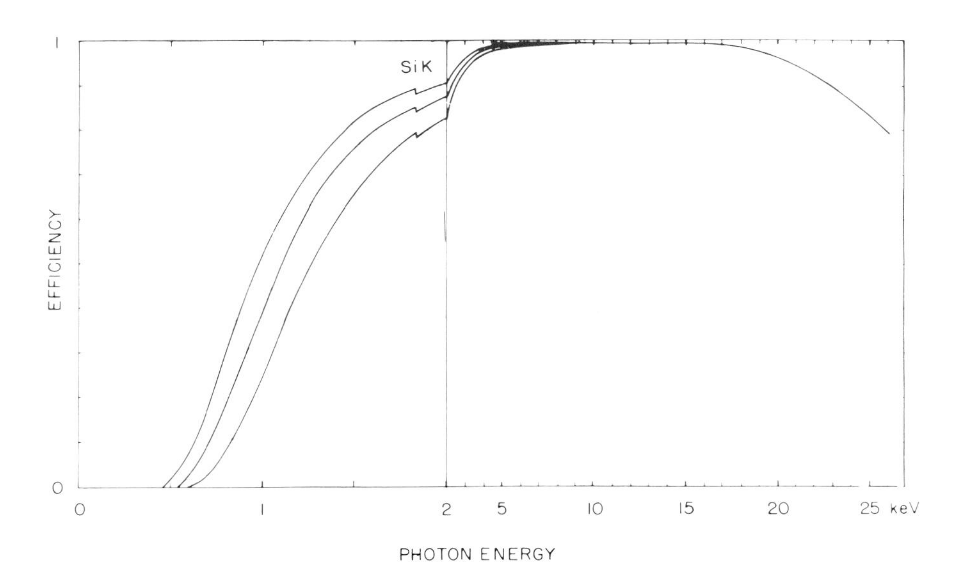

Fig. 5.31. Efficiency of the Si(Li) detector. Curves correspond, from left to right, to beryllium windows of 7, 10, and 15 μm thickness. The scale of photon energy changes at 2 keV, so that efficiency at energies below 2 keV can be seen more accurately. SiK is the silicon K-edge.

predicted by the Poissonian distribution. This is expressed by a factor F (Fano factor), as follows:

$$F = \sigma_n^2/\overline{n}. \tag{5.3.8}$$

If the distribution were Poissonian, the Fano factor would be equal to one. In real counters, this factor is smaller (about 0.22 for argon). The statistics of both the primary ionization and the avalanche formation were studied by Campbell [5.25] who indicated that an optimistic prediction from the theory of the total pulse spread gives

$$\sigma_V/\overline{V} = 0.11E^{-1/2} \tag{5.3.9}$$

for an argon detector while measurements, by Mulvey and Campbell [5.26] give

$$\sigma_V/\overline{V} = 0.15E^{-1/2} \tag{5.3.10}$$

where E is expressed in kiloelectron volts (keV) and $\overline{V}$ is the mean pulse height of the distribution. Since this parameter is proportional to the photon energy of the detected radiation, E, it follows that the statistical spread of the observed photon energy of a spectral line is:

$$\sigma_E/E = \sigma_V/\overline{V} = 0.15E^{-1/2}, \qquad \sigma_E = 0.15E^{1/2}. \tag{5.3.11}$$

The resolution of detectors is frequently expressed in terms of "full width at half maximum" (FWHM). A Gaussian distribution attains a height equal to half its maximum at a distance of 1.18 times its standard deviation. Hence, the FWHM of a curve close to Gaussian is equal to twice this distance (2.36σ). Assuming that the pulse height distribution observed at the detector output were Gaussian, we would obtain: FWHM $\simeq 0.35E^{1/2}$. Real pulse height distributions from gas detectors are not Gaussian, and have an observable skewness. Furthermore, imperfections in design and operation may produce, at the tails of the bell-shaped distribution curves, values considerably higher than those expected from a Gaussian of the same FWHM. Hence, the above equation is only approximative. FWHM's of distributions obtained in the detectors of electron-probe microanalyzers are shown in Fig. 5.20, in which Eq. (5.3.11) is also represented. In this figure, the photon energy E is, however, expressed in electron volts, so that the numerical coefficients of Eqs. (5.3.9) to (5.3.11) must be multiplied by $1000^{1/2} = 31.6$. We obtain:

$$\text{FWHM} = 11E^{1/2} \tag{5.3.12}$$

where E is measured in electron volts (eV).

The energy resolution of the silicon detector is determined by both the intrinsic statistics of electron-pair formation, and by the noise contribution of the first amplification stage. The first contribution can be obtained by the equation:

$$(\text{FWHM}) = 2.36(FE\epsilon)^{1/2} \text{ eV} \cong (2.5E)^{1/2} \text{ eV} = 1.6E^{1/2} \text{ eV} \qquad (5.3.13)$$

in which E (measured in electron volts) is the energy of the absorbed photon, F is the Fano factor (which typically is 0.115 to 0.135), and ϵ is the energy required to create an electron-hole pair (for silicon, 3.6 eV). If we call the noise contribution to energy spread E_n, then we obtain for the total spread

$$\Gamma = (2.5E + E_n{}^2)^{1/2} \text{ eV}. \qquad (5.3.14)$$

At present, the values of E_n in typical detectors vary from 80 to 120 eV (Fig. 5.20). These values depend on the techniques used to minimize the electronic noise (pulse shaping). The noise contribution of the first stage of amplification diminishes with increasing duration of the shaped pulse. However, with increasing pulse-shaping time, the dead-time also increases. There is thus a tradeoff between speed (i.e., short dead-time) and resolution of the detector, which can be regulated by means of the pulse-shaping constant.

5.3.7 Detector Artifacts

The Escape Peak. As was discussed in Section 4.3, the photoelectric absorption, with ionization of the inner orbital levels, is the main mechanism for the attenuation of X-rays. This is also true of the absorption of X-rays in the detector. Ionization of the K-level is frequently followed by emission of a K-line of the absorbing element (e.g., argon, and silicon, in the respective detectors). The K-lines are located at the low side of the absorption edge of the emitting element; i.e., the detector is fairly transparent to its own K-lines. Therefore, the escape of such an X-ray photon from the active volume of the detector is a fairly frequent event. The eventual fate of such a photon (exit from the detector, or absorption in the detector walls or other inactive detector parts) has no effect on the detector process, in which, in the primary ionization, the energy of the escaping photon is missing. Therefore, the height of the corresponding pulse—the escape pulse—is proportionally smaller:

$$V_e/V = (E_1 - E_2)/E_1 \qquad (5.3.15)$$

where V_e is the height of the escape pulse, V is the height of the normal pulse, E_1 is the energy of the exciting photon, and E_2 is the energy of the escaping photon.

Theoretically, one X-ray line produces more than one escape peak when interacting with the detector gas. For instance, in the detection of Ti $K\alpha$

photons in an argon detector, the escape peak of energy corresponding to $(E_{\text{Ti K}\alpha} - E_{\text{Ar K}\alpha})$ and that corresponding to $(E_{\text{Ti K}\alpha} - E_{\text{Ar K}\beta})$ are produced. However, these two peaks differ little in energy and cannot be separated by pulse height analysis (Figs. 5.22 and 5.33).

The relative frequency of escape peak production depends on the fluorescence yield for the escaping radiation, its mass attenuation coefficient in the detector gas, and the shape and size of the detector. The conditions for escape pulse formation are particularly favorable for K-lines of the elements 19-25 (or other lines in the same wavelength region) received by an argon detector, and of the elements of atomic number 15 to 20 for the silicon detector (Fig. 5.32). For the silicon detector, escape of Si Kα photons other than through the front face of the detector can be neglected. Reed and Ware [5.28] showed that for an incident radiation having an absorption coefficient in silicon μ_i the fraction ϵ of emitted Si Kα radiation which escapes from the dectector is:

$$\epsilon = \frac{1}{2}\left\{1 - \frac{\mu_{Si}}{\mu_i} \ln\left(1 + \frac{\mu_i}{\mu_{Si}}\right)\right\} . \tag{5.3.16}$$

On the energy scale, the distance between two escape peaks is the same as between the normal peaks, but their statistical pulse height spread is smaller, because fewer individual events contribute to the pulse formation. Therefore,

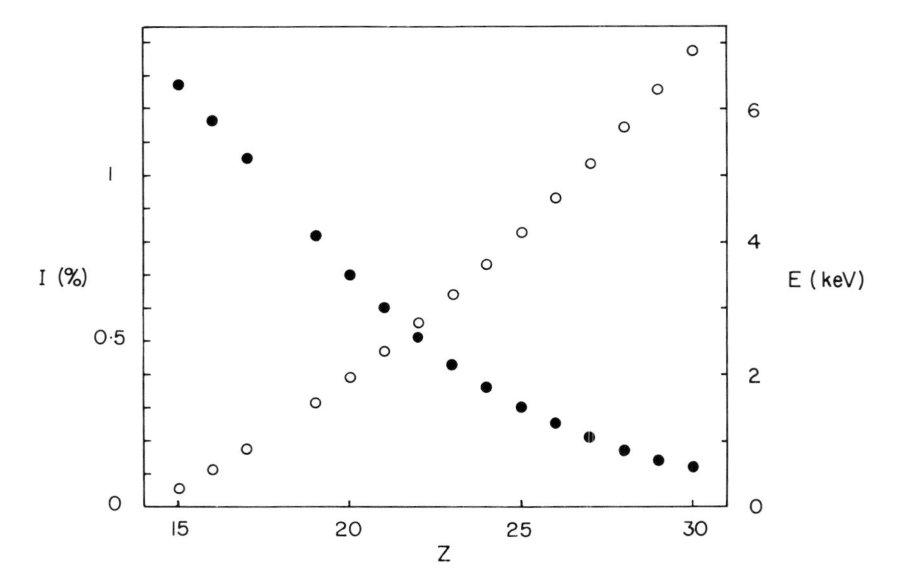

Fig. 5.32. Intensity, relative to the main Kα peak (left scale, full circles), and energy (right scale, empty circles) of escape peaks produced by a silicon detector (after Reed and Ware [5.28]). Z:atomic number.

escape peaks of adjacent elements can be resolved better than the corresponding main peaks. Notwithstanding this property, escape peaks are, in general, an unwanted complication.

In detectors containing elements of higher atomic numbers, such as krypton and xenon, analogous escape peaks appear due to the escape of L-lines; since there are more L-lines than K-lines, these escape peaks greatly complicate the interpretation of energy-dispersive spectra obtained with such detectors.

In calculating the detector efficiency, corrections should be applied for the loss through escape peak formation, unless these pulses can be counted together with the main pulses.

Incomplete Charge Collection. If a photon is absorbed in the inactive regions of the detector, no charge is collected. At the fringes of these regions, only part of the charge may reach the collectors; pulses so generated are abnormally small. The charge loss varies from almost complete charge collection to almost total loss of charge; unlike escape pulses, incomplete charge collection pulses have a very wide pulse distribution (Fig. 5.33). This artifact occurs both in gas and in solid-state detectors, but it can be particularly well demonstrated in the Si(Li) detector, because of its low background, high resolution, and the highly symmetrical peak shape, which, except for the artifact under discussion, is extremely close to Gaussian, in the energy region of use in electron-probe microanalysis (Fig. 5.34).

Incomplete charge collection occurs preferentially at shallow depths within the detector and is therefore strongest with low-energy radiation which is absorbed close to the detector window. The intensity and energy distribution of incomplete charge collection depend on the detector shape and must be experimentally determined for each detector.

The *silicon fluorescence peak* (Fig. 5.35), $E_{Si\,K}$, is produced by excitation of silicon atoms following absorption of a photon in the inactive surface layer (p-layer, dead layer) of the Si(Li) detector, and detection of the emitted silicon K photon in the active part of the detector. The intensity of this peak depends on the thickness of the inactive layer.

Pulse pile-up peaks are produced when the electrons produced by the absorption of two photons reach the collector of the detector simultaneously, and are interpreted as being originated by the same photon. The energy of the pile-up pulse is the sum of the individual pulse energies:

$$E_{pu} = E_1 + E_2 .$$

The number of pile-up pulses can be reduced if the multichannel pulse height analyzer used contains a pile-up rejector (Fig. 5.36).

Pulse Coincidence Losses of X-Ray Detector Systems. During the passage of an electric pulse through an electronic device such as a detector, an amplifier, or a pulse height analyzer, the device becomes temporarily incapable of

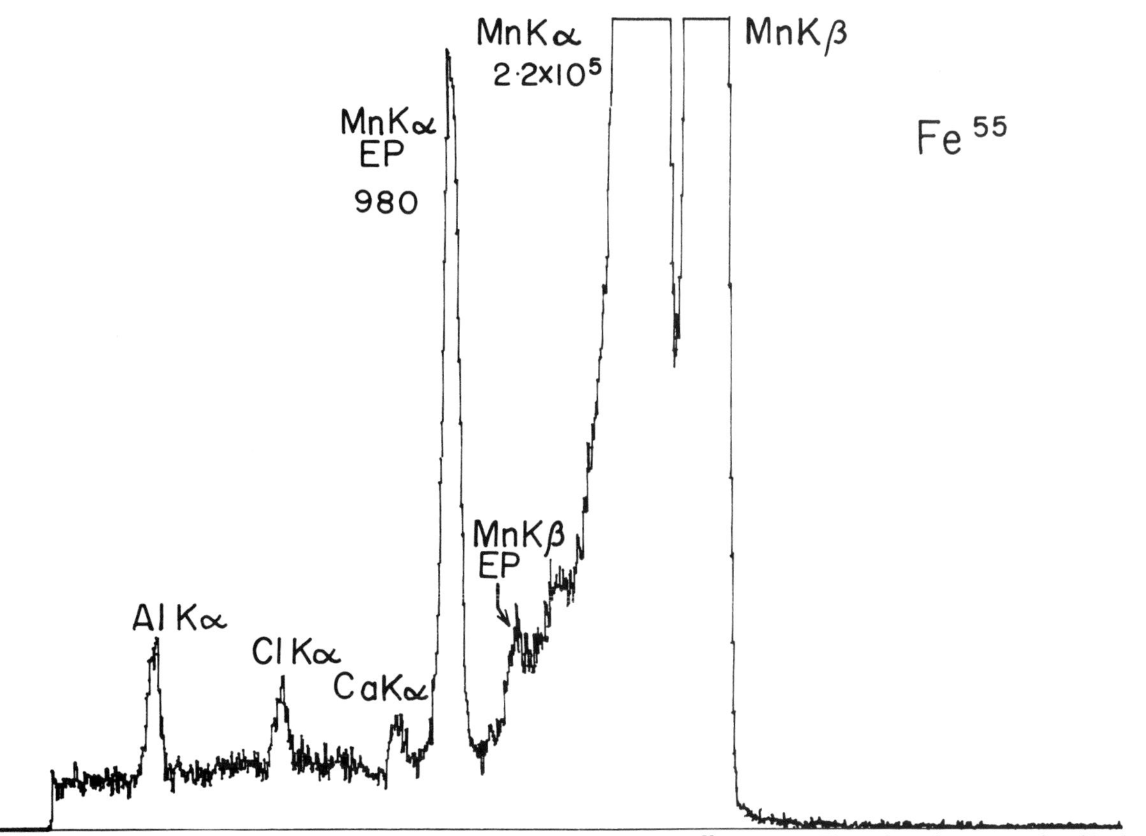

Fig. 5.33. Energy-dispersive spectrum obtained with a Si(Li) detector from a ^{55}Fe source. The main and escape peaks of both Mn Kα and Mn Kβ can be seen. The shelf formed by the incomplete charge collection extends from the main peaks to zero photon energy. Peaks of aluminum (Al), chlorine (Cl), and calcium (Ca) are impurities excited indirectly.

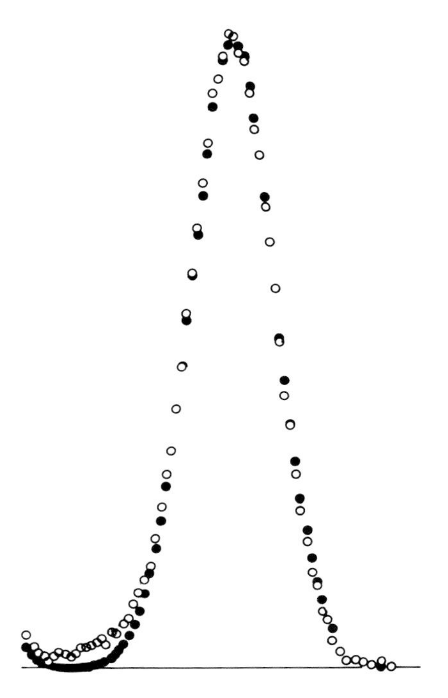

Fig. 5.34. Fe Kβ line obtained with a Si(Li) detector (empty circles), and Gaussian fit to the line (full circles). The shoulder of the Fe Kα-line is seen at the extreme left. The misfit to the left of the peak is due to incomplete charge collection.

transmitting another pulse. The length of this period of inactivity is called the dead-time or paralysis time, τ. The effect produces a reduction of observed count rates, particularly if these rates are high, since the probability of pulse coincidence increases with the counting rate. Therefore, the values of relative intensities (i.e., ratios of count rates) measured by real systems are affected by coincidence losses, and for quantitative measurements these losses must be estimated.

A pulse arriving at a detector system during its paralysis can interact with the system in two different ways. Either it can be simply rejected, without affecting the state of the system, and, in particular, the length of the dead-time period caused by a previous event. In this case, the dead-time is called nonextendible. At an infinitely high count rate, a device with nonextendible dead-time would produce an output of $1/\tau$ pulses/sec. Or else, the pulse arriving during system paralysis, although rejected, may initiate its own paralysis period, thus extending that triggered by the previous event. A device with such an extendible dead-time would, at an infinitely high count rate, produce no output signal, since the initial period of paralysis would be continuously extended by newly arriving photons.

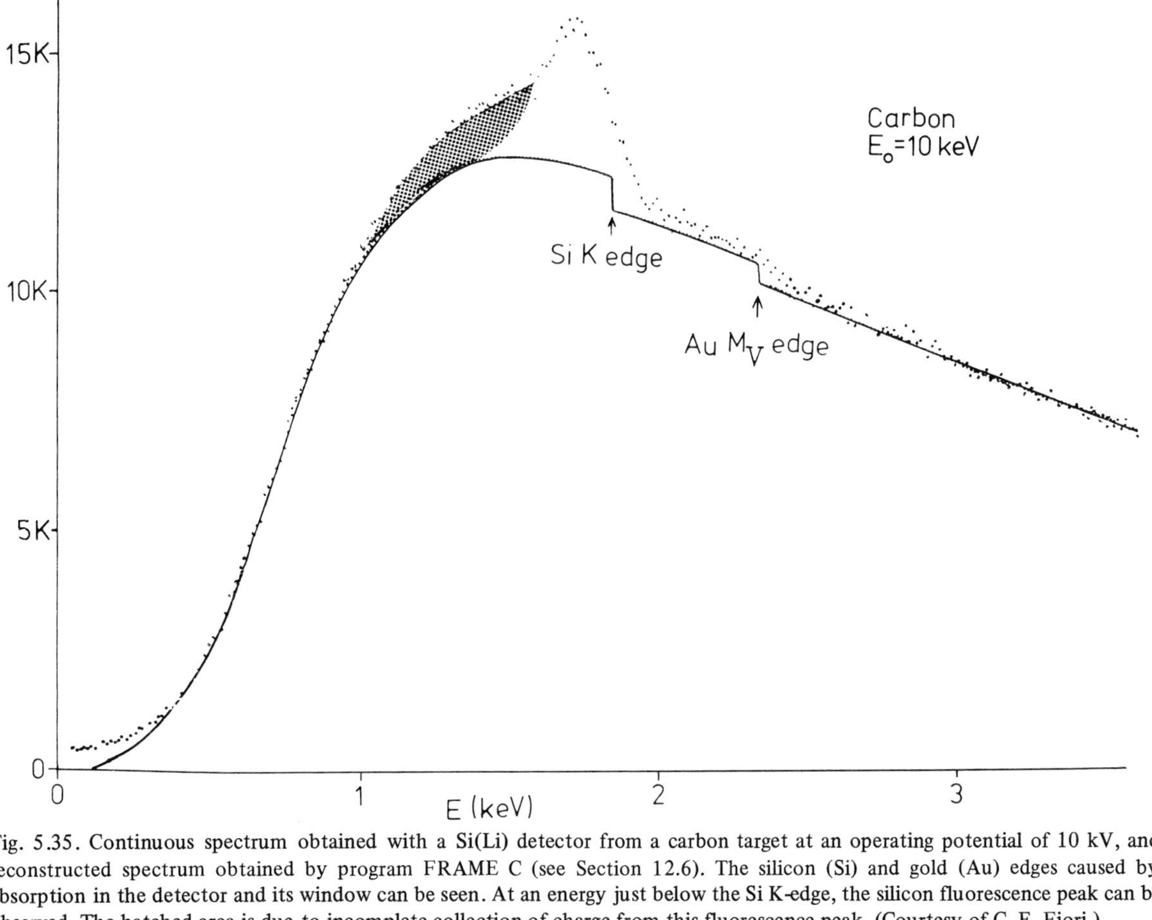

Fig. 5.35. Continuous spectrum obtained with a Si(Li) detector from a carbon target at an operating potential of 10 kV, and reconstructed spectrum obtained by program FRAME C (see Section 12.6). The silicon (Si) and gold (Au) edges caused by absorption in the detector and its window can be seen. At an energy just below the Si K-edge, the silicon fluorescence peak can be observed. The hatched area is due to incomplete collection of charge from this fluorescence peak. (Courtesy of C. E. Fiori.)

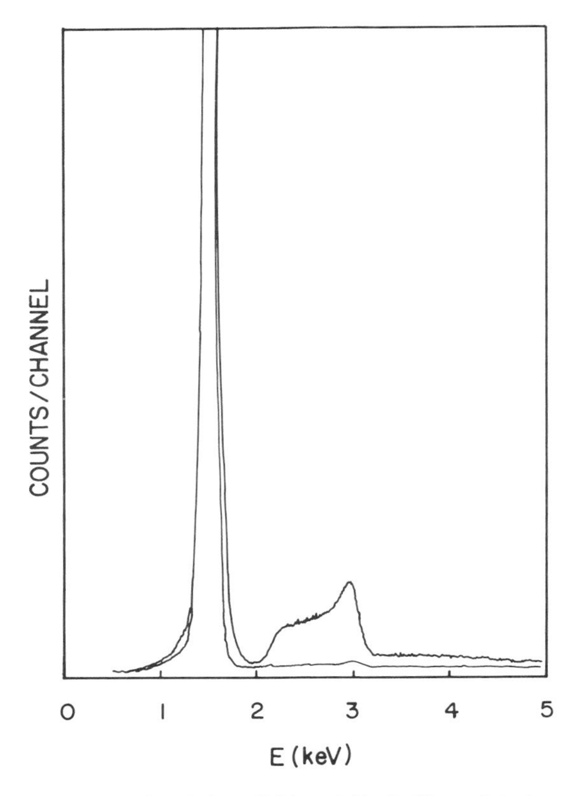

Fig. 5.36. Pile-up pulses produced in a lithium-drifted silicon detector system receiving Al Kα radiation at a high count rate. Upper curve: without pile-up rejection. Lower curve: with pile-up rejection. The pile-up pulse distribution peaks at twice the energy of the main peak.

The loss in the apparent count rate due to coincidence losses in systems with extendible dead-time is, according to Schiff [5.29] equal to:

$$Y = Y' \exp \tau \tag{5.3.17}$$

while for nonextendible coincidence losses the expression of Ruark and Brammer [5.30] is valid:

$$Y = Y'(1 - Y'\tau)^{-1}. \tag{5.3.18}$$

In both equations, Y' is the observed count rate (counts/sec), and Y is the true count rate which would have been observed in the absence of coincidence losses.

In most proportional detector systems, the nonextendible dead-time of the electronics used for amplification exceeds the extendible dead-time of the detector, which is shorter than 1 μsec. Such an arrangement is desirable, since the dead-time of the detector may vary with operating conditions. Therefore, the equation of Ruark and Brammer is applicable [5.31].

For moderately high counting rates ($2\text{-}4 \times 10^4$ counts/sec for typical proportional detector systems), the differences between the effects of Eqs. (5.3.17) and (5.3.18) are within experimental error (Fig. 5.37).

The classical experiment for determining coincidence losses consists of inserting in the path of X-rays a variable number of absorbers of identical thickness. If the coincidence losses are significant, an apparent deviation from Beer's law (Eq. 4.3.2)] will be observed, and this deviation can be used to calculate the dead-time.

In most electron-probe microanalyzers the path to the X-ray detector is not easily accessible, and therefore the method of determining coincidence losses by means of filters is not practical. If the target current can be measured with pre-

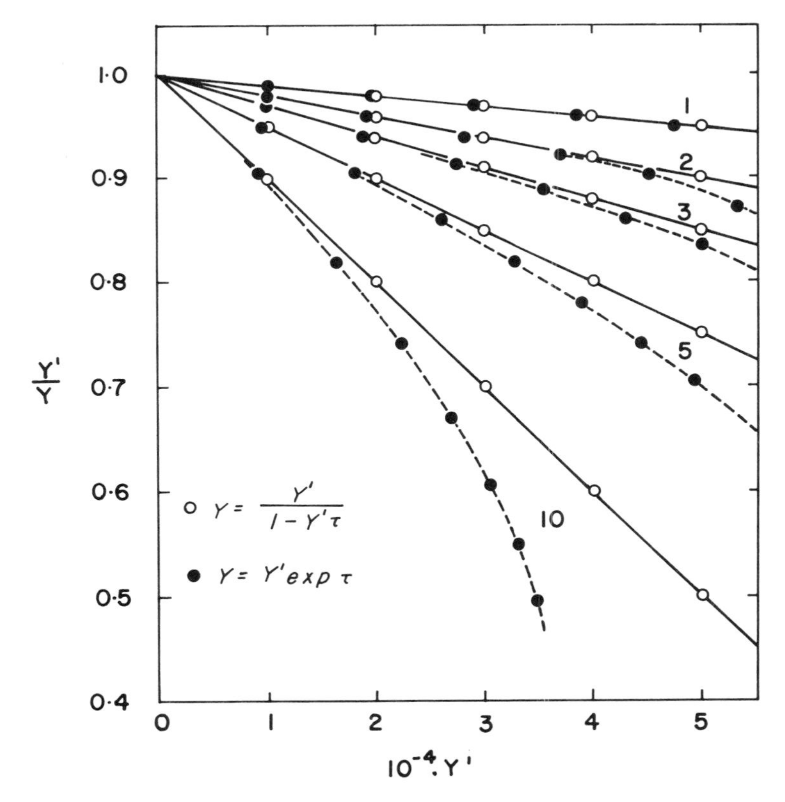

Fig. 5.37. Comparison of the equations of Schiff [5.29] and of Ruark and Brammer [5.30].

cision over a wide range of beam intensities, then it is possible to measure both this current and the X-ray signal over such a range of intensities [5.31]. For any given set of experimental conditions, the flux of photons reaching the detector is proportional to this current:

$$Y = \kappa i. \tag{5.3.19}$$

From Eq. (5.3.18), it follows that

$$Y'/i = \kappa(1 - \tau Y'). \tag{5.3.20}$$

Hence, a plot of Y'/i as a function of Y' should give a straight line, if Eq. (5.3.18) applies (Fig. 5.38). The value of κ can be obtained by extrapolating this line to negligible count rate ($Y' = 0$), and τ can be obtained from any point on the line by

$$\tau = [1 - (Y'/i)/\kappa]/Y'. \tag{5.3.21}$$

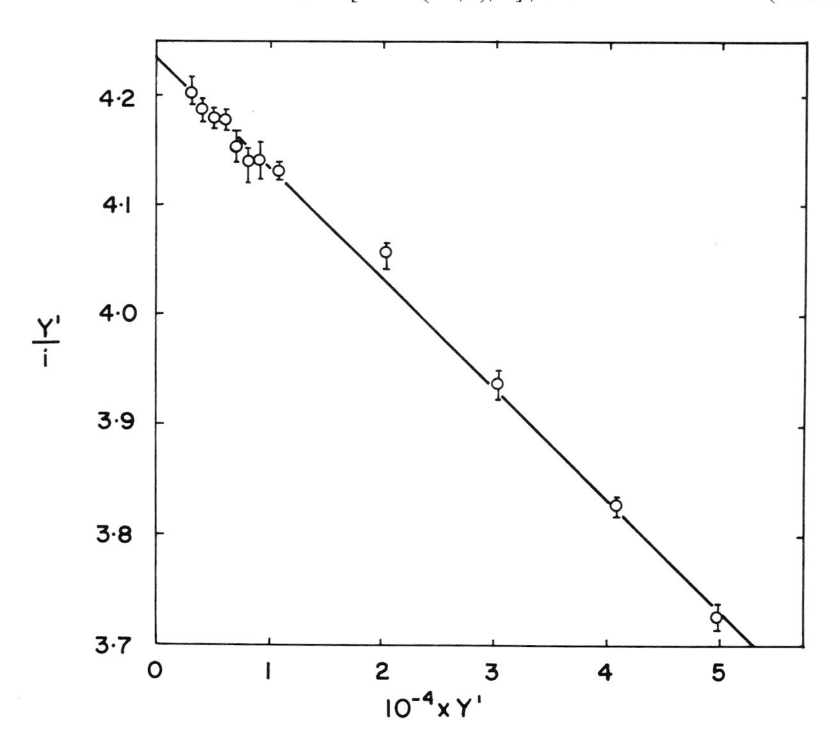

Fig. 5.38. Determination of the dead-time constant with the help of target current measurements.

If the target current cannot be measured accurately over a wide range of intensity, the dead-time can be determined by means of simultaneous X-ray measurements on two X-ray spectrometers. In a first series of measurements, the count rate in one spectrometer is held much higher than in the second spectrometer; this series of measurements chiefly reflects the count losses in the first spectrometer. In a second series of measurements, the intensities observed are chosen in such a fashion that the count rate in the second spectrometer is substantially higher.

Using for the two detector systems the subindices 1 and 2, we obtain:

$$Y_1 = Y_1'(1 - Y_1'\tau_1)^{-1} \quad \text{and} \quad Y_2 = Y_2'(1 - Y_2'\tau_2)^{-1}. \tag{5.3.22}$$

Calling $C = Y_1/Y_2$, we obtain

$$\begin{aligned} 1/Y_2' - \tau_2 &= C(1/Y_1' - \tau_1) \\ Y_1'/Y_2' &= C + Y_1'(\tau_2 - C\tau_1). \end{aligned} \tag{5.3.23}$$

If we plot Y'_1/Y'_2 as a function of Y'_1, we thus obtain a straight line, and by extrapolating to the point ($Y'_1 = 0$), we obtain the value of C.

Considering now the two sets of measurement, we will call C_1 and C_2 the values of C obtained from the first and the second series of measurements. For the values of the observed count rates Y', the first subscript will denote the spectrometer, and the second subscript the series of measurement. The dead-times of the two detectors can be obtained as follows:

$$\begin{aligned} \tau_1 &= \frac{1}{C_2 - C_1}\left[\frac{1}{Y'_{11}}\left(\frac{Y'_{11}}{Y'_{21}} - C_1\right) - \frac{1}{Y'_{12}}\left(\frac{Y'_{12}}{Y'_{22}} - C_2\right)\right] \\ \tau_2 &= \frac{1}{C_2 - C_1}\left[\frac{C_2}{Y'_{11}}\left(\frac{Y'_{11}}{Y'_{21}} - C_1\right) - \frac{C_1}{Y'_{12}}\left(\frac{Y'_{12}}{Y'_{22}} - C_2\right)\right]. \end{aligned} \tag{5.3.24}$$

The values of C_1, C_2, Y'_{11}, Y'_{12}, Y'_{11}/Y'_{21}, and Y'_{12}/Y'_{22} can be read from the graphs (Fig. 5.39).

On the plots we have described, we frequently observe deviations from linearity in the region of low counting rates. These are due to spurious signals and can be neglected. To the contrary, deviations from linearity at high counting rates indicate that at the conditions of the measurements the equation of Ruark and Brammer does not apply. Apparent increases of coincidence losses at high count

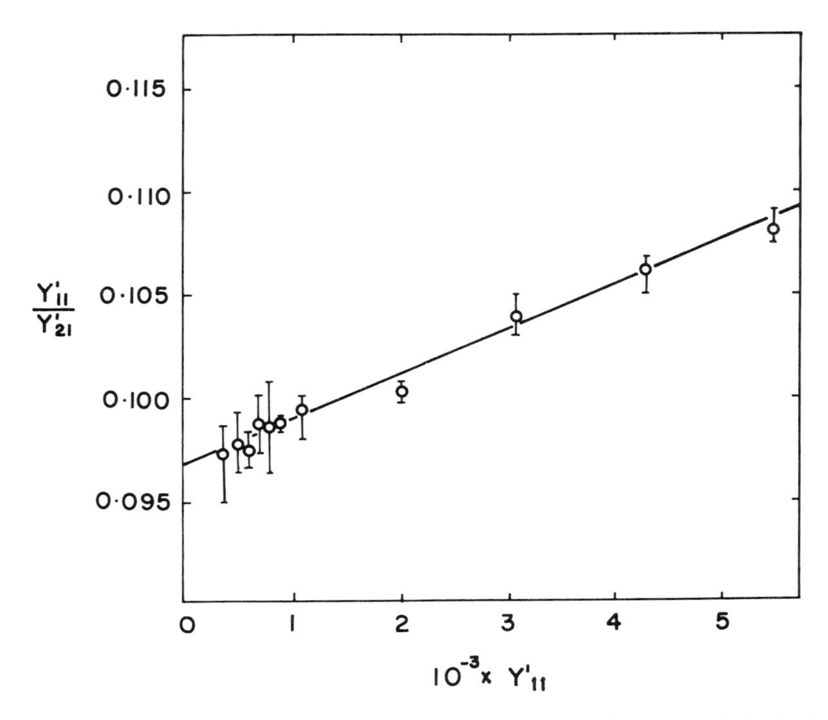

Fig. 5.39. Dead-time determination with two detectors. An analogous graph is obtained from plotting Y'_{12}/Y'_{22} versus Y'_{12}.

rates may be due to a combination of pulse shrinkage and improperly high discriminator settings.

Whenever quantitative measurements are performed, the losses due to coincidence should be determined experimentally, and the observed count rates should be corrected accordingly. It is a common fallacy that coincidence losses can be neglected except for very high count rates. On the other hand, it is not necessary to frequently recalibrate the detector circuits unless a malfunction is suspected or unless changes were made in them.

It should also be considered that the concepts underlying the equation of Ruark and Brammer are simplifications, and that the dead-time correction cannot be expected to eliminate the effects of large coincidence losses. Even if the device were to exactly follow the behavior predicted by theory, it is difficult to accurately determine the dead-time. The formulas of propagation of error (see the Appendix) indicate that the relative error in the count rate Y due to an error in the estimate of the dead-time $\epsilon(\tau)$ is given by the equation

$$\epsilon(Y)/Y = Y \cdot \epsilon(\tau). \tag{5.3.25}$$

Hence, the omission of the correction for a count rate of 10^4 counts/sec and a dead-time of 3 μsec results in an error of 3% in the measured count rate. A partial compensation takes place in the calculation of intensity ratios. If $k = Y_1/Y_2$, then the error in k resulting from an error in the estimate of τ is obtained by

$$\epsilon(k)/k \simeq -(Y_1' - Y_2')\,\epsilon(\tau). \tag{5.3.26}$$

In view of the uncertainties in the dead-time correction, it is good practice to limit the count rates to levels at which the coincidence losses do not exceed a few percent, even if the dead-time has been determined and the correction is performed.

Interference by High-Energy Electrons. Electrons having an energy in the kilovolt range can also produce signals in Si(Li) spectrometers. Normally, backscattered electrons do not reach the active part of the detector because they are deflected by the magnetic shields already mentioned, or absorbed by the beryllium window. However, at operating potentials above 20 kV, and in the absence of magnetic shielding, they can enter the detector and cause considerable distortion and increase of the spectral background (Fig. 5.40).

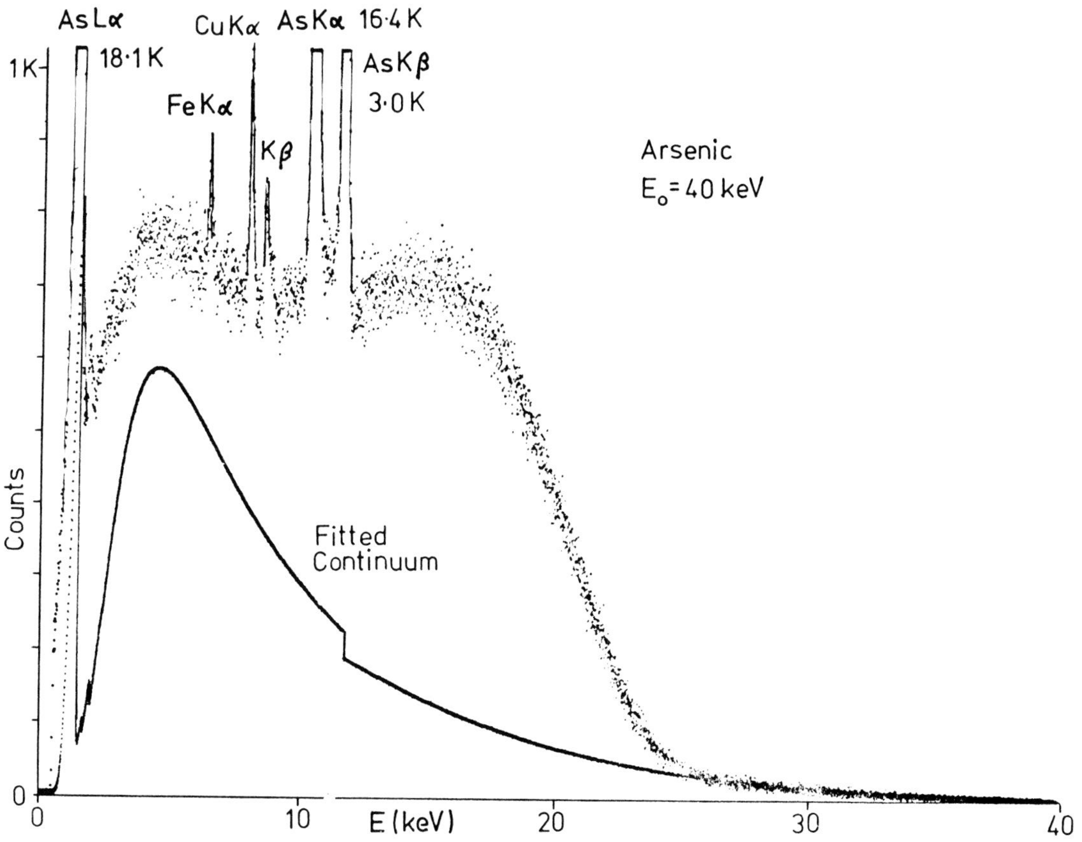

Fig. 5.40. Interference of high-energy electrons with the Si(Li) spectrum. Target of arsenic, irradiated with 40-keV electrons. The full line shows the spectrum obtained in the absence of electron interference (characteristic lines omitted). Because of the high operating voltage, a large absorption edge exists in the continuum at the As L_{III} level. The background above the fitted continuum is due to backscattering electrons entering the detector, after crossing the beryllium window.

5. REFERENCES

5.1 Jenkins, R., *X-Ray Spectrometry* **1**, 23 (1972).

5.2 Birks, L. S., *Electron Probe Microanalysis*, 2nd ed., John Wiley & Sons, New York, 1971, p. 41.

5.3 Allison, S. K., *Phys. Rev.* **41**, 1, (1932).

5.4 Ross, P. A., *J. Opt. Soc. Amer.* **16**, 433 (1921).

5.5 Castaing, R. and Pichoir, F., Res. Aérosp. No. 108, 3 (1965); also *Proc. 4th Int. Conf. on X-Ray Optics and Microanalysis*, Hermann, Paris, 1966, p. 454.

5.6 Pichoir, F., Res. Aérosp. No. 4, 225 (1973); also *Proc. 5th Int. Conf. on X-Ray Optics and Microanalysis*, Springer, Berlin, 1969, p. 373.

5.7 Herglotz, H., *J. Appl. Phys.* **38**, 4565 (1967).

5.8 Borovsky, I. B., Private commun. (1974).

5.9 Reed, S. J. B., *Electron Microprobe Analysis*, Cambridge University Press, Cambridge, Mass., 1975.

5.10 Fergason, L., *Rev. Sci. Instr.* **41**, 696 (1956).

5.11 Burkhalter, P. G., Brown, J. D., and Myklebust, R. L., *Rev. Sci. Instr.* **37**, 1266 (1966).

5.12 Beamann, D. R., *Anal. Chem.* **38**, 599 (1966).

5.13 Jenkins, R., *J. Sci. Instrum.* **41**, 696 (1956).

5.14 Delattes, R. D., Simson, B. G., and La Villa, R. E., *Rev. Sci. Instr.* **37**, 596 (1966).

5.15 Poen Sing Ong, in *The Electron Microprobe*, McKinley, T. D., Heinrich, K. F. J., and Wittry, D. B., Eds., John Wiley & Sons, New York 1966, p. 43.

5.16 Bowman, H. R., Hyde, E. K., Thompson, D. G., and Jared, R. C., *Science* **151**, 562 (1966).

5.17 Fitzgerald, R., Keil, K., and Heinrich, K. F. J., *Science* **159**, 528 (1968).

5.18 Russ, J. C. (Ed.), *Energy Dispersion X-Ray Analysis: X-Ray and Electron Probe Analysis*, ASTM Special Tech. Publ. 485, American Society for Testing and Materials, Philadelphia, Pa., 1971.

5.19 Gedke, D. A., *X-Ray Spectrometry* **1**, 129 (1972).

5.20 Woldseth, R., *X-Ray Energy Spectrometry*, Kevex Corp., Burlingame, Calif., 1973.

5.21 Dolby, R. M., *Proc. Phys. Soc. London* **73**, 81 (1959).

5.22 Dolby, R. M., *J. Sci. Instrum.* **40**, 345 (1963).

5.23 Birks, L. S. and Batt, A., *Anal. Chem.* **35**, 778 (1963).

5.24 Myklebust, R. L., Fiori, C. E., and Heinrich, K. F. S., NBS Technical Note 1106, (FRAME C) U. S. Department of Commerce, Washington, D.C., 1979.

5.25 Campbell, A. J., *Norelco Reporter* **14**, 103 (1967).

5.26 Mulvey, T and Campbell, A. J., *Brit. J. Appl. Phys.* **9**, 406 (1958).

5.27 Shimizu, R. and Shinoda, G., *Japan. J. Appl. Phys.* **4**, 543 (1965).

5.28 Reed, S. J. B and Ware, N. G., *J. Phys. E* **5**, 582 (1972).

5.29 Schiff, L. J., *Phys. Rev.* **50**, 88 (1936).

5.30 Ruark, A. and Brammer, F. E., *Phys. Rev.* **52**, 322 (1937).

5.31 Heinrich, K. F. J., Vieth, D., and Yakowitz, H., *Adv. X-Ray Analysis* **9**, 208 (1966).

6. Measurement of the Intensity of X-Ray Emission

6.1 SCALES OF INTENSITY OF X-RAY EMISSION

The intensity of X-ray generation can be measured in two ways. If the energy transmitted by the generated X-rays is collected and measured in a calorimeter, we obtain a rate of flow of energy, Y_e; this is the expression of intensity preferred in the past in physics. If the X-rays are produced as a consequence of the impact of an electron beam, the ratio of the X-ray energy produced to the expended energy of the electrons is called the *energy* efficiency, ϵ. Or, we may express the X-ray intensities by means of the rate at which X-ray photons are produced, Y. We can divide this intensity by the number of electrons impinging on the specimen, and thus obtain a *quantum* efficiency, I, which is the probability that the impact of an electron will produce a photon of the type under consideration. Calling E_{qm} the energy of a photon of line m, produced by ionization of the shell q (e.g., $E_{K\alpha}$), E_0 the initial energy of the exciting photon, and I_e the average photon energy produced by the impact of one electron, we can relate the efficiencies by

$$\epsilon = \frac{I_e}{E_0} = I \cdot \frac{E_{qm}}{E_0} \,. \tag{6.1.1}$$

Since quantitative electron probe analysis is based on ratios of X-ray intensities, the distinction between the two scales of intensities is unimportant when the two intensities are of the same line, and hence, of the same energy per quantum. The distinction becomes significant, however, when we compare lines of different wavelength, and is therefore to be considered in the discussion of the weights of lines and of continuous radiation. *In electron probe analysis we always count photons rather than measuring energies.* Therefore, in the discussion of the generation of X-rays (Chapter 8), the term "intensity" will always denote quantum efficiency. Concerning information from the literature on relative intensities, we must ascertain which of the two scales were employed, particularly in older references.

In Chapter 9 we will distinguish *observed* X-ray count rates, of radiation emitted within a solid angle covered by the spectrometer, and affected by the absorption losses of the X-rays within the target and the detector efficiency, from intensities *generated within the target*, into all directions. In some early references to line intensity ratios the absorption losses may not have been taken into account; such ratios are therefore dependent of the physical parameters of the experiment, such as X-ray emergence angle and operating voltage. Furthermore, it was often not considered that relative line intensity measurements should be performed by measuring integral line intensities rather than peak intensities. The distinction is significant when the spectrometer resolution is narrower than the natural line widths.

Depending on the purpose of the measurement, we may either measure an ideally unchanging count rate while keeping the conditions of the measurement as constant as possible, or we may observe the variation of count rate in time as some parameter of the experiment–such as the observed wavelength–is varied. The latter procedure is used to obtain wavelength spectra, such as that of Fig. 2.6, by varying in time the Bragg angle of a crystal spectrometer. In the performance of quantitative analyses (Chapters 8 through 12), one usually measures constant X-ray intensities. The rate of arrival of the amplified pulses produced by the detector is counted for a known time interval. Usually, this operation is performed several times in order to control statistical variability. The count rate, Y, is the ratio of the number of observed pulses, N, to the length of time of observation, t:

$$Y = N/t \quad \text{counts/sec.} \tag{6.1.2}$$

The analytical determination of the specimen composition is based on the ratio of count rates obtained from both the specimen and the standard. Factors which enter both count rates–such as the spectrometer efficiency P and the solid angle Ω covered by the spectrometer–cancel in these ratios. The precision of the determination of the ratio is limited, however, by factors which may affect independently the values of individual measurements. The most important of these factors are:

1. the statistical fluctuations of measurements under identical conditions;
2. the degree to which instrumental conditions can be maintained constant (stability);
3. the production of pulses which are not caused by X-rays (electronic noise),

or which are caused by X-ray photons not belonging to the X-ray line which is being measured (background and line interferences);
4. the coincidence losses of the detector system, which were discussed in Subsection 5.3.7.

6.2 STATISTICS OF THE MEASUREMENT OF COUNT RATES

If we perform repeatedly a counting experiment under virtually identical conditions, the number of events observed (count) will, in general, differ between experiments. If the differences are small, we use the average of counts as a measure of the true count rate, and determine the dispersion of the individual values of the measurements by the statistical procedures described in the Appendix. A set of measurements of count rates can be characterized by the sample mean, $\overline{x}$, and the estimate of the standard deviation of the sample count, s, or of the standard error of the mean, $s_{\overline{x}}$. We can also calculate confidence limits for the mean value obtained; if the number of measurements is below ten, these should be obtained by means of Student's t-distribution.

The experimentally observable dispersion of the measured values is due to a combination of the effects of minor changes in the conditions of the experiment, and of the statistical variations inherent to counting events which are generated independently, such as the emission of X-ray photons. The inherent counting statistics are described by the properties of the Poissonian distribution (see the Appendix). The most interesting property of this distribution is the equality between the variance of a large set of counts, and the mean of the counts:

$$\sigma^2_{\text{Poisson}} = \overline{N}. \tag{6.2.1}$$

The total standard deviation of a set of counts can thus be obtained by combining the respective contributions as follows:

$$\sigma_{\text{total}} = \sqrt{\overline{N} + \Sigma\sigma_i^2} \tag{6.2.2}$$

where $\overline{N}$ is the mean count obtained in the experiment, and the variances σ_i^2 are those due to small variations of parameters of the experiment. Depending on the stability of the system and the number of pulses counted, the dispersion of the results may be practically Poissonian (when $\overline{N} \gg \Sigma\, \sigma_i^2$), or much larger. But clearly, the total standard deviation cannot be smaller than Poissonian. The property of the Poissonian distribution shown in Eq. (6.2.1) thus provides a limit for precision, as a function of counts, which cannot be surpassed even in an ideally stable counting experiment.

By combining the Poissonian variances of several measurements in an analytical procedure, the theoretical standard deviation of the analytical result can be derived. This standard deviation can, in turn, lead to a theoretical limit of detection. To assess the validity and significance of this concept, we must, however, consider the remaining factors which affect the precision of the measurement. These factors, denoted $\Sigma\sigma_i^2$ in Eq. (6.2.2), can be greatly reduced by improving the quality of the experimental setup. However, whenever extreme care is taken to reduce the counting statistics by increasing the beam intensity and signal-collecting time intervals, then the non-Poissonian statistical errors will become predominant. Therefore, the Poissonian estimate of the precision of a measurement is always optimistic, and may become quite unreliable when efforts are made to reduce the Poissonian uncertainty.

The effects of combined Poissonian statistics on the accuracy of analysis are described in detail in Section 12.3. There, we also discuss the errors which are committed in the interpretation of the signal and in its use for the determination of concentrations.

6.3 THE GENERATION OF CONTINUOUS X-RAYS

The continuous background which, together with the characteristic lines, forms the X-ray emission spectrum (Fig. 2.6), is called *continuous radiation,* or *bremsstrahlung* (from the German: deceleration radiation). This emission is caused by the deceleration of an electron which passes through the electron cloud of an atom. The electrostatic field of the necleus deflects the electron, and the energy freed is emitted as an X-ray photon. As the charge of the nucleus, $Z \cdot e^-$, increases with the atomic number, Z, the quantum efficiency of emission of continuous radiation also increases with Z. The energy of the emitted photons varies continuously from zero to the initial energy of the exciting electron. The wavelength, λ_0, and the frequency, ν_0, of a photon which has received all of the energy of the electron are called the Duane-Hunt limits of wavelength and frequency. As the energy of the exciting electrons, E_0, is increased, these limits shift to shorter wavelengths and higher frequencies, according to Eq. (4.2.1), with $(E = E_0)$.

Kramers [6.1] derived a theoretical equation which predicts the intensity of continuous radiation from a thick target as a function of the observed frequency of the X-rays and the atomic number of the absorber. He obtained that for a thin film the generated continuum intensity, $I_e(\nu)d\nu$ is proportional to $Z^2 \cdot d\nu$, and does not depend upon the value of ν. For the calculation of the energy losses of electrons within a thick target, Kramers used the Thomson-Whiddington law [Chapter 9, Eq. (9.2.30)]. With this approximation, he obtained, for a thick target, the following expression, generally known as Kramers' law:

$$I_e(\nu)d\nu = h^2k \cdot Z(\nu_0 - \nu)d\nu \qquad (6.3.1)$$

Where $I_e(\nu)d\nu$ is the average energy produced in the continuous spectrum, between the frequencies ν and $\nu + d\nu$, by one incident electron, Z is the atomic number, and h^2k (J $\cdot$ sec^2 or keV $\cdot$ sec^2) is a constant (h is Planck's constant). The theory underlying Kramers' law also predicts that the emission of the continuum is strongly anisotropic with respect to the direction of the primary electron beam. The directional changes of electron paths within the thick target reduce the anisotropy which is strongest in a thin target.

Kramers' equation can also be expressed in terms of wavelength. From $\nu = c/\lambda$ we derive that $d\nu = -c\lambda^{-2}\,d\lambda$. Since $I_e(\lambda)d\lambda$ is, in terms of wavelength, equal to $I_e(\nu)d\nu$, we obtain:

$$I_e(\lambda)d\lambda = h^2k \cdot c^2 \cdot Z\left(\frac{1}{\lambda} - \frac{1}{\lambda_0}\right)\frac{1}{\lambda^2}\,d\lambda. \tag{6.3.2}$$

It can be shown by differentiation that the maximum for this function occurs at $\lambda = 3/2\lambda_0$. From Eq. (6.3.1) and $E = h\nu$ we obtain Kramers' equation in terms of energy ranges:

$$I_e(E)dE = k \cdot Z(E_0 - E)dE. \tag{6.3.3}$$

The constant k has the dimension of kV^{-1}. Integrating over the entire spectrum we obtain the total continuum energy per electron:

$$I_e(\text{total}) = k \cdot Z \int_0^{E_0} (E_0 - E)dE = \frac{1}{2}k \cdot Z \cdot E_0^2\,. \tag{6.3.4}$$

The energy efficiency of the generation of bremsstrahlung, $\epsilon = I_e/E_0$, is therefore equal to $\frac{1}{2}k \cdot Z \cdot E_0$. This efficiency, and the term $\frac{1}{2}k$, which is called by Green the continuous efficiency constant, can be determined experimentally, as is discussed in Green's thesis [6.2, p. 62].

Because we measure X-ray intensities by counting photons, we must transform Eqs. (6.3.1) to (6.3.3) into equivalent expressions giving the probabilities of photon production. The reader should be cautioned that in several places in the electron probe literature this transformation has been overlooked. By dividing I_e by the energy of one photon, E, we obtain:

$$I(\nu)d\nu = hk \cdot Z \cdot (\nu_0/\nu - 1)d\nu \tag{6.3.5}$$

with a maximum intensity at $\nu = 0$, and

$$I(\lambda)d\lambda = hck \cdot Z \cdot (\lambda^{-2} - \lambda_0^{-1}\lambda^{-1})d\lambda \tag{6.3.6}$$

with a maximum at $\lambda = 2\lambda_0$. Finally,

$$I(E)dE = k \cdot Z \cdot (E_0/E - 1)dE \tag{6.3.7}$$

with a maximum at $E = 0$.

Compton and Allison [4.14] averaged early measurements of the continuous efficiency constant, $k/2$, obtaining a value of 1.1×10^{-6} keV^{-1}. Kirkpatrick and Wiedmann [6.3] proposed a value of 1.3×10^{-6}. Measurements for several elements by Dyson [6.4] were reported in the thesis of Green, who also added several measurements made by himself. He indicated that the constant, rather than being independent on the atomic number, exhibits a dependence on Z as shown in Fig. 6.1.

Extensive measurements of the shape of the continuum made in 1922 by Kulenkampff [6.5] gave results in good agreement with Kramers' equation. Kulenkampff recognized the importance of the absorption of continuous radiation within the target. He performed measurements at various X-ray emergence angles in order to estimate the loss of the continuum by absorption (compare with the absorption of characteristic radiation, Chapter 10). His results indicated that after correction for absorption, the values of $I_e(\nu)$ diminished linearily with ν, as predicted by Eq. (6.3.1), but that at frequencies close to ν_0 ($E_0 - E < 500$ eV) the negative slope of $I_e(\nu)$ versus ν increased rapidly. To take this observation into account, Kulenkampff added another term to Kramers' equation:

$$I_e(\nu)d\nu = c \cdot Z\,[(\nu_0 - \nu) + Z \cdot b]\,d\nu. \tag{6.3.8}$$

If the constant b vanishes, and with appropriate adjustment of the constant c, we would obtain Kramers' equation. Kramers attributed the effect expressed in Kulenkampff's equation to the backscattering of primary electrons [6.6].

Kirkpatrick and Wiedmann [6.3] recalculated the theoretical cross section for continuum production from Sommerfeld's theory [6.7]. Reed [6.8] used their

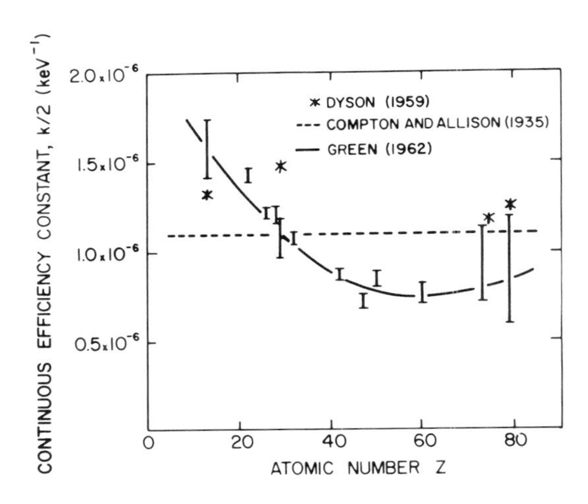

Fig. 6.1. Variation with Z of the "constant" in Kramers' equation, according to Green [6.2].

results, as well as Bethe's law for electron deceleration [Eq. (9.2.2)], instead of the Thomson-Whiddington law, to calculate the production of the continuum. He concluded that Kramers' "constant" varies, not only with the atomic number of the target, but also with the wavelength at which the continuum is measured, and, to a lesser degree, the operating voltage.

The form of the continuum distribution was further investigated by Rao-Sahib and Wittry [6.9]. These authors performed measurements with crystal spectrometers on several elements, excitation potentials, and wavelengths. They applied absorption corrections, assuming that the depth distribution of generation of the continuum is virtually identical to that of characteristic X-rays (see Chapter 10). They also proposed that the effects of anisotropy of generation can be ignored. Their measurements indicated that the Kramers' equation was not accurate, and they found that for any given wavelength the atomic number dependency can be expressed as a proportionality of the intensity to Z^n, where the value of the exponent n varies from 1.15 to 1.38, depending on Z, E_0, and E. They also performed theoretical calculations using Sommerfeld's cross sections, Bethe's retardation law, and the calculation of energy loss through backscattered electrons which had been given by Duncumb and Reed (see Section 9.3). The results of these calculations were in agreement with the measurements by Rao-Sahib and Wittry. Reed [6.8] concluded from similar considerations that

$$I(E)dE = \text{constant} \times \frac{E_0 - E}{E^{1+a}}\, dE \; (a \sim 0.21) \qquad (6.3.9)$$

for $13 \leqslant Z \leqslant 29$ and $15 \text{ kV} \leqslant E_0 \leqslant 20 \text{ kV}$. This algorithm, with an additional correction for continuum generated by backscattered electrons in the window of the Si(Li) detector, is also used by Statham [6.10] who presents a table of values for a as a function of the atomic number of the target. Hehenkamp and Böcker [6.11] also made measurements similar to those of Rao-Sahib with similar conclusions.

A different approach to the continuum function was taken by Lifshin et al. [6.12] who expanded Kramers' law, from the form taken in Eq. (6.3.7), to

$$I(E)dE = \frac{Z}{E} \cdot [k_0 + k_1 (E_0 - E) + k_2 (E_0 - E)^2]\, dE. \qquad (6.3.10)$$

Lifshin et al. as well as Fiori et al. [6.13] used this equation to obtain estimates of the background under characteristic lines from measurements, with the same target, of continuum regions free of characteristic lines, as will be described in Section 12.6. In this application, the dependence of $I(E)dE$ on Z is therefore unimportant. Smith et al. [6.14] found that Eq. (6.3.10) agreed with measurements they performed on various minerals. The first term of the equation k_0, is small, and can be omitted.

Measurements of the continuous radiation emitted from composite targets were performed by Moreau and Calais [6.15], who reported that the continuum changed proportionally to the weight fractions of the constituent elements. These results were confirmed in our laboratory (see Fig. 6.2).

The reader should be aware that all intensities mentioned here refer to emission into all directions per incident electron and that absorption within the target has not been taken into account. The intensity observed in a detector would be equal to $i_b \cdot I \cdot f \cdot (\Omega/4\pi) \cdot P$ [see Eq. (5.3.7)], where f is the attenuation factor for the X-rays, and i_b is the beam current (electrons/sec).

6.4 LINE AND BACKGROUND

Background is that part of the signal which is not due to line emission. In a well constructed and properly working spectrometer, background contributions other

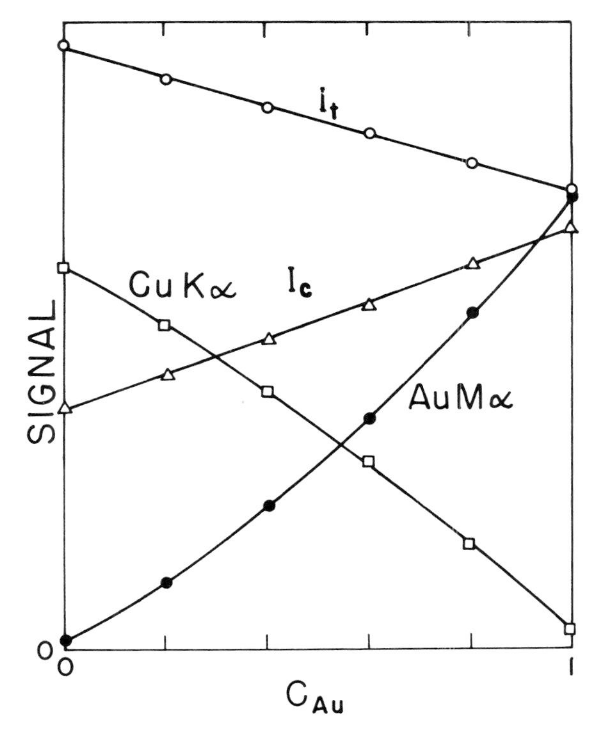

Fig. 6.2. Variation of several signals from gold-copper (Au-Cu) alloys, as a function of composition. Operating potential: 20 kV. Cu Kα and Au Mα are the intensities of X-ray signals, I_c is the intensity of the continuous radiation at 2.29 Å, and i_t is the intensity of the target current. All intensity scales are arbitrary.

than continuous radiation should be insignificant. In practice, however, scattered electrons and X-rays as well as electric noise may contribute to the background. Unless pulse height discrimination is used, higher orders of reflections of the continuum may also be found in the background produced by crystal spectrometers.

The purpose of the background measurement is to obtain a correction for background emission in the determination of the intensity of a line. This line intensity, Y_l, is obtained by subtracting the background intensity, Y_b, from the total intensity, Y_t:

$$Y_l = Y_t - Y_b. \tag{6.4.1}$$

The rule of propagation of errors (see the Appendix) indicates how the standard deviation of Y_l, σ_l, can be obtained if the standard deviation of Y_t, σ_t, and that of Y_b, σ_b, are known:

$$\sigma_l = (\sigma_t{}^2 + \sigma_b{}^2)^{1/2}. \tag{6.4.2}$$

We assume that the background can be obtained without significant systematic error. If this is the case, the accuracy of the line measurement would depend only on the precision with which Y_t and Y_b can be measured. If the Poissonian counting statistics apply, we obtain:

$$\sigma_l \simeq (N_t + N_b)^{1/2}. \tag{6.4.3}$$

The background contribution to σ_l becomes important at low concentrations of the emitting element, because $N_t \simeq \sigma_t^2$ becomes, for equal counting periods, comparable in size to σ_b^2. The detection of an element depends, in fact, on the operator's ability to prove that Y_t is larger than Y_b. To obtain conditions conducive to the detection of small concentrations of an element, the analyst must measure a line of this element in such a fashion that the background contribution is as small as possible. This condition can be expressed by means of the ratio of line to background for the pure element, which should be as large as possible. This ratio, and hence the sensitivity, increases with the resolution of the spectrometer. The total number of events (count) accumulated also codetermines the precision of the measurement, since the relative standard deviation of the total intensity, ν_t is equal to $N_t^{-1/2}$. This subject is further pursued in Section 7.3.

The direct observation of the background beneath a line is impossible since we cannot suppress the emission of the line without affecting the background emission. As is illustrated in Fig. 6.3, the background contribution can be obtained by either of two methods. One can measure the signal intensity from the emitting target at a different wavelength, and calculate the background of inter-

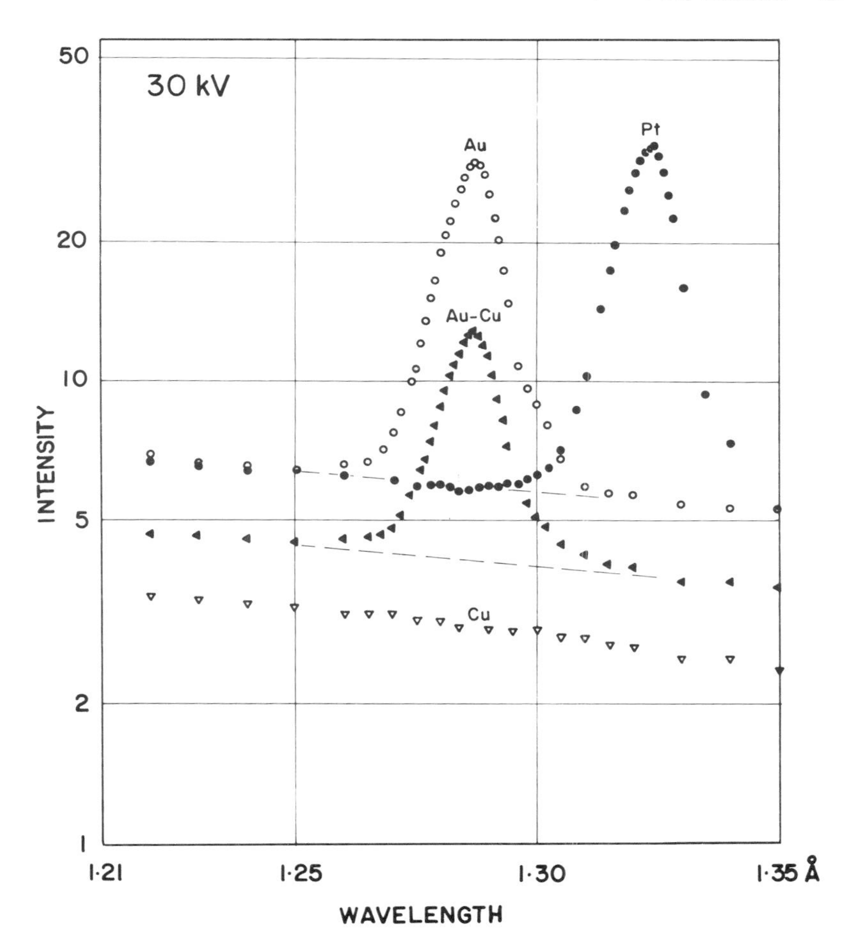

Fig. 6.3. In the absence of lines, the continuous X-ray intensities vary with the atomic number of the emitter as indicated by Eq. (6.3.6). Therefore, the level of radiation from a platinum (Pt.) target ($Z = 78$) is almost identical with the extrapolated background under the Au Lα line from a gold (Au) target ($Z = 79$). However, the intensity at the same wavelength from a copper (Cu) target ($Z = 29$), and the extrapolated background from a copper-gold (Cu-Au) alloy, are much lower.

est by means of a function relating the continuum emission to energy, or wavelength. Or, the emission of the continuum at the wavelength or energy of the line is measured from another target which does not emit the line of interest. In this case, a function is needed which relates the emitted intensities of continuous radiation to the composition of the target. Both these functions are contained in Kramers' law [Eqs. (6.3.1)-(6.3.7)], or in refined versions of this law. For the observed continuous background, the absorption of the continuum on emergence, and the variation of detector efficiency with wavelength, must also be considered, particularly if an absorption edge of either the target or the detector material is within the observed wavelength region.

The choice of the method of background correction depends on the technique of signal production and recording, on the line-to-background ratio, and on the available time. The most accurate and time consuming is the interpolation with respect to wavelength or energy of the background. This technique is particularly indicated when the ratio of signal to background is low. It requires, however, the measurement of X-ray intensities at two or more wavelengths, and is therefore easier to perform in energy-dispersive analysis, where several wavelength regions can be measured simultaneously, than with crystal spectrometers. If the spectrometers are manually driven, the frequent change of setting is time consuming and may induce error. Attention must also be paid to the change of background intensity at the positions of the absorption edge, which may be hidden by a nearby line such as the Kβ-line (Fig. 6.4).

When the background is measured at a wavelength close to that of a line, the tail of this line may produce a slight overlap with the background position. The resulting displacement of the background is proportional to the height of the peak being measured, and its effects cancel in the formation of the relative intensity. Therefore, and in view of the high peak-to-background ratios and spectral resolution obtainable with crystal spectrometers, it is usually sufficient to perform only one background measurement for each specimen and peak, close to the line of interest. Careful measurement at several points on both sides of the peak position, with graphic or mathematical interpolation, may be required for trace determinations. The accuracy achievable with a good crystal spectrometer by this procedure permits the determination of the background signal at a trace concentration level that cannot be matched with an energy-dispersive spectrometer.

Line Interference. Serious systematic errors can, however, arise as a consequence of overlap of the line of interest with a line of another element present in the specimen. In the analysis with crystal spectrometers, the problem can usually be resolved by using another line.

For manually driven crystal spectrometers, the method of calculating the background on the unknown specimen from measured background intensities on other targets is attractive because it avoids the repeated readjustment of the wavelength setting. The simplest situation is present in a binary system (Fig. 6.5). Let us assume that a line of element a is measured in the system consisting of elements a and b. It is a safe practice in such a case to use throughout the binary system the background measured at the pure element b. If the calibration curves for both the line and the background are linear, the total observed intensity at the concentration C_a is

$$\begin{aligned} Y &= C_a Y_a + Y_{Bb} - (Y_{Bb} - Y_{Ba})C_a \\ &= (Y_a + Y_{Bb} - Y_{Ba})C_a + Y_{Bb}. \end{aligned} \tag{6.4.4}$$

We obtain a linear relationship with respect to C_a, with background Y_{Bb}. The argument remains valid if the curves are not linear, provided that it can be assumed that the variations of the net line signal and of the background as a function of C_a are proportional between themselves. Any error arising from this assumption will be small when C_a, and hence the ratio of line-to-background, are small. Our experience indicates that the inaccuracies arising from this simple procedure cannot normally be detected. This simple technique can be extended to a system of more than two elements, if the backgrounds on all elements which do not emit the line in question can be measured and averaged in weight proportions (or proportions of their characteristic signals). If a line of every element present is observed by means of one spectrometer, no additional measurements, besides those on the elementary standards, are needed. A similar system can be based in oxydic materials if all standards are the corresponding oxides. However, in complex materials, such as minerals, not all necessary measurements can be made simultaneously, or not all oxide standards may be available. For such cases it is best to use a function such as Kramers' law, to relate background intensities emitted from different targets.

This simple procedure is surprisingly effective, often even at the trace level, so long as crystal spectrometers are used. In energy-dispersive systems, however, the line-to-background ratios are much lower (by a factor of about ten), and therefore the background correction must be performed with greater care. Measurements at a single energy, and even linear interpolation, may be too inaccurate, and it is preferable to reconstruct the entire background spectrum. This can be achieved by collecting a spectrum of a material producing no X-ray lines, such as carbon, and scaling this spectrum, according to the atomic number dependence previously discussed [6.14]. Absorption corrections must be performed, but the spectrometer characteristics (detector efficiency) cancel between line and background, and therefore the detector need not be characterized. I prefer the measurement of the background intensity at two or more regions free of lines, from the specimen to be analyzed. The algorithm used for the continuum is that proposed by Lifshin [Eq. (6.3.10)]. This procedure was incorporated in a program (FRAME C) which will be described in Chapter 12. It should be noted that an accurate background correction for energy-dispersive spectra must include a calculation of the discontinuities at the absorption edges, which requires the knowledge of the specimen composition and must therefore be incorporated in the iterative analytical procedure to be described in Chapter 12. The use of as few as two measurements of continuum intensity is unavoidable when the target spectrum contains many lines (Fig. 6.6), and is justified by the excellent back-

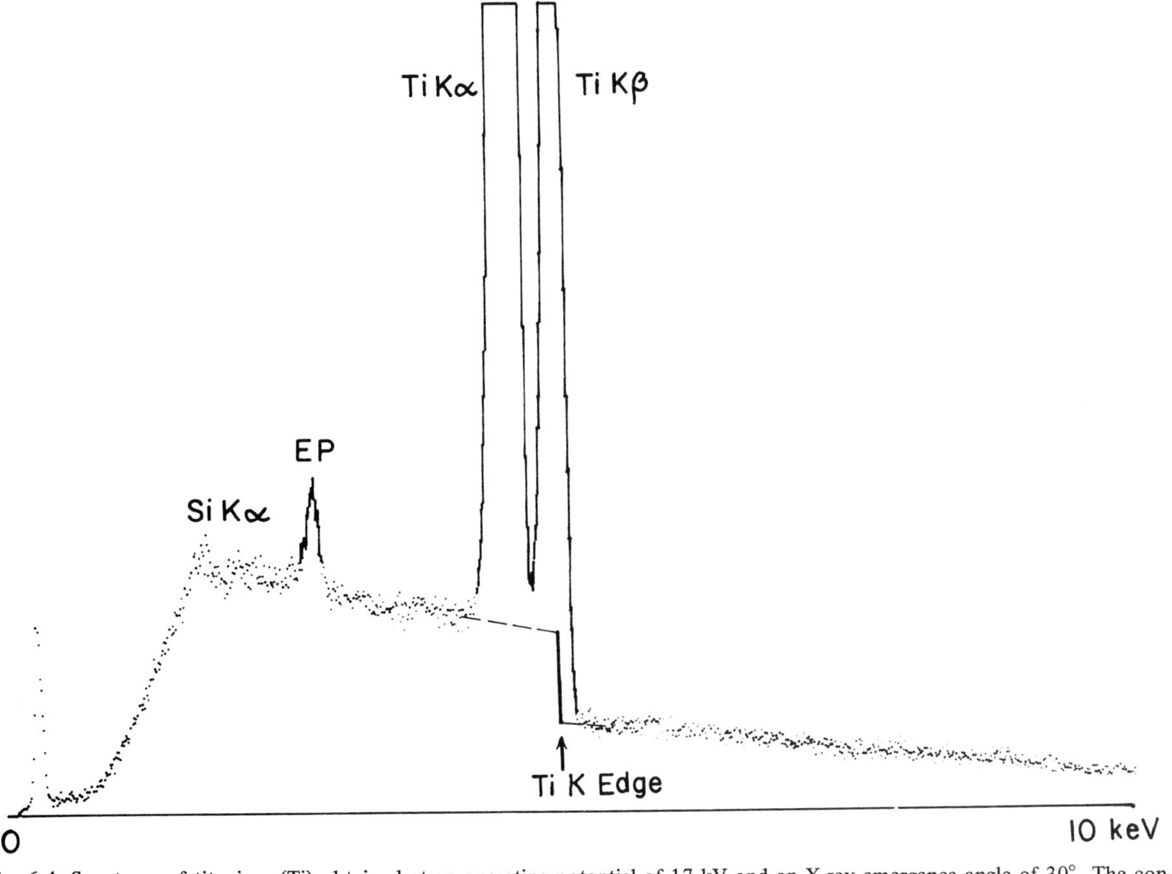

Fig. 6.4. Spectrum of titanium (Ti) obtained at an operating potential of 17 kV and an X-ray emergence angle of 30°. The continuum shows a strong change of intensity at the Ti K-edge, which is located below the broadened Ti Kα peak. The Ti Kα escape peak and the silicon (Si) fluorescence peak (see Subsection 5.3.7) can be seen in this spectrum.

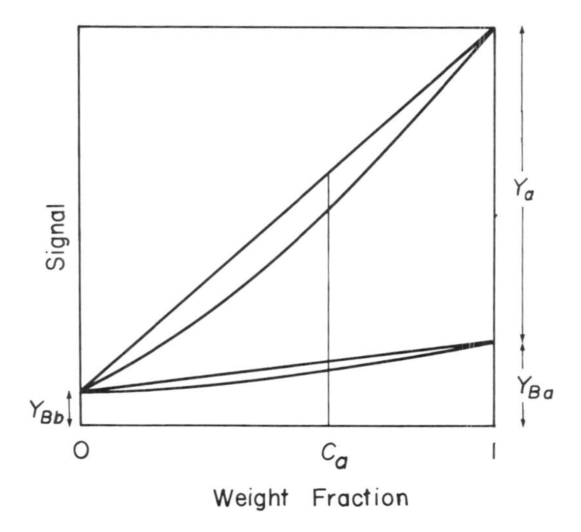

Fig. 6.5. No error is committed if the background at the zero level of a binary system is used throughout, provided that the variation of background as a function of C_a is proportional to the variation of the line signal. This property is also shared by systems in which the calibration curves are nonlinear (lower curves).

ground fits obtainable. The Duane-Hunt limit provides another point for the construction of the background distribution curve. The detector parameters which must be known in this procedure (mainly the window thickness) can be empirically determined and do not require frequent recalibration.

6.5 THE RECORDING OF VARIABLE COUNT RATES

In electron probe microanalysis, variable count rates are mainly observed in the performance of wavelength scans, and of topographic line scans and in alignment procedures. The purpose of the wavelength scan is to record the X-ray spectrum which contains the positions and relative intensities of the X-ray lines emitted from the specimen (Fig. 6.2). Wavelength scans are usually obtained by recording the signal intensity as a linear function of the wavelength (angstroms) or the Bragg angle (degrees). The topographic line scan (Fig. 6.7) records the variations in intensity of one or more X-ray lines as a function of the beam position with respect to the specimen. In either scan one must obtain the variations of X-ray intensity in time. A signal proportional to the rate of detection of photons is obtained by feeding the amplified and normalized pulses into an electronic ratemeter. The wavelength scans obtained by the outlined procedure are distorted by several effects, some of which reside in the detector and spectrometer, and others in the ratemeter. Distorting effects due to the detector and spectrometer include the *coincidence losses,* which are significant only if a qualita-

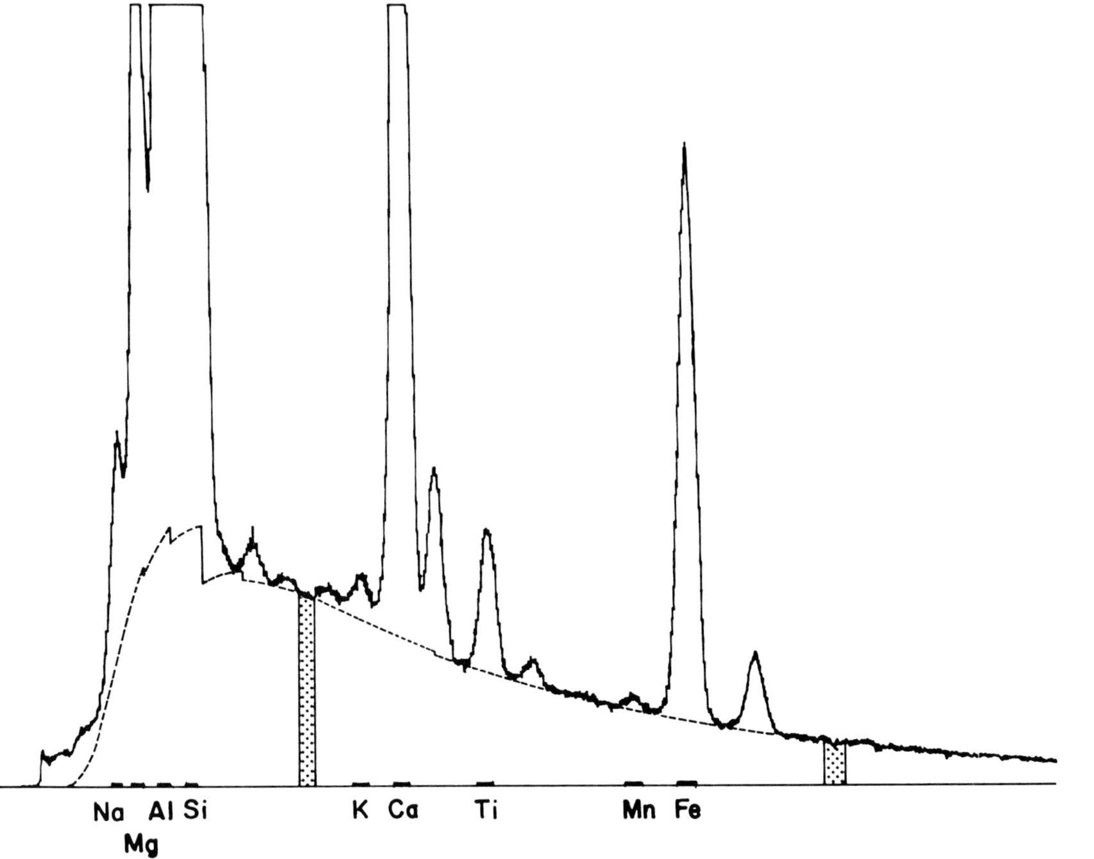

Fig. 6.6(a). Background correction for the energy-dispersive analysis of Juan de Fuca basaltic glass (see Table 12.5). The baseline of this spectrum goes from 0 to 10 keV. The operating potential was 20 kV. The regions of interest for the background correction are marked as dotted areas. Those for the elements to be determined are marked on the baseline. The reconstructed continuous background is drawn as a broken line. Absorption edges of magnesium (Mg), aluminum (Al), gold (Au), and calcium (Ca) can be seen. The convolution of the background signal by the detector resolution has not been applied to the curve shown in the graph. See Fig. 6.6(b).

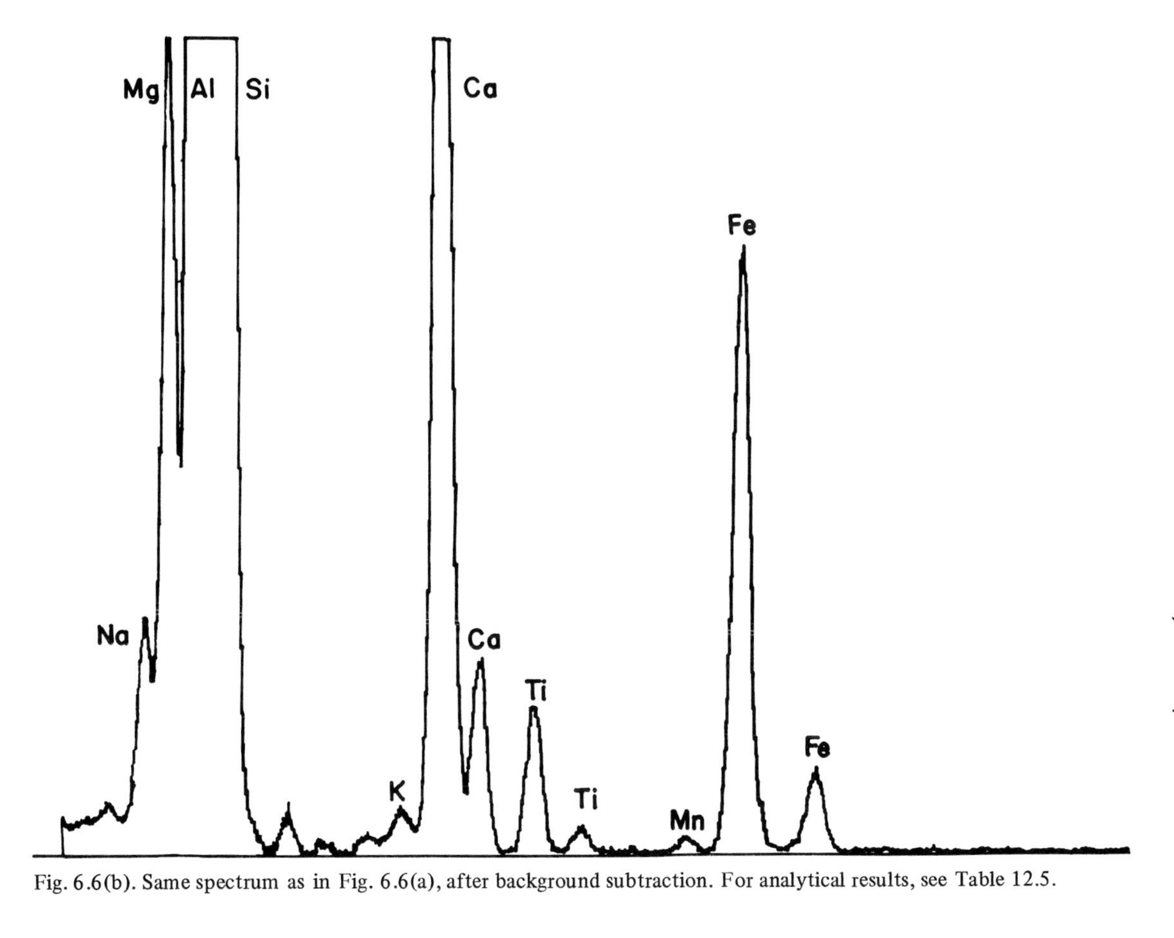

Fig. 6.6(b). Same spectrum as in Fig. 6.6(a), after background subtraction. For analytical results, see Table 12.5.

tive evaluation of the recorded intensities is intended, the *efficiency variations* with wavelength of detectors and spectrometers, which modify the intensity relations of lines at different wavelengths, and the limited *spectral resolution* of single-crystal spectrometers, which broadens and distorts the natural lines, lowers the ratios of line-to-background, and therefore reduces the sensitivity of detection of lines. The line broadening also causes or aggravates line interferences, such as that of the titanium (Ti) Kα line with the vanadium (V) Kβ line (Fig. 6.8), or the arsenic (As) Kα line with the lead (Pb) Lα_1 line (Fig. 6.9).

Further distortions of the X-ray emission spectrum are due to the counting statistics and to the dynamic characteristics of the ratemeter. The problem caused by the presence of statistical count rate fluctuations is fundamental. It is impossible to fully distinguish and separate them from spectrally significant variations in signal intensity. Since the relative importance of the Poisson fluctuations diminishes as more counts are considered, in a given measurement, their effect can be reduced by limiting the speed of response of the ratemeter to rapid

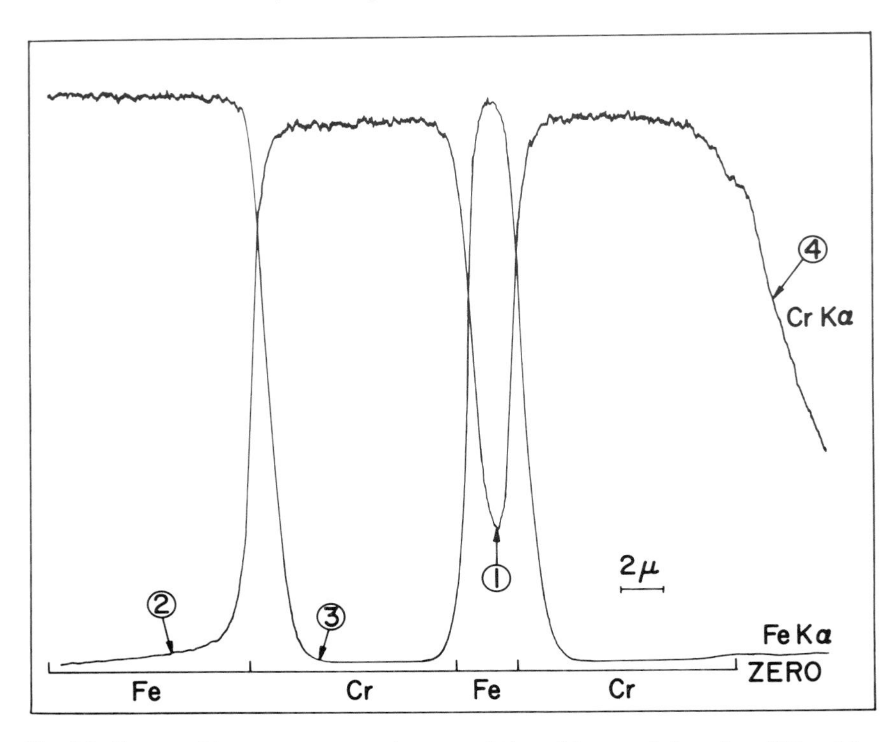

Fig. 6.7. Topographic scan over a specimen consisting of layers of chromium (Cr) and iron (Fe). At the interface the iron signal decays more abruptly ③ than the chromium signal ① and ②; the difference is due to fluorescent excitation of chromium K-radiation by the iron K-radiation. Point ④ is the edge of the specimen.

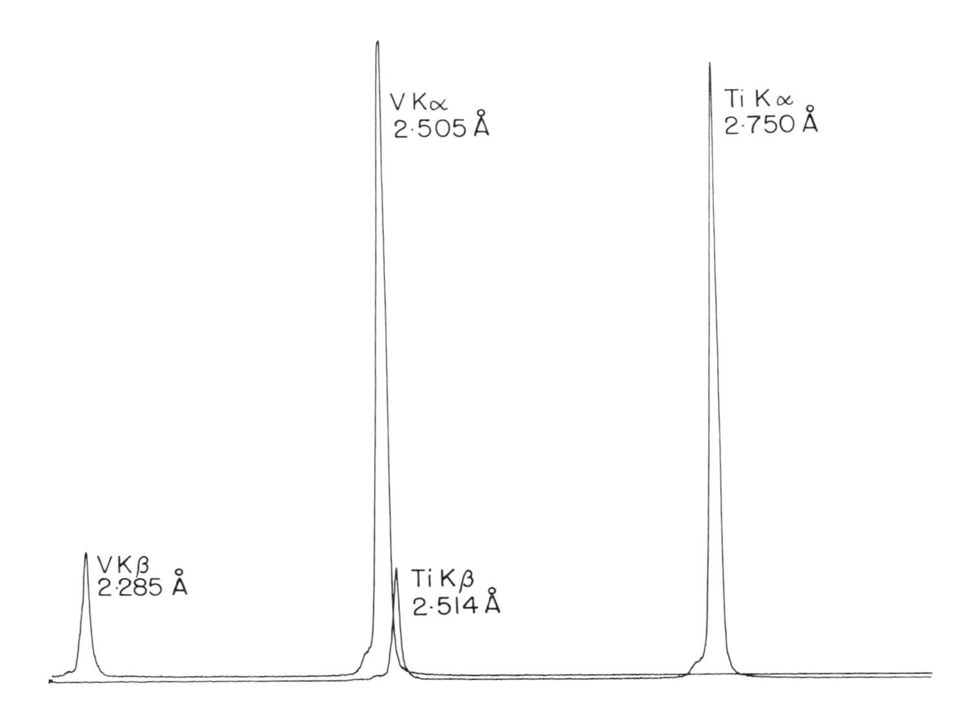

Fig. 6.8. Partial overlap between the vanadium (V) Kα and the titanium (Ti) Kβ lines; spectra were obtained with the same spectrometer as in Fig. 6.9.

changes in the count rate. Such a limitation, however, also distorts or masks significant rapid variations in the signal. The only ways to remove or diminish the uncertainty which is reflected by this tradeoff are to increase the signal intensity level, or to slow down the significant changes (i.e., the rate of spectrum scanning), so that time is allowed for every significant element of the signal to be represented in the output by a sufficiently large number of photons.

The Characteristics of the Ratemeter. The ratemeter transforms a pulsed input signal into an output voltage proportional to the rate of arrival of the pulses. The most essential parts of the ratemeter are a storage capacitor, C, and a resistor, R, which is connected across the capacitor C (Fig. 6.10). The arriving pulses are all normalized with respect to height, shape, and duration, and then applied to the capacitor. If the charge across C is not excessive, each of the arriving pulses increases the voltage across C by the same amount. This voltage originates a current across the resistor R, which diminishes the charge across the storage capacitor. If the rate of arrival of the pulses is constant, the potential across C tends to an equilibrium value, V_0. Let us now assume that at the time t_0 the

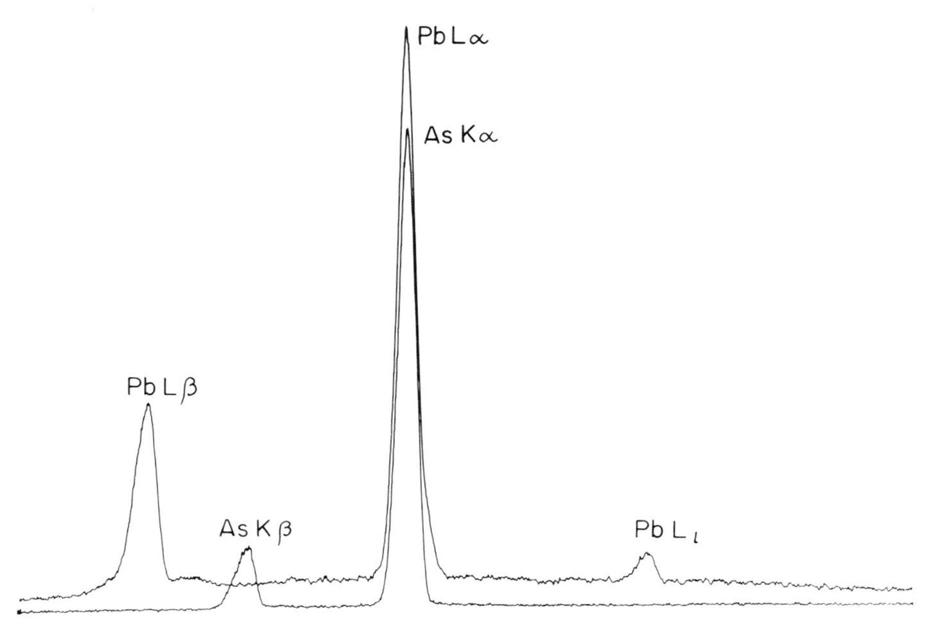

Fig. 6.9. Complete overlap of the lines arsenic (As) Kα (1.177 Å) and lead (Pb) Lα (1.176 Å). The spectra were obtained from the elements by a curved-crystal lithium fluoride (LiF) spectrometer of 15-cm radius. $E_0 = 20$ kV.

count rate suddenly changes, and remains constant thereafter at a new value. The potential across the capacitor will asymptotically approach a new equilibrium value, V. The potential at the time $t_1 = t_0 + RC$ is equal to

$$V_1 = V - \frac{\Delta V}{e}. \tag{6.5.1}$$

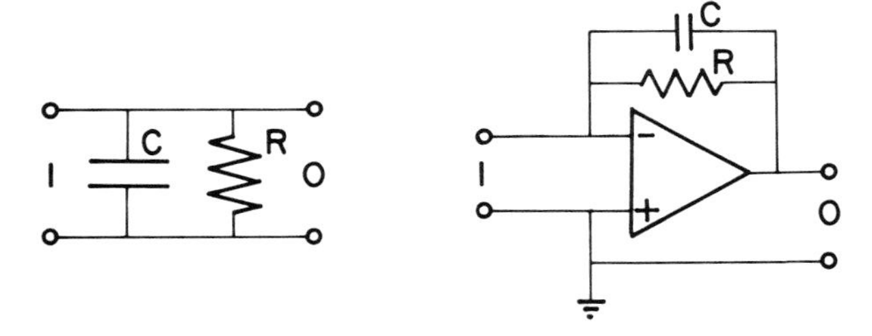

Fig. 6.10. The ratemeter. Left: principle: a capacitor is charged by the incoming pulses and its charge is drained through the resistor. C:capacitor, R:resistor. Right: application of the ratemeter principle to a circuit with operational amplifier.

In this equation, ΔV is equal to $V - V_0$, and e is the base of the natural logarithms. The resistance R is measured in ohms, and the capacitance C is measured in farads. Similarly, at any arbitrary time

$$t_n = t_0 + nRC$$

the potential is equal to

$$V_n = V - \frac{\Delta V}{e^n}. \tag{6.5.2}$$

Calling k the fraction of the voltage change—from V_0 to V—by which the voltage differs from the equilibrium value V:

$$k = \frac{V - V_n}{\Delta V} = e^{-n}$$

and calling $\Delta t = t_n - t_0$, we obtain:

$$k = e^{-n} = e^{-\Delta t/RC}. \tag{6.5.3}$$

With this equation we can calculate the time necessary for approaching to any desired degree the equilibrium output voltage of the ratemeter (Fig. 6.11).

The output of the ratemeter can be displayed on a voltmeter, calibrated in counts per second, or it can be applied to chart recorders and other output devices. Some ratemeters also have an output showing the logarithm of the count rate, which can be used to show wide ranges of count rates.

The lag of the ratemeter output with respect to sudden signal intensity variations can be varied, according to Eq. (6.5.3), by varying either the value of the resistor, R, or of the capacitor, C. As the product RC (called the *time constant* of the ratemeter) is changed, the response to statistical count rate fluctuations also changes. For low count rates, and when sudden changes in the signal are not expected, a high value for RC is used, while for high count rates a lower value can be selected; in such a case, rapid signal changes will be more accurately displayed in the ratemeter output (Fig. 6.12).

Graphic Correction of Ratemeter Distortion. Tournarie [6.16] proposed a simple method for graphic correction of the distortion of a count rate trace due to the ratemeter. This method does not, however, correct for distortions due to other reasons, such as the limited speed of the recorder.

Let us assume that at an initial time, t_0, the ratemeter has reached an equilibrium condition, and the charge of the storage capacitor will be called Q_0. At a time t, which follows after t_0, the charge, Q_t, differs from Q_0 for two reasons:

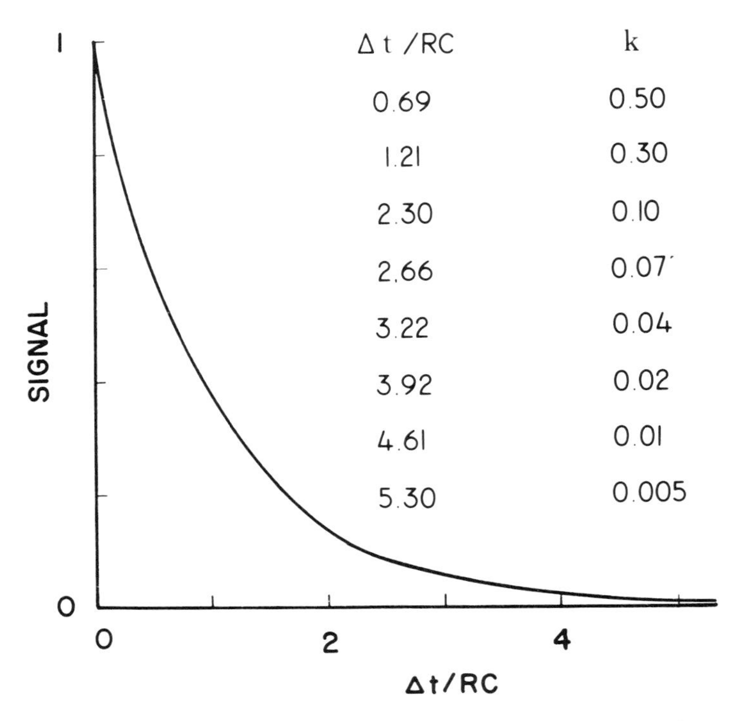

Δt /RC	k
0.69	0.50
1.21	0.30
2.30	0.10
2.66	0.07
3.22	0.04
3.92	0.02
4.61	0.01
5.30	0.005

Fig. 6.11. Decay of a ratemeter signal after the input to the ratemeter stops. The symbols of the table and graph are those of Eq. (6.5.3).

because of the discharge of the capacitor through the resistor, R, and because of the charging of the capacitor by the input current, i. The charge increase of the capacitor due to the input current during dt_1 at the time t_1 is:

$$dQ(t_1) = i(t_1) \cdot dt_1 .$$

At the time t, after t_1, the charge increment $dQ(t_1)$ has diminished exponentially to

$$dQ(t) = dQ(t_1) \cdot \exp \quad -\frac{t - t_1}{RC} \quad .$$

Hence, the increase in potential across the capacitor at time t which is due to the input received at time t_1 is

$$dV(t) = dQ(t)/C = \frac{i(t_1)dt_1}{C} \exp\left(-\frac{t - t_1}{RC}\right).$$

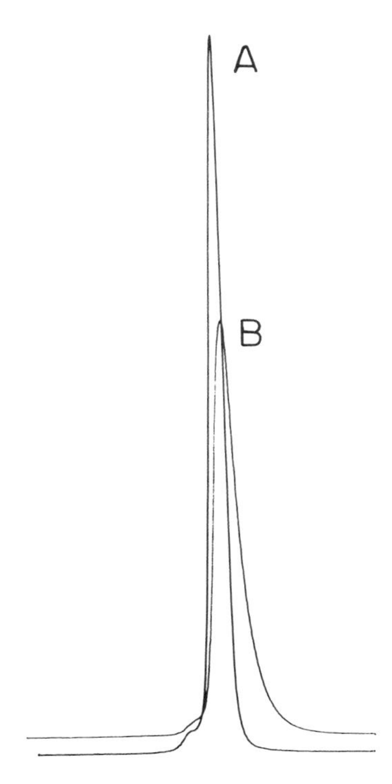

Fig. 6.12. Profiles of a titanium (Ti) Kα line, obtained with a lithium fluoride (LiF) spectrometer, at two scanning speeds. Spectrum B is distorted due to speed limitations in the recorder and associated electronics.

The current through the resistor R due to this potential increase is

$$dj = \frac{dV(t)}{R} = \frac{i(t_1)dt_1}{RC} \exp\left(-\frac{t - t_1}{RC}\right).$$

Since the total output current is determined by the effects of all input current fractions which have arrived since t_0, we must integrate over the time t_1, in the above equation, from t_0 to t. Calling $T = RC$, we obtain:

$$j(t) = \int_{t_0}^{t} \frac{dj(t)}{dt_1} dt_1 = \frac{1}{T} e^{-t/T} \int_{t_0}^{t} j(t_1) e^{t_1/T} dt_1 .$$

This current produces the potential drop across the resistor R which determines the output signal. To relate $j(t)$ (the output) to $i(t)$ (the input at the same time), we differentiate the above equation:

$$\frac{dj(t)}{dt} = \frac{1}{T}\left[-\frac{1}{T}e^{-t/T}\int_{t_0}^{t} i(t_1)e^{t_1/T}\,dt_1 + e^{-t/T}\,i(t)e^{t/T}\right] = \frac{1}{T}[-j(t) + i(t)]$$

and we obtain a differential equation:

$$i(t) = j(t) + T\frac{dj(t)}{dt}. \tag{6.5.4}$$

The equation links the observed ratemeter signal j, its slope with respect to the time axis dj/dt, the time constant T, and the undistorted count rate i. The latter can therefore be obtained (Fig. 6.13) if one prolongs the tangent to the observed signal at the point A corresponding to the time t, until reaching, on the time axis, the value $t + T$ (point B). To reconstruct the graph, it is now necessary to translate the value of i back to the time t (point C). A reconstruction of a square-wave input signal distorted by a ratemeter with a very large time constant is shown in Fig. 6.14. It indicates that the technique is very effective in restoring the original shape of the signal. It should be noted, however, that the effects of counting statistics, which may have been smeared out by the ratemeter distortion, will also become visible when the curve is restored.

Due to the availability of fast electronic circuits, Tournarie's method is not used in practice. However, Eq. (6.5.4) is a useful description of the dynamic distortion of ratemeter traces which the reader should recognize and minimize.

The Periodic Integrator. In most ratemeter signals, the counting statistics, effects of the ratemeter time constant, and often also the dynamic characteristics of the chart recorder combine in modifying the recorded trace. It is therefore difficult to ascribe limits of accuracy or even significance to rapid excursions or changes of level in the ratemeter record. The interpretation of the signal is easier if a periodic ratemeter is used [6.17]. This electronic device accumulates

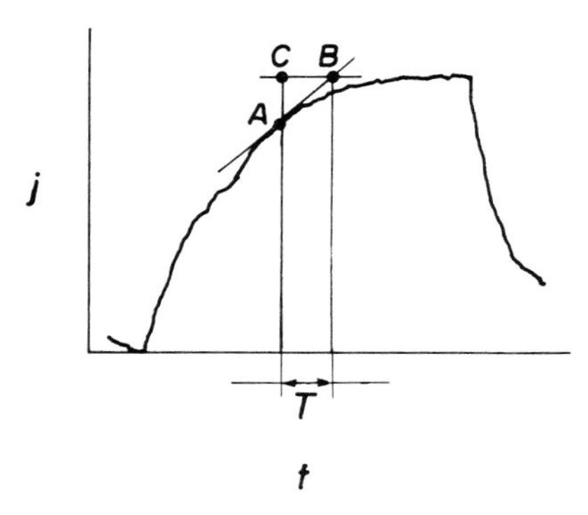

Fig. 6.13. Graphic correction of ratemeter distortion.

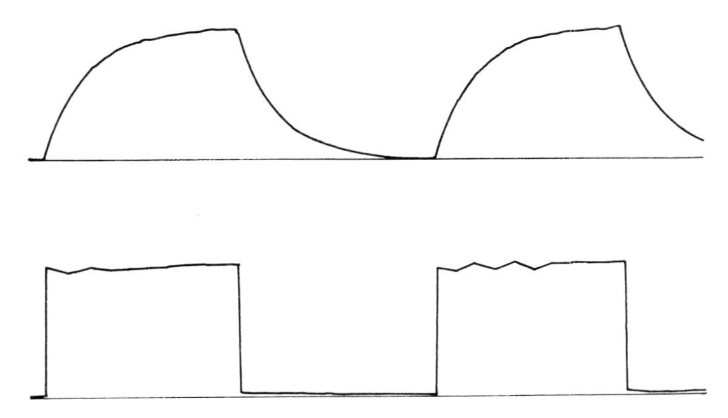

Fig. 6.14. Ratemeter signal of a square-wave X-ray signal, with large time constant (above), and signal rectified by Tournarie's method (below).

the signal pulses arriving during a short adjustable time interval (the period), and emits, at the end of the period, a voltage proportional to the number of pulses, which is maintained until the signal from the next period can be displayed. The device is equivalent to a fast scaler with analog output, working in an uninterrupted sequence of count periods of equal time. Early periodic integrators were based on analog circuits, and thus prone to drift or poor calibration. At present, digital techniques can be used to advantage in building periodic integrators of high accuracy, with optional subtraction of a fixed count level.

The output of the periodic integrator is shown in Fig. 6.15. Poisson statistics can be used to establish confidence limits for the excursion of the output signal from a steady input (e.g., the average number of counts per period is $\overline{N}$, the limits of 95% confidence for each period are $\overline{N} \pm 2\overline{N}^{1/2}$. If the input rate is steady, only 1 period in 20 should, in average, fall outside these limits). Periodic ratemeter traces with such confidence limits can be used to test instrument stability and—by moving the specimen during the measurements—homogeneity of the target materials (Fig. 6.16). Another useful property of the periodic integrator is the fact that the effects of a sudden large excursion of the input signal do not affect the output beyond the integration period comprising the change. In contrast, in the standard ratemeter the effects of any change, though exponentially attenuated, theoretically persist forever.

The periodic ratemeter cannot, of course, eliminate the dynamic characteristics of the recorder, but it clearly shows the resulting deformation in Fig. 6.17. It also cannot eliminate the uncertainties inherent in counting statistics. The effects are merely shifted, but in a way which is frequently advantageous to the observer. While the carry-over of signal beyond one period is eliminated, there is no information whatsoever left concerning the signal changes *within* one period.

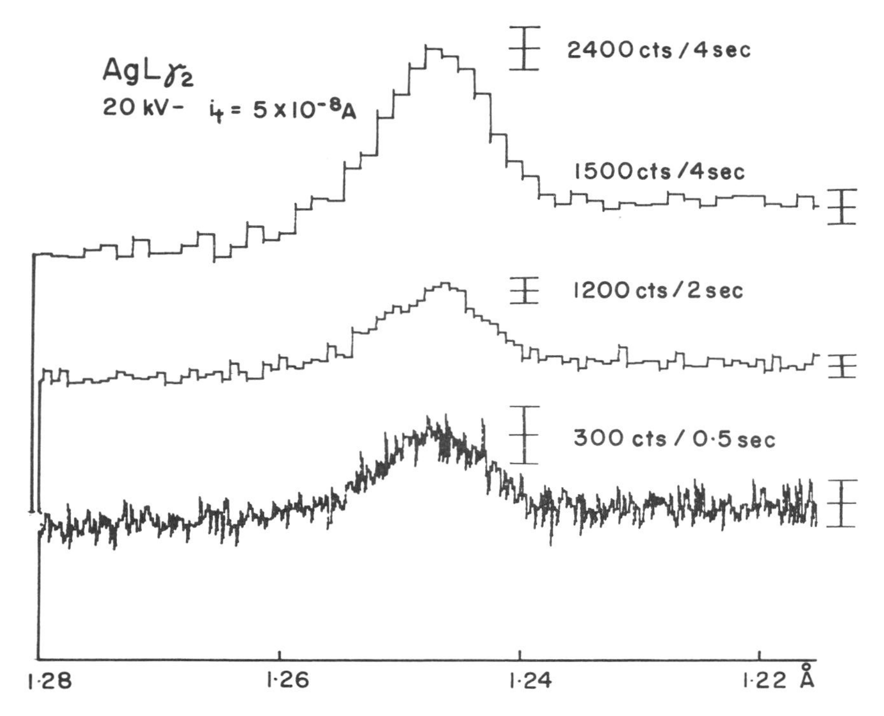

Fig. 6.15(a). Periodic ratemeter traces over a small peak (Ag $L\gamma_2$). The vertical bars indicate Poissonian limits of $\pm\ 2\ \sqrt{N} \simeq \pm\ 2\sigma$, where N is the number of counts collected in one counting period. In the absence of non-Poissonian deviation, 1 period out of 20 will fall outside these limits. Bars at the left indicate the ratemeter range from zero counts to the indicated count rate. Note overshoot of the recorder pen.

In an abrupt signal change (e.g., a square-wave signal), the signal level of the period during which the change took place depends on the time relations and cannot be interpreted in terms of signal intensity.

The length of the period plays the same role as the time constant in the conventional ratemeter, and for similar effect, should have the same value as RC of the latter device. The period of the periodic ratemeter should be adjustable within wide limits (e.g., from 10 msec to 10 sec). The signal height must be accurately calibrated to count rate, and a wide choice of well-calibrated signal subtraction levels is very useful.

Acoustic Monitoring of the X-Ray Count Rate. It is often useful to acoustically monitor the output of an X-ray spectrometer, particularly for the tuning of the spectrometer, during which the operator may not be able to continuously observe a meter or chart recorder. Monitors in the form of loudspeakers which receive the amplified detector pulses were frequently used in the past. Such an arrangement, however, produces a noise, due to the random distribution of the

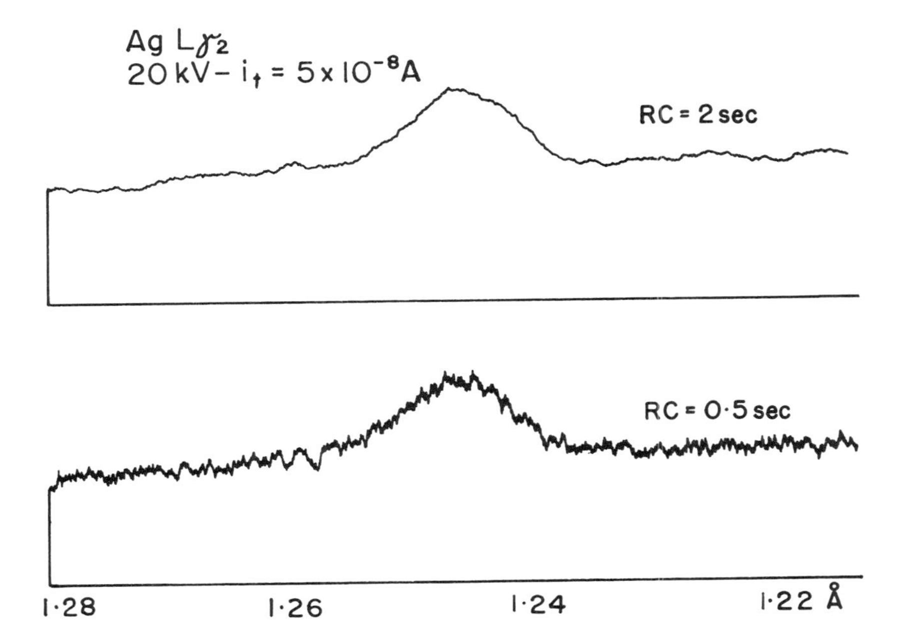

Fig. 6.15(b). Normal ratemeter scans with two different time constants, through the same peak as shown in Fig. 6.15(a).

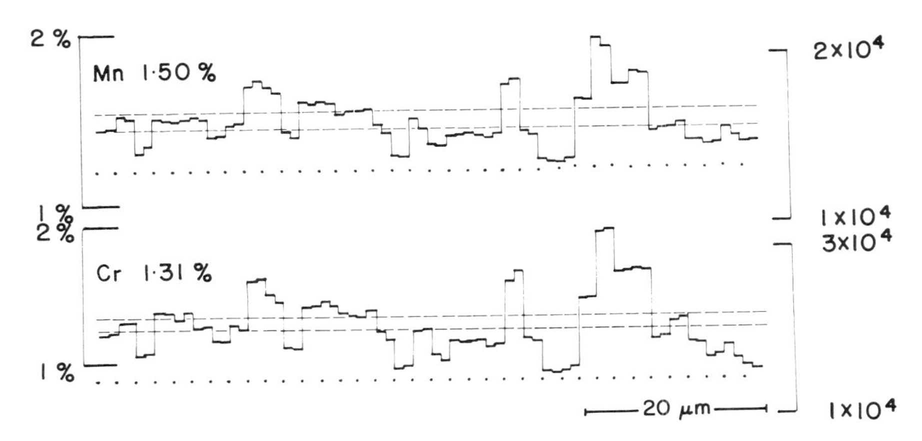

Fig. 6.16. Homogeneity test on a low-alloy steel. The scales at the left indicate mass percents. The scales on the right are in counts collected per counting period. The beam diameter was 2 μm and the specimen was moved 2 μm after every other counting period (marked by dots below the traces). Inhomogeneity is recognized because the changes in intensity after moving the specimen frequently exceed the Poissonian limit of $\pm 3N^{1/2}$ (broken lines), but they stay within the Poissonian limit when the specimen is not moved. Note a strongly positive correlation between the concentrations of chromium (Cr) and manganese (Mn).

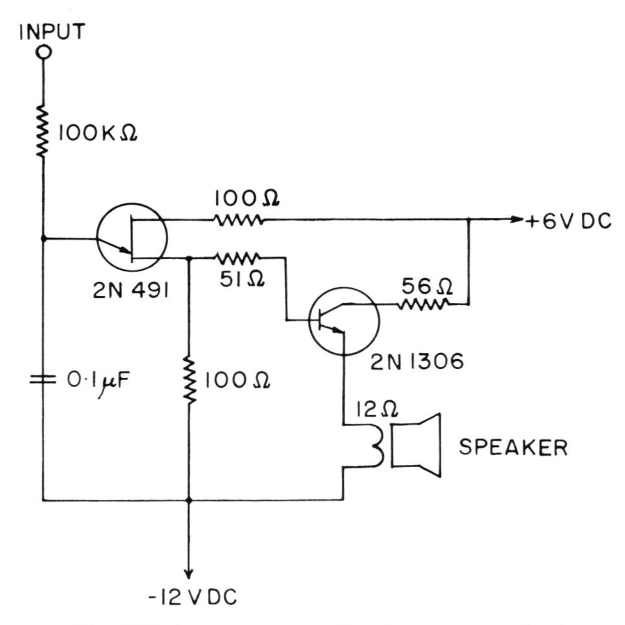

Fig. 6.17. Acoustic rate monitor for ratemeter signals.

pulses, which cannot be used effectively to judge small intensity variations. It is much more preferable to use a device which produces a sound, the pitch of which is modulated by the count rate, because the ear is much more sensitive to changes in pitch than to changes in intensity. The acoustic monitor is particularly effective when coupled to the output of a periodic integrator, because the well-defined pitches of the individual integration periods can be compared with good sensitivity.

A schematic of a circuit usable for acoustic monitoring is shown in Fig. 6.17 [6.17].

6.6 DRIFT

We call a system of measurement stable if we can repeat a measurement as many times as we wish, without observing a systematic change in the time of the results. In our previous discussion concerning the statistics of the measurement of count rates we have presumed that such systematic changes in time do not occur. Such an assumption is required in applying Poisson statistics to counting experiments. Similarly, the statistical contributions called $\Sigma\sigma_i^2$, in Eq. (6.22) represent perturbations which do not establish a trend in time.

There are, however, parameters which affect the measurement in the electron probe microanalyzer, and which change gradually over long periods of instrument operation. Some of these parameters drift away from an optimum value and must be periodically readjusted by alignments. Other characteristics, such as those due to the wear of the filament, change irreversibly during a set of measurements of long duration. Stability is an ideal condition which can only be realized imperfectly in practice. The degree to which the quality of an analysis is affected by drift depends on the analytical problem, on the characteristics and conditions of the instrument, and on the technique selected by the operator. When the series of measurements is of short duration, the effects of drift may not be observable. However, as the duration of the experiment is increased, and as the Poissonian uncertainty diminishes, a point will be reached beyond which further improvement of the accuracy of analysis is limited by drift.

The drift problem is particularly insidious in electron probe analysis because the switching from standards to the specimen regions of interest usually requires time and involves delicate operations of the stage. There is, therefore, always a resistance to frequent changes from specimens to standards. Yet, in quantitative procedures, the operator is well advised not to perform too many measurements without such a switch, and to measure the intensities of X-ray emission from the standards both before and after the measurements on the specimen. A comparison of the two sets of standard measurements may reveal the presence and the extent of drift.

It is usually assumed that the intensity variations of the signal during the period of measurement are close to linear with time. Such an assumption may lead to a correction procedure which will reduce the effect of drift; however, unless the drift is small, the assumption is often not valid (Fig. 6.18). It is therefore advisable to reject measurements if the differences between the bracketing standard measurements exceed preestablished tolerances, such as the allowable range of measurement error.

In view of the limited effectiveness of mathematical drift corrections, the operator should make all efforts to understand the mechanisms operating in the instruments, and thus the diverse sources of instability. He should maintain or modify his instrument in order to reduce drift; he should know the magnitude of the residual drift in his instruments, and he should choose the conditions of analysis so as to reduce the effects of drift on the accuracy of the analysis. Such measurements require skill, experience, and an observant attitude. The measures to be taken frequently depend on the characteristics of particular instruments. We will, nontheless, describe in a general way some potential sources of drift.

Causes of Drift: The High-Voltage Supply. The stability of the operating voltage between the filament and the anode plate affects in various ways the performance of the instrument. While high-frequency components of the operating voltage are mainly a nuisance in scanning image formation with secondary electrons or target current, and errors in the value of the operating voltage may cause errors in the calculation of the corrections, a drift of the operating poten-

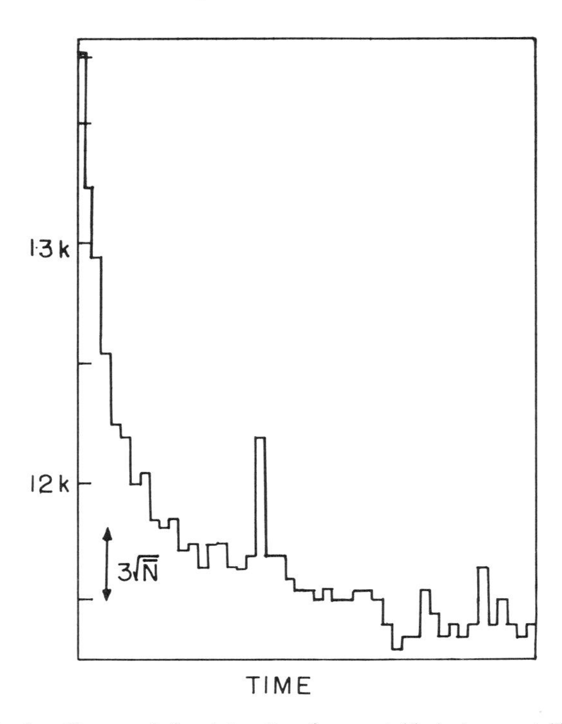

Fig. 6.18. Drift of an X-ray emission intensity of an unstable instrument. The graph was obtained with the aid of a periodic integrator; the integration periods were 10 sec long. The arrow to the left is equal to three standard deviations, according to Poissonian statistics. The real drift greatly exceeds this prediction, and is markedly nonlinear.

tial can cause serious inaccuracies in the measurement of ratios of X-ray intensities.

Concerning the effects of high-voltage drift, we first note that the number of photons emitted per incident electron increases with increasing operating potential (Section 9.2), except for soft radiation at high operating potentials. In the latter case the absorption loss with increasing depth of excitation more than compensates for the increase of X-ray emission within the specimen (see Fig. 6.19). These observations are valid in the case that the beam current has not changed as a consequence of the change of operating potential. However, according to Eq. (3.2.6), the focal length of an electromagnetic lens is proportional to the acceleration potential of the electrons which are being focused. As can be seen in Figs. 2.1 and 3.24, a change in focal length of the first (condenser) lens varies the fraction of the electron beam which passes through the aperture beneath it, and thus the intensity of the beam which impinges upon the specimen. Furthermore, the gun emission also increases with increasing acceleration potential [Eq. (3.2.14)]. Hence, unless the beam current is monitored and regulated, a change in operating potential also affects the emitted X-ray intensity through changes in the beam current (Fig. 6.20). For these reasons the power supply used to accelerate the electrons emitted by the gun should be very stable.

Monitoring the Beam Current. The fluctuation of accelerating voltage is not the only possible cause for variation in the beam current. A drift of the beam intensity may also be due to misalignment of the electron optics caused by thermal expansion (Subsection 3.2.3), or instability of the strength of the con-

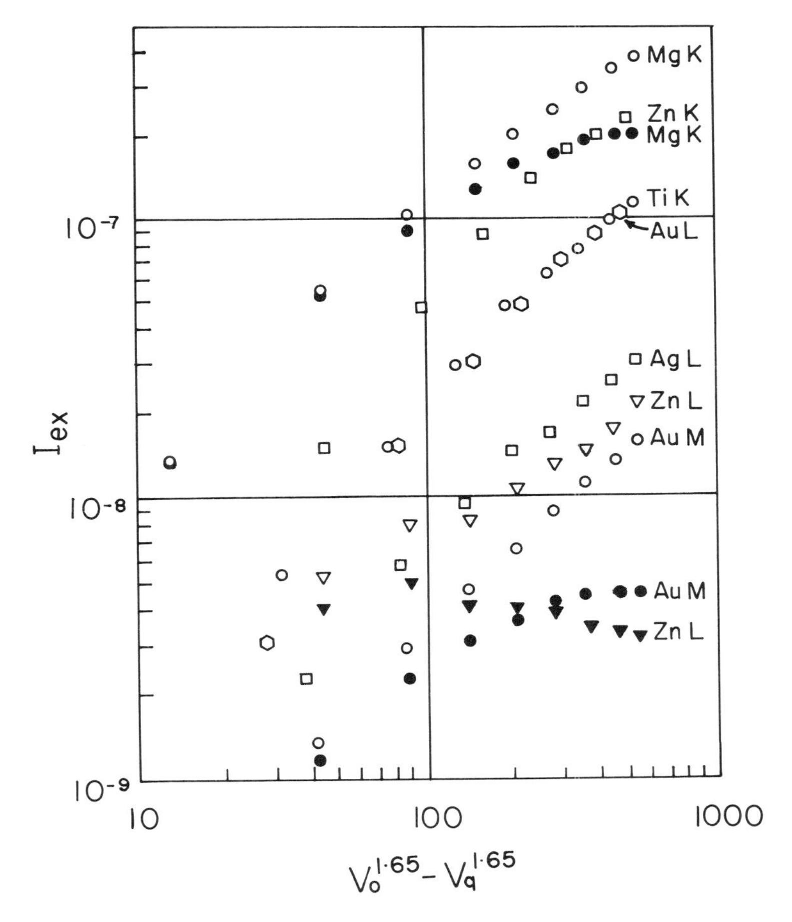

Fig. 6.19. Observed X-ray intensities, obtained from a curved-crystal spectrometer of a radius of 10 cm, as a function of $(V_0{}^{1.65} - V_q{}^{1.65})$. The intensity is normalized to photons/incident electron. Empty symbols represent intensities corrected for absorption within the target.

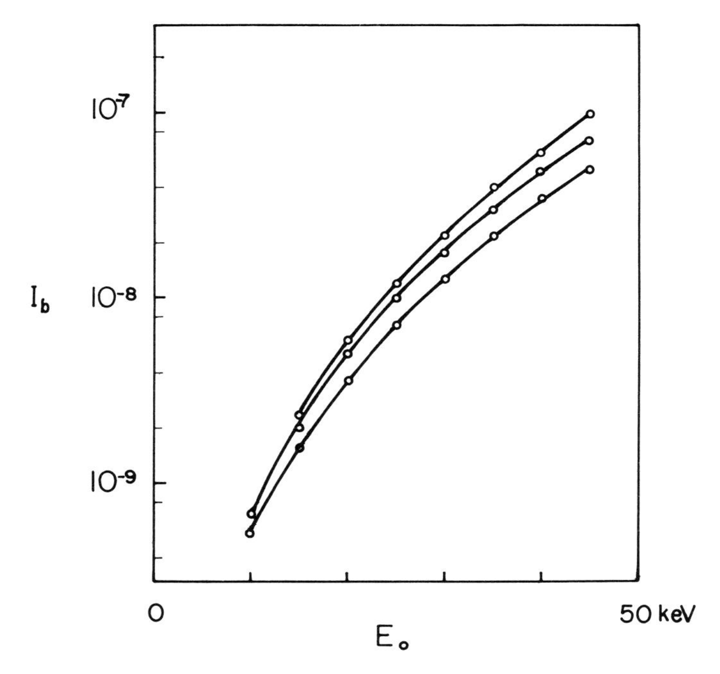

Fig. 6.20. Change of beam current due to changing the operating voltage, and hence the electron energy, E_0.

denser lens. The first defect must be corrected periodically by realignment of the anode plate and/or the filament position. Another source of instability is the evaporation of material from the filament. As the diameter of the filament decreases, the temperature produced by the same heating voltage rises; therefore, the output of electrons also increases. This effect is particularly noticeable towards the end of the filament life. Further instability can be produced by faulty setting of the saturation level of the gun, or by electrostatic charging of contamination deposits in lens gaps and other parts of the electron optics.

It is desirable to observe these changes in beam current, and to compensate for them. But the only direct way of measuring the beam current is to trap the beam in a cavity (Faraday cage). Such a trap can be constructed by covering a small drill hole in a target with an aperture. However, the measurement cannot be performed during the analysis of a specimen.

The specimen current (target current) which flows (in the classical sense) from the ground connection to the specimen cannot be used to monitor the beam current, since its magnitude depends of the composition and topography of the specimen, and of the emission of secondary electrons. Monitoring of the beam current can be performed, however, if the limiting aperture is replaced by a pair of concentric apertures (Fig. 3.24). The lower aperture determines the fraction of the electron beam which reaches the specimen. The upper aperture is wider, and therefore the lower aperture diaphragm receives an annular cross section of the beam. The current received by this diaphragm can be collected and used to monitor variations in the beam current. The proportionality between this monitor current and the beam current is limited, and should be experimentally verified. It depends on the uniformity of the current density within the beam which reaches the lower aperture. At high currents the proportionality does not hold; if the entire beam is made to pass the lower aperture, the monitor current vanishes entirely. For these reasons, the monitor current cannot be used to compensate for large fluctuations of beam current due to misalignment of the electron optics, although it is effective in the correction for minor drifts.

It is possible for a small fraction of backscattered electrons from the specimen to be focused back upon the aperture diaphragm. In such an event, the monitor current would not be totally independent of the topography and composition of the specimen. It is easy to test if the monitor current varies when the beam passes from a specimen of low atomic number to one of high atomic number. If the difference is significant, a correction may be advisable.

The proportionality of the monitor and beam currents may also be affected by the collection of secondary electrons on the monitoring aperture. In one such case, an improvement of the proportionality was observed when the collecting aperture was biased to −75 V.

It should be noted that the current monitoring technique can compensate for variations in the beam current, but not in the operating voltage, although the latter affects the efficiency of X-ray production.

The Spectrometer Assembly. A well-built and aligned X-ray spectrometer should not produce observable drift due to variation in its efficiency, even for lengthy operation. Yet, under unfavorable conditions, such drift is occasionally observed. The causes for the erratic behavior of the spectrometer may be mechanical deficiencies (backlash), thermal expansion of components [6.12], [6.18], large pressure variations of the detector gas in flow-proportional detectors, marginal alignment conditions of the spectrometer parts or of the electronic components (amplifier and pulse height analyzer), or severe variations in the detector voltage.

The coefficients of thermal expansion of crystals differ considerably, and accordingly their sensitivity to changes in temperature also varies greatly. The following values are reported in [6.12]:

	coefficient/°C
sodium chloride	40×10^{-6}
lithium fluoride	34×10^{-6}
quartz	$8\text{-}13 \times 10^{-6}$
topaz	$5\text{-}8 \times 10^{-6}$

Causes of Drift: The Specimen. Frequently, the source of instability of X-ray emission measurements is in the specimen itself. Specific reasons for instability in such a case are poor electrical conductivity, absorption of electrons or X-rays in contamination deposits which grow on the specimen surface, or evaporation and alteration of the specimen itself. Such effects should be suspected if stability is restored when the beam is directed towards a solid metal target such as brass or aluminum. Problems related to the nature of the specimen will be discussed in Chapter 16.

6. REFERENCES

6.1 Kramers, H. A., *Phil. Mag.* **46,** 6th ser., 836 (1923).
6.2 Green, M., Ph.D. Thesis, Univ. of Cambridge, Cambridge, England, 1962.
6.3 Kirkpatrick, P. and Wiedmann, L., *Phys. Rev.* **67,** 321 (1945).
6.4 Dyson, N. A., *Proc. Phys. Soc. London* **73,** 924 (1959).
6.5 Kulenkampff, H., *Ann. Phys. (Leipzig)* **69,** 548 (1922).
6.6 Kramers, H. A., *Phil. Mag.* **46,** 6th ser., 869 (1923).
6.7 Sommerfeld, A., *Ann. Phys. (Leipzig)* **11,** 257 (1931).
6.8 Reed, S. J. B., *X-Ray Spectrometry* **4,** 14 (1975).
6.9 Rao-Sahib, T. S. and Wittry, D. B., *J. Appl. Phys.* **45,** 5060 (1974).
6.10 Statham, P. J., *X-Ray Spectrometry* **5,** 154 (1976).
6.11 Hehenkamp, Th. and Böcker, J., *Mikrochim. Acta,* Suppl **5,** 29 (1974).
6.12 Lifshin, E., Ciccarelli, M. F., and Bolon, R. B., in *Practical Scanning Electron Microscopy,* Yakowitz, H. and Goldstein, J. I., Eds., Plenum, New York, 1975, p. 291.
6.13 Fiori, C. E., Myklebust, R. L., Heinrich, K. F. J., and Yakowitz, H., *Anal. Chem.* **48,** 172 (1976).
6.14 Smith, D. G. W., Gold, C. M., and Tomlinson, D. A., *X-Ray Spectrometry* **4,** 149 (1975).
6.15 Moreau, G. and Calais, D., *J. Phys.* **25,** Suppl. **6,** 83 (1964).
6.16 Tournarie, M., *J. Phys. Radium,* Suppl. **15,** 16A (1954).
6.17 Heinrich, K. F. L., *Adv. X-Ray Analysis* **7,** 393 (1964).
6.18 Davies, T. A., *J. Sci. Instrum.* **35,** 407 (1958).

7. Qualitative Analysis

The qualitative elemental microanalysis of a specimen is a very important function of the electron probe microanalyzer. The characteristic line spectra used for this purpose can be obtained either by wavelength-dispersive or by energy-dispersive spectrometers. In addition, signals other than X-rays can also convey useful information on the specimen composition which must not be disregarded.

In spite of the simplicity of the X-ray spectrum, the interpretation of spectra from composite materials is not always straightforward. The presence of elements of low atomic numbers is easily overlooked. Even elements of atomic number from 11 to 30 emit only two strong lines useful for identification. If these lines are interfered with by strong lines of other elements, a serious problem may arise. In crystal spectrometers, further complications stem from the presence of higher-order lines. Misinterpretation of spectral features is, particularly for the inexperienced analyst, a surprisingly frequent accident. Therefore, the analyst is well advised to give to this task the attention it deserves. Orderly procedures of line identification, rather than haphazard operations, result in considerable gain in time and accuracy, particularly if the analyst has acquired experience with them and strictly adheres to an invariable scheme of analysis and interpretation. This is not an area in which virtuosity and improvisation pay off.

7.1 ELEMENT IDENTIFICATION BY MEANS OF WAVELENGTH-DISPERSIVE SPECTRA

The crystal spectrometer offers the advantage of higher line resolution than the energy-dispersive detector. This feature produces higher line-to-background ratios, and, more importantly, reduces the possibility of line overlaps. The disadvantage of the crystal spectrometer resides in its sequential mode of operation, which reduces the signal collection time for each line to a few seconds, and in the limited wavelength range covered by any of the available crystals (Table 5.1, Fig. 5.2). This disadvantage is increased by the fact that the analysis of an unknown specimen often involves—or should involve—the qualitative investigation of composition at several sites of the specimen surface.

When the specimen is a complete unknown, the analyst should obtain wavelength spectra over as large a spectral range as possible. Emission of wavelengths from 1 to 100 Å can be covered, in typical spectrometers, with four crystals (e.g., lithium fluoride, ammonium dihydrogen phosphate, rubidium phthalate, and lead stearate). With such an array of spectrometers, all elements of atomic number five or higher can be detected (Fig. 5.2).

When intense lines are emitted, their higher orders of reflection also appear in the spectrum and thus complicate its interpretation (Fig. 7.1). If the analyst were to attempt the identification of all lines of such a spectrum without a systematic procedure, the time involved would be considerable, and the interpretation would frequently be ambiguous. Several investigators [7.1]-[7.3] have proposed schemes involving the use of pulse height analyzers to mark or eliminate higher-order lines. In practice, however, such techniques are cumbersome and

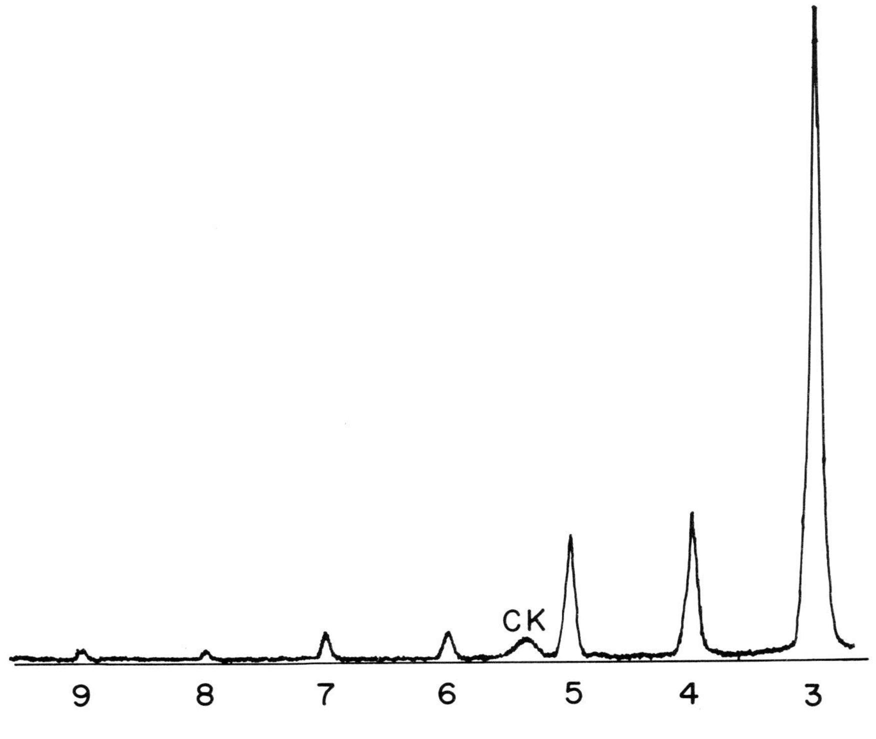

Fig. 7.1. High orders of Al Kα radiation observed by means of a lead stearate pseudocrystal. The numbers indicate the order of reflection. The C K-band, due to contamination with hydrocarbon polymers of the aluminum target, is distinctly broader than the high-order aluminum peaks.

are seldom used. We must therefore discuss the interpretation of spectra in which no distinction is made instrumentally of the order of the reflection of lines.

For the systematic interpretation of spectra the analyst needs a graph or table in which the emitted lines of all elements appear as a function of their wavelength, or Bragg angle, depending on the calibration of the spectrometers. If a spectrometer can be used with more than one monochromator crystal, it may be necessary to use graphs or tables with the "equivalent wavelengths" in which the spectrograph is calibrated. These graphs and tables are used to identify unknown lines, starting with the most intense ones. Once an element has been tentatively identified, another table should be used in which all observable lines, for all crystals and orders of diffraction, are tabulated for each element. By means of this table, the identification of the element in question can be confirmed, and all lines emitted by this element are checked off, so that they cannot be misinterpreted later. The strongest remaining unidentified line is investigated next, and the procedure is continued until all lines have been identified. If the presence of one or more elements in the specimen is known to the analyst, he will identify, and check off, all observable lines of these elements, in all orders of diffraction, before attempting to identify lines of other elements. The systematic procedure we have described not only results in a considerable saving of time, but significantly reduces the possibility of errors in the interpretation of lines.

The identification of the order of diffraction can be aided by the simple device of observing, on the screen of an oscilloscope, the amplified pulses which enter the single-channel pulse height analyzer or discriminator. The height of these pulses can be compared with the height of background pulses—which are mostly in the first order of diffraction—or the height of pulses of a known line (Fig. 7.2). Since the observation of the amplified pulses is also very useful in the discovery and diagnosis of electronic failures, or the occurrence of electronic noise, it is advisable to have such an oscilloscope permanently installed for the monitoring of pulses.

The following observations may be of help in the identification of spectral lines of unknown origin.

1. In the search for high-order lines, one should take into account that the emission of short-wavelength lines is limited by the operating potential. Therefore, the probability of the appearance of higher-order lines increases with the wavelength. In the region of wavelengths above 10 Å, the patterns of higher-order K-lines, with sharp peaks, which repeat at intervals corresponding to the wavelength of the first-order line, are particularly conspicuous. They contrast with the bands of K-emission of elements of low atomic number, which are substantially broader than the spectrometer resolution (Fig. 7.3).
2. The patterns of lines within series are characteristic. The ratio of intensity of

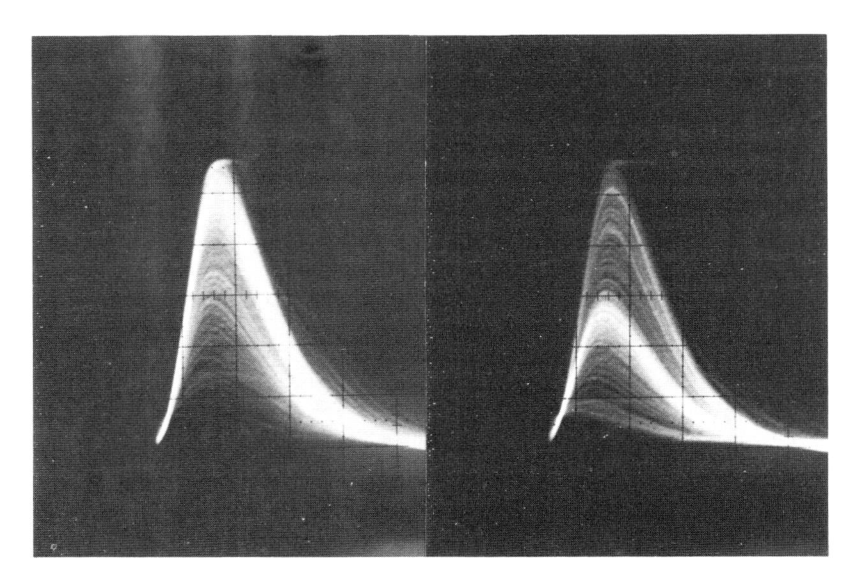

Fig. 7.2. Pulse height distribution of Cu K reflected in the second order from an LiF crystal (left), and of the adjacent background, which is mostly in the first order of reflection (right).

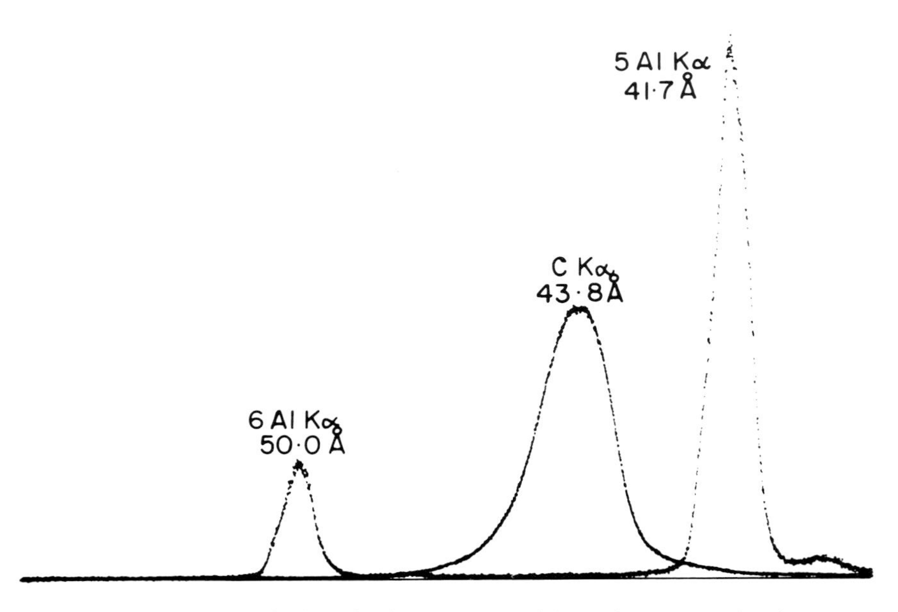

Fig. 7.3. Spectra obtained with a lead stearate crystal from aluminum and carbon targets. The peak obtained from the carbon band is broader than the high-order aluminum reflections.

the Kβ-line to the Kα-line increases slowly with the atomic number. The $K\alpha_1$-line is twice as intense as the $K\alpha_2$-line; this pair, however, is usually unresolved in the first order. The pattern of the L-lines is more complex and varies with atomic number. The most conspicuous M-lines, the $M\alpha_1$ and $M\beta_1$-lines, are approximately equal intensities.

3. It is important to keep in mind that, within the relations of intensities characteristic of each series (see Subsection 4.2.2), all lines of an element will be generated if the element is present and the shell is excited. Hence, if a tentative identification of a Kα-line is made, this identification must be in error if the corresponding Kβ-line does not appear at the expected intensity level. The same is, of course, true of other levels. Furthermore, within the limits given by the excitation potentials and the spectrometer sensitivity, all series of an element should be present if the element is a constituent of the specimen.
4. The operator should choose a standard set of conditions (voltage, beam current, crystals, and spectrometer slew speeds) and strictly adhere to them whenever a qualitative analysis is performed.

A wavelength scan with three crystals of a phosphate glass usable as a standard is shown in Fig. 7.4.

7.2 QUALITATIVE ANALYSIS BY ENERGY DISPERSION

Because the resolution of the energy-dispersive detectors is inferior to that of diffracting spectrometers (Figs. 5.21 and 5.25), some line interferences cannot be resolved satisfactorily with the solid-state detector, although they are avoided if the crystal spectrometer is used. Moreover, even in the absence of line interferences, the ratios of line to background are lower in energy-dispersive spectra; this tends to reduce the sensitivity of the method. In spite of these limitations, the lithium-drifted silicon detector has proven to be a valuable tool for qualitative electron probe microanalysis. It permits the identification of major constituents of the specimen in a few seconds or, at most, minutes, while taking the complete spectrum with diffractive spectrometers requires the irradiation of the specimen for about half an hour. The difference is due to the collection of photons within the complete data collection time over the entire wavelength range in the energy-dispersive detector. Due to the larger dead-time, the total (lines and background) count rate in the solid-state detector must be kept below 10^4 counts/sec. But in the crystal spectrometer the observation of any single line extends over only a few seconds, while in the energy-dispersive detector all lines are observed during the entire signal collection period. Therefore, in the absence of line interferences, the sensitivity of detection with the energy-dispersive detector is similar to that of the crystal spectrometer in the wavelength scan, in spite of the unfavorable line-to-background ratios. Because of its high efficiency, the solid-state detector is particularly useful when the generated spectral intensity is small, as in the analysis of thin films and small particles.

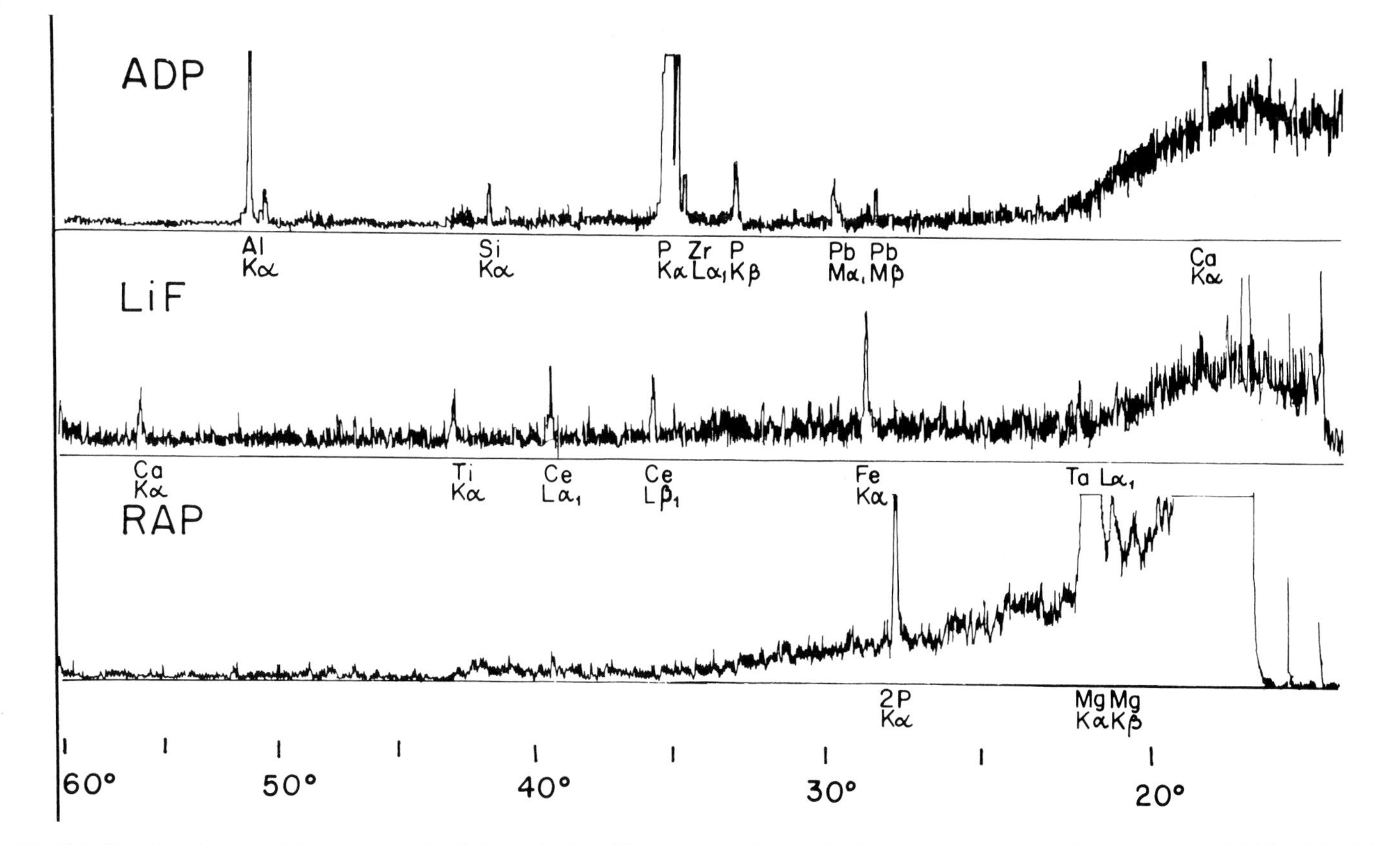

Fig. 7.4. Wavelength scan with three crystals (0.2 Å/min for LiF) of a glass of the following composition (mass fractions): MgO:0.086, Al_2O_3:0.109, SiO_2:0.003, P_2O_5:0.760, TiO_2:0.0035, Fe_2O_3:0.0035, ZrO_2:0.0054, CeO_2:0.0076, Ta_2O_5:0.0098, PbO:0.0099. Truncated peaks of major components exceed scale of recorder. For the composition of the crystals see Table 5.1.

The energy-dispersive use of flow-proportional detectors in qualitative analysis is less common, although they are quite useful in the range from 10 to 50 Å. In this region one observes the emissions of elements such as oxygen, nitrogen, and carbon, with surprising intensities (Fig. 7.5). Therefore, carbides, oxides, and similar compounds can be quickly identified. The interferences are not severe, since lines of shells above the K-level in this wavelength region are weak. Although lithium-drifted silicon detectors with a removable window can also be obtained commercially, their use is not advantageous, since in this region their resolution is about equal to that of a flow-proportional detector. They are sensitive to contamination; however, they do not adversely affect high vacuum, as flow-proportional detectors with permeable windows will.

Several manufacturers of Si(Li) detectors have issued slide rules giving the energies of the principal lines and of the edges which are very useful in practice. A table containing more lines (some of which cannot be observed in practice) has been prepared by Johnson and White [7.5].

7.3 THE LIMIT OF DETECTION

The limit of detection is the lowest concentration of an element at which it can be identified in a given type of specimen. In electron probe microanalysis, to identify an element at such a level of concentration is equivalent to distinguishing its characteristic X-ray emission from the background. Once we have determined which is the smallest signal that can be distinguished from the background, we must determine a calibration function, such as the one shown graphically in Fig. 7.6, which relates the concentration of an element in the specimen to the

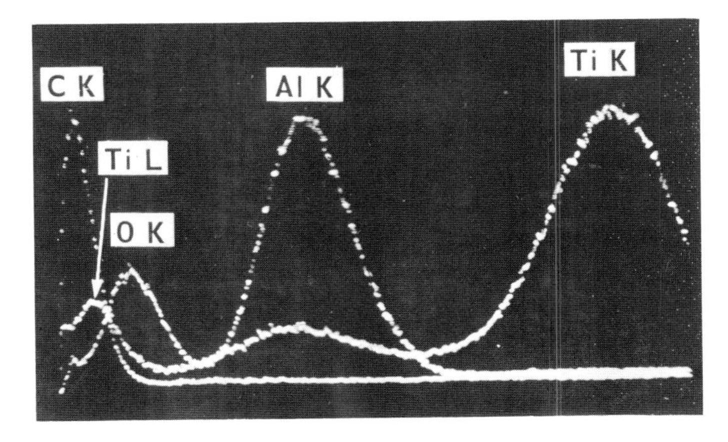

Fig. 7.5. Energy-dispersive spectra obtained with an Ar-CH_4 flow-proportional detector with a parlodion window, at an operating voltage of 20 kV. The targets used were titanium, carbon, and aluminum oxide.

signal this element produces. By means of this function, we can then determine the lowest concentration, or limit of detection, C_{ld}, which corresponds to the smallest detectable signal.

As will be discussed in detail in Chapter 9, the signal typically used in electron probe microanalysis is the ratio of characteristic emitted X-ray intensities from the specimen and from a standard. We will superscript the intensity from the specimen with an asterisk, and that from the standard with st; these intensities were corrected for background and for coincidence losses. Calling the ratio k, we obtain:

$$k = Y^*/Y^{st}.$$

Because these intensities were corrected for background, it follows that the signal for zero concentration must be equal to zero: $k(0) = 0$. Although in general the calibration function is not linear, calibration functions for concentrations up to 1-2% can always be assumed to be virtually linear for a given matrix (see Fig. 7.6):

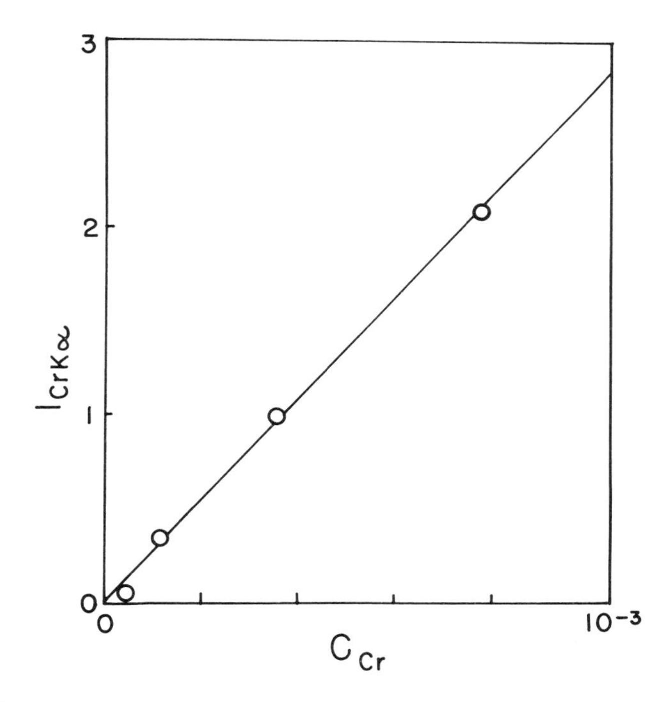

Fig. 7.6. Calibration curves for trace concentrations in a constant matrix are linear. Cr Kα measured in steel, at 30 kV, ψ = 52.5°. Ordinate:background-corrected intensity of Cr Kα, on an arbitrary scale.

$$C = \alpha k \tag{7.3.1}$$

so that $C_{ld} = \alpha k_{ld}$. The factor α is identical with the factor used in Chapter 8 to define the hyperbolic approximation.

The criterion of a limit of detection was initially coined as a measure of the sensitivity of qualitative chemical reactions. Such reactions typically lead to a binary decision: the element sought is either detected, or not detected. The concept of a limit of detection would seem to imply at first sight that the element in question can always be detected when present above this limit, and is never found if its concentration is below the limit of detection. But, even in qualitative reactions, there exists a region of uncertain detection. Within this region, the element is sometimes, but not always, detected. The probability of detection for a given concentration and test can be expressed by statistics. The definition of a limit of detection thus also defines statistical risks of a false positive or negative result when decisions are made for specimens in which the element to be identified is close to the limit.

A statistical uncertainty is also observed when the detection of an element depends on a measurement, or measurements, which produce numerical information. In general, several measurements are combined in an analytical procedure. The standard deviation of the combined result can be obtained by well-known statistical procedures (see Appendix, Eq. 14). To the extent that the detectability of traces of elements depends on these statistical fluctuations, Kaiser [7.6] states that "... the concept of the limit of detection of a chemical analytical procedure can be defined precisely with the aid of mathematical statistics." Kaiser [7.7] also concluded that "... only a statistical definition will hold and will give an objective number for a limit *which is independent of any personal judgment.*"

However, the distinction between line signal and background is limited not only by the statistical uncertainties in the measured X-ray intensities, but also by the systematic errors inherent in the procedure used to determine the background level which were discussed in Section 6.4. These systematic errors affect the accuracy of the value of k; and the more carefully the analyst tries to eliminate the counting errors, the more prevalent the systematic errors become in determining the uncertainty of the relative intensity ratio. Hence, it is incorrect in principle and in practice to define a limit of detection solely on the basis of statistical considerations as is so frequently proposed.

Both the statistical and the systematic components which define the limit of detection depend on the conditions of the measurement. Such conditions would include the instrument and its alignment, the time of measurement, the operating potential and beam current, the element and the line chosen to measure it, and the composition of both specimen and standard. To detect, for instance, a minor constituent, such as chromium in a low-alloy steel, we may perform a continuous wavelength scan and look for the Kα line of chromium, or we may collect an energy-dispersive spectrum with a solid-state detector. Alternatively, we can perform fixed-time counting measurements of the emission of the characteristic line with a crystal spectrometer. In each case the sensitivity, and hence the limit of detection, will be different. Further differences may arise from increasing the operating voltage, in which case we trade X-ray intensity for spatial resolution, or by using another line, such as Cr Lα, when the state of the specimen surface is of interest. Each of the mentioned procedures has its particular advantages. The wavelength scan and the energy-dispersion spectrum contain information about a large number of elements. The scan has relatively high resolution (hence the information is very specific); a complete scan, however, requires 20-40 min, while an energy-dispersion spectrum, of much lower resolution, can be obtained in 2 min. The fixed-time counting operation, which is the most sensitive alternative, only informs us about one single element. With each technique, the limit of detection can be lowered as more time is invested in the analysis. It is therefore impossible to define the sensitivity or the limit of detection of any analytical procedure without specifying all conditions which unequivocally define a complete analytical procedure. Common sense also dictates that a meaningful limit of detection correspond to instrumental conditions such as can be found in a laboratory dedicated to analytical service and to procedures and precautions such as can be followed in practice.

Let us first consider only the statistical aspects of the problem. Separating for each intensity the total count rate, Y_t, and the background count rate, Y_b, we obtain:

$$k = \frac{Y_t{}^* - Y_b{}^*}{Y_t^s - Y_b{}^s} . \tag{7.3.2}$$

If we assume that the same time interval is spent in obtaining each of the four count rates, we obtain:

$$k = \frac{N_t{}^* - N_b{}^*}{N_t{}^s - N_b{}^s} = \frac{u - v}{w - z} \tag{7.3.3}$$

where u, v, w, and z are the numbers of photons detected for each of the counting rates which define k. The variance of k can be obtained by combining the variances of these four variables [7.8]

$$\sigma_k^2 = \sigma_u^2 \left(\frac{\partial k}{\partial u}\right)^2 + \sigma_v^2 \left(\frac{\partial k}{\partial v}\right)^2 + \sigma_w^2 \left(\frac{\partial k}{\partial w}\right)^2 + \sigma_z^2 \left(\frac{\partial k}{\partial z}\right)^2$$

$$\frac{\partial k}{\partial u} = -\frac{\partial k}{\partial v} = (w - z)^{-1}; \qquad \frac{\partial k}{\partial w} = -\frac{\partial k}{\partial z} = (u - v)(w - z)^{-2} \tag{7.3.4}$$

hence

$$\sigma_k^2 = (w - z)^{-4} \; [(\sigma_u^2 + \sigma_v^2)(w - z)^2 + (\sigma_w^2 + \sigma_z^2)(u - v)^2]. \quad (7.3.5)$$

As the concentration of the element in question approaches zero, the total count becomes equal to the background count, and the standard deviations of these counts also become equal. Hence

$$u \to v, \qquad \sigma_k^2 \to (\sigma_u^2 + \sigma_v^2)(w - z)^{-2} \to 2\sigma_u^2 (w - z)^{-2}. \qquad (7.3.6)$$

If the background count z is small in comparison with the total count on the standard, w, we get:

$$\sigma_K(0) \simeq \sqrt{2}\,\frac{\sigma_b{}^*}{N_t{}^s}. \qquad (7.3.7)$$

If we can assume that the distribution of $N_b{}^*$ is Poissonian, we obtain:

$$\sigma_K(0) \simeq \frac{\sqrt{2N_b{}^*}}{N_t{}^s}. \qquad (7.3.8)$$

If we call $1/\alpha$ the initial slope of the calibration curve ($k/C \xrightarrow[C \to 0]{} 1/\alpha$), we can obtain an expression for the standard deviation of the measured concentration at the zero level of concentration, $\sigma_C(0)$:

$$\sigma_C(0) = \alpha\frac{\sqrt{2}\sigma_b{}^*}{N_t{}^s} \simeq \alpha\frac{\sqrt{2N_b{}^*}}{N_t{}^s}. \qquad (7.3.9)$$

The coefficient α is identical with that of the hyperbolic approximation which will be introduced in Section 8.2 [Eq. (8.2.4)]. The subject will be further pursued at the end of Chapter 8.

Several authors [7.8]-[7.12] have proposed limits of detection which are defined by a constant with which the standard deviation of the measured concentration at the zero level, $\sigma_C(0)$, is multiplied. Thorough analyses of the statistics of trace analysis leading to such definitions were performed by Kaiser [7.6], [7.7] and by Currie [7.12]. We will briefly describe their concepts, as they apply to electron probe microanalysis. Both authors assume that no systematic error has been committed in the determination of the signal at the zero concentration level, $k(0)$. Hence, for an analytical signal to signify detection, it must exceed $k(0)$ by a multiple of the standard deviation $\sigma_k(0)$. The value of this multiplication factor, ν, determines the confidence level of the detection. The one-sided confidence levels for various values of ν are as follows:

ν :	0.85	1.28	1.64	2.00	2.32	3.00
confidence level (%) :	80	90	95	97.9	99	99.9

The authors propose that, in order to define a limit of detection, all parts concerned should agree to a certain confidence level, and hence to the corresponding statistical risk of erroneous decision. Such an agreement may not be universally acceptable since the risk the investigator is willing to accept may depend on the penalties incurred in case of an erroneous decision. Kaiser proposes, however, that a reasonable value is: $\nu = 3$. Therefore, a value of the signal, k, is considered significant if it exceeds the background by at least $3\sigma_k(0)$

$$k \geqslant k_{\text{ld}} \equiv 3\sigma_k(0).$$

The value k_{ld} above which the signal is considered genuine is called by Kaiser the measure at the limit of detection. The corresponding limit of detection can be obtained by means of the analytical calibration function:

$$C_{\text{ld}} = \alpha k_{\text{ld}} = 3\alpha\sigma_k(0) \simeq 3\alpha(2N_b{}^*)^{1/2}/N_t{}^s. \tag{7.3.10}$$

Let us now assume that we analyze a specimen that contains a concentration equal to the limit of detection, C_{ld}, of the element in question. In the absence of a systematic bias the measurements of such a specimen will center around the measure at the limit of detection $k_{\text{ld}} = 3\sigma_k(0)$. Hence, half of the analyses of the specimen will yield a "positive" result ($C > C_{\text{ld}}$), while the other half will indicate "no detection" ($C < C_{\text{ld}}$). Such a low probability at the limit of detection is puzzling: this limit is certainly not "the lowest concentration at which the element in question can be identified." At what concentration level would the risk of a false negative be negligible (at the 99.7% confidence level)? As indicated in Fig. 7.7, this level, C_g, would be higher than C_{ld} by three times its standard deviation, $\sigma_C(g)$, which we assume not to differ drastically from $\sigma_C(0)$:

$$C_g = 3\sigma_C\,(0) + 3\sigma_C\,(g) \simeq 6\sigma_C\,(0). \tag{7.3.11}$$

In terms of X-ray counts, and admitting Poissonian statistics, we obtain for this second level:

$$C_g = 6\sqrt{2}\alpha\frac{\sqrt{N_b{}^*} + 3/2}{N_t{}^s} \simeq 6\sqrt{2}\alpha\frac{\sqrt{N_b{}^*}}{N_t{}^s}. \tag{7.3.12}$$

Kaiser calls this second limit the "limit of guarantee of purity," because a signal smaller than $k(0) + 3\sigma_k(0)$ guarantees, within the chosen confidence level, that the concentration of the element in question is below C_g. Currie, instead, pro-

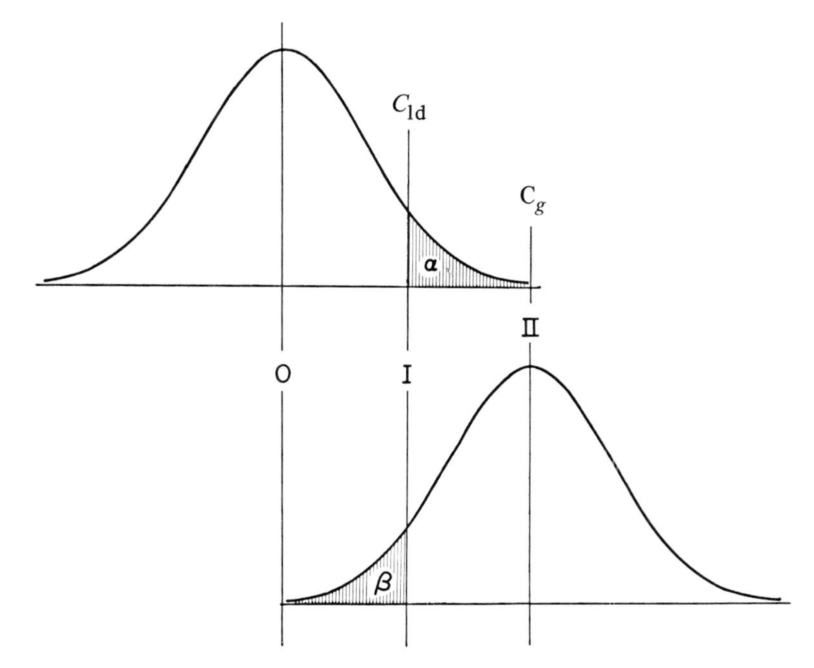

Fig. 7.7. Limit of detection. The horizontal axis represents concentration. If a specimen has zero concentration of the element which is being measured, the measured values have a distribution centering about zero, unless the method is biased. If we accept a chance of a false positive equal to the area α, the level C_{ld} is determined by the distribution around zero. If the specimen has a real concentration of C_g, the probability of obtaining a false negative ($C_{measured} < C_{ld}$) is given by the area β.

poses that this limit be called limit of detection, and that C_{ld} be called the "decision limit."

Currie [7.12] reviewed a series of proposed definitions for the limit of detection. The range of these, expressed in terms of the multiplier ν, goes from 1 to 20. One must conclude that there is no consensus as to what should be a reasonable statistical limit of detection.

As efforts are made to lower the limit of detection of an analytical procedure, by increasing counts, the effect of Poissonian statistics diminishes, and other random and systematic factors exert a significant effect on the precision and accuracy of the measurement. The Poissonian limit of detection must therefore be considered a theoretical and optimistic limit, which could not be surpassed even in ideal conditions of calibration and stability, but which may not be attainable in practice. Some authors have proposed including the effects of non-Poissonian error by arbitrarily increasing the value of the multiplier ν. There is, however, no good way to decide how large this increase should be. For any realistic evaluation, the estimate of the standard deviation of the measurement must be obtained by comparing the results of repetitions of the measurement. Abbreviated procedures based on the range of a small (2-10) number of measurements are useful, and certainly preferable to mere prediction of Poissonian statistics (see the Appendix). But even if a realistic estimate of *precision* (i.e., of the standard deviation) has been obtained, the possibility of systematic bias (such as in the background estimate in energy-dispersive analysis) must be considered in the estimate of *accuracy*.

In view of these difficulties it seems preferable not to use the concept of "limit of detection." Instead one should, as usual in other measurements, give estimates of both the standard deviation of the concentration at the zero level, and of the systematic error [7.13]. The second estimate, however, requires a personal judgment on the part of the analyst. The facts of life which apply here were succinctly stated by Lundell: ". . . the analyst sells an intangible commodity, an opinion" and ". . . all results are matters of opinion rather than fact, and so the true result is never known" [7.14].

7.4 ANCILLARY TECHNIQUES

The experienced analyst will frequently obtain information from signals other than X-rays. One of the most important ancillary tools of electron probe microanalysis, the use of reemitted secondary and scattered electrons, will be discussed in detail in Sections 9.2 and 14.3.

The observation of the illuminated specimen through the optical microscope, with polarized light where applicable, provides an indispensable adjunct and preliminary step. The visual observation is frequently the guide for selecting the areas to be investigated with the electron beam, with collection of X-rays (scanning electron microscopy is an alternative frequently used); it is particularly useful in the study of minerals and rocks. For this reason, the requirements for the quality of the microscope are particularly stringent when such specimens are investigated.

The optical microscope also permits the observation of the emission of light in the visible range as a consequence of the electron beam impact on the specimen (*cathodoluminescence,* see color plates [7.15]). Such radiation can also be emitted in the infrared or ultraviolet regions. While the visual observation permits a rapid qualitative evaluation of the bright and frequently beautiful light emissions in various colors, a more detailed study, particularly of the emission outside the visible range, requires the use of light detectors, and, usually, of monochromators. The optics of the light microscope can be used to collect the cathodoluminescent emission over a wide angle or a lightpipe can be employed for the same purpose. Many investigations of cathodoluminescence are performed with scanning electron beams. Therefore, the subject will be pursued further in Chapter 14.

Another interesting ancillary technique used by Zähringer [7.16] involves the use of a residual gas analyzer to observe gases emitted from the specimen as a consequence of irradiation with the electron beam.

7. REFERENCES

7.1 Weber, K. and Marchal, J., *J. Sci. Instrum.* **41,** 15 (1964).
7.2 Heinrich, K. F. J., *Adv. X-Ray Analysis* **4,** 370 (1961).
7.3 Riggs, Jr., F. B., *Rev. Sci. Instr.* **34,** 912 (1963).
7.4 Heinrich, K. F. J. and Giles, M. A. M., NBS Tech. Note 406, National Bureau of Standards, U.S. Dept. of Commerce, Washington, D.C., 1967.
7.5 Johnson, Jr., G. G. and White, E. W., ASTM Data Series DS 46, American Society for Testing and Materials, Philadelphia, Pa., 1970.
7.6 Kaiser, H., *Z. Anal. Chemie.* **209,** 1 (1965).
7.7 Kaiser, H., in *Trace Characterization,* Meinke, W. W. and Scribner, B. F., Eds., NBS Monograph 100, National Bureau of Standards, U.S. Dept. of Commerce, Washington, D.C., 1967, p. 153.
7.8 Heinrich, K. F. J., *Adv. X-Ray Analysis* **3,** 95 (1960).
7.9 Kaiser, H. and Specker, H., *Z. Anal. Chemie* **149,** 46 (1956).
7.10 Ziebold, T. O., *Anal. Chem.* **39,** 858 (1967).
7.11 Liebhafsky, H. A., Pfeiffer, H. G., and Zemany, P. D., in *X-Ray Microscopy and X-Ray Microanalysis,* Engström, A., Cosslett, V., and Pattee, H., Eds., Elsevier, Amsterdam, 1960, p. 321.
7.12 Currie, L. A., *Anal. Chem.* **40,** 586 (1968).
7.13 Eisenhart, Ch., *Science* **60,** 1201 (1968).
7.14 Lundell, G. E. F., *Anal. Educ., Ind. Eng. Chem.* **5,** 221 (1933).
7.15 Smith, J. V. and Stenstrom, R. C., *J. Geology* **73,** 627 (1965).
7.16 Zähringer, J., *Earth Planet. Sci. Lett.* **1,** 20 (1966).

PART IV.
QUANTITATIVE ANALYSIS

8. The Empirical Approach to Quantitation

Unlike the classical analytical procedures based upon stoichiometry, or the colorimetric techniques which rely on Beer's law, X-ray spectrometry cannot be treated by a simple theoretical model. The absorption of X-rays follows Beer's law, but it depends strongly upon the specimen composition, due to the wide variations of the X-ray attenuation coefficients. Further important complications arise in the generation of X-rays, which depend on the balance between the deceleration of the impinging electrons and the ionization of the target atoms. In view of the complexity of the process of X-ray emission, it is easy to conclude that considerable simplicity would be gained if the theoretical considerations could be dispensed with, and the analysis could be based entirely upon empirical calibration. If appropriate standard reference materials are available, empirical calibration is a viable and powerful procedure. Furthermore, the empirical study of the analytical calibration curve is useful even if the lack of standards forces us to perform the theoretical data reduction procedures first advocated by Castaing [1.1]. For these reasons we will start our discussion of quantitative electron probe microanalysis with the empirical approach.

8.1 THE CALIBRATION FUNCTION

The use of weight as a measure of quantity of material originated in the time-honored practice of weighing merchandise. In science, the weight fraction of an element is a convenient measure of its concentration when we discuss phenomena related to the mass of the atoms (which resides almost entirely in the nuclei). Molar fractions are more appropriate when dealing with interactions involving orbital electrons. Since the generation of characteristic X-rays is related to such interactions, the properties of characteristic X-rays are related to the atomic number, Z, rather than to the atomic weight, A [see, for instance, Moseley's law, Eq. (4.2.3)]. For the same reason, it will be simpler in principle to relate the generation of characteristic X-rays in multielement targets to molar fractions rather than to mass fractions. In this sense, X-ray spectrochemical analysis shows its chemical character.

From the preceding considerations it should be clear that any law according to which emitted X-ray intensities should be proportional to mass fractions cannot have solid physical foundations. It is important to underline this fact, since, due to fortuitous compensations, approximate linearity is often, but not always, observed. In the past it was sometimes erroneously believed that such a law could be postulated for the generation or even for the emission of characteristic X-rays, and due to this tradition, most electron probe analysts discuss and report concentrations in weight fractions rather than in atomic or molar fractions.

The relation between the observable signal and the corresponding concentration is called the *analytical calibration function*, and its graphic representation is called the *analytical calibration curve*. Once the X-ray intensities emitted by the specimen have been measured, as described in the previous chapter, the problem of quantitation is reduced to finding the analytical calibration function and then to calculating the concentration of the emitting element.

To be useful for the analytical calibration of any binary system, the calibration function must be monotonic, so that there is one, and only one, value of the concentration for any experimental value of the signal. In electron probe microanalysis, the concentration is usually expressed as a mass fraction. (If C is the mass fraction, $0 \leqslant C \leqslant 1$.) The signal is usually the relative X-ray intensity, k, mentioned in the previous chapter. With these parameters a continuous monotonic calibration function is always achievable. Elements are often used as standards, and even if this is not possible for practical reasons, we can calculate from theory, as will be explained in Chapter 9, the intensity we would have obtained had the pure element been used as a standard. Hence, in the definition of the *relative X-ray intensity*, k,

$$k = \frac{Y^*}{Y^s}$$

we will henceforth use the symbol k, without superscript, for a ratio which refers to the elementary standard. We recall that count rates (Y) are corrected for dead-time and background. With the above definition, and if the analytical function is monotonic, we obtain for k the range $0 \leqslant k \leqslant 1$. In summary, the analytical calibration function $k = f(C)$ can be represented by a monotonic curve going from the origin (0,0) to the value (1,1). A straight line connecting these points is called a linear calibration curve ($k = C$). Real calibration curves are either convex or concave (Fig. 8.1). Although perfectly linear curves cannot be obtained, it is desirable that the analytical curve should approach linearity, so that it contains no regions of shallow slope, and hence low sensitivity. (Sensi-

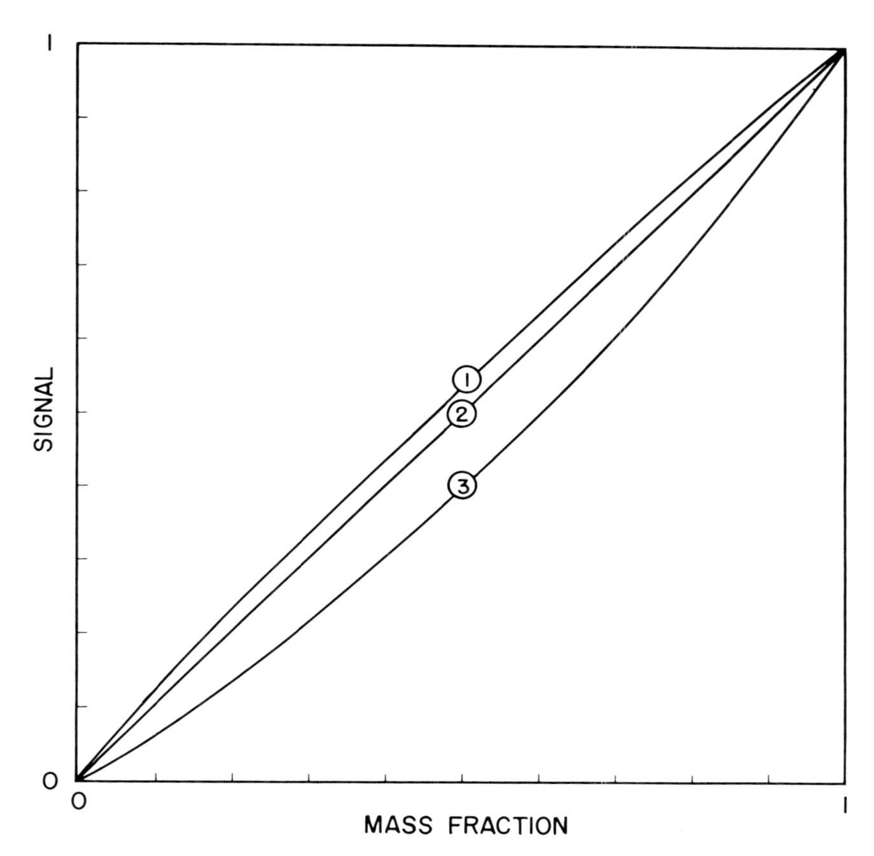

Fig. 8.1. Possible shapes of the calibration curve. The convex curve (1) can result from atomic number effects. Concave curves (3) are usually due to absorption of primary X-rays in the specimen. Linear calibration curves (2) are a limiting case hardly ever found in practice.

tivity can be defined as the first derivative of the calibration function dk/dC [7.7].)

In a ternary system constituted by the elements *a*, *b*, and *c* we obtain for each element (e.g., *a*) a family of calibration curves which fall between those for the binary systems (*a-b*, and *a-c*). Unless these curves happen to coincide, the signal of an element depends at any given concentration on the mass fractions of the other elements present. This dependence is called matrix effect, or interelement effect. It should be noted that for a ternary or higher system, the ratio of the weight fraction to the atomic fraction of any element is also composition dependent. Hence, the magnitude of the matrix effect depends on the way in which compositions are defined.

For systems containing more than two or three elements, graphic representation of the calibration equations is impractical, and algebraic models ("corrections") become indispensable. The term "correction" reflects the traditional belief in an ideal law of linear calibration curves, but it is impossible to eradicate this misnomer. The number of standards required for empirical characterization of a system also increases rapidly with the number of elements present. However, concentrations of constituent elements up to 0.01 (weight fraction) can be neglected in empirical or theoretical calibration corrections.

8.2 THE HYPERBOLIC APPROXIMATION

If it could be assumed that in a given system the relative signal k_a emitted by element a were equal to the concentration (i.e., weight fraction) of this element, C_a, then the calibration equation could be formulated as follows:

$$k_a = \frac{C_a}{\Sigma C_i} = C_a \text{ since } \Sigma C_i = 1. \tag{8.2.1}$$

The summation over the i elements present in the system includes the emitting element, a. This linear calibration algorithm can be simply extended to nonlinear calibration curves if it can be assumed that, for the particular conditions of analysis, the concentration of each element (in whatever scale has been chosen) must be multiplied by an efficiency factor, α_i, characteristic for this element but independent of the specimen composition. We obtain [8.1]:

$$k_a = \frac{\alpha_a C_a}{\Sigma \alpha_i C_i}. \tag{8.2.2}$$

This model is widely used in X-ray fluorescence spectrometry [8.2], and is formally identical to the so-called third approximation of Castaing [3.22] which was, however, developed for generated rather than emerging X-ray intensities. Equation (8.2.2) does not fully define the values of the factors α_i, since k_a will not change when all factors are multiplied by a constant.

For a binary system (a,b), Eq. (8.2.2) transforms into

$$\begin{aligned} C_a/k_a &= \alpha + (1 - \alpha)C_a, \qquad \alpha = \alpha_b/\alpha_a \\ k_a/C_a &= 1/\alpha + (1 - 1/\alpha)k_a \end{aligned} \tag{8.2.3}$$

or

$$\frac{C_a}{1 - C_a} = \alpha \frac{k_a}{1 - k_a}. \tag{8.2.4}$$

Mathematically, these equations represent hyperbolas [8.2, Appendix]; therefore, the assumption concerning the calibration equation implicit in Eqs. (8.2.2) through (8.2.4) is called the *hyperbolic approximation*. The symmetry of this approximation is conserved if one passes from weight fractions to molar fractions. Since $\Sigma m_i = 1$, we obtain:

$$m_a = \frac{m_a}{\Sigma m_i} = \frac{C_a/A_a}{\Sigma C_i/A_i}. \tag{8.2.5}$$

If

$$k_a = \frac{\alpha_a C_a}{\Sigma \alpha_i C_i} \quad \text{and} \quad C_a = m_a \cdot A_a \cdot \Sigma C_i/A_i$$

then

$$k_a = \frac{\alpha_a m_a A_a \Sigma C_i/A_i}{\Sigma \alpha_i m_i A_i \Sigma C_i/A_i} = \frac{\alpha_a A_a m_a}{\Sigma \alpha_i A_i m_i}. \tag{8.2.6}$$

In passing from the weight fractions to molar fractions we have merely changed the hyperbolic coefficient from α_a to $\alpha_a \cdot A_a$. Hence, if the hyperbolic approximation is admissible for use with weight fractions, it will be equally acceptable when molar fractions are used.

The hyperbolic approximation, when applicable, is particularly useful in extended investigations within a binary composition range. It is not unusual to perform hundreds of determinations on a single binary diffusion couple, and to analyze many couples in the course of a study. In such a case it may be uneconomical to apply the full correction procedure (to be described in Chapters 9 and 10) to all analyses. Once the coefficient α has been determined, a simple correction calculation can be performed:

$$C = \frac{\alpha k}{1 + k(\alpha - 1)}. \tag{8.2.7}$$

The factor α can be determined experimentally if one or several dependable standards pertaining to the system are available. This is a reasonable requirement when a laboratory is dedicated to an extensive study of a binary system. If such standards are not available, the factor can be calculated by obtaining k for a given value of C, by means of the theoretical procedures which will be described in Chapters 9 through 12.

Although we have, for simplicity, omitted double subscripts to the factor

α, its value depends on two elements (the element a which is being determined, and the element i which affects the intensity of emission of the line of element a). It also depends on the line which is being measured, on the operating voltage, and on the angles of electron beam incidence and X-ray emergence. For these reasons, caution must be observed when applying coefficients determined with one instrument to measurements performed with another.

The shape of the analytical curve is related to the value of α. If this factor is below one, the curve is convex ($k > C$). If it is larger than one, the curve is concave ($k < C$). Equations (8.2.3) suggest a useful alternative to plotting the analytical curve (k as a function of C). These equations indicate that C_a/k_a is a linear function of C_a, varying from the value of α for concentrations approaching zero, to unity when C_a approaches one. Analogously, k_a/C_a changes linearly from $1/\alpha$ to one, as a function of k_a. Therefore, on graphs showing either C_a/k_a as a function of C_a, or k_a/C_a as a function of k_a, nonlinear analytical curves which follow the hyperbolic model are represented by straight lines, and the vertical axis of such graphs can be expanded to any desired degree so that small deviations from the curve can be seen accurately (Fig. 8.2). For the analysis of binary targets the graph representing k_a/C_a as a function of k_a is the more useful. We will therefore describe it in more detail (Fig. 8.3). Any point such as $(k_a, k_a/C)$ on the lower graph represents a concentration C on which the intensity ratio k_a was obtained. If the binary system under consideration follows the hyperbolic approximation, the coordinates for all points of the system define a straight line passing through the point (1,1). The line passing from point (0,1) to (1,1) corresponds to a linear calibration curve in the upper graph ($k = C$). For a convex calibration curve (such as k_a in the graph), the corresponding calibration line in the lower graph is above the line from (0,1) to (1,1). For a concave calibration curve, the line is below that connecting (0,1) with (1,1). The intercept of the calibration line of the lower graph with the vertical at the origin ($k = 0$) gives the value of $1/\alpha$. The abscissa of the point $(k, k/C)$ is equal to k. At the intercept with ($k/C = 1$) of the line connecting the points (0,0) and $(k, k/C)$ (point P of Fig. 8.3) we can read the value of C on the same linear scale used for k. Hence we can obtain from the same graph the values for k, C, and $1/\alpha$. The graph of Fig. 8.3 shows these relations both for $k < C$ and for $k > C$. The same considerations are valid for the graph of Fig. 8.2, if C and k are interchanged, and $1/\alpha$ is replaced by α.

Figures 8.2. and 8.3 further illustrate the relations between the conventional plot of k versus C and those corresponding to Eqs. (8.2.3). Within the upper square, the point (C, k) of Fig. 8.2 has the coordinates C (abscissa) and k (ordinate). By tracing a line at 45° from point 2 of Fig. 8.2, we obtain point 3 which is distant from the origin (0,0) by the measure of k. A vertical line projects this distance into the lower graph. If the vertical axis of the lower graph represents C/k, then the connection between point 4, on the line ($C/k = 1$), and

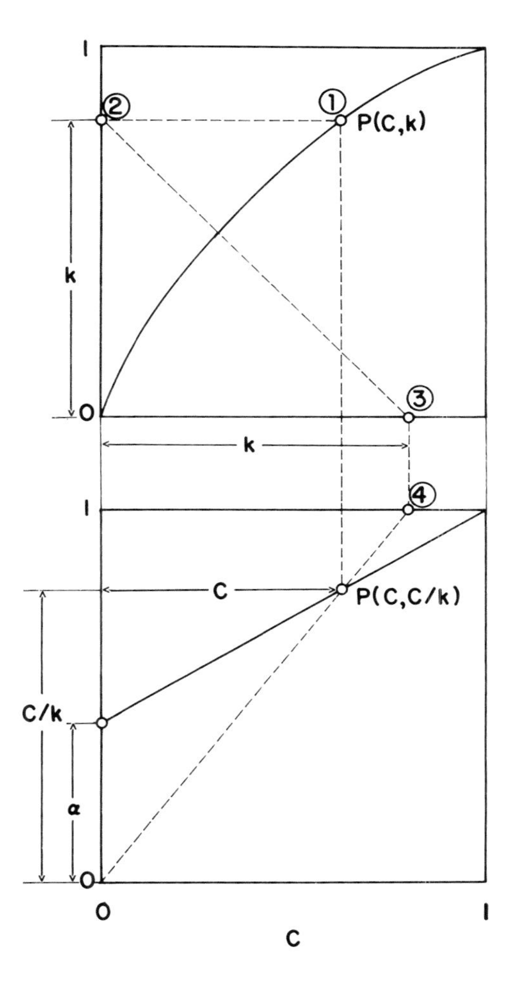

Fig. 8.2. Analytical curves of a binary system in which the hyperbolic approximation is accurate. Above: the conventional plot of k as a function of C. Below: plot of C/k versus C. The calibration curve is a straight line through the points $(0, \alpha)$, $(C, C/k)$, and $(1,1)$. Dotted lines explain the passage from one plot to the other.

the origin (0,0), contains all points corresponding to the value k. A vertical dropped from the upper graph through point (C,k), intersects this line and so defines the point $(C,C/k)$. If, to the contrary, the vertical scale of the lower graph represents k/C, as in Fig. 8.3, then the line connecting the points (0,0) and $(C,1)$ contains all points corresponding to C. The intersection of this line with the vertical dropped from point 3 (upper graph) defines the point $(k,k/C)$.

How accurate is the hyperbolic approximation? Before answering this ques-

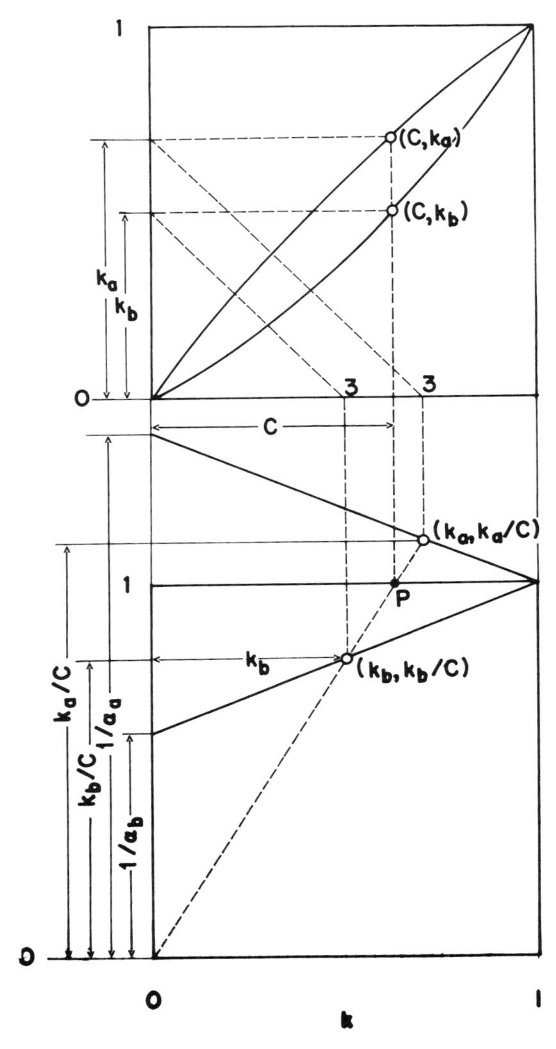

Fig. 8.3. Two calibration curves, plotted conventionally, and as k/C versus k, for binary systems in which the hyperbolic function is accurate. A convex calibration curve (k_a) produces in the k/C plot a straight line which falls with increasing k, while for a concave calibration curve (k_b) the line rises with increasing k. The broken line from (0,0) to (k_a,k_a/C) in the lower graph is the locus of all calibration curves for the concentration C.

tion, we must consider the factors which determine the intensity of emitted characteristic radiation. If we assume that the contribution of the continuous radiation has been taken into account by means of a background correction, then the number of photons, N, observed in a time period t by a detector of efficiency of detection P which covers a solid angle $\Omega/4\pi$ of the emitted radiation, is equal to

$$N = Y't = \frac{\Omega}{4\pi} \cdot P \cdot i_b \cdot t \cdot (I_p f_p + \Sigma I_f f_f). \tag{8.2.8}$$

In this equation, i_b is the beam current intensity, (electrons/sec)[1], I_p is the probability of production of a primary X-ray photon per incident electron, I_f is the probability of production of a fluorescent X-ray photon per incident electron, and the factors f_p and f_f (absorption factors) are the probabilities that a photon (primary or fluorescent) emitted towards the angle covered by the detector escapes absorption and thus enters the detector. The summation sign indicates that the fluorescent excitation may be due to more than one primary line, or also to continuous radiation.

The solid angle and efficiency of the detector are the same for the specimen and the standard, as is—within limits of stability—the beam current, and these parameters thus cancel in the calculation of a relative intensity. The time intervals of data collection are well known. The accurate determination of the parameters I_p, I_f, f_p, and f_f will be extensively discussed in Chapters 9 through 11. At this point it is sufficient to know that the values of the generated intensities I_p determine to a large extent the observed total intensities. The largest deviations from linearity in the calibration curve are due to the absorption factor of primary radiation, f_p. For this practical reason, the "absorption correction" is the most important (see Chapter 10). Differences among the atomic numbers of the constituents usually cause moderate nonlinearity of the calibration curve. Nonlinearity due to the generation of secondary X-rays is significant only in specimens which emit hard radiation. It occurs mainly in the presence of elements of atomic number between 20 and 30. The conditions conducive to fluorescent emission will be discussed in Chapter 11.

Both the theory and experiments indicate that the effects of primary absorption, which may be very significant, can be described with high accuracy by the hyperbolic approximation (Figs. 8.4 and 8.5). The approximation is somewhat less successful in the case of large differences of atomic number (Fig. 8.6), and fails significantly in the presence of strong fluorescent excitation (Fig. 11.11). For these reasons, the approximation can be used with great success in the analysis of materials of relatively low atomic number such as silicates and many other minerals. On the other hand, it does not produce accurate results in the analysis of materials such as ferrous alloys, in which strong fluorescent excitation is frequently present.

---

[1] 1 A = 6.25×10^{18} electrons/sec.

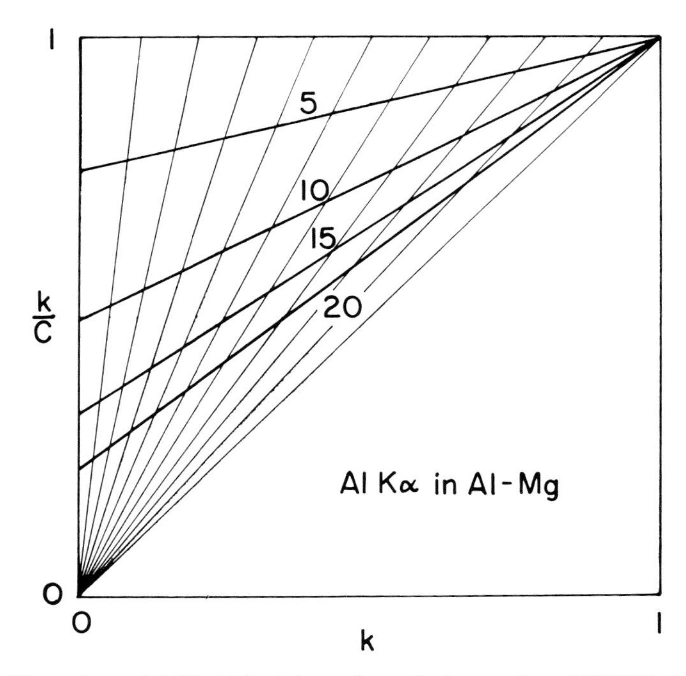

Fig. 8.4. The values of k/C obtained by a theoretical procedure (MULTI8, [4.42]) are linear with respect to k, as predicted by the hyperbolic approximation. (X-ray emergence angle $\psi = 30°$.) The numbers above the calibration curves indicate the operating potentials in kilovolts.

From a practical point of view it is important to obtain all the coefficients α_i required in Eq. (8.2.2). Their empirical determination is particularly cumbersome in complex targets, such as natural silicates in which several elements are present. Ziebold and Ogilvie [8.1], who first advocated the use of the hyperbolic approximation in electron probe microanalysis showed that the theoretical models of X-ray generation are compatible with the form of hyperbolic correction. Bence and Albee [8.4] described how the coefficients α_i can be obtained, for the analysis of minerals, by means of theoretical formulations. Further studies of the empirical method, with the introduction of additional correction terms, were performed by Laguitton et al. [8.5].

Internal Consistency of Analytical Measurements. Figure 8.5 shows how the internal consistency of a series of measurements of X-ray intensities from specimens of presumably known compositions can be tested. Obviously, whatever effects exert an influence on the shape of the calibration curve, they will not abruptly change at any concentration level. Therefore, even if a calibration curve of k/C versus k is not linear, it must be continuous. In the example shown, the discontinuities of the calibration curves between the 0.500 and 0.519 alloys

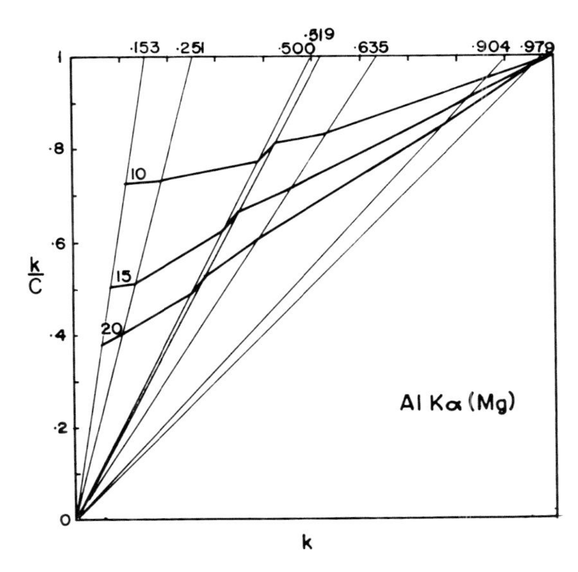

Fig. 8.5. Same system as in Fig. 8.4. Experimental results, obtained by Goldstein et al. [8.3]. Numbers above the graph indicate the composition of the alloys used in the experiments, in weight fractions.

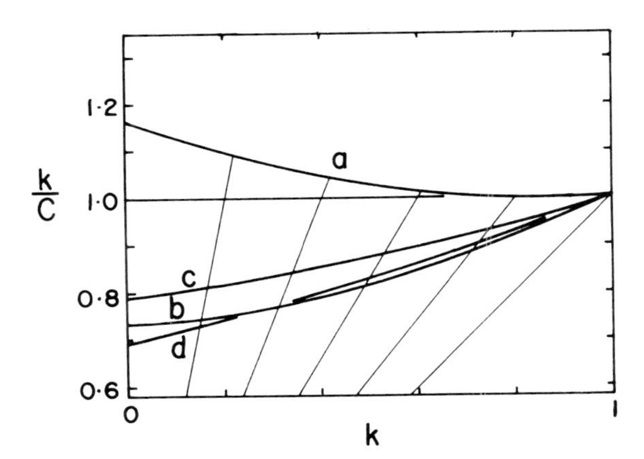

Fig. 8.6. Analytical calibration curves predicted by the theoretical program MULTI8 for binary aluminum-gold alloys. a:Al Kα at 10 kV, b:Al Kα at 20 kV, c:Au Lα at 30 kV, and d:Au Lα at 20 kV. The X-ray emergence angle ψ is 52.5°.

are similar at all three operating potentials. It may be concluded that they are not the result of faulty measurement or statistical factors, but that at least one of the postulated concentrations must be in error. Similar consistency tests are shown in [8.6] in which the proven inconsistency of sets of postulated compositions is used to exclude suspect measurement results from a statistical study of error in electron probe microanalysis. The tests can also be performed by plotting k/C as a function of operating voltage, as shown in [8.6] and in Fig. 8.7 (see also Fig. 12.9).

Further consideration is given to the empirical method in Chapter 12, where its accuracy is discussed in more detail.

Standard Deviation of the Concentration on the Zero Level of Concentration. The consequences of Eq. (7.3.8) can now be further investigated. We can express the intensity of the characteristic emission from the standard by means of Eq. (8.2.8). Ignoring the contribution from the background, we have:

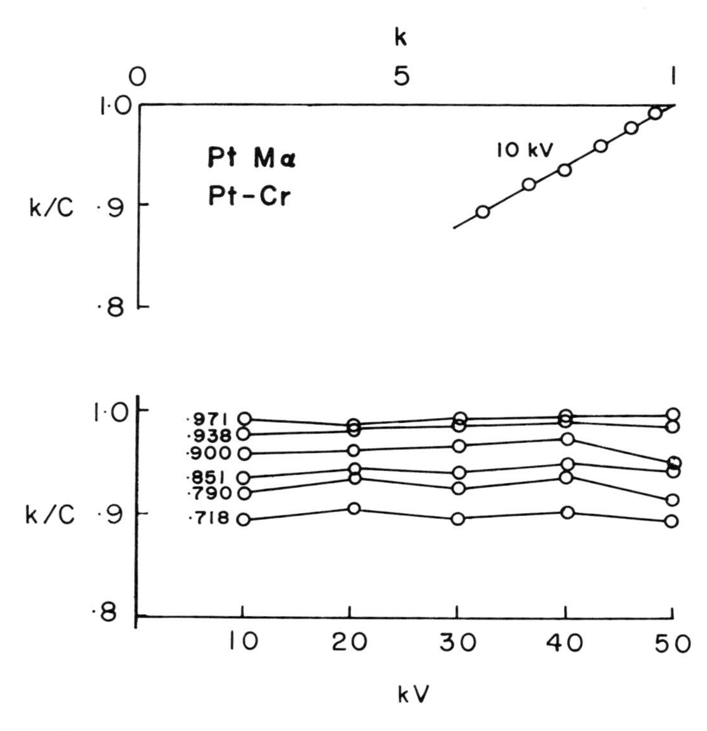

Fig. 8.7. Consistency tests on measurements of the Pt Mα line emitted by several binary Pt-Cr alloys. The consistency is proven in comparisons of measurements of several alloys at one operating potential (above) as well as by plotting for each alloy the ratio k/C versus the operating potential (below).

$$N_t^s = \frac{\Omega}{4\pi} P \cdot i_b \cdot t \cdot I^s f^s.$$

Analogously, for the background,

$$N_b{}^* = \frac{\Omega}{4\pi} P \cdot i_b \cdot t \cdot I_b{}^* f_b{}^*.$$

We thus obtain for the standard deviation of C at the zero level:

$$\sigma_{C=0} = \alpha \left[\frac{\Omega}{4\pi} P \cdot i_b \cdot t\right]^{-1/2} [2I_b{}^* f_b]^{1/2} [I^s f^s]^{-1}. \tag{8.2.9}$$

According to Kramers' Eq. (6.3.7) the continuum intensities, and hence, ideally, the background intensities of two targets are related, for any wavelength, to the mean atomic numbers of these targets by

$$I_b{}^*/I_b{}^s = \bar{Z}^*/Z^s. \tag{8.2.10}$$

We will also introduce in the expression for $\sigma_{c=0}$ the line-to-background ratio of the emission from the elemental standard, r^s, which is defined by

$$r^s = \frac{I^s f^s}{I_b{}^s f_b{}^s}. \tag{8.2.11}$$

By combining Eqs. (8.2.9), (8.2.10), and (8.2.11), we obtain the following expression:

$$\sigma_{C=0} \simeq \alpha \left(\frac{1}{2} \cdot \frac{\Omega}{4\pi} P \cdot i_b \cdot t \cdot r^s \cdot I^s f^s \cdot \frac{Z^s}{\bar{Z}^*} \cdot \frac{f_b{}^s}{f_b{}^*}\right)^{-1/2}. \tag{8.2.12}$$

This equation provides useful indications concerning the factors which determine the sensitivity of detection of an element. The parameters α and $(f_b{}^s/f_b{}^*)^{1/2}$ are usually fairly close to unity. The ratio $Z^s/\bar{Z}^*$ reflects the dependence of the sensitivity upon the background level due to the continuum which grows with the atomic number. The sensitivity is seen to improve when the solid angle subtended by the detector, the efficiency of the detector, the beam current, and the duration of each measurement, increase. Furthermore, the sensitivity improves as the line-to-background ratio for the element of interest increases, and hence with the resolution of the spectrometer. Finally, the statistical uncertainty of the measurement diminishes as the probability of emission of a photon per electron for the standard element, $I^s f^s$, increases. This relation explains the increase in sensitivity which is usually obtained if the operating voltage, and hence I^S, increases. For very soft radiation, however, the absorption factor, f^S, may decrease more rapidly with voltage increase than the increase of I^S (see Fig. 6.19). Equation (8.2.12) can also be transformed, with the simplifications indicated above, to

$$\sigma_C(0) \simeq \sqrt{2}\,\alpha\left(N^S \cdot r^S \cdot \frac{Z^S}{\bar{Z}^*}\right)^{-1/2} \tag{8.2.13}$$

which establishes a simple relationship between the standard deviation of the concentration at the zero level, the intensity generated in the element, the signal-to-line ratio in the element, and the respective atomic numbers of standard and specimen. As discussed in Chapter 7 [e.g., Eq. (7.3.10)], limits of detection can be derived from the expressions (8.2.12) and (8.2.13) by multiplying them with the appropriate factors. The caveats expressed in the last lines of Section 7.3 are, of course, equally appropriate for these limits of detection.

8. REFERENCES

8.1 Ziebold, T. O. and Ogilvie, R. E., *Anal. Chem.* **36**, 322 (1964).

8.2 Rasberry, S. D. and Heinrich, K. F. J., *Anal. Chem.* **46**, 81 (1974).

8.3 Goldstein, J. I., Majeske, F. J., and Yakowitz, H., *Adv. X-Ray Analysis* **10**, 431 (1967).

8.4 Bence, A. E. and Albee, A. L., *J. Geology* **76**, 382 (1968).

8.5 Laguitton, D., Rousseau, R., and Claisse, F., *Anal. Chem.* **47**, 2174 (1975).

8.6 Heinrich, K. F. J., *Adv. X-Ray Analysis* **11**, 40 (1968).

9.
Theory of Quantitative Electron Probe Microanalysis: Primary Emission

9.1 INTRODUCTION

In his doctoral thesis, Castaing showed the results of quantitative analyses of various alloys, and there, as well as in subsequent publications [9.1, 3.22] he developed a coherent and detailed theory of quantitative microanalysis. The later work of many investigators has lead to a better understanding of the theoretical foundations of quantitation with the electron probe microanalyzer, and several modifications were introduced in Castaing's correction scheme. We will now give a description of the present state of the art, reserving for later a discussion of some of the earlier procedures.

The rationale and limitations of the procedure are as follows. It is assumed that the analyst does not always have access to appropriate matching standards, so that a procedure for data reduction must be provided that permits the calculation of the analytical calibration function for any particular case. If this can be achieved, then simple standards, including pure elements, can be used to establish the relative X-ray intensities for all elements which emit observable lines. Moreover, assumptions of stoichiometry or analysis by difference permit the inclusion of one or more elements which have not been observed through X-ray emissions, in the data reduction scheme.

The relative X-ray intensities depend on many parameters, and these parameters must be known if the "correction scheme" is to succeed. Such parameters include physical constants such as mass absorption coefficients and fluorescent yields, but also a series of instrumental parameters including the operating potential and the angles of electron beam incidence and of X-ray emergence. These instrumental parameters must be accurately known. Therefore, the method is restricted as to specimen geometry: unless the analyzed region is homogeneous in composition and bounded by a flat surface of known orientation with respect to the electron beam and the X-ray detector, the theoretical scheme cannot be applied. The above requirements are frequently stated by saying that the specimen must be flat and semi-infinite in depth.

Before presenting the impressive array of "laws" and derivations on which the theory is based, we should make some general assessment as to the nature of the evidence presented. The processes which are involved in X-ray production by electrons are complex; few physicists would say that our knowledge of the subject suffices to establish a theoretical quantitative calculation procedure the accuracy of which will satisfy the analyst. To the contrary, we believe that many of the partial aspects of the theory are due to, and valid because of, accommodation to partial experiments, simplification of very complex matters, and a trial and error evaluation procedure on specimens of known composition. The purpose of the theoretical model is to indicate which aspects of X-ray generation dominate in determining the emitted intensities for a given specimen and experimental conditions, and hence to permit an adjustment of the corresponding aspects of the procedure to the experimental evidence.

Our first objective is to determine the average number of characteristic primary X-ray photons of the line of interest produced within the specimen by one impinging electron (I_p). The calculation of this number is called the *generation calculation*, or–for historical reasons–the *atomic number correction.* The intensity of secondary characteristic radiation is determined in the *fluorescence calculations* or *fluorescence corrections.* Depending on the source of excitation for the fluorescence, we distinguish *fluorescence due to characteristic lines* and *fluorescence due to the continuum.* The latter is frequently omitted in the data processing schemes, since it is usually of negligible consequences. The attenuation of both primary and secondary radiations during the passage towards the specimen surface is calculated in the *absorption calculations* or *absorption corrections.* The degree of attenuation depends on the distribution in depth of the generation of X-rays. Since fluorescent radiation is generated at a larger mean depth, its attenuation is larger, and must be calculated separately from that of primary radiation. The most important correction in quantitative analysis is usually that for attenuation of primary X-rays. In theoretical discussions, references to the absorption correction which do not specify the type of emission usually refer to attenuation of primary characteristic X-rays. However, experimental values of the absorption correction usually refer to total emission and can be equated to the primary absorption corrections only to the degree to which secondary emission can be neglected.

Let us recall that the fraction of X-rays emitted towards the detector is characterized by the solid angle, Ω, which is subtended by the detector, that the X-ray intensities are measured in terms of count rates, and that we have defined the probability, P, of detection of a photon which has reached the detector. The theoretical scheme presupposes that corrections were applied to the characteristic X-ray intensities for their background contributions, and for the variation of detector efficiency as a function of count rate (dead-time correction), as described in previous chapters. We will now relate the emitted intensity, Y^*, to the processes of generation and absorption within the specimen. We will use the subscripts p, c, and f to distinguish the parameters related to primary radiation, to fluorescence excited by the continuum, (continuum fluorescence), and to fluorescence excited by characteristic lines (characteristic fluorescence). The superscripts $*$ and s relate, respectively, to the specimen and the standard which, in the general case, need not be a pure element. If the probability of generation of a photon by an electron is I, the probability that this photon will be emitted within the solid angle subtended by the spectrometer, and that it will not be absorbed, is

$$I' = \frac{\Omega}{4\pi} \cdot I \cdot f. \tag{9.1.1}$$

In this equation f is the probability that a photon generated within the target and traveling toward the detector, within the solid angle Ω, will reach the specimen surface without being absorbed (absorption factor). The emergent intensity, I', is defined by Eq. (9.1.1). It is usually defined by other authors without considering the spectrometer aperture, by $I' = I \cdot f$. This parameter would then have only limited physical significance since it would be determined by the mean angle of emergence of the X-rays, but be independent of the solid angle subtended by the detector.

If we separate the mechanisms of X-ray generation, we obtain:

$$Y^* = I'^* P\, i_b = \frac{\Omega}{4\pi}(I_p{}^* f_p{}^* + I_c{}^* f_c{}^* + \Sigma I_f{}^* f_f{}^*)\, P\, i_b. \tag{9.1.2}$$

An analogous expression, with superscripts s, is obtained for the intensity of emission from the standard. We note that different absorption factors are applied to primary and fluorescent radiations, since the distributions of depth of generation, and hence the probability of absorption, differ.

The parameter in Eq. (9.1.2) which we wish to know is $I_p{}^*$, since this probability is directly related to the unknown concentration of the emitting element and thus leads to its determination. However, the factor $(\Omega/4\pi)\, Pi_b$—which is not a function of the specimen composition—is difficult to determine with sufficient accuracy to permit an accurate calculation of the value of $I_p{}^*$. If we calculate, instead, the ratio of the emissions from specimen and standard, $k_s = Y^*/Y^s$, obtained under identical experimental conditions, the above-mentioned factor cancels:

$$\frac{Y^*}{Y^s} = \frac{I'^*}{I'^s} = k_s = \frac{I_p{}^* f_p{}^* + I_c{}^* f_c{}^* + \Sigma I_f{}^* f_f{}^*}{I_p{}^s f_p{}^s + I_c{}^s f_c{}^s + \Sigma I_f{}^s f_f{}^s}. \tag{9.1.3}$$

Therefore, after correcting for background and dead-time, we obtain a *relative X-ray intensity,* k_S, from which, after applying the absorption, fluorescence, and atomic number "corrections," we calculate the concentration of the emitting element in the specimen. If the standard is a pure element, the term $I_f^S f_f^S$ can be omitted from Eq. (9.1.3), since pure elements do not emit a significant intensity of characteristic fluorescence.

To obtain the ratio of primary emitted intensities, we must first calculate the fluorescent contributions to the emergent intensities from both specimen and standard, and subtract them from the total emitted intensities. Then, the primary absorption factors, $f_p{}^*$ and $f_p{}^S$, must be calculated and the emergent primary intensities divided by them. But the calculation of all these parameters requires the knowledge of the composition of specimen and standard. For this reason, we must start with a rough estimate of the specimen composition which is used in an initial calculation of these parameters. The final estimates of the concentrations are obtained by means of an iterative procedure, in successive approximations. In our description we will first deal with the calculation of the intensity of X-rays emitted from a specimen of known composition, leaving for later the discussion of the iterative process required in the analysis of unknown specimens.

The primary characteristic emergent intensity is almost always much larger than the sum of all emergent fluorescent contributions. As was first noted by Castaing [1.1], the primary generated intensities are *roughly* proportional to the respective mass fractions of the emitting element *Castaing's first approximation):*

$$\frac{I_p{}^*}{I_p{}^S} \simeq \frac{C^*}{C^S} \,. \tag{9.1.4}$$

If the fluorescent contributions are negligibly small, and if the primary absorption factors are close to unity, the observed intensity ratios between specimen and standard are thus roughly equal to the ratios of the respective concentrations (mass fractions) of the emitting element [9.2]:

$$k_S \simeq \frac{C^*}{C^S} \quad \text{or} \quad k \simeq C^*. \tag{9.1.5}$$

For observations in which no high accuracy is required—for instance in line scans (Chapter 15), the validity of Eq. (9.1.5) is often assumed. But neither Eq. (9.1.5) nor Eq. (9.1.4) are sufficiently accurate for quantitative analysis.

The traditional though unfounded assumption of an ideal law of linear calibration [such as the one expressed by Eqs. (9.1.4) and (9.1.5)] has also determined the form in which the "corrections" are usually applied; i.e., as multiplicative "correction factors" for atomic number effects, absorption (of primary radiation), and fluorescence. This procedure is usually called the "*ZAF* procedure." It assumes that elemental standards are used, and that the fluorescent emission due to the continuum can be neglected, for both the specimen and the standard. Therefore, Eq. (9.1.3) reduces to

$$k = \frac{I_p{}^* f_p{}^* + \Sigma I_f{}^* f_f{}^*}{I_p{}^s f_p{}^s}. \tag{9.1.6}$$

The deviations from Castaing's first approximation are described by an "atomic number correction factor," F_Z, so that

$$C = \frac{I_p{}^*}{I_p{}^s} F_Z. \tag{9.1.7}$$

An "absorption correction factor" is defined as the ratio of the absorption factors for specimen and standard:

$$F_A{}^* = \frac{f_p{}^s}{f_p{}^*}. \tag{9.1.8}$$

Finally, a "fluorescence correction factor," for the fluorescence due to characteristic lines, is defined by

$$F_F{}^* = \frac{1}{1 + \dfrac{\Sigma I_f{}^* f_f{}^*}{I_p{}^* f_p{}^*}} = \frac{I_p{}^* f_p{}^*}{I_p{}^* f_p{}^* + \Sigma I_f{}^* f_f{}^*} \tag{9.1.9}$$

With these definitions, Eq. (9.1.6) can now be recast into an expression in which the deviations from the "ideal" linearity are taken into account by multiplicative factors (*ZAF* equation) [9.3]:

$$C = k \cdot F_Z{}^* F_A{}^* F_F{}^*. \tag{9.1.10}$$

This apparently simple equation may give the impression that the concentration of the element in question, C, can be calculated directly. This, however, is not possible since the correction factors are composition dependent (as indicated by the asterisks), and thus contain implicitly the unknown variable C. Therefore, the concentration must be determined by the iterative procedure previously mentioned. Moreover, as shown by Criss [9.4], the use of multiplicative correction factors leads to undesirable complications when fluorescence emission is present in the standard, or when the continuum fluorescence is taken into account.

Even in the simple case presented by Eq. (9.1.6) (elemental standard, no continuum fluorescence), the proposed *ZAF* procedure lacks logical rigor. As can be seen from the definitions, the absorption correction factor does not contain the terms for absorption of fluorescent contributions, which are present in the "fluorescence correction." The fluorescence correction factor also contains the terms $I_p{}^*$ and $f_p{}^*$ which describe the generation and absorption of primary X-rays. This lack of separation of effects has led in practice to inaccurate approximations for the terms $I_p{}^*$ and $f_p{}^*$ in the fluorescence correction.

Philibert [9.3] noted that in the calculation of relative X-ray intensities the sequence Z-A-F must be maintained–i.e., one must start with the generation calculation, and end with the calculation of the fluorescent contributions. In the calculation of unknown compositions, the order should be inverted to $F \rightarrow Z$. Such a sequence is logical if Eq. (9.1.3) is used: the product $I_p f_p$ must be obtained before the fluorescent contributions are added. It is, however, incomprehensible how the sequence of the multiplicative factors in Eq. (9.1.10) should affect the outcome of the calculation.

It is clear from the preceding that the use of the ZAF procedure for theoretical calculations is unnecessary and didactically disadvantageous, particularly if composite standards are used and if the fluorescence due to the continuum is taken into account. However, for traditional reasons, it is used where its application does not create complications, e.g., in the FRAME data reduction procedure. For a rigorous discussion of the problem, the use of Eq. (9.1.3) is much to be preferred.

Before entering into the details of the process of X-ray generation, we wish to recall that in our nomenclature, subscripts and superscripts are frequently omitted when such an omission does not cause ambiguities. This applies, in particular, to the expressions used for the measure of depth and path length. In the discussion of electron-target interactions, as well as in absorption (Section 4.4), it is cumbersome to express dimensions in metric units of length because such a choice renders the descriptions of the interactions dependent on density, and hence on physical states such as temperature, allotropic state, and pressure. Depth within the target is more conveniently scaled in mass cross sections (g/cm^2). In publications in which z and x denote, respectively, depth within the target and path lengths, in metric units of length, these symbols are invariably combined with the density of the material, ρ. To avoid the repetition of cumbersome formulas, in this book the path length, x, and the depth within the target, z, are scaled throughout in units of grams per square centimeter (g/cm^2). Hence our symbols dx and dz replace the symbols $d(\rho x)$ and $d(\rho z)$, used in most publications. This departure from the customary notation seems justified in view of the significant simplification.

9.2 THE GENERATION OF CHARACTERISTIC PRIMARY X-RAYS

The intensity of X-ray emission from a target bombarded by electrons depends on the interrelationship of the processes of electron deceleration and scattering and ionization. The ionization of an inner shell which is required for the emission of an X-ray photon is a relatively rare event. Most of the electron energy is spent in interactions with outer, weakly bound orbital electrons. The primary electron thus suffers a large number of small energy losses which are usually described as a quasi-continuous deceleration. Only a small fraction of the primary electrons produce an X-ray photon (see Fig. 11.4).

The probability of ionization of an inner shell depends on the energy of the penetrating electron, which diminishes along its path within the specimen. After the energy of the primary electron has dropped below the critical excitation potential of a given inner shell, no further ionization of this shell can occur. Therefore, the X-ray lines produced by this shell can no longer be excited. Interactions of the beam electrons with an atom also include large changes in direction of the electron (large-angle scattering). Some of the scattered electrons reemerge from the specimen surface, before their energy has dropped below the critical excitation potential. The backscattering of electrons reduces the number of inner-shell ionizations with respect to that which would be observed if all primary electrons had remained within the target (Fig. 9.1).

Finally, as we have discussed previously, the probability of emission of an X-ray photon of a given wavelength depends on the corresponding fluorescent yields, weights of lines, and, for shells above the K-level, of the Coster-Kronig transitions (see Section 4.2).

In order to calculate the average number of primary characteristic photons of a line produced by one electron along its trajectory within the specimen, $I_p{}^*$, we must therefore quantitatively describe:

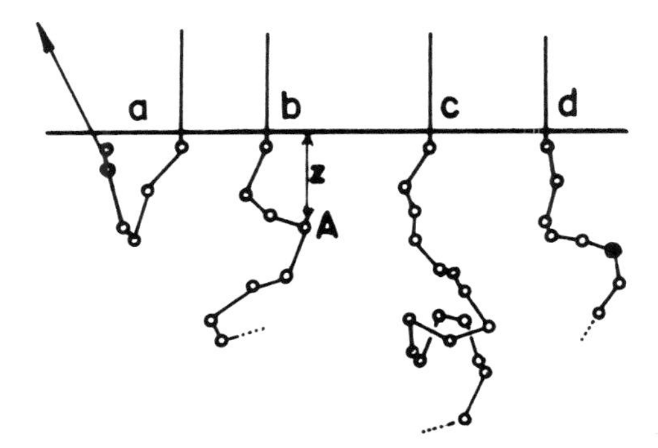

Fig. 9.1. Schematic representation of collisions within the target. The first electron at the left shows large-angle scattering at the third collision and subsequent backscattering. The dark circle at the path at the right represents an ionization in an inner orbit which may cause the emission of a characteristic X-ray photon. Note (second path from left) that the path from the surface to point A (denoted x) is larger than the depth of penetration, z. Dotted paths correspond to electrons which do not have sufficient kinetic energy left to produce ionization of the inner orbit of interest.

a. the loss of energy of the electron along its path in the target (stopping power);
b. the probability of ionization of the inner shells of target atoms, as a function of the energy of the ionizing atom (ionization cross section);
c. the loss of X-ray production caused by electron backscatter (electron backscatter correction factor);
d. the probability of photon emission of the line of interest subsequent to ionization of the corresponding shell (fluorescent yield, weight of X-ray line).

We will start with a discussion of the energy losses within the target.

9.2.1 Stopping Power

If we regard the deceleration of the beam electron within the target as a continuous process, we can express the average rate of energy loss by means of the equation:

$$-dE = Sdx. \tag{9.2.1}$$

The symbol E denotes the energy of the electron at any point of the path. (The initial energy of the electron before the first collision will be called E_0.) This energy of the electron, in electron volts, is numerically equal to its potential, V, in volts. For a small energy loss, the energy diminishes proportionally to the number of atoms in the way of the electron, and thus to the mass thickness, dx (g/cm^2). The proportionality factor, S, is called *stopping power* (eV g^{-1}cm^2). We note that Eq. (9.2.1) relates energy loss to the actual length of path, x, rather than to depth beneath the surface, z. This distinction is necessary, since, due to scattering, the path of the electron is larger than the depth z (Fig. 9.1).

Bethe [9.5] derived from theory the following equation for stopping power:

$$S = 2\pi e^4 \cdot N_{av} \cdot \frac{Z}{A} \cdot \frac{1}{E} \ln \frac{\epsilon E}{J}. \tag{9.2.2}$$

where e is the charge of the electron, N_{av} is Avogadro's number (the number of atoms in one gram-atom, or mole, of material), Z and A are the atomic number and atomic weight, respectively, of the target material, E, as before, is the instantaneous energy of the decelerating electron, ϵ is the square root of half the base of natural logarithms (ϵ = 1.166), and J is the logarithmic mean excitation energy, i.e., the logarithmic mean loss of energy of an electron when interacting with an orbital electron of the target atoms.

Since the targets of interest in electron probe microanalysis contain more than one element, we must obtain an expression of Bethe's law applicable to such multielement targets. For this purpose, we recall that the number of atoms in 1 g of substance is equal to N_{av}/A. If the target contains more than one element, the number of atoms of an element a in 1 g of substance is $m_a N_{av}/\Sigma m_i A_i$. It can be assumed that in a solid target the stopping effects of the orbital electrons of atoms of different elements are combined additively (Bragg's rule [9.6]). Therefore, the following form of Bethe's equation is applicable to multielement targets:

$$S^* = \frac{2\pi e^4 N_{av}}{E} \cdot \frac{\Sigma m_i Z_i \ln (\epsilon E/J_i)}{\Sigma m_i A_i}. \tag{9.2.3}$$

In this equation, the concentrations of the elements in the target are expressed in atomic fractions. Since mass fractions and atomic fractions are related by

$$C_a = \frac{m_a A_a}{\Sigma m_i A_i},$$

Bethe's equation for multielement targets can also be written

$$S^* = \frac{2\pi e^4 N_{av}}{E} \cdot \Sigma C_i \frac{Z_i}{A_i} \ln \left(\epsilon \frac{E}{J_i}\right). \tag{9.2.4}$$

Equations (9.2.2), (9.2.3), and (9.2.4) contain the atomic weight, A_i, which is not a smooth function of the atomic number, Z (Fig. 9.2). For this reason, S is not a smoothly varying function of the atomic number.

In Eq. (9.2.1), the unit of mass cross section of material traversed by the electron beam is 1 g/cm^2. It is equally possible to define as a unit of layer thickness

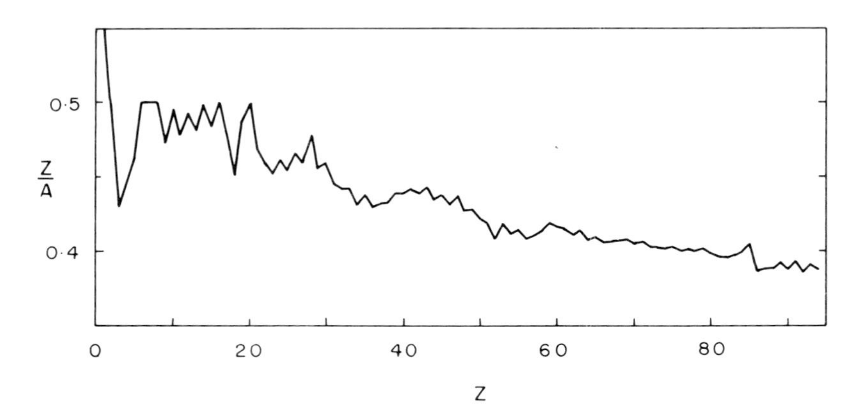

Fig. 9.2. Ratio of atomic number, Z, and atomic weight, A, as a function of atomic number.

that of a layer containing one atom per square centimeter of absorber (or, if preferred, one atom per barn). The atomic stopping power is equal to

$$S_{at}^* = S^* \frac{\Sigma m_i A_i}{N_{av}} = \frac{2\pi e^4}{E} \Sigma m_i Z_i \ln (\epsilon E/J_i) \tag{9.2.5}$$

and for pure elements

$$S_{at}(a) = \frac{2\pi e^4}{E} Z \ln (\epsilon E/J_a). \tag{9.2.6}$$

If the barn were to be used as the unit of surface, the values of S_{at} must be multiplied by 10^{24}. Equations (9.2.5) and (9.2.6) do not contain atomic weights; thus, the atomic stopping power varies smoothly with atomic number. A minor effect, the dependence of J upon the atomic number—which is almost smooth, and enters in a logarithmic term—can safely be neglected in this consideration. The atomic-mass dependence of the stopping power S in Eq. (9.2.2) is thus merely a consequence of the definition of the length of path in terms of mass per area of the absorber.

The variation of stopping power with electron energy is shown in Fig. 9.3 for the elements of atomic numbers 10, 30, and 70.

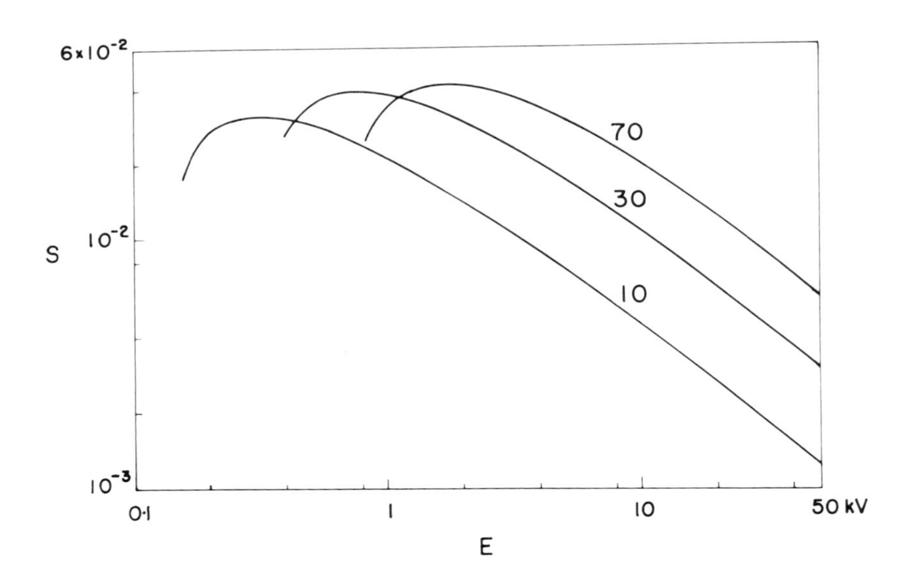

Fig. 9.3. Variation of stopping power with energy, for elements of atomic numbers 10, 30, and 70. The values at low energies are uncertain, because Bethe's assumption of the invariance with energy of the mean excitation potential does not hold in this region.

9.2.2 Mean Excitation Energy

Bethe's equation has proven in practice to provide a satisfactory basis for calculating the electron deceleration in the treatment of electron probe measurements of X-ray intensities. However, some objections can be raised to its application, which we must now consider. The stopping power equation (9.2.3), as used in the data reduction scheme which we are describing, implies the following assumptions:

a. that the loss of energy of a primary electron can be uniquely defined as a function of the pathlength (i.e., there is no "electron straggling" or spread in energy at a given pathlength);
b. that no mechanisms contribute to the energy loss of the primary electron, except the interactions with orbital electrons of individual atoms which are considered in Bethe's equation;
c. that the mean excitation energy, J_i, is well defined, known, and independent of chemical states and of the electron energy E, and that Bragg's rule of additivity is valid.

In reality, electrons which have traveled for the same pathlength, have not, in general, lost exactly the same amount of energy. The value of the mean excitation energy is a logarithmic average of the energy losses suffered by the electron in individual independent interactions. The individual energy losses vary, and the distances between collisions also scatter around an average (the mean free path). There is, therefore, no unique function relating the distance covered by an electron and its residual energy. However, since in the performance of the analysis a very large number of electrons interact with the specimen or standard, we may use a mean value of energy loss, rather than considering the chain of events caused by each single electron.

The mean excitation energy, J, "is a well-defined parameter of the theory . . . . From its definition, J depends only upon the ground- and excited-state wave functions of a stopping material, independent of the energy and other characteristics of an incident particle . . . . In practice, J is determined empirically from stopping power and/or range measurements" [9.6]. In most cases, such measurements were performed at energies much higher than those employed in electron probe microanalysis, in experiments involving the stopping of protons. The results show considerable scatter (Fig. 9.4), and there is evidence that the mean excitation energy is not a smooth function of the atomic number, particularly for light elements ($Z < 10$). The value of J may also vary as a function of chemical binding [9.8]. From a theoretical viewpoint, it is clear that the interaction of electrons with solid matter cannot be described rigorously in terms of collisions of the electron with free atoms, particularly at low electron energies. It would therefore be unreasonable to expect that the mean excitation voltage should not vary significantly as E approaches J. From a formal point of view, we note that according to Eq. (9.2.4), S tends to zero as $(\epsilon E)/J$ approaches one. In this region of energies, the Bethe equation is not a valid model.

Several authors have proposed simple empirical formulas for the approximate calculation of the mean excitation energies of the elements. Some of these expressions are listed in Table 9.1 and are shown in Fig. 9.4. The equation published by Berger and Seltzer—but attributed to Sternheimer (Table 9.1)—is widely used. This equation is a fit to experimental data. Another expression widely employed in electron probe microanalysis is the expression of Duncumb and DaCasa (Table 9.1). Duncumb's fit differs from all other fits in that it was not obtained from the particle deceleration experiments mentioned previously; rather, it represents an effort to empirically adjust the parameter J so as to minimize the analytical errors of electron probe analyses of materials of known composition. It therefore has the advantage of being derived from experiments which were similar to the analytical measurements of an unknown specimen.

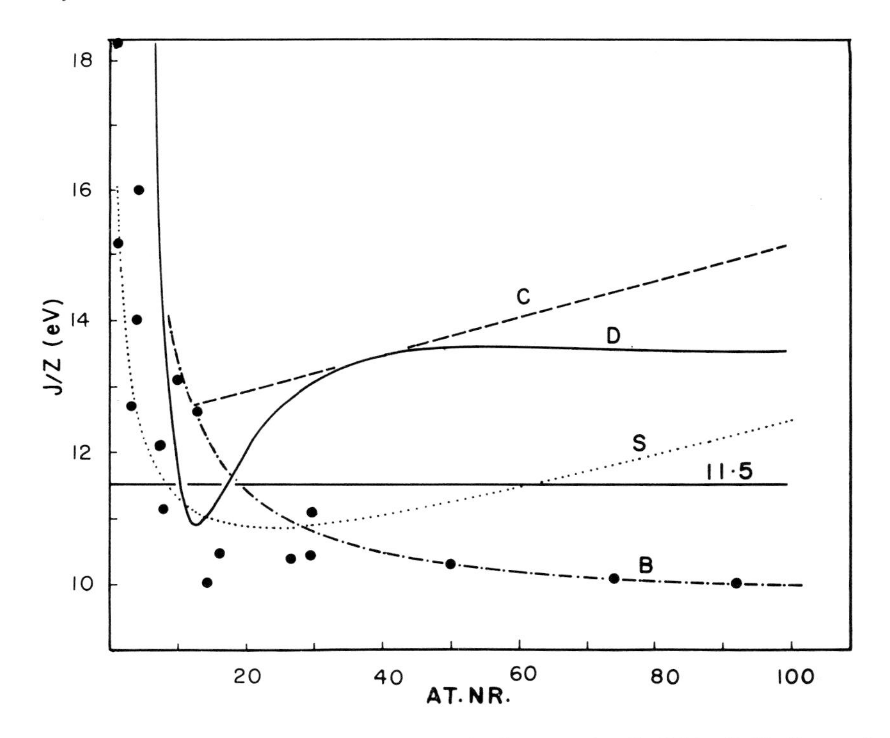

Fig. 9.4. Proposed formulas for the mean excitation energies. C: Caldwell, D: Duncumb-DaCasa, S: Springer, B: Sternheimer, reported by Berger and Seltzer. Solid circles are experimental results reported by Turner [9.8]. For sources see Table 9.1.

Table 9.1. Proposed Equations for J.

EQUATION	SOURCE
$J/Z = 13.5$	F. Bloch, *Z. Phys.* **81**, 363 (1933).
$J/Z = 11.5$	R. R. Wilson, *Phys. Rev.* **60**, 749 (1941)
$J/Z = 9.76 + 58.82 \cdot Z^{-1.19}$	R. M. Sternheimer, quoted by M. J. Berger and S. M. Seltzer, National Research Council Publ. 1133, National Academy of Sciences, Washington, D.C., 1964, p. 205.
$J/Z = 9.0(1 + Z^{-2/3}) + 0.03 \cdot Z$	G. Springer, *Neues Jahrbuch Fuer Mineralogie,* Monatshefte. (1967), 9/10, p. 304.
$J/Z = 12.4 + 0.027 \cdot Z$	K. F. J. Heinrich and H. Yakowitz, *Mikrochim. Acta* (1970), p. 123. [Fit of Caldwell's data, from D. O. Caldwell, *Phys. Rev.* **100**, 291 (1955).]
$J/Z = 14.0[1 - e^{-0.1Z}] + 75.5/Z^{Z/7.5} - Z/(100 + Z)$	P. Duncumb and C. DaCasa, cited in P. Duncumb, P. K. Shields-Mason, and C. DaCasa, *Proc. 5th Int. Congr. on X-Ray Optics and Microanalysis,* Springer, Berlin, 1969, p. 146.
$J/Z = 10.04 + 8.25 \cdot e^{-Z/11.22}$	C. Zeller, cited in [12.29] by J. Ruste and M. Gantois, *J. Phys. D., Appl. Phys.* **8**, 872 (1975)

(all in electron volts)

Duncumb's procedure yields values for the main excitation energy which differ significantly from those obtained by other means, particularly for elements of atomic number below ten. It has been claimed [9.9] that Duncumb's equation yields smaller analytical errors than others, but this claim is not uncontested. The main objection that can be raised against it is that corrections for effects other than electron stopping were applied to the data on which the formula is based. Hence, errors in the ionization cross section, in the X-ray absorption correction or in the backscatter correction, or the omission of the correction for continuum fluorescence, could have affected the procedure. Further investigations will be needed to definitely settle this argument. In fact, if the effects of chemical state and other uncertainties in the value of J prove to be significant, they may set an ultimate limit to the accuracy of quantitative electron probe microanalysis by the theoretical approach.

The failure of the equation by Bethe for stopping power to adequately describe energy loss at low electron energies has been discussed by Love et al. [9.10]. These authors proposed a modified expression for stopping power:

$$-dE/dx = (Z/A \cdot J)/[1.18 \times 10^{-5}(E/J)^{1/2} + 1.47 \times 10^{-6}(E/J)] \quad (9.2.7)$$

which produces values similar to that of Eq. (9.2.2), except for low values of the electron energy. It remains to be seen what the effects of the proposed change are, since at very low values of E the production of X-rays is absent or low.

9.2.3 Ionization Cross Section

If a thin layer of an elemental target material is crossed perpendicularly by i electrons, the number of ionizations, n, of electrons in the shell q, of the target atoms is proportional to the number of atoms present, N, to that of the electrons which cross the layer, and inversely proportional to the area a over which the N atoms are uniformly distributed:

$$n = Q_q N i a^{-1}. \quad (9.2.8)$$

The atomic ionization cross section, Q_q, for the shell q has the dimensions of an area. Consider now a layer of material of thickness dx (g/cm^2). The number of atoms in this layer is equal to

$$\frac{N_{av}}{A} \cdot a \cdot dx$$

so that the number of ionizations occurring when i electrons cross the layer is equal to

$$dn = Q_q i \frac{N_{av}}{A} dx. \quad (9.2.9)$$

Most experimental and theoretical investigations of ionization cross sections concern the K-shell. This subject has been discussed in detail in Green's thesis [6.2], and more recently by Powell [9.11]. As discussed by Powell, Bethe provided a theoretical formula which can be brought into agreement with experimental data:

$$Q_q = \frac{4\pi e^4}{E \cdot E_q} \cdot z_q \cdot b_q \cdot \ln \frac{4E}{B_q}$$

$$= \frac{6.51 \times 10^{-14}}{E \cdot E_q} \cdot z_q \cdot b_q \cdot \ln \frac{c_q E}{E_q} \text{ cm}^2 \quad (9.2.10)$$

or

$$Q_q E_q^2 = 6.51 \times 10^{-14} \cdot z_q \cdot b_q \cdot \frac{\ln (c_q \cdot U)}{U} \text{ cm}^2 \cdot \text{eV}^2.$$

The term z_q denotes the number of orbital electrons present in the filled shell—or subshell—q, the ionization of which leads to the emission of the line of interest. The values of z_q for the shells of main interest are given in Table 9.2.

Table 9.2. Numbers of Electrons in the Filled Shell

Level	Important Lines	z_q
K	$K\alpha_{1,2}$, $K\beta_1$	2
L_I	$L\beta_3$	2
L_{II}	$L\beta_1$	2
L_{III}	$L\alpha_{1,2}$	4
M_{IV}	$M\beta$	4
M_V	$M\alpha$	6

The overvoltage, $U = E/E_q$, is the ratio of the instantaneous energy of the electron at each point of the trajectory to that required to ionize the level q. (The critical ionization energy, E_q, is related to the critical ionization potential, V_q by $E_q = e \cdot V_q$.) The values of B_q, b_q, and c_q are constants which presumably are independent of the atomic number of the target. Powell fitted Eq. (9.2.10) to the experimental and calculated cross sections for overvoltages above five from several investigators. He obtained fits for pairs of values of b_q and c_q, for the K- and L-shells, which are, in general, within the following limits:

	b_q	c_q
K-shell	0.7-1.0	0.6-1.0
L-shell	0.5-1.0	0.5-0.9

A simplification is achieved if c_q is assumed to be equal to one; such an assumption, according to Powell's results, is not seriously in error. This simplified Bethe equation, also cited by Philibert and Tixier [9.3], will be used in the development of the atomic number calculation of primary emitted X-ray intensities which follows:

$$Q_q = \frac{b\pi e^4 z_q}{E_q^2} \cdot \frac{\ln U}{U}. \tag{9.2.11}$$

This equation is similar to the equation given by Green and Cosslett [9.12], and derived from the work of Worthington and Tomlin [9.13]:

$$Q_q = 7.92 \times 10^{-20} E_q^{-2} \frac{\ln U}{U}. \qquad (9.2.12)$$

The variation with U of the energy-dependent term $(\ln U)/U$ is shown in Fig. 9.5. Q_q has a maximum at $(U = e \simeq 2.7)$. The constant b in Eq. (9.2.11) is postulated to be independent of the atomic number of the target. Henoc et al. [9.14] assumed that the value of b was equal to 0.76, for all shells and subshells. Burhop [9.15], on the contrary, calculated that the value of b for the L-shell should be approximately two-thirds of that of the K-shell. Green published in his thesis values for the factor b obtained experimentally for the L_{III}-shell which were about twice as high as predicted by Burhop.

Worthington and Tomlin [9.13] observed that Eq. (9.2.12) was inaccurate for small overvoltages. They suggested instead that

$$Q_K = \frac{2\pi e^4}{E \cdot E_k} \cdot \ln\left[0.35 \cdot \frac{4U}{(1.65 + 2.35 \exp(1 - U))}\right]. \qquad (9.2.13)$$

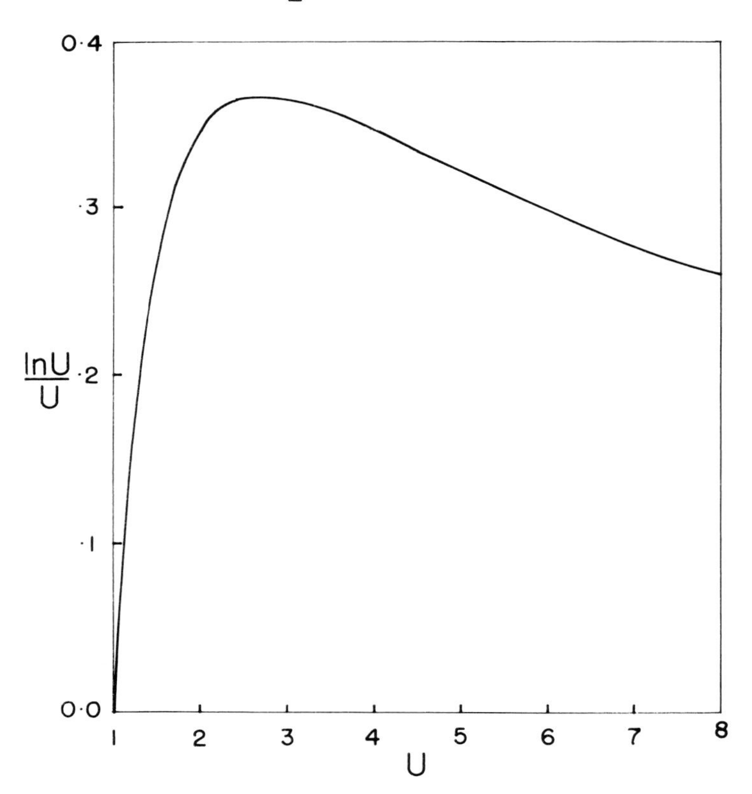

Fig. 9.5. Equations (9.2.11) and (9.2.12) rise to a maximum located at $U \simeq 2.7$.

The experimental evidence on Q is scanty (see [9.16]). Therefore, the use of expressions for Q introduces a considerable uncertainty in our calculations. Fortunately, the uncertainties in the value of the ionization cross section tend to compensate when the relative primary X-ray intensities are calculated. They are, however, significant in the computation of intensities of secondary emission (see Sections 11.1 and 11.2), and of X-ray generation in thin films and in small particles. The importance of this parameter for quantitation has apparently been underestimated in the past.

Powell [9.16], in a recent review of expressions for the ionization cross section recommends, for overvoltages from 1.5 to 25, a modification by Fabre [9.17] of the Bethe equation:

$$Q_q \cdot E_q{}^2 = \frac{6.51 \times 10^{-14} z_q \cdot \ln(U_q)}{a_q(U_q + b_q)} \text{ cm}^2\text{eV}^2 \qquad (9.2.14)$$

in which $a_k = 1.18$ and $b_k = 1.32$. [The numerical fit has been made only for the K-shell ($q = k$).] This expression fits recent cross-section measurements within 6%.

9.2.4 Probability of Ionization Per Incident Electron

Let us assume that a target containing several elements is bombarded by electrons having the energy E_0. We are interested in the ionization of the shell q of element a, which is present at the atomic fraction m_a. Consider a hypothetical layer of target material which contains 1 atom/cm^2. The probability of ionization, according to Eq. (9.2.11), is

$$m_a Q_{qa} = m_a b \pi e^4 z_q \frac{\ln U}{U} E_q{}^{-2}. \qquad (9.2.15)$$

The mean deceleration of the electron passing through this layer can be obtained from Bethe's Eq. (9.2.5). An integration over the energy range within which ionization is possible (from E_0 to E_q) yields the average number of ionizations produced by an electron which remains within the target for this energy range:

$$n_a = m_a \int_0^{x_q} Q dz = m_a \int_{E_0}^{E_q} \frac{Q_q}{-S_{at}{}^*} dE. \qquad (9.2.16)$$

where x_q is the effective range of the electron path, at the end of which the electron has the energy E_q. We can now substitute the values for the ionization cross section and the stopping power from Eqs. (9.2.11) and (9.2.5). We obtain:

$$n_a = \frac{m_a \cdot b \cdot z_q}{2} \int_{E_q}^{E_0} \frac{\frac{E}{E_q{}^2} \cdot \frac{\ln U}{U} dE}{\Sigma m_i Z_i \ln \frac{\epsilon E}{J_i}} = \frac{m_a \cdot b \cdot z}{2} \int_1^{U_0} \frac{\ln U dU}{\Sigma m_i Z_i \ln \frac{\epsilon E}{J_i}}. \tag{9.2.17}$$

The limit U_0 denotes E_0/E_q. The formal integration of Eq. (9.2.17) is feasible, as shown by Worthington and Tomlin [9.13], and for composite targets by Philibert and Tixier [9.3]. To simplify the notation we introduce the following auxiliary variables:

$$L = \Sigma m_i Z_i \quad \text{and} \quad M = \exp \frac{\Sigma m_i Z_i \ln \left(\frac{\epsilon E_q}{J_i}\right)}{L}$$

so that

$$\Sigma m_i Z_i \ln \left(\frac{\epsilon E_q}{J_i}\right) = L \cdot \ln M. \tag{9.2.18}$$

The summations are over all elements present in the specimen. Equation (9.2.17) now becomes

$$n_a = \frac{m_a b z_q}{2} \int_1^{U_0} \frac{\ln U}{L \ln (UM)} dU. \tag{9.2.19}$$

We now call $t = \ln (UM)$. Therefore, $\ln U = t - \ln M$, $M \cdot dU = e^t dt$, and $dU = (e^t/M)dt$. We obtain:

$$\int_1^{U_0} \frac{\ln U}{L \ln (UM)} dU = \frac{1}{L} \int_{\ln M}^{\ln(U_0 M)} \frac{t - \ln M}{t} \cdot \frac{e^t}{M} dt$$

$$= \frac{1}{L} \left[\frac{1}{M} \int_{\ln M}^{\ln(U_0 M)} e^t dt - \frac{\ln M}{M} \int_{\ln M}^{\ln(U_0 M)} \frac{e^t}{t} dt \right]$$

$$= \frac{1}{L} \left\{ U_0 - 1 - \frac{\ln M}{M} \left[\int_{-\infty}^{\ln(U_0 M)} \frac{e^t}{t} dt - \int_{-\infty}^{\ln M} \frac{e^t}{t} dt \right] \right\} \tag{9.2.20}$$

$$\equiv \frac{1}{L} \left\{ U_0 - 1 - \frac{\ln M}{M} [\mathrm{li}\,(U_0 M) - \mathrm{li}\,(M)] \right\}.$$

The symbol li denotes a function called the logarithmic integral, defined by

$$\text{li}\ (x) \equiv \int_{-\infty}^{\ln x} \frac{et}{t} dt \tag{9.2.21}$$

which can be calculated numerically [9.3] as follows:

$$\text{li}\ (x) = \text{constant} + \ln |\ln x| + \sum_{S=1}^{\infty} \frac{(\ln x)^s}{S \cdot S!}\ . \tag{9.2.22}$$

The constant in Eq. (9.2.22) cancels in the difference of logarithmic integrals in Eq. (9.2.20). The probability of ionization per incident electron is finally:

$$n_a^* = m_a \frac{bz_q}{2L} \left\{ U_0 - 1 - \frac{\ln M}{M} [\text{li}\ (U_0 M) - \text{li}\ (M)] \right\}\ . \tag{9.2.23}$$

Worthington and Tomlin pointed out that this equation does not take into account the loss of ionizations caused by the backscattering of electrons at energies above the critical excitation potential. If we introduce a *backscatter correction factor,* R^*, which represents the loss of ionization due to backscattering, we finally obtain:

$$n_a^* = R_a^* m_a \frac{bz_q}{2L} \left\{ U_0 - 1 - \frac{\ln M}{M} [\text{li}\ (U_0 M) - \text{li}\ (M)] \right\}\ . \tag{9.2.24}$$

Probability of Production of Primary X-Ray Photons per Primary Electron. Postponing further discussion of the backscattering correction, we will now calculate the average number of photons produced by an incident electron. To obtain this number, we must multiply the number of ionizations by the corresponding fluorescence yield, ω_q, and by the weight of the line of interest, p_{qm} :tion 4.2). In shells other than the K-shell, the effective yields, corrected ter-Kronig transitions, apply. We obtain:

$$I_p^* = n_a^* \cdot p_{qm} \cdot \omega_q = \frac{1}{2} R^* \omega_q \cdot p_{qm} \cdot m_a \cdot b \cdot z_q \int_1^{U_0} \frac{\ln\ U dU}{\Sigma m_i z_i \ln\left(\frac{\epsilon E}{J_i}\right)}$$

$$= \frac{1}{2} R^* \omega_q p_{qm} \frac{m_a}{\Sigma m_i Z_i} bz_q \left\{ U_0 - 1 - \frac{\ln M}{M} [\text{li}\ (U_0 M)\ -\ \text{li}\ (M)] \right\}\ .$$

(9.2.25)

If the target is an element, $m_a = 1$, $L = Z_a$, $M = \epsilon E_q / J_a$, and

$$I_p(a) = \frac{1}{2} R_a \, \omega_{qa} p_{qma} b z_q \int_1^{U_0} \frac{\ln U}{Z_a \ln \dfrac{\epsilon E}{J_a}} \, dU \tag{9.2.26}$$

$$= \frac{1}{2} R_a \, \omega_{qa} p_{qma} b z_q \frac{1}{Z_a} \left\{ U_0 - 1 - \frac{\ln\left(\dfrac{\epsilon E_q}{J_a}\right)}{\dfrac{\epsilon E_q}{J_a}} \left[\mathrm{li}\left(\frac{\epsilon E_q}{J_a}\right) - \mathrm{li}\left(\frac{\epsilon E_0}{J_a}\right) \right] \right\}.$$

Note that neither Eq. (9.2.25) nor Eq. (9.2.26) contain the atomic weights of the elements present in the targets. If we prefer expressing the results in weight fractions rather than atomic fractions, Eq. (9.2.25) can be transformed with the aid of Eq. (8.2.5):

$$I_p^* = \frac{1}{2} R^* \omega_q p_{qm} \frac{C_a}{A_a} b z_q \int_1^{U_0} \frac{\ln U}{\Sigma C_i \dfrac{Z_i}{A_i} \ln \dfrac{\epsilon E}{J_i}} \, dU \tag{9.2.27}$$

and since

$$M = \exp \frac{\Sigma C_i \dfrac{Z_i}{A_i} \ln\left(\dfrac{\epsilon E_q}{J_i}\right)}{\Sigma C_i \dfrac{Z_I}{A_i}} \tag{9.2.28}$$

we obtain

$$I_p^* = \frac{1}{2} R^* \omega_q p_{qm} \frac{\dfrac{C_a}{Aa}}{\Sigma C_i \dfrac{Z_i}{A_i}} b z_q \left\{ U_0 - 1 - \frac{\ln M}{M} \left[\mathrm{li}\,(U_0 M) - \mathrm{li}\,(M) \right] \right\}. \tag{9.2.29}$$

For an elementary target, this equation again reduces to Eq. (9.2.26). The formal integration by means of the logarithmic integral can easily be performed by a computer.

In the calculation of the retardation of electrons and production of primary X-rays, no assumption was made concerning the electron beam incidence angle and the configuration of the specimen surface. This was possible as long as it was assumed that all incident electrons remain within the specimen. The above factors do, however, affect the value of the backscatter correction factor, R, as well as the attenuation of the emerging primary X-rays, which will be discussed in Chapter 10.

Simple Approximations for Primary X-Ray Intensities. For certain practical uses such as in the fluorescence correction for secondary emission due to characteristic lines [Eq. (11.1.36], the approximations to the primary intensity given by Eqs. (9.2.25), (9.2.26), and (9.2.29) may seem too complex. In other instances we may wish to predict the variation of the X-ray generation with the operating voltage only, or to obtain a rough estimate of absolute intensities. In such cases simple models may have limited usefulness. Whiddington [9.18] proposed the following model for the energy of a penetrating electron:

$$S = -dE/dx = \text{constant}/E. \tag{9.2.30}$$

Hence,

$$E_0^2 - E^2 = c \cdot x \tag{9.2.31}$$

with x given in grams per square centimeter. It was later found that the constant c changes slowly with the energy of the penetrating electron as shown in Table 9.3.

Table 9.3.

E (keV)	1	2	5	10	20	50
c (keV^{2}cm^{2}g^{-1})	1×10^5	1.3×10^5	1.8×10^5	2.3×10^5	2.9×10^5	4.4×10^5

Combining Eq. (9.2.31) with Bethe's law for ionization cross sections [Eq. (9.2.9)] and considering that the number of ionizations per electron is equal to

$$n_q = -R \int_{E_q}^{E_0} \frac{N_{av}}{A} \cdot Q \cdot \frac{dx}{dE} \, dE$$

we obtain the following expression [9.12]:

$$n_k = 9.54 \times 10^4 \frac{R}{Ac} \left[U_0 \ln U_0 - (U_0 - 1)\right] \tag{9.2.32}$$

for the generation of K-lines; the number of photons of a given line K1 is $n_K \cdot \omega_K \cdot p_{K1}$. Duncumb showed in his thesis [9.19] that the approximation

$$U_0 \ln U_0 - (U_0 - 1) \simeq 0.365\,(U_0 - 1)^{1.67} \tag{9.2.33}$$

is applicable for an overvoltage range from 1.5 to 16, with less than 10% error.

As early as 1917, Webster and Clark [9.20] had proposed that

$$I = k(V_0 - V_q)^{1.5}.$$

This relation, with a slightly higher exponent, is widely applied (see, for instance, Cosslett [9.21]). It appears that 1.65 is a generally usable scaling power for energy dependence of various parameters related to X-ray generation (see, for instance, Fig. 6.19). A simpler relation obtained by Castaing [1.1] by a radical simplification of a model for ionization by Rosseland [9.22] was used to predict the ratio of intensities of lines from two elements as follows:

$$\frac{I(a)}{I(b)} = \frac{E_{qb}}{E_{qa}} \cdot \frac{A_b}{A_a} \cdot \frac{\omega_a}{\omega_b} \cdot \frac{p_a}{p_b}. \tag{9.2.34}$$

This approach was proven to be inadequate [9.23] and should not be used. Instead, one can derive from the formulas developed by Duncumb and Cosslett the relation

$$\frac{I(a)}{I(b)} = \frac{A_b}{A_a} \cdot \left(\frac{U_b - 1}{U_a - 1}\right)^{1.67} \frac{\omega_a p_a}{\omega_b p_b}. \tag{9.2.35}$$

This equation is frequently used in abbreviated formulations of the fluorescence correction [Eq. (11.1.42)].

If the speed of the computer used for the data reduction procedure is limited, it may be impractical to perform the iteration of Eqs. (9.2.20) and (9.2.21) because the logarithmic integrals require the calculation of the infinite series in Eq. (9.2.22). Poole and Thomas [9.24] proposed a simplification in which it is assumed that the stopping power for the entire energy range is, in average, equal to that obtained from Eqs. (9.2.4) to (9.2.6), with an average energy

$$\overline{E} = \frac{1}{2}(E_0 + E_q). \tag{9.2.36}$$

Except for extreme atomic number differences and excitation conditions, this substitution will not cause serious inaccuracies in the analytical result (see Section 12.3). Equation (9.2.36) has been used in many simplified *ZAF* procedures, and the principles common to them will be explained here, since they concern mainly the stopping power part of the atomic number correction. The simplified procedure is described in detail by Duncumb and Reed [9.26] who also recommended the use of Eq. (9.2.36). We recall that the stopping power is usually defined as in Eq. (9.2.1), and that concentrations are commonly expressed in weight fractions. From Eq. (9.2.16) and the relation between C_a and m_a shown in Eq. (8.2.5), inverting the limits of integration, and introducing the backscatter correction factor, we obtain:

$$n_a = R_a^* \frac{N_{av}}{A_a} C_a^* \int_{E_q}^{E_0} \frac{Q}{S^*} dE \tag{9.2.37}$$

Let us assume a case in which the absorption corrections and indirect excitation can be neglected ($F_A = 1, F_F = 1$). We obtain for the ratio of primary intensities:

$$k_a = \frac{I_p^*}{I(a)} = C^* \frac{R_a^*}{R(a)} \cdot \frac{\int_{E_q}^{E_0} Q/S^* \cdot dE}{\int_{E_q}^{E_0} Q/S(a) \cdot dE} . \tag{9.2.38}$$

If we now replace the stopping powers by average stopping powers $\overline{E}$, and since the ionization cross section is not dependent upon specimen composition [Eqs. (9.2.9)-(9.2.13)], the ionization cross sections cancel, and we obtain:

$$k_a = C_a \cdot \frac{R_a^*}{R(a)} \cdot \frac{\overline{S}(a)}{\overline{S}_a^*}. \tag{9.2.39}$$

In terms of the *ZAF* correction, this is equivalent to

$$F_Z = \frac{\overline{S}_a^*/R_a^*}{\overline{S}(a)/R(a)} . \tag{9.2.40}$$

A refined model in which the mean voltage is adjusted to the value of the overvoltage E_0/E_q was proposed by Heinrich and Yakowitz [9.25] (Fig. 9.6).
It is also frequently assumed that the average stopping power is additive in mass fractions. This, according to Eq. (9.2.4), is only approximately true. If the absorption and fluorescence corrections are significant, the corresponding *ZAF* factors are applied, according to Eq. (9.1.10). The procedure we have described is the basis of the program FRAME which we have, with the aid of analyses of a large number of known specimens, compared with the more elaborate program COR, and found to be very accurate. These programs are described in Chapter 12, where an assessment of the accuracy of the simplified procedure is also given.

A variety of related data reduction schemes were developed which are extensively described in a review by Martin and Poole [9.27]. Since I do not believe that any of the variants is preferable to that given by Duncumb and Reed, these variants will not be described here.

9.3 THE BACKSCATTERING OF ELECTRONS

The electrons which are reemitted with energies above 100 eV from the specimen are called backscattered. The emission of these electrons can be used for the

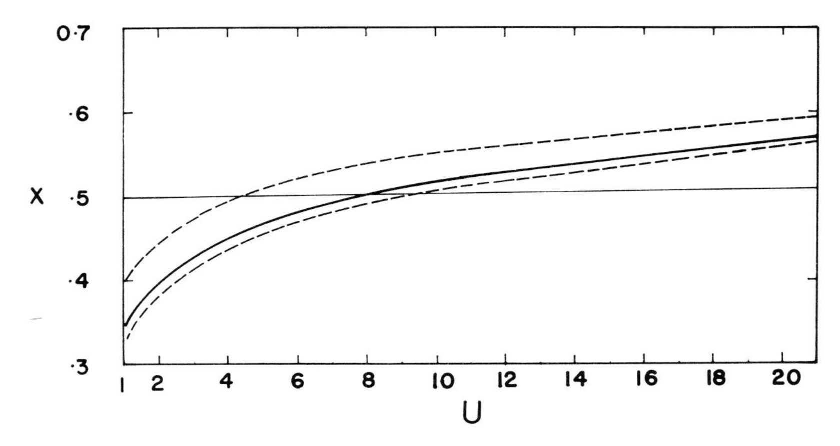

Fig. 9.6. The equivalent voltage for the atomic number correction can be determined from the above graph as a function of overvoltage, $U = E_0/E_q$. The central curve gives the suggested value of x and the broken lines give its variation with atomic number from 4 to 79. The value of the equivalent voltage is: $E_{eq} = x \cdot E_q + (1 - x) \cdot E_0$.

formation of scanning images in scanning electron microscopy (Chapter 14), or for the characterization of the mean atomic number of the bombarded specimen region. Furthermore, the electron backscattering affects the primary X-ray intensities generated in the target, and this effect, which is expressed by the backscatter correction factor R mentioned in the previous chapter, must be taken into account in the data reduction procedure.

The energy distribution of the backscattered electrons is broad (Fig. 9.7) and depends somewhat upon the atomic number of the target. The ratio of backscattered to primary electrons,

$$\eta = i_r/i_b \tag{9.3.1}$$

is called the *electron backscatter coefficient.* Unless otherwise specified, this coefficient is defined for an electron beam incidence normal to the flat surface of a target of depth larger than the maximum electron range. It comprises all electrons emitted into the hemisphere above the specimen surface.

The backscatter coefficient for elementary targets varies strongly with atomic number, approaching zero for the lightest elements and reaching values close to 0.5 for the heaviest elements (Fig. 9.8). It does not, however, vary substantially with the energy of the primary electrons. The backscattering coefficients of heavy elements slightly increase with increasing energy, while those of light elements decrease moderately (Fig. 9.9).

The first accurate measurements of electron backscatter coefficients were performed by Palluel [9.28]. More recent measurements by Bishop [9.29] and

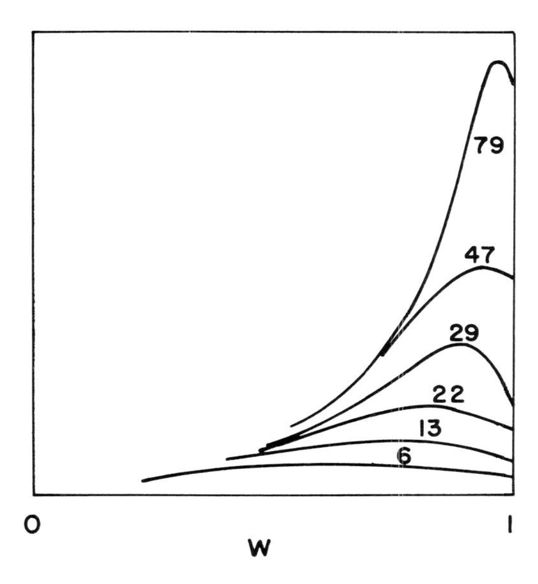

Fig. 9.7. Energy distribution of the backscattered electrons (schematic). The curves mainly follow Bishop's data [9.29]; for the determination of R, such distributions must be averaged over all angles of backscatter emission. The value $w = 1$ corresponds to the energy of primary electrons.

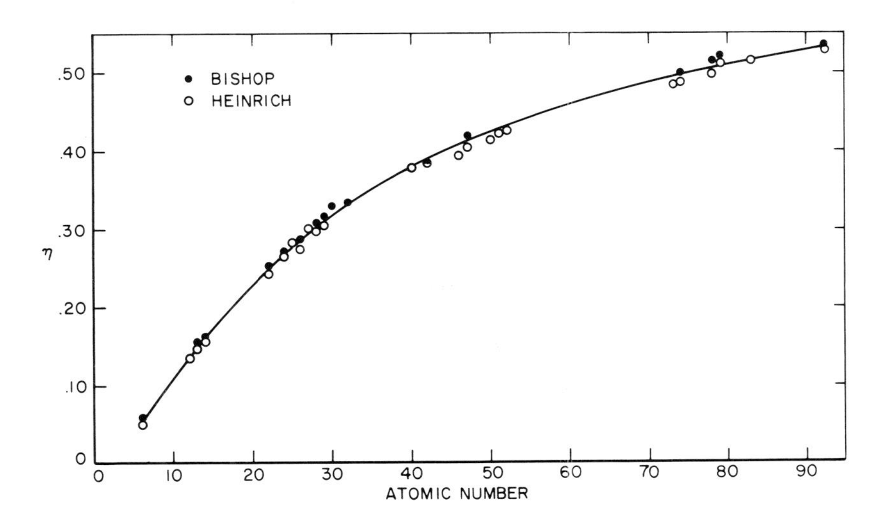

Fig. 9.8. The electron backscatter coefficient for normal electron beam incidence, according to Bishop [9.29] and Heinrich [9.30], at 30 kV.

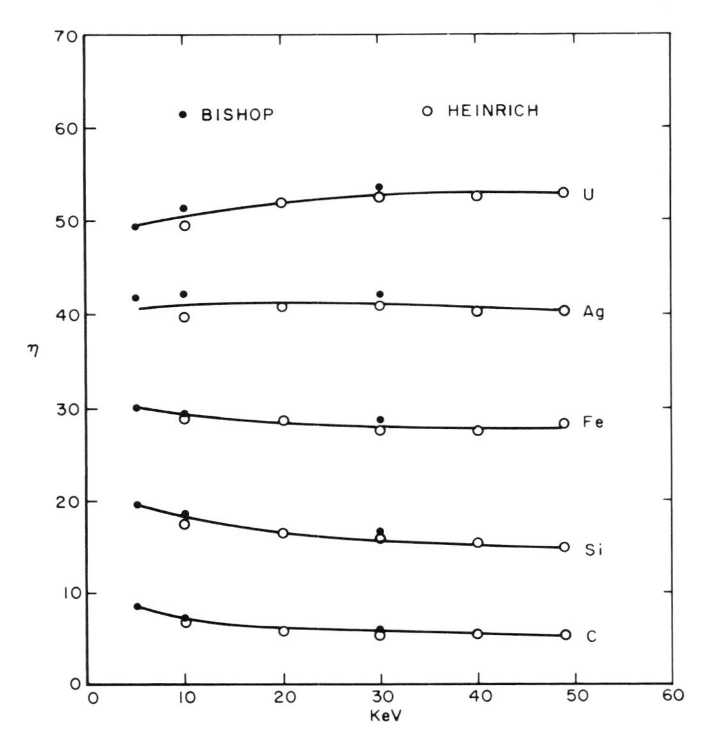

Fig. 9.9. Variation of the backscatter coefficient with the incident electron energy, E_0. Data are from [9.29] and [9.30].

Heinrich [9.30], in the energy range from 5 to 50 keV, are in good agreement between themselves and with Palluel; values obtained by Weinryb and Philibert [9.31] are about 10% lower. Other authors also obtained values close to those of Heinrich and Bishop. We may thus assume that the values of the backscatter coefficient are fairly well known (Table 9.4). They can be approximated by the following empirical formula:

$$\eta = 0.5 - 0.228 \times 10^{-4} (80 - Z) \times (|80 - Z|)^{1.3}. \qquad (9.3.2)$$

It is usually assumed that the backscatter coefficient is a smooth and monotonic function of the atomic number. Our measurements [9.30] indicate, however, that it also depends on the ratio of atomic weight to atomic number (Fig. 9.10), and that it is neither smooth nor monotonic with Z. A dependence on the atomic weight is also obtained in a theoretical calculation by Borovsky and Rydnik [9.33]. It is related to the role which the mass of the nucleus plays in the scattering of electrons.

Table 9.4. Electron Backscatter Coefficients Obtained From Massive Targets [9.30].[a]

Element	Atomic Number	10 keV	20 keV	30 keV	40 keV	49 keV
C	6	0.069	0.060	0.052	0.054	0.052
Mg	12	0.145	0.140	0.136	0.132	0.130
Al	13			0.149		
Si	14	0.174	0.164	0.159	0.153	0.150
Ti	22	0.262	0.253	0.243	0.245	0.243
Cr	24	0.273	0.268	0.265	0.259	0.256
Mn	25	0.292	0.286	0.284	0.277	0.276
Fe	26	0.289	0.287	0.275	0.276	0.282
Co	27	0.309	0.302	0.302	0.297	0.292
Ni	28	0.307	0.301	0.298	0.289	0.286
Cu	29	0.318	0.309	0.306	0.303	0.297
Zr	40	0.376	0.379	0.379	0.381	0.376
Mo	42	0.384	0.382	0.386	0.384	0.377
Pd	46	0.403	0.405	0.397	0.402	0.398
Ag	47	0.397	0.408	0.407	0.403	0.402
Sn	50		0.394	0.416	0.417	0.415
Sb	51	0.420	0.423	0.423	0.426	0.427
Te	52	0.425	0.427	0.427	0.426	0.431
Ta	73	0.481	0.491	0.487	0.501	0.503
W	74	0.482	0.494	0.489	0.499	0.506
Pt	78	0.486	0.504	0.499	0.503	0.510
Au	79	0.483	0.506	0.512	0.510	0.511
Bi	83	0.498	0.513	0.516	0.519	0.521
U	92	0.495	0.520	0.527	0.527	0.530

[a]The ranges of duplicate measurements fall within 0.003. The accuracy of the values is believed to be within 0.01.

The measurement of the electron backscattering of binary target materials by various authors [9.24], [9.29], [9.30], [9.34] shows that the backscatter coefficient of a multicomponent target can be expressed with good accuracy by the equation

$$\eta^* = \Sigma C_i \eta_i. \tag{9.3.3}$$

The additivity of the backscatter coefficients in mass fractions appears to hold to 1% or better, at electron energies of 20 keV or higher. For the purpose of calculating the effect of backscattering on the primary X-ray generation it may be considered to be valid at all electron energies of interest. Therefore, the measurement of the backscatter coefficient by means of target current measurements can serve as a base for accurate analyses of binary targets if the atomic numbers of the constituent elements differ sufficiently. Results obtained with this method, which is described in detail in [9.30], are shown in Table 9.5.

The spatial distribution of backscattering of electrons as a function of the specimen inclination, ϵ, is shown in Fig. 9.11 [9.35].

With increasing inclination of the electron beam with regards to the specimen

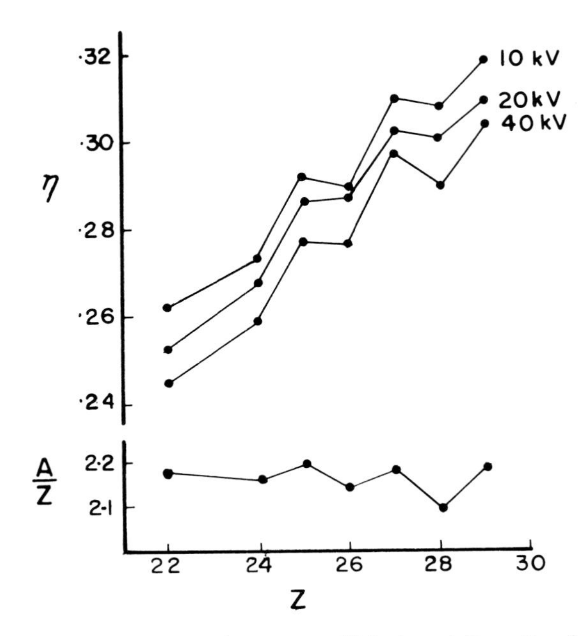

Fig. 9.10. Comparison of electron backscatter coefficients and the ratio of atomic weight and atomic number [9.32].

Table 9.5. Quantitative Analyses of Binary Specimens by Target Current Measurement With the ARL Electron Probe Microanalyzer.

SPECIMEN	INVESTIGATOR	WEIGHT FRACTION OF HEAVIER ELEMENT (THEORETICAL)	(FOUND)
GaP	Heinrich (1963)	0.692	0.688
UB_4	Heinrich (1963)	0.836	0.853
UB_{12}	Heinrich (1963)	0.647	0.679
USi_3	Heinrich (1963)	0.739	0.736
ZrB_2	Heinrich (1963)	0.808	0.807
ZrB_{12}	Heinrich (1963)	0.413	0.458
U_3Si	Colby (1964)	0.962	0.956
UP	Colby (1964)	0.885	0.884
UC	Colby (1964)	0.908	0.912
US_2	Colby (1964)	0.881	0.883
UN	Colby (1964)	0.944	0.936
U_6Fe	Colby (1964)	0.962	0.957

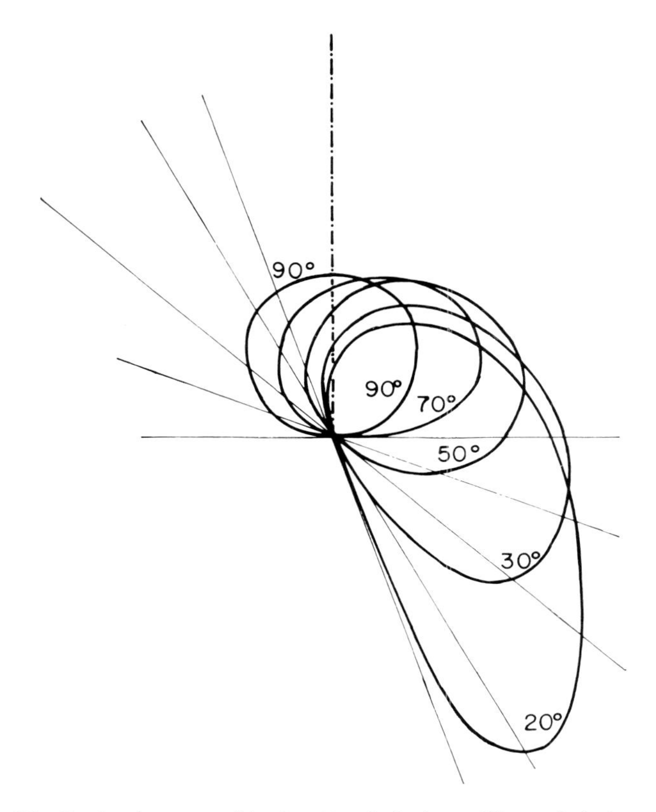

Fig. 9.11. Distribution in space of backscattered electrons. The angle between the impinging electron beam and the specimen surface ϵ is indicated for each distribution curve. The target is tantalum and the operating potential is 10 kV. (After date from [9.35].)

surface, the backscatter coefficient also increases (Fig. 9.12). This variation of the coefficient with the electron beam impact angle is of importance in the quantitative interpretation of X-ray intensities measured with inclined electron beam impact, and in the formation of target current and backscatter images, in scanning electron microscopy [9.30].

At electron energies below 5 keV the atomic number dependence of the backscatter coefficient diminishes, and the coefficients for all elements converge towards approximately 0.5. The behavior of backscattered electrons in this energy region was discussed by Darlington [9.37].

9.3.1 The Backscatter Correction Factor, *R*

The effects of electron backscattering on the generated primary X-ray intensities were recognized by Webster et al. in 1931 [9.38]. They described a

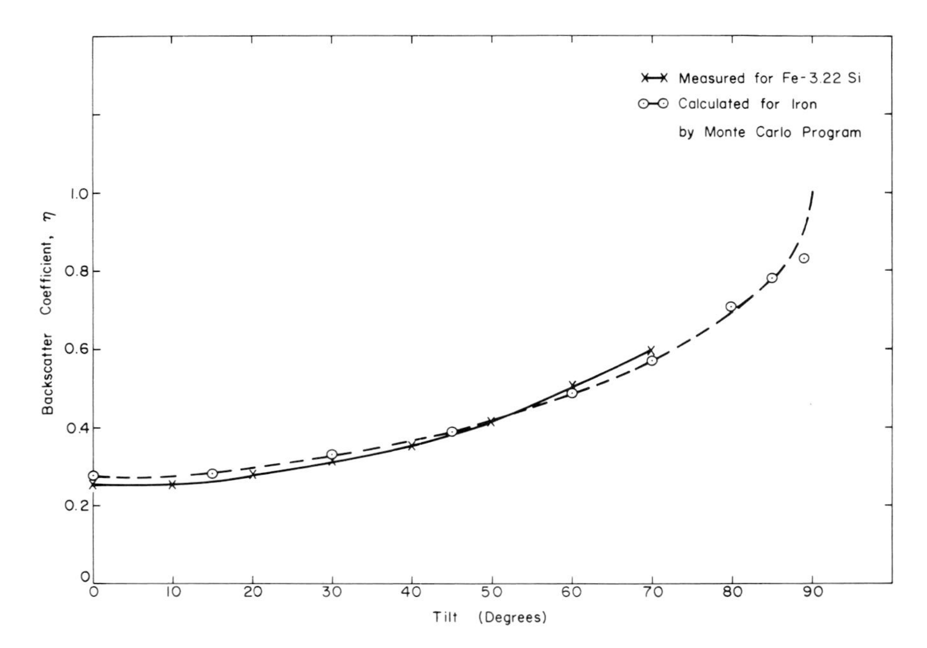

Fig. 9.12. Variation of the backscatter coefficient of iron as a function of specimen tilt, $90° - \epsilon$ [9.36].

method of calculating the correction factor R from the backscatter coefficient, η, and from information on the energy distribution of the backscattered electrons. Archard and Mulvey [9.39] indicated that in cases of large atomic number differences the reduction of generated X-ray intensity due to backscattering may approach 50%. The effect was taken into account by Castaing [3.22] in 1960, as one of the factors determining the value of α in his third approximation:

$$\frac{Ia}{I(a)} = \frac{\alpha_a \, C_a}{\Sigma \alpha_i \, C_i}. \tag{9.3.4}$$

(In [3.22], this approximation is called the "second approximation"; however, there exists another, in [9.1], which was then called the "second approximation.")

It should be noted that, unlike Eq. (8.2.2), this one refers to generated, rather than emitted, intensities. Both the stopping power effect and the effect of electron backscattering are embedded in a single "correction." The same approach was taken by Thomas [9.40]. Most investigators preferred to keep these two effects separate, as we have done in Eqs. (9.2.24) through (9.2.29).

Webster's calculation is based on the fact that each electron leaving the target, after scattering, with a residual energy, E_b, could have produced had it remained in the target, a number of ionizations equal to

$$n_b = \int_{E_q}^{E_b} \frac{Q}{S} dE.$$

To obtain the full number of ionizations which would have occurred additionally, had there been no backscattering, this probability must be integrated over the energy range of the backscattered electrons:

$$n_b = \int_{w_q}^{1} \frac{d\eta}{dw} \int_{Eq}^{wE_0} \frac{Q}{S} dE \cdot dw. \tag{9.3.5}$$

where w denotes the ratio of the energy of the backscattered electron to that of the primary electron impinging the specimen. The limit w_q corresponds to the value $(E_b = E_q)$.

$$w = \frac{E_b}{E_0} \qquad w_q = \frac{E_q}{E_0} = \frac{1}{U_0}.$$

Hence, the relative loss of ionizations caused by electron backscattering, R, is equal to

$$R = 1 - \frac{\int_{w_q}^{1} \frac{d\eta}{dw} \int_{E_q}^{wE_0} \frac{Q}{S} dE dw}{\int_{E_q}^{E_0} \frac{Q}{S} dE}. \tag{9.3.6}$$

Duncumb and Reed [9.26] performed the calculation of R according to Eq. (9.3.6). He used in this calculation the experimental results on the energy distribution of backscattered electrons obtained by Bishop [9.29] (Fig. 9.13).

Duncumb also proposed an algebraic expression which fits the numeric results of his computation:

$$\begin{aligned} R = \; & 1.000 + (-0.581 + 2.162 w_q - 5.137 w_q^2 + 9.213 w_q^3 \\ & - 8.619 w_q^4 + 2.962 w_q^5) \times 10^{-2} Z + (-1.609 - 8.298 w_q \\ & + 28.791 w_q^2 - 47.744 w_q^3 + 46.540 w_q^4 - 17.676 w_q^5) \\ & \times 10^{-4} Z^2 + (5.400 + 19.184 w_q - 75.733 w_q^2 + 120.050 w_q^3 \\ & + 110.700 w_q^4 + 41.792 w_q^5) \times 10^{-6} Z^3 + \end{aligned}$$

$$(-5.725 - 21.645w_q + 88.128w_q^2 - 136.060w_q^3 + 117.750w_q^4 - 42.445w_q^5) \times 10^{-8}Z^4 + (2.095 + 8.947w_q - 36.510w_q^2 + 55.694w_q^3 - 46.079w_q^4 + 15.851w_q^5) \times 10^{-10}Z^5. \quad (9.3.7)$$

This equation, as well as Duncumb's curves, and similar curves proposed by other authors, are based on the assumption that the electron backscatter coefficients vary smoothly with atomic number. As we have seen (Fig. 9.10), this assumption is not quite correct. A more accurate function for R could be obtained if the dependence of backscattering on the atomic weight could be included.

In the absence of experimental evidence, it is not obvious how the backscatter loss of a composite target can be related to those of pure elements. Duncumb

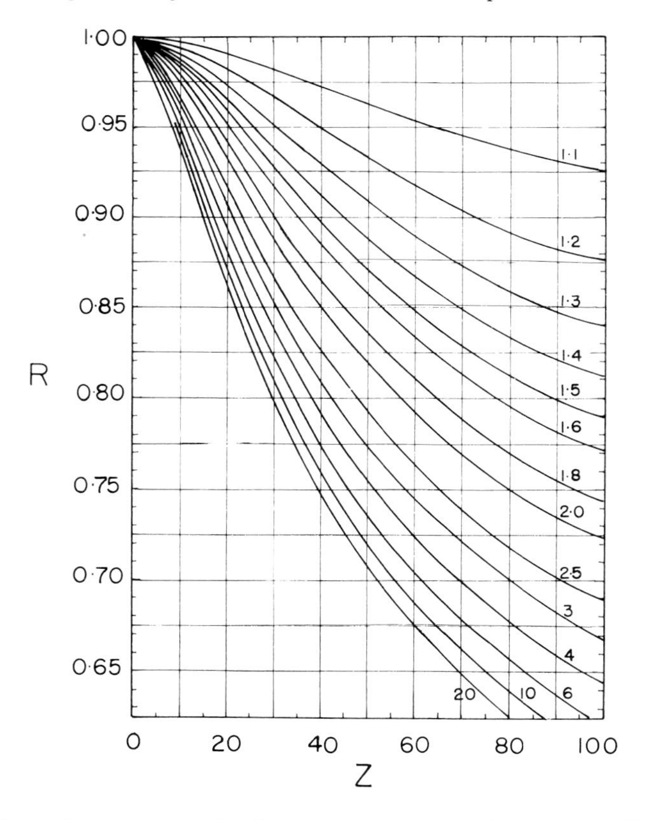

Fig. 9.13. The backscatter correction factor R as a function of overvoltage (figures inside graph) and atomic number, Z, from Duncumb. Data on energy distribution of backscattered electrons are from Bishop.

and Reed [9.26] proposed that the backscatter correction factors R can be considered additive, in proportions of the mass fractions of the elements present:

$$R^* = \Sigma C_i R(i). \tag{9.3.8}$$

This assumption is plausible since the same additivity holds for the backscatter coefficients [Eq. (9.3.3)]. Theoretical calculations by Bishop [9.41] and by Brown [9.42] indicate fairly good agreement with this summation rule.

When Duncumb and Reed determined the values of the backscatter correction factors, several conflicting sets of values for the energy distribution of backscattered electrons were available. The authors used the values published by Bishop, as well as those from Kulenkampff and Spyra [9.43]. Springer [9.44] performed similar calculations on the base of the values of Kulenkampff and Spyra. Other proposed values for R have been generated by Green [6.2] and by Thomas. All these alternatives to the values by Duncumb and Reed based on Bishop, are in disagreement with the experimental values for R obtained by Dérian which will be discussed next [9.45], [9.46].

The experimental method for determining the backscatter correction factor used by Dérian is shown schematically in Fig. 9.14. The electron beam is directed upon a flat semi-infinite surface of the target material of interest. The backscattered electrons emitted from this target are absorbed in a thin foil of the same material as the main target, and located at a short distance above it. A small hole in this foil permits the passage of the electron beam towards the main target. The thickness of the foil is chosen in such a fashion that all backscattered electrons are collected, while the absorption of X-rays generated within both the main target and within the foil is sufficiently low to be calculated with reasonable accuracy. By comparing the X-ray intensity generated by this target with that obtained from a conventional flat target of the same element, the fraction of X-rays which is generated within the foil, and hence R, can be determined. With this method, Dérian provided experimental values of R for aluminum, copper, and gold, at several primary electron energies. It would also be possible to obtain with his method values of R^* for multielement targets, but such experiments have not yet been performed.

Figure 9.15 indicates the considerable uncertainty by which Dérian's measurements are affected. Notwithstanding, the values obtained experimentally are very useful, since the spread of the data obtained by the theoretical calculations [Eq. (9.3.6)] is even larger.

The expression (9.3.7) is widely used in programs for electron microprobe data reduction. It is obvious, however, from Fig. 9.15, that the values of R are not as well known as would be desirable for accurate analyses, particularly if the standard and the specimen differ considerably in atomic number, and if the specimen contains large concentrations of elements of high atomic number.

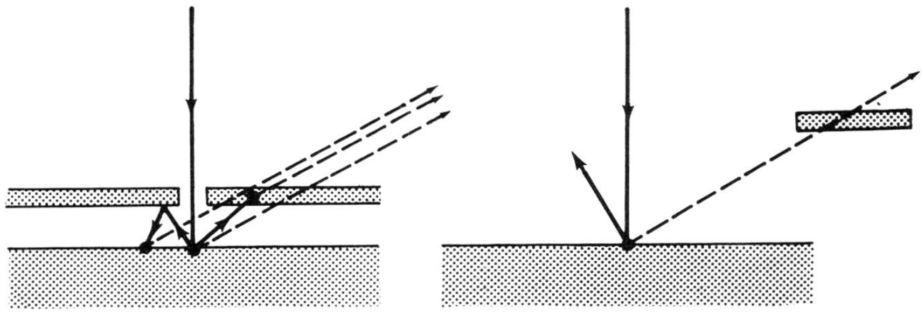

Fig. 9.14. Determination of the backscatter correction factor, R, by Dérian [9.45]. Left: the backscattered electrons (solid paths) are absorbed in a thin foil of the target material. X-rays are emitted from both the solid target and the foil (broken paths). Right: the loss of X-radiation on passage through the thin foil is determined separately and the results of the first experiment are corrected accordingly.

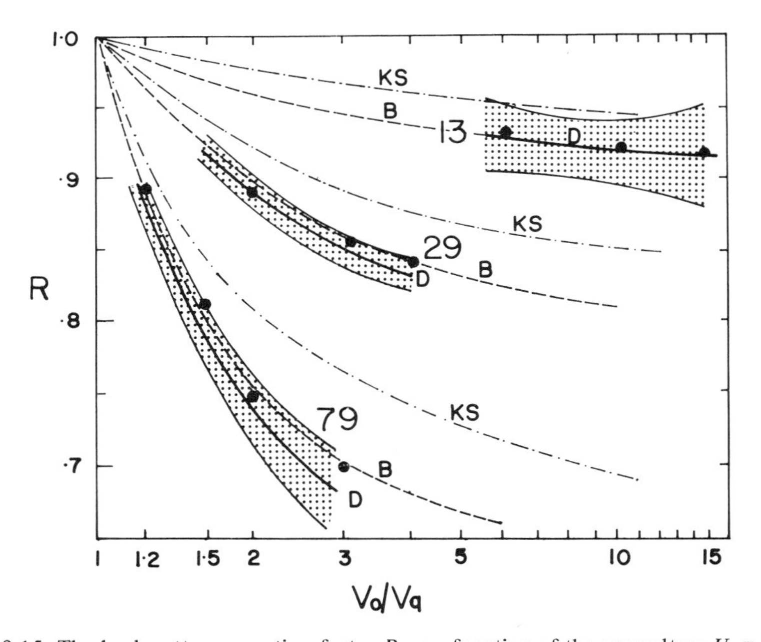

Fig. 9.15. The backscatter correction factor R, as a function of the overvoltage $U_0 = V_0/V_q$ and the atomic number (large numerals in the graph). Heavy lines: Dérian [9.45], [9.46]. Shaded area: uncertainty of Dérian's results. Full circles: Bishop [9.29]. B:generalization from Bishop's experiments by Duncumb [9.26] and Eq. (9.3.7). KS:generalization from Kulenkampff and Spyra [9.43], [9.44].

The limits of accuracy of the summation rules [Eqs. (9.3.3) and (9.3.8)] should be reevaluated as well. The validity of Eq. (9.3.8) has been challenged by Love et al. [9.10], who derived an expression for R as a function of the backscatter coefficient of the target.

9. REFERENCES

9.1 Castaing, R. and Descamps, J., *J. Phys. Radium* **16**, 304 (1955).

9.2 Brown, J. D., *Adv. X-Ray Analysis* **7**, 340 (1964).

9.3 Philibert J. and Tixier, R., in *Quantitative Electron Probe Microanalysis,* Heinrich, K. F. J., Ed., NBS Special Publ. 298, National Bureau of Standards, U.S. Dept. of Commerce, Washington, D.C., 1968, p. 13.

9.4 Criss, J., in *Quantitative Electron Probe Microanalysis,* Heinrich, K. F. J., Ed., NBS Special Publ. 298, National Bureau of Standards, U.S. Dept. of Commerce, Washington, D.C., 1968, p. 53.

9.5 Bethe, H. A., *Ann. Phys. (Leipzig)* **5**, 325 (1930).

9.6 Barkas, W. H. and Berger, M. J., in *Studies in Penetration of Charged Particles in Matter,* National Research Council Publ. 1133, U.S. National Academy of Sciences, 1964, p. 205.

9.7 Jorgensen, T., in *Penetration of Charged Particles in Matter,* National Research Council Publ. 752, U.S. National Academy of Sciences, 1960, p. 20.

9.8 Turner, J. E., in *Studies in Penetration of Charged Particles in Matter,* National Research Council Publ. 1133, U.S. National Academy of Sciences, 1964, p. 99.

9.9 Sweatman, T. R. and Long, J. V. P., *Proc. 5th Int. Conf. on X-Ray Optics and Microanalysis,* Möllenstedt, G. and Gaukler, K. H., Eds., Springer, Berlin, 1969, p. 432.

9.10 Love, G., Cox, M. G., and Scott, V. D., *J. Phys. D., Appl. Phys.* **11**, 7 (1978).

9.11 Powell, C. J., *Rev. Mod. Phys.* **48**, 33 (1976).

9.12 Green M. and Cosslett, V. E., *Proc. Phys. Soc. London* **78**, 1206 (1961).

9.13 Worthington, C. R. and Tomlin, S. G., *Proc. Phys. Soc. A,* **69A**, 401 (1956).

9.14 Hénoc, J., Heinrich, K. F. J., and Myklebust, R. L., NBS Tech. Note 769, National Bureau of Standards, U.S. Dept. of Commerce, Washington, D.C., 1973.

9.15 Burhop, E. H. S., *Proc. Cambridge Phil. Soc.* **36**, 43 (1940).

9.16 Powell, C. J., in *Use of Monte Carlo Calculations in Electron Probe Microanalysis and Scanning Electron Microscopy,* Heinrich, K. F. J., Newbury, D. E., and Yakowitz, H., Eds., NBS Special Publ. 460, National Bureau of Standards, U.S. Dept. of Commerce, Washington, D.C., 1976, p. 97.

9.17 Fabre de la Ripelle, M., *J. Phys. (Paris)* **10**, 319 (1949).

9.18 Whiddington, R., *Proc. Roy. Soc., Ser. A* **86**, 360 (1911-1912); also **89**, 554 (1914).

9.19 Duncumb, P., Ph.D. Thesis, Univ. of Cambridge, Cambridge, England, 1957.

9.20 Webster, D. L. and Clark, H., *Phys. Rev.* **9**, 571 (1917).

9.21 Cosslett, V. E., *Proc. 2nd Int. Symp. on X-Ray Optics and Microanalysis,* Engström, E., Cosslett, V. E., and Pattee, H. H., Eds., Elsevier, Amsterdam, 1960, p. 346.

9.22 Rosseland, S., *Phil. Mag.* **45**, 65 (1923).

9.23 Reed, S. J. B., *Brit. J. Appl. Phys.* **16**, 913 (1965).

9.24 Poole, D. M. and Thomas, P. M., *J. Inst. Metals* **90**, 228 (1961-1962).

9.25 Heinrich, K. F. J. and Yakowitz, H., *Mikrochim. Acta,* 123 (1970).

9.26 Duncumb, P. and Reed, S. J. B., in *Quantitative Electron Probe Microanalysis,* Heinrich, K. F. J., Ed., NBS Special Publ. 298, National Bureau of Standards, U.S. Dept. of Commerce, Washington, D.C., 1968, p. 133.

9.27 Martin, P. M. and Poole, D. M., Review 150, Metallurgical Reviews, The Metals & Metallurgy Trust, England, 1971, p. 19.

9.28 Palluel, M. P., *C. R. Acad. Sci.* **224,** 1492 (1947); also **224,** 1551 (1947).

9.29 Bishop, H. E., *Proc. 4th Int. Congr. on X-Ray Optics and Microanalysis,* Castaing, R., Deschamps, P., and Philibert, J., Eds., Hermann, Paris, 1966, p. 153.

9.30 Heinrich, K. F. J., *Proc. 4th Int. Congr. on X-Ray Optics and Microanalysis,* Castaing, R., Deschamps, P., and Philibert, J., Eds., Hermann, Paris, 1966, p. 159.

9.31 Weinryb, E. and Philibert, J., *C. R. Acad. Sci.* **258,** 4535 (1964).

9.32 Heinrich, K. F. J., in *Quantitative Electron Probe Microanalysis*, Heinrich, K. F. J., Ed., NBS Special Publ. 298, National Bureau of Standards, U.S. Dept. of Commerce, Washington, D. C., 1968, p. 7.

9.33 Borovsky, I. B. and Rydnik, V. I., in *Quantitative Electron Probe Microanalysis,* Heinrich, K. F. J., Ed., NBS Special Publ. 298, National Bureau of Standards, U.S. Dept. of Commerce, Washington, D. C., 1968, p. 35.

9.34 Colby, J. W., in *The Electron Microprobe,* McKinley, T. D., Heinrich, K. F. J., and Wittry, D. B., Eds., John Wiley & Sons, New York, 1966, p. 95.

9.35 Hart, D. M., Records of the 10th Symp. on Electron Ion and Laser Beam Technology, National Bureau of Standards, U.S. Dept. of Commerce, Washington, D.C., May 1969, p. 473.

9.36 Myklebust, R. L., Newbury, D. E., and Yakowitz, H., in *Use of Monte-Carlo Calculations in Electron Probe Microanalysis and Scanning Electron Microscopy,* NBS Special Publ. 460, National Bureau of Standards, U.S. Dept. of Commerce, Washington, D.C., 1976, p. 105.

9.37 Darlington, E. H., *J. Phys. D., Appl. Phys.* **8,** 85 (1975).

9.38 Webster, D. L., Clark, H., and Hansen, W. W., *Phys. Rev.* **37,** 115 (1931).

9.39 Archard, G. D. and Mulvey, T., *Proc. 3rd Int. Conf. on X-Ray Optics and Microanalysis,* Pattee, H. H., Cosslett, V. E., and Engström, A., Eds., Academic Press, New York, 1963, p. 393.

9.40 Thomas, P. M., *Brit. J. Appl. Phys.* **14,** 397 (1963).

9.41 Bishop, H. E., Ph.D. Thesis, Univ. of Cambridge, Cambridge, England, 1965.

9.42 Brown, D. B., Ph.D. Thesis, Massachusetts Institute of Technology, Cambridge, Mass., 1965.

9.43 Kulenkampff, H. and Spyra, W., *Z. Phys.* **137,** 416 (1954).

9.44 Springer, G., *Mikrochim. Acta,* 587 (1966).

9.45 Dérian, J. C., Ph.D. Thesis, CEA Rep. R3052, Univ. of Paris, Paris, France, 1966.

9.46 Castaing, R. and Dérian, J. C., *Proc. 4th Int. Congr. on X-Ray Optics and Microanalysis,* Castaing, R., Deschamps, P., and Philibert, J., Eds., Hermann, Paris, 1966, p. 193.

10.
The Absorption of Primary X-Rays

We will now discuss the attenuation of the primary photons which travel in the direction toward the detector, from their points of origin to the surface of the specimen. The absorption of primary X-rays is defined in Eq. (9.1.2) by the factor f_p which is equal to the fraction of the radiation directed toward the spectrometer which remains after emergence from the surface; in other words, the fraction absorbed by this radiation is equal to $(1-f_p)$. We will call f_p the primary absorption factor. The value of this factor is strongly affected by the penetration of the electron beam, which in turn depends on the operating potential, V_0, the angle of observation of the X-rays (X-ray emergence angle ψ), and the composition of the target which determines its X-ray absorption coefficient. The absorption loss can be negligibly small ($f_p \simeq 1$), or, depending on the above-mentioned parameters, it can be severe, and thus introduce serious uncertainty in the analytical result. Errors in the estimate of f_p are, in fact, the most significant cause for inaccuracy of the analysis. For this reason, the accurate determination of the primary X-ray absorption factor is of great importance.

Figure 10.1 shows schematically the emergence of the primary X-rays which, we recall, are emitted from a roughly spherical region. The intensity distribution of the generation of primary X-rays as a function of depth z, is shown at the left side of the figure. We call the function which expresses this depth distribution $\phi(z)$. (Depth is usually measured in linear units such as micrometers (μm). Therefore, the depth distribution is noted, by many authors, including Castaing, as $\phi(\rho z)$. In this book, however, z (mg/cm^2) denotes mass thickness, so that we write for the depth distribution $\phi(z)$, and we can henceforth ignore the density, ρ, as was discussed in Subsection 9.2.)

The characteristic X-rays generated within the specimen propagate isotropically in all directions. Hence, if the spectrometer or the energy-dispersive detector subtend an angle Ω, then a fraction of primary X-rays generated at any given point, equal to $(\Omega/4\pi)$. I_p, moves toward the spectrometer. The mean angle of emergence of the X-rays is ψ, but in reality the emergence angles cover

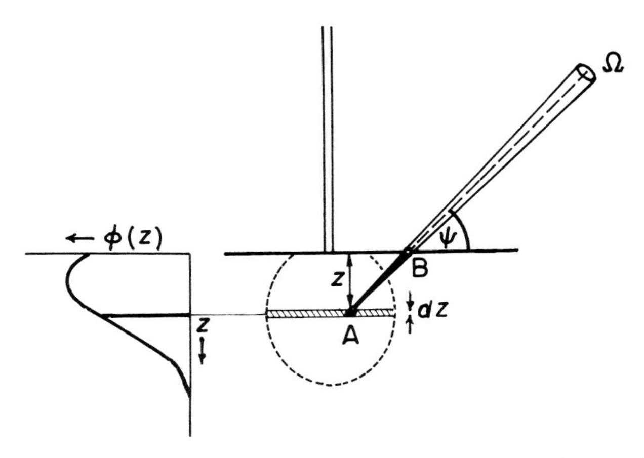

Fig. 10.1. Distribution in depth of generation of primary X-ray photons. The electrons diffuse within the target and produce X-rays within a roughly spherical volume. The probability distribution of primary X-ray generation, $\phi(z)$, is shown at the left, as a function of depth within the specimen, z (g/cm^2). The X-rays emerging at an angle ψ from the specimen surface, and contained within a small solid angle, Ω, reach the detector. On their path within the target (from A to B), some photons are absorbed.

a range which is determined by the aperture of the detector. The probability of absorption of an X-ray photon from any point within the excited region depends on the distance of this point from the specimen surface. Because of the distribution in depth of the sites of X-ray generation, the absorption loss cannot be computed directly from Beer's law, and it is not sufficient to determine for this purpose a mean depth of generation. Rather, the calculation requires the knowledge of the distribution in depth. The length of the path from a point at a depth z to the specimen surface depends on the X-ray emergence angle with respect to the specimen surface, ψ. In a rigorous calculation, the absorption loss would be obtained by integrating over both depth and range of X-ray emergence angles. In practice, it is assumed that no significant error arises if we stipulate that all photons emerge at the angle ψ. This assumption is not strictly valid when both the detector aperture and the absorption corrections are large, as could be the case with a Si(Li) detector very close to the specimen.

The distribution in depth of the primary X-ray generation depends on the deceleration and the scattering of the primary electrons, as well as on the variation of the ionization cross section with the electron energy. The scattering of electrons varies greatly with the atomic number of the scattering atoms, and the knowledge of the interactions of the penetrating electron with the target matter is affected by large uncertainties. Theoretical calculations of the absorption loss correctly predict trends with atomic number and acceleration potential, but they are not as yet sufficiently accurate to provide a good model for quantitative analysis. Therefore, theoretically derived models must be tested by experiments, and, if necessary, adjusted empirically to conform with the results of experimental evidence. Even after such an adjustment, the models are affected by a significant uncertainty. Fortunately, the operator can, by the choice of experimental conditions, almost always minimize the absorption correction and hence the corresponding uncertainty.

10.1 THE ABSORPTION FACTOR, f_p, AND THE DISTRIBUTION IN DEPTH, $\phi(z)$

To define the parameters used in discussing X-ray absorption, we will consider here the most common and best documented case: The electron beam impinges at a right angle upon the flat surface of a homogeneous specimen which is thicker than the range of the electrons (normal beam incidence and flat semi-infinite homogeneous specimen).

Consider a thin layer of target material of thickness, dz, parallel to the specimen surface, at a distance z (Fig. 10.1). Call $I(z)dz$ the probability that one primary electron impinging upon the specimen will produce a photon within this layer. The probability of production of an X-ray photon anywhere within the target is equal to

$$I = \int_0^\infty I(z)dz. \qquad (10.1.1)$$

Let a spectrometer of aperture Ω observe the specimen at a mean angle of X-ray emergence ψ. Because the X-ray emission after ionization is isotropically distributed, in all directions, the intensity emitted from the layer at depth z which reaches the detector is, after absorption, equal to

$$I'(z) = \frac{\Omega}{4\pi} I(z)dz \quad \exp(-\mu z \cdot \csc\psi). \qquad (10.1.2)$$

To obtain the probability of detection of a photon from anywhere within the specimen, we must integrate from the surface to the maximum depth of X-ray emission (or infinite):

$$I' = \frac{\Omega}{4\pi} \int_0^\infty I(z) \exp(-\mu z \cdot \csc\psi)dz. \qquad (10.1.3)$$

If there had been no X-ray absorption ($\mu = 0$), the radiation arriving at the detector would have been

$$I'(\mu = 0) = \frac{\Omega}{4\pi} \int_0^\infty I(z)dz = \frac{\Omega}{4\pi} I. \qquad (10.1.4)$$

The absorption factor, f, is thus equal to

$$f(\mu, \csc\psi) = \int_0^\infty I(z) \exp(-\mu z \cdot \csc\psi)dz \Bigg/ \int_0^\infty I(z)dz \qquad (10.1.5)$$

If in this definition only the absorption of primary X-rays were considered, the parameters f, $I(z)$, and I' in the previous equations, would carry the subscript p.

We note that μ and ψ always appear in the combination $\mu \csc \psi$, for which the following notation is common:

$$\mu \csc \psi = \chi. \tag{10.1.6}$$

Therefore, the absorption factor is frequently called $f(\chi)$, although it depends on several variables other than χ. The notation $f(\chi)$ is often ambiguous, since in theory it usually refers to primary radiation only, while in experiments as a rule it denotes the attenuation of all characteristic radiation. If the measured radiation contains a significant fraction of fluorescent radiation, the difference is not negligible; it is therefore preferable to denote the absorption factor of primary radiation with the subscript (f_p). In experiments in which the contributions of primary and secondary excitation are not separated, an overall experimental attenuation factor, f, (without subscript) can be defined.

Equation (10.1.5) indicates that the absorption factor f_p decreases with increasing values of the mass absorption coefficient, μ, and with decreasing X-ray emergence angle, ψ (a decreasing absorption factor means increased absorption losses). The absorption factor also decreases with increasing operating potential, V_0, due to the change in electron range, which increases approximately as $V_0^{5/3}$. There is also a dependence on the critical excitation potential, V_q, which sets a lower limit to the energy of an electron which can produce an X-ray photon of the line of interest. Since the average electron energy diminishes with depth, a higher critical excitation potential—other conditions being equal—produces a shallower primary emission and hence a smaller absorption loss.

The distribution in depth of primary radiation should also vary with the atomic number of the absorber, since the scattering parameters depend strongly upon the atomic number. However, due to a compensation between electron scattering and deceleration, this effect is not very large.

In summary, an accurate model for the calculation of the primary absorption factor should, in principle, have the form:

$$f_p = f(\mu, \psi, V_0, V_q, Z) \quad \text{or} \quad f_p = f(\chi, V_0, V_q, Z). \tag{10.1.7}$$

This function will be valid for one given electron beam incidence angle only. If this angle is considered a variable, it must also enter the list of parameters which modify f_p. This aspect is covered at the end of this chapter. Unless specified otherwise, we will always assume that the angle of electron beam incidence is normal.

It is practical to express the depth distribution of X-rays on a relative scale.

In so doing we avoid the inclusion in its definition of the primary excitation probability, $I(z)dz$, which is affected by the considerable uncertainties of the generation equation, and cannot be determined with great accuracy by a direct experiment. Castaing and Descamps proposed for this reason an intensity scale in which the intensity of X-rays generated by a thin target layer at any depth z is divided by the intensity produced by an unsupported layer of the same thickness of target material. This intensity is proportional to the ionization cross section of the target material at the energy E_0:

$$I_u(0) \cdot dz = i_b \cdot Q_q \cdot dz \cdot \omega \cdot p.$$

We will call the ratio of generated intensities $\Phi(z)$:

$$\Phi(z) = I(z) \cdot dz \ / \ I_u(0) \cdot dz.$$

This relative scale was used in the publications of the authors' tracer experiments for determining the distribution in depth, which will be described later. Castaing and Descamps [9.1] also defined an integral absorption function, $F(\chi)$:

$$F(\chi) = \int_0^\infty \Phi(z) \cdot e^{-\chi z} dz. \tag{10.1.8}$$

This function is a Laplace transform [10.1] of $\Phi(z)$. A Laplace transform is defined by:

$$\mathcal{L}\,[f(t), p] = \int_0^\infty e^{-pt} f(t) dt. \tag{10.1.9}$$

If the absorption is omitted ($\chi = 0$), one obtains the corresponding function

$$F(0) = \int_0^\infty \Phi\,(z) dz,$$

and the absorption factor can be obtained by

$$f_p = F(\chi)/F(0). \tag{10.1.10}$$

Since $F(0)$ is a measure of the generated primary intensity, it contains the generation equation, or "atomic number correction," being proportional to our generated intensity I_p. Several authors tried to derive analytical expressions for the practical use of $F(\chi)$ which would include the effects of both primary X-ray generation and absorption in the same algorithm. In the interest of clarity, and to avoid error compensations from one effect to the other, it seems preferable to separate the primary generation (expressed by I_p) from the primary attenuation (expressed by f_p). We will therefore present the depth distribution function in a normalized form in which $\phi(z)$ is the fraction of the total generated primary radiation which is produced in the layer of thickness dz, at depth z. The normalizing condition is thus

$$\int_0^\infty \phi(z)dz = 1 \tag{10.1.11}$$

and f_p can now be obtained simply as

$$f_p = \int_0^\infty \phi(z) \exp(-\chi z)dz. \tag{10.1.12}$$

The values of the normalized function can be obtained experimentally, provided that corrections for secondary emission are applied. The corresponding values of $I(z)$ would be:

$$I(z) = I_p \cdot \phi(z). \tag{10.1.13}$$

The relation between $\Phi(z)$ and $\phi(z)$ is:

$$\Phi(z) = \phi(z) \frac{I_p}{Q_q(E_0) \cdot \omega \cdot p}.$$

The uncertainties inherent in the determination or calculation of absolute X-ray intensities, which affect the values of I, $I(z)dz$, and I', are of no consequence to practical analysis of electron-opaque ("thick") specimens, which is based on intensities relative to those obtained from a known standard. In contrast, the uncertainties in f_p can be of serious consequences.

10.2 THE EXPERIMENTAL DETERMINATION OF THE DEPTH DISTRIBUTION OF X-RAY EMISSION

The experimental information needed to determine the absorption losses of X-rays in a target can be obtained in the following types of experiments:

1 the measurement of X-ray emission from tracers implanted at various depth levels in an experimental target (tracer method);
2. the measurement of the variation of X-ray emission from a target as a function of the angle of X-ray emergence (variable emergence angle method);
3. the observation of electron absorption, transmission, and backscattering, as well as of X-ray emission, from targets which are thinner than the depth of electron penetration.

Observations on the behavior of electrons interacting with thin films were among the first experiments of the above list. They were frequently summarized in "laws" of attractive simplicity, such as the exponential laws of electron attenuation of Lenard [10.2] or those of Williams [10.3] and Webster et al. [10.4]. The reason for their simplicity is the absence, in very thin films, of significant effects of multiple interactions, diffusion, and backscattering. For this very reason, the relations valid for thin films cannot be extended in a simple fashion to thick targets. On the other hand, the theoretical and experimental study of single interactions furnishes us with the bases for the Monte Carlo method, which will be discussed in Chapter 13.

10.2.1 The Tracer Method (Fig. 10.2)

Castaing and Descamps [9.1] described in 1955 the experiments with tracers which enabled them to determine the distribution function, $\Phi(z)$. They deposited by vacuum evaporation a layer of 0.03 mg/cm^2 of zinc on a copper block. By the addition of layers of copper of various thicknesses, they created a target in which, in effect, a uniformly thick layer of the zinc tracer was buried within the copper target at various levels of depth. Simultaneously, an identical zinc layer was deposited on a thin foil of an organic material; this layer served as a unit of mass of the tracer, hence it was unnecessary to accurately measure the mass cross section of the tracer. The authors postulated that, to give an adequate representation of the X-ray intensities which would be emitted at the same depth by the target element, the tracer should meet the following conditions:

a. It should be close in atomic number to the target element so that the electron distribution would not be significantly altered by the presence of the tracer.
b. Fluorescent excitation of the observed tracer line by a strong line of the target element must be avoided.

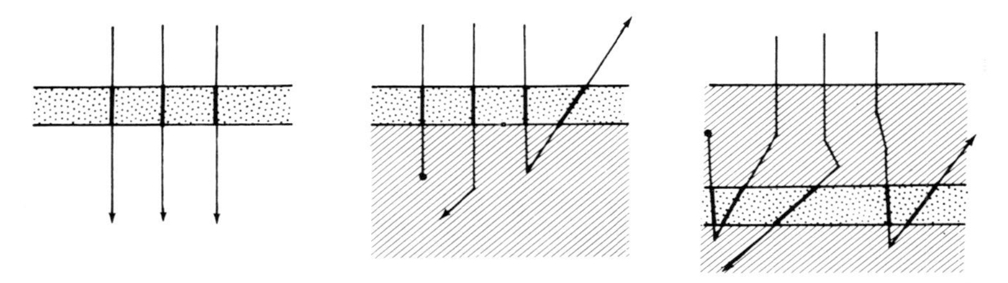

Fig. 10.2. The tracer method. In the passage through a thin unsupported tracer (left), the paths are in the original beam direction. The X-ray production of supported films (center) is enhanced by backscattered electrons which traverse the tracer in the reverse direction. At greater depth (right), the intensity of generation of X-ray increases further due to the randomization of the path directions. The absorption of the emitted X-rays also increases with depth; for the purpose of the determination of $\phi(z)$ it can be taken into account by means of Beer's law.

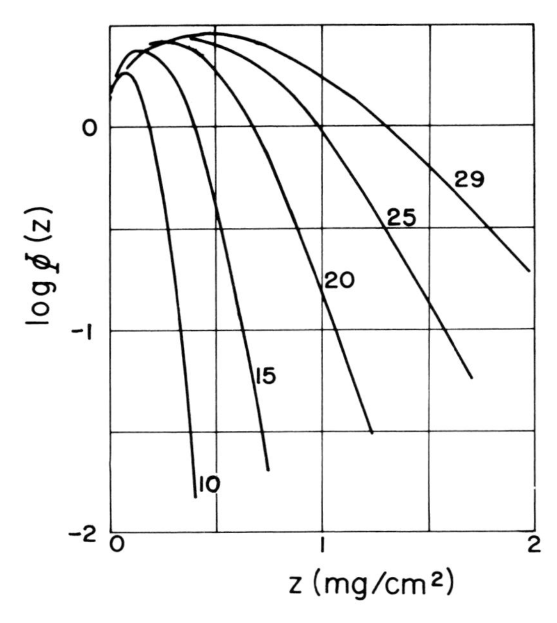

Fig. 10.3. Depth distribution in aluminum of the generation of X-rays [10.5]. The tracer was Mg K. Numbers close to the curve indicate the operating potential (kV). The logarithm (base 10) of the generated intensity divided by that of an unsupported layer is plotted versus depth (mg/cm^2).

c. The emission from the tracer should not be strongly absorbed by the target material so that the absorption loss can be accurately calculated.

The experiments were performed at an operating potential of 29 kV. The observed X-ray emission was corrected by absorption in the target, by means of Beer's law. The logarithm of the depth distribution function $\Phi(z)$ was plotted as a function of depth (g/cm^2) of the tracer within the target, (z).

The authors, and later Castaing and Hénoc [10.5], measured the X-ray emission in tracer experiments covering an array of target materials and tracers. A list of these experiments, and of similar tracer experiments by Vignes and Dez [10.6], Shimizu et al. [12.82, p. 127], and Parobek and Brown [10.7] is given in Table 10.1.

The observed distributions (Figs. 10.3 through 10.5) have the following common characteristics:

a. The general shape of the function $\phi(z)$ is similar for all targets and tracers.
b. Where the tracer is directly at the specimen surface ($z = 0$), the value of $\phi(z)$ is larger than that obtained from the "unsupported" tracer layer (deposited

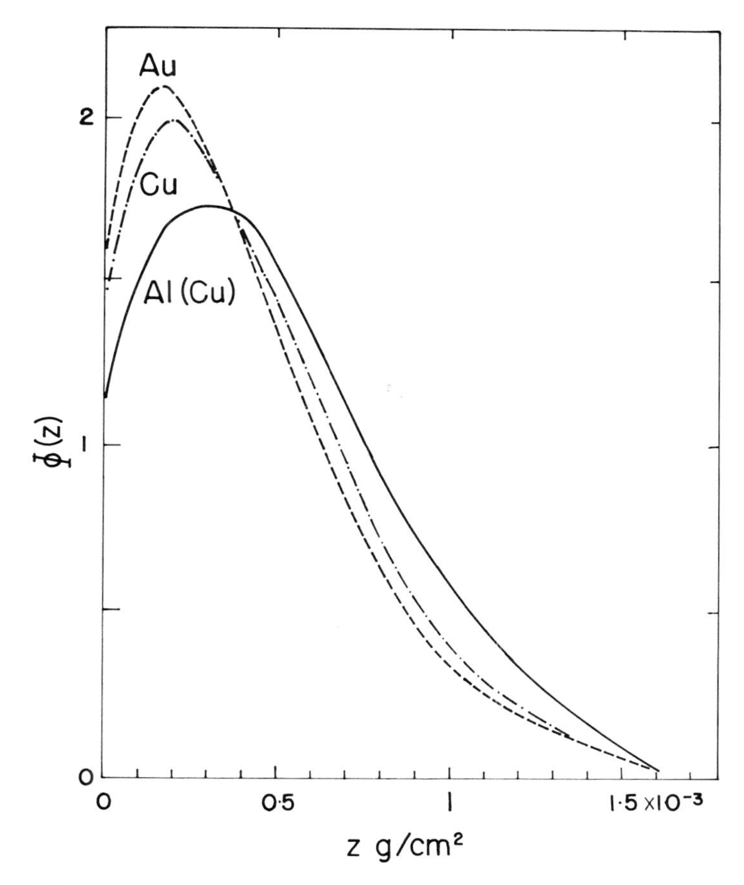

Fig. 10.4. Depth distribution $\Phi(z)$ of the generation of X-rays in aluminum (tracer: Cu Kα), copper, (Zn Kα), and gold (Bi Lα_1), at 29 kV, after Castaing and Descamps [9.1]. The intensity scale is linear and the intensity emitted by the unsupported traces is the unit of scale.

on the organic film) ($\Phi(0) > 1$). With increasing depth, $\phi(z)$ further increases to a maximum which, at an operating potential of 29 kV, is located close to 0.2 mg/cm^2. From there on, the decrease is rapid and approaches an exponential curve.

c. The penetration increases rapidly with accelerating voltage.
d. The value of the depth distribution at depth zero, $\phi(0)$, and that of the maximum, increase with increasing atomic number of the target; the depth at which the maximum occurs decreases with increasing atomic number.

The increase of emission from a layer at the target surface over the emission

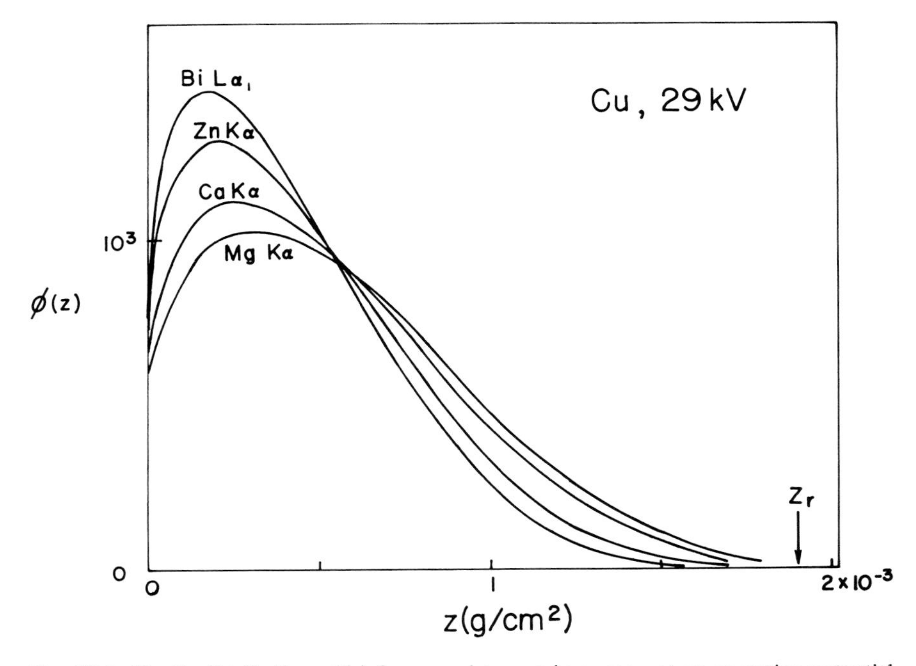

Fig. 10.5. Depth distributions $\phi(z)$ for several tracers in copper, at an operating potential of 29 kV, obtained by the NBS Monte Carlo program. The depth of excitation decreases with increasing excitation potential. z_r is the electron depth range, Eq. (13.1.5), with $V_q = 0$.

from an unsupported layer is caused by electrons which return from the substrate to the layer, having changed direction due to scattering (backscattered electrons, see Fig. 10.2). A small contribution may also be due to fluorescence of the tracer by continuous radiation generated within the target. Both effects increase with the atomic number of the target. The further increase of $\phi(z)$ with depth is due to changes in direction of the downward electrons. Initially, all primary electrons travel along paths perpendicular to the tracer layer. After onset of the electron diffusion, most electron paths within the layer are oblique, and therefore longer than the layer thickness. Hence, the probability of ionization per primary electron increases until such depth that the effects of energy loss of the primary electrons become predominant.

The depth distribution curve for the gold target shows a peculiar flattening at large depth. This abnormally high level of X-ray production at large depth is caused by fluorescence due to the continuum, which will be discussed in Chapter 12.

If the tracers are thin enough, their presence will not significantly affect the depth distribution of the electrons. The differences in depth distribution

Table 10.1. Tracer Experiments.[a]

Castaing and Descamps, 1955 [9.1]			
Cu(ZnKα)	29	Al(ZnKα)	29
Au(BiLα)	29	Al(CrKα)	29
Al(CuKα)	29	Al(BiLα_1)	29
Castaing and Hénoc 1966 [10.5]			
Al(MgKα)	10, 15, 20, 25, 29		
Vignes and Dez, 1968 [10.6]			
Ti(VKα)	17, 20, 24, 29, 35	(also gives values of $\Phi(0)$)	
Ni(CuKα)	17, 20, 24, 29, 35		
Pb(BiLα_1)	29, 33		
Schmitz, Ryder, and Pitsch (wedge technique, see Fig. 11.17), 1969 [11.13]			
Cu(NiKα)	20, 25.5, 30		
Shimizu, Murata, and Shinoda, 1972 [12.82, p. 127]			
Cu(ZnKα)	23.8, 25.1, 26.5, 28.5, 30.0, 31.7, 33.1, 34.7, 37.0		
Brown and Parobek, 1972 [12.84, p. 163] ($\epsilon = 60°$)			
Cu(ZnKα)	13.4, 18.2, 23.1, 27.6		
Cu(NiKα)	18.2, 23.1, 27.6		
Cu(CoKα)	18.2, 23.1, 27.6		
Cu(FeKα)	18.2, 23.1, 27.6		
Cu(DyLβ_1)	18.2, 23.1, 27.6		
Au(DyLα_1)	18.2, 23.1, 27.6		
Brown and Parobek 1976 [10.7] ($\epsilon = 90°$, $\psi = 75°$)			
Al(SiKα)	6, 8, 10		
Ni(SiKα)	6, 8, 10		
Ag(SiKα)	6, 8, 10		
Au(SiKα)	6, 8, 10		
Al(CuKα)	6, 8, 10		
Ag(CuKα)	12, 15		
Ni(CuKα)	12, 15		

[a]Numbers are operating potentials in kilovolts.

observed in the same matrix with different tracer elements (Fig. 10.5) are therefore attributed to the respective critical excitation potentials.

When Castaing and Descamps performed the first tracer experiments in an aluminum matrix, they were forced to use a tracer which emitted hard radiation, because of limitations in the wavelength range of their spectrometers. In practice, experiments with tracers of widely different atomic numbers are useful. If we were, for instance, to analyze for traces of copper in an aluminum matrix, the appropriate $\phi(z)$ distribution for such a target would be that of a copper tracer in aluminum, while for a standard of pure copper the zinc tracer in copper would be appropriate. Fortunately, the effect of the atomic number of the target on the distribution in depth is moderate, particularly in view of other uncertainties which affect the primary absorption correction.

The most important result of the tracer experiments was the determination of the absorption factor, f, which can be obtained by Eq. (10.1.5) or Eq. (10.1.12). It should be noted, however, that the *primary* absorption factor, f_p, can be obtained only if the experimental depth distributions have been corrected for the contributions, if significant, of secondary emission.

Figures 10.6 and 10.7 show the absorption factors derived from some of the tracer experiments on logarithmic, linear, or inverse linear scales. The relative merits of these scales will be discussed later.

10.2.2 The Variable Emergence-Angle Method for the Determination of the Absorption Factor

Attention was drawn toward the absorption of X-rays within the target when Webster observed the attenuation of emerging radiation at glancing angles [10.9]. In 1922, Kulenkampff [6.5] developed a method for determining the absorption within the X-ray target, on the basis of measurements of X-rays at varying emergence angles. This method was further developed, and extensively used, by Green [10.10]. The principles of the method are seen in Fig. 10.8. A

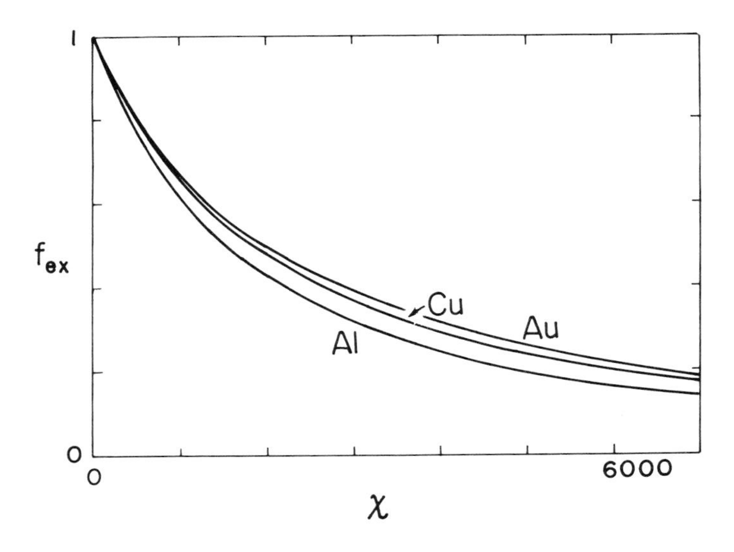

Fig. 10.6(a). Linear presentation of the absorption factors obtained from the tracer experiments of Castaing and Descamps [9.1]. All measurements were made with an operating potential of 29 kV. Tracers were: Cu Kα for aluminum, Zn Kα for copper, and Bi Lα for gold (see Table 10.1).

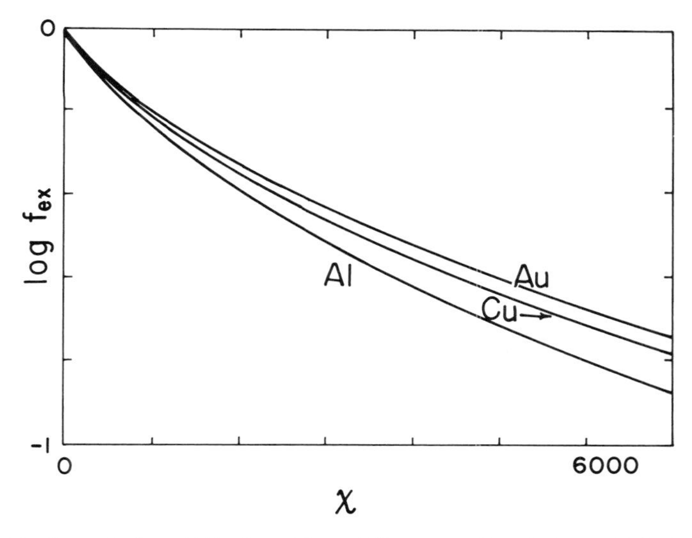

Fig. 10.6(b). Presentation of the logarithm of the absorption factors obtained from the tracer experiments of Castaing and Descamps [9.1].

flat semi-infinite target is bombarded by a stable electron beam impinging at a fixed (usually normal) angle. The emission of a characteristic X-ray line is measured by a spectrometer of narrow aperture, Ω, at the average emergence angle ψ. The mass absorption coefficient for the target and radiation of interest is known, and hence we can compute the parameter $\chi = \mu$ cosec ψ. At high values of the emergence angle, the absorption is relatively small. We call the highest emergence angle ψ_1. After further measurements at smaller values for ψ, over the full range available with the instrument, we calculate the ratios of emergent intensities, using angle ψ_1 as a reference (i.e., we calculate I'_i/I'_1). According to Eq. (9.1.1), the following relation holds:

$$I'_i/I'_1 = f(\chi_i)/f(\chi_1). \tag{10.2.1}$$

At the highest theoretically possible angle ($\psi = 90°$), χ would be equal to μ (Fig. 10.9). To obtain the absolute value of $f(\chi_i)$, we must extrapolate the ratios I'_i/I'_1 on a graph which shows, on some appropriate scale, the variation of $f(\chi_i)/f(\chi_1)$, to the hypothetical value $1/f(\chi_1)$, which would occur where χ_i is zero. Once this extrapolated value is obtained, $f(\chi_1)$, and therefore all values for $f(\chi_i)$, can be calculated. The extrapolation, however, is not trivial, since the absorption factor, f, is not a linear function of χ. Green uses for this purpose a drastically simplified model for primary X-ray generation which assumes that all radiation is generated at the same depth level, the "mean depth of X-ray produc-

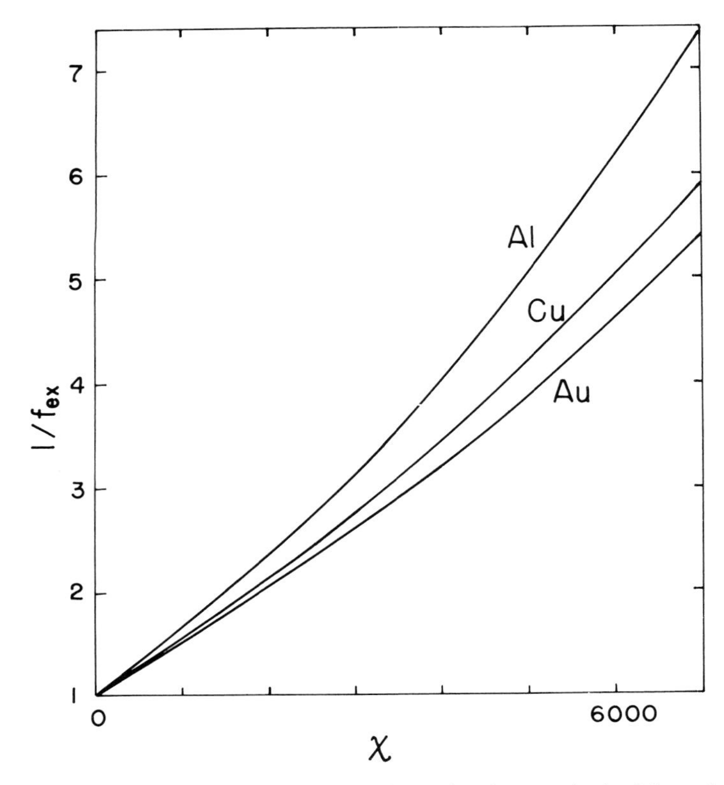

Fig. 10.6(c). Inverse linear presentation of the absorption factors obtained from the tracer experiments of Castaing and Descamps [9.1].

tion," $\bar{z}$. With this assumption, the attenuation within the target follows simply from Beer's law:

$$-\ln f_p = \bar{z} \cdot \csc \psi \cdot \mu = \bar{z}\chi. \qquad (10.2.2)$$

It would therefore be best to extrapolate, not the linear value of I'_i/I'_1—which is proportional to the absorption factor—but its logarithm. However, the implied simplification, and hence the extrapolation, are not accurate. Furthermore, some of the mass absorption coefficients for X-rays used by Green in the evaluation of his experiments were significantly in error. For these reasons, the values of f_{ex} proposed by him are not fully reliable. However, they have very significantly extended our knowledge about absorption of X-rays in a target, and have therefore greatly stimulated the development of generalized analytic solutions such as that of Philibert (see p. 272) for the absorption factor.

Unfortunately, the performance of Green's experiment by means of conven-

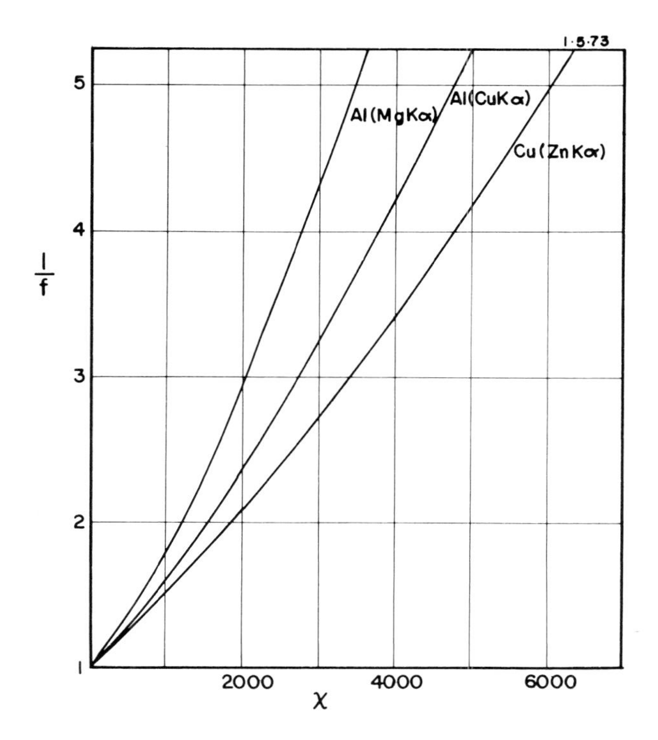

Fig. 10.7. Absorption factors from tracer experiments by Castaing and Descamps [9.1] and Castaing and Hénoc [10.5]. All curves shown are for an operating potential of 29 kV. In this graph the inverse of the absorption factor, $1/f$, is plotted as a function of χ.

tional electron probe microanalyzers is not possible, since in such instruments the angle between the electron beam and the X-ray path into the spectrometer is fixed. Hence, for normal electron beam incidence upon the target, the X-ray emergence angle is uniquely defined. The instrument used by Green (Fig. 10.8) had a conventional electron beam column, except for the elongated objective lens which did not contain a pole piece. The electron beam could be focused at a working distance of 8 cm. This arrangement provided the space for an assembly in which the X-ray emergence angle of the spectrometer could be varied from 1 to 45°. The X-ray spectrometer was located outside the vacuum enclosure; for the observation of the soft carbon K-band, the crystal spectrometer was replaced by a flow-proportional detector operating in the energy-dispersive mode. All other measurements were performed with the aide of a single-crystal spectrometer, with flat lithium fluoride and gypsum crystals. The angular divergence of the spectrometer was restricted to less than 1°. The specimens were positioned normally to the electron beam and could be displaced to expose fresh surfaces as desired. The accelerating voltage covered a continuous

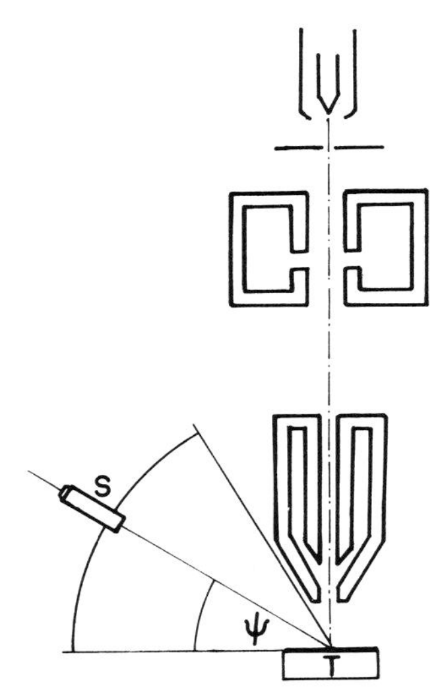

Fig. 10.8. Schematics of the instrument used for the variable emergence-angle method. The spectrometer, S, can be energy or wavelength dispersive. ψ:X-ray emergence angle, which is variable over a large range. T:target.

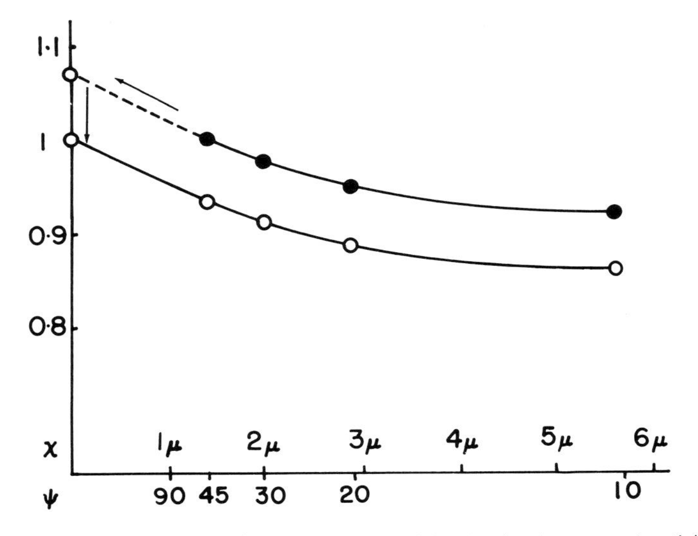

Fig. 10.9. Extrapolation in Green's determination of f_{ex}. Full points are ratios $I'_i/I'_{45°}$. The curve must be extrapolated to $\chi = 0$, to obtain the value of $1/f(\chi_1)$. The experimental ratios are divided by this value, in order to obtain the corresponding $f(\chi)$ values. The uncertainties of the procedure include the shape of the extrapolated curve, and the mass absorption coefficient μ. To reduce the curvature of the function to be extrapolated, Green used a logarithmic scale.

Table 10.2. Table of f_{ex} Curves Obtained by Green [10.10].

ELEMENT	Z	LINE	E_q (kV)	λ_{q1} (Å)	ENERGIES (kV)
C	6	$K\alpha$	0.283	44.7	0.57, 2.5, 10
Al	13	$K\alpha$	1.559	8.34	3.1, 4.4, 5.9, 7.8, 9.8, 15.3, 20.5, 40.0
Ti	22	$K\alpha$	4.964	2.75	10.0, 40.0
Fe	27	$K\alpha$	7.111	1.94	10.0, 14.2, 40
Cu	29	$K\alpha$	8.980	1.54	13.3, 16.8, 20.3, 24.0, 27.4, 30.2, 35.3, 39.9, 44.5, 49.7
Ge	32	$K\alpha$	11.103	1.25	22.2, 40.0
Mo	42	$K\alpha$	20.002	0.71	40.0
Ag	47	$K\alpha$	25.517	0.56	29.8, 34.8, 39.5, 44.5, 49.7
Nd	60	$L\alpha_1$	6.215	2.37	10.0, 12.4, 40.0
Ta	73	$L\alpha_1$	9.876	1.52	19.7, 40.0
Au	79	$L\alpha_1$	11.919	1.28	16.1, 20.4, 25.4, 30.3, 35.3, 39.9, 44.5, 49.7
		$M\alpha_1$	2.123	5.84	4.44, 10

range up to 50 kV. Table 10.2 lists the elements, lines, emergence angles, and electron energies used in Green's experiments.

The variable emergence-angle technique is potentially applicable to homogeneous multielement targets, and it could be simplified and made more accurate with the use of a silicon X-ray detector. Unfortunately, there has been no further research effort in this direction.

10.3 GENERALIZED MODELS FOR PRIMARY ABSORPTION

10.3.1 Philibert's Absorption Model

The same year in which Green presented in his thesis his experimental absorption factors (1963), Philibert published a generalized model for the absorption factor of primary radiation, which is widely and successfully used in analytical practice. In his paper, Philibert mentioned that the same equation had been obtained independently also by Tong. A full derivation of the model was presented in the same proceedings of a conference in Stanford [10.11] which also contains the report of the experimental results of Green on absorption factors. A similar, though less accurate, version was published by Theisen in 1965 [4.39, p. 11]. We follow here the derivation given by Philibert.

The quantitative description of the target phenomena is complicated by the gradual change in the characteristics of the electron penetration. The electrons start as a well-oriented stream, moving in parallel and at uniform speed, and change into a diffused state of movement in which the direction of the electrons is almost random, and their energies spread over a wide range. The change of direction was explained either as due to a gradual diffusion, or to single large-angle scatter. Archard [10.12] pointed out that the first mechanism prevails in elements of low atomic number, and the second in heavier elements.

The transition into a state of diffusion was described by Bethe et al. [10.13] as follows:

"A beam of fast electrons will in the beginning of its path suffer energy loss but very little scattering and thus move in almost a straight line. With decreasing energy, scattering will become more important until finally the stage of diffusion is reached where the direction of motion becomes almost random. *Both the limiting cases of straight motion and diffusion . . . are easy to treat whereas the intermediate region is not*. We therefore approximate the problem by assuming a direct transition from straight motion into diffusion. As the transition point we take the point at which the average cosine of the angle between the direction of motion and the primary beam becomes $1/e$."

Philibert's derivation starts with the calculation of the number of ionizations of the emitting element a within a layer at depth z and of thickness dz. Calling ζ the angle of the electron trajectory at depth z with respect to the initial trajectory—assumed to be normal to the surface—we obtain:

$$d^2 n(\zeta, z) = N_{\mathrm{av}} \cdot \frac{C_a}{A_a} \cdot Q \cdot \frac{dz \cdot d\zeta}{\cos \zeta}. \tag{10.3.1}$$

Let us assume that a large number, n_0, of electrons impinge upon the specimen surface. At the depth z they are reduced to n_z electrons traveling in the downward direction. In addition, some of the electrons which were backscattered at depths below z return and cross the layer at depth z in the upward direction. We call the number of these electrons $(r - 1)n_z$. The total number of electrons crossing the layer is thus $r \cdot n_z$; the average number of ionizations caused in the layer by one impinging electron is

$$dn(z) = \frac{1}{n_0} C_a \frac{N_{\mathrm{av}}}{A_a} Q \sum_{i=1}^{rn_z} \frac{dz}{\cos \zeta_i}. \tag{10.3.2}$$

The summation in Eq. (10.3.2) includes the backscattered electrons and the corresponding directions are expressed by the respective values of ζ.

We now define a mean electron pathlength within the layer dz, at depth z, $R(z)dz$. This term includes the trajectories of the returning backscattered electrons:

$$\sum_{i=1}^{rn_z} \frac{1}{\cos \zeta_i} = n_z R(z). \tag{10.3.3}$$

The number of ionizations per electron within the layer of the shell q is:

$$dn(z) = N_{\text{av}}(C_a / A_a) Q_q (n_z / n_0) R(z) dz. \tag{10.3.4}$$

We assume that at great depth the number of electrons traveling upward is equal to that moving downward ($r = 2$), and that the distribution of angles of trajectory at this depth is such that $R = 4$. This compares to a value of R_0—at the surface—only slightly above unity. Strictly, such a state of complete diffusion is not attainable since, at equilibrium, there is at any level of depth an excess of downward moving electrons which replace those stopped below this depth.

In order to calculate $dn(z)$ as a function of any depth, z, Philibert makes the following additional simplifying assumptions:

1. The ratio n_z/n_0 is given by the following law (Lenard's law [10.2]):

$$n_z = n_0 \exp(-\sigma z). \tag{10.3.5}$$

2. The value of $R(z)$ is assumed to vary from R_0 to R_∞ according to the following exponential relation:

$$R(z) = R_\infty - (R_\infty - R_0)\exp(-kz). \tag{10.3.6}$$

3. The ionization cross section, Q_q, is independent of the electron energy, and therefore of the depth z. This assumption is quite inaccurate (see Eq. (9.2.10) and Fig. 9.5).

With these assumptions, we obtain:

$$dn(z) = R_\infty \frac{N_{\text{av}}}{A_a} C_a Q_q \left\{ \exp(-\sigma z) - \left[\left(1 - \frac{R_0}{R_\infty}\right) \exp - (\sigma + k) z \right] \right\} dz. \tag{10.3.7}$$

The X-ray emission at the depth z is thus:

$$I(z) = p_{qm} \omega_q dn(z) = R_\infty \frac{N_{\text{av}}}{A_a} C_a Q_q p_{qm} \omega_q \cdot \left\{ \exp(-\sigma z) - \left[\left(1 - \frac{R_0}{R_\infty}\right) \exp - (\sigma + k) z \right] \right\} dz. \tag{10.3.8}$$

A new variable, h, is now defined by

$$h = \frac{\sigma}{k}. \tag{10.3.9}$$

Philibert obtains an expression for h from an equation by Bothe [10.14] which gives the average scattering angle $\bar{\zeta}$ (not its standard deviation, as indicated in reference [10.11]) of an electron at depth z:

$$\bar{\zeta} \simeq \frac{400}{E_0}\left(\frac{Z^2}{A}z\right)^{\frac{1}{2}}. \tag{10.3.10}$$

Philibert defines the "depth of complete diffusion," z_d, as that at which the average value of $\bar{\zeta}$ is equal to $\pi/4$:

$$z_d \simeq \frac{\pi^2}{16} \cdot \frac{E_0{}^2}{400^2} \cdot \frac{A}{Z^2} \simeq 4 \times 10^{-6} \frac{A}{Z^2} E_0{}^2 \tag{10.3.11}$$

and he assumes that at this point the value of $R(z)$ is such that

$$R_\infty - R(z) = 0.1\,(R_\infty - R_0) \tag{10.3.12}$$

so that $\exp(-kz_d) = 0.1$. Therefore, he finds that

$$k = 0.58 \times 10^{-6} \frac{Z^2}{A} \frac{1}{E_0{}^2}. \tag{10.3.13}$$

Since Lenard's coefficient σ is found to be approximately proportional to E_0^{-2} as well, Philibert assumes that the value of h is independent of the electron energy:

$$h = \text{const}\,\frac{A}{Z^2}. \tag{10.3.14}$$

For best results, the constant of the above expression must be fitted to experimental results. Introducing h into Eq. (10.3.7), we obtain:

$$I(z)dz = K\left[\exp(-\sigma z) - \left(1 - \frac{R_0}{R_\infty}\right)\exp\left(-\frac{1+h}{h}\sigma z\right)\right] dz \tag{10.3.15}$$

where

$$K = R_\infty \frac{N}{A_a} C_a Q_q P_{qm} \omega_q.$$

The total primary generated intensity is thus

$$I_p = K\int_0^\infty \left[\exp(-\sigma z) - \left(1 - \frac{R_0}{R_\infty}\right)\exp\left(-\frac{1+h}{h}\sigma z\right)\right] dz$$

$$= \frac{K}{\sigma}\left[1 - \left(\frac{R_0}{R_\infty}\right)\frac{h}{1+h}\right]. \tag{10.3.16}$$

From Eq. (10.1.12) we obtain:

$$\phi(z) = \frac{I(z)}{I_p} = \frac{\sigma\left[\exp(-\sigma z) - \left(1 - \frac{R_0}{R_\infty}\right)\exp\left(-\frac{1+h}{h}\sigma z\right)\right]}{1 - \left(1 - \frac{R_0}{R_\infty}\right)\frac{h}{1+h}}. \tag{10.3.17}$$

By applying to Eq. (10.3.16) the X-ray attentuation terms, according to Beer's law, we obtain for the emergent primary radiation:

$$I'_p = \frac{\Omega}{4\pi} K\int_0^\infty \left\{\exp[-(\sigma + \chi)z] - \left(1 - \frac{R_0}{R_\infty}\right)\exp\left[-\left(\frac{1+h}{h}\sigma + \chi\right)z\right]\right\} dz$$

$$= \frac{\Omega}{4\pi}\frac{K}{\sigma}\frac{\frac{1}{1+\chi/\sigma} + h\frac{R_0}{R_\infty}}{1 + h\left(1 + \frac{\chi}{\sigma}\right)}. \tag{10.3.18}$$

Finally, the primary absorption factor, f_p, is obtained:

$$f_p = \frac{4\pi I'_p}{\Omega I_p} = \frac{\frac{1}{1+\chi/\sigma} + h\frac{R_0}{R_\infty}}{\left[1 - \left(1 - \frac{R_0}{R_\infty}\right)\frac{h}{1+h}\right]\left[1 + h\left(1 + \frac{\chi}{\sigma}\right)\right]}. \tag{10.3.19}$$

Because the formula was still cumbersome, Philibert proposed a final simplification, which consisted in assuming that the initial value for the "mean electron path," R_0, instead of being slightly above one, was equal to zero. We obtain:

$$I_p = \frac{K}{(1+h)\sigma} \tag{10.3.20}$$

$$\phi(z) = \sigma(1 + h)\left[\exp(-\sigma z) - \exp\left(-\frac{1+h}{h}\sigma z\right)\right] \qquad (10.3.21)$$

$$I'_p = \frac{\Omega}{4\pi}\frac{K}{\sigma}\frac{1}{\left(1 + \frac{\chi}{\sigma}\right)\left[1 + h\left(1 + \frac{\chi}{\sigma}\right)\right]}. \qquad (10.3.22)$$

And, at last, the equation which is commonly known as Philibert's absorption correction:

$$f_p = \frac{1}{\left(1 + \frac{\chi}{\sigma}\right)\left(1 + \frac{h}{1+h}\frac{\chi}{\sigma}\right)}. \qquad (10.3.23)$$

The factor $h/(1 + h)$ (see Fig. 10.10) introduces in Eq. (10.3.23) a dependence of f_p on Z. Actually, Philibert presented in lieu of f_p the corresponding value for the function $F(\chi)$, which in our nomenclature would be equal to

$$F(\chi) \equiv \frac{4\pi}{\Omega} \cdot \frac{\sigma}{K} I'_p.$$

Since this function does not have a well-defined physical meaning, we will not consider it further.

The change introduced in the equations by assuming that R_0 is zero is partic-

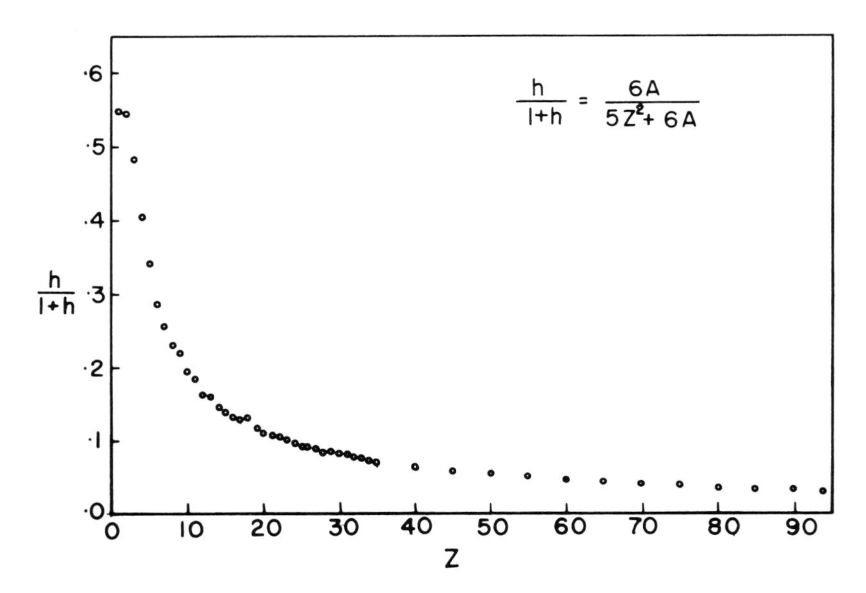

Fig. 10.10. The parameter $h/(1 + h)$ used in Philibert's Eq. (10.3.23).

ularly significant when the X-ray excitation close to the specimen surface is of importance, as in cases of serious absorption of primary radiation because, according to Eq. (10.3.21), the value of $\phi(z)$ at the specimen surface is zero, contrary to both theory and experimental evidence (see Fig. 10.4). Philibert also suggests that a further empirical adjustment be made for the use of Eqs. (10.3.20)-(10.3.23), so that the value of h becomes equal to $1.2(A/Z^2)$; finally he presents a set of adjusted values of the "Lenard coefficient," σ.

The fact that the model developed for f_p by Philibert has been very successful in spite of the poor fit of the corresponding $\phi(z)$ function indicates that the function f_p is insensitive to small changes in the depth distribution $\phi(z)$. This observation will encourage us to speculate about further simplifications of the model for the absorption factor. But even the derivation of the complete model [Eqs. (10.3.17)-(10.3.19)] is not rigorous. Its use is legitimated, not by the underlaying theory, but by the good fit which Philibert obtained to Castaing's tracer experiments, and to Green's variable emergence-angle experiments.

As shown in Eq. (10.1.7), the model for the absorption factor should contain the separate effects on the factor of the X-ray absorption coefficients, μ, the emergence angle, ψ, the operating and minimum excitation potentials, V_0 and V_q, and the composition of the specimen. Philibert's model accomplishes most of this, since χ depends only on the parameters μ and ψ, while the parameter h is dependent only on the atomic number and atomic weight of the elements present in the specimen. The "Lenard coefficient" σ, related to the deceleration of the electrons, is, in Philibert's original version, dependent only on the operating potential, V_0, and independent of the specimen composition. There arises a question as to the appropriate value to be used for h in the case of a composite target (i.e., one which contains more than one element). Philibert's suggestion to use the h corresponding to the average atomic number of the specimen, is suspect, since the atomic weights, and hence h, are not a smooth function of the atomic number, nor natural constants. In practice, one may average either h or $1/(1 + h)$ in weight proportions. The uncertainty resulting from the choice of averaging procedures is insignificant, unless the attentuation of X-rays is very strong.

Reuter [10.15] critically reviewed Philibert's procedure, and introduced in it several modifications, some of which are drawn from the experimental work of Cosslett and Thomas [10.16]. Reuter accepts the formulation of the variation of the mean electron pathlength, $R(z)$, given in Eq. (10.3.6), and the definition of the onset of "complete diffusion" according to Eq. (10.3.12). He introduces, however, the variation of the ionization cross section with energy [see Eq. (9.2.9)] and he changes the expression for the average scattering angle $\overline{\zeta}$ and hence of the "depth of complete diffusion," z_d [Eqs. (10.3.10) and (10.3.11)]. The pathlength at the surface, R_0, is defined by the following equation:

$$R_0 \simeq \Phi(0) = 1 + 2.8(1 - 0.9/U)\eta \qquad (10.3.24)$$

which is based on Reuter's measurements of X-ray intensities from thin foils on substrates of varying thickness [10.15]. The equality $R_0 = \Phi(0)$ is obtained if it is assumed (as Philibert does) that the ionization cross section for the backscattered electrons is equal to that for the primary electrons. From Heinrich's measurements of electron backscatter coefficients, Reuter derives an algebraic expresssion for η and inserts it in Eq. (10.3.24). The changes also include a linear electron transmission law, instead of an exponential attenuation [Eq. 10.3.5)], for the initial segments of the electron trajectories. They render impractical the development of a simple formula for $\phi(z)$ or f_p; rather, these parameters are obtained with the aid of a computer program. Although some of the remaining, or new, assumptions may still be questionable, the author obtains good fits to experimental results.

10.3.2 The Effect of the Critical Excitation Potential on the Primary X-Ray Attenuation

In his thesis, Green [6.2] observed that Philibert's model does not take into account the critical excitation potential of the observed radiation. Because, on average, the energy of primary electrons diminishes with increasing depth of penetration, and since the critical excitation potential limits the energy of the electrons at which they can excite the radiation of interest, it follows that, other conditions being unchanged, the mean depth of X-ray production diminishes as the critical excitation potential increases. As the operating potential approaches the critical excitation potential, the X-ray generation becomes very shallow. Such a trend is not, however, followed by Eqs. (10.3.17)-(10.3.23), if the values for σ tabulated by Philibert are applied to them. As mentioned on p. 267, Green discussed the depth of X-ray production with the aid of a model which assumes that all X-ray photons are generated at the same depth level so that the absorption factor can be calculated by Beer's law [see Eq. (10.2.2)]. Green concluded from his experiments that his main depth of X-ray production was nearly proportional to $(V_0 - V_q)$. He obtained for copper:

$$\bar{z} = 2.7 \times 10^{-5}(V_0 - V_q) \tag{10.3.25}$$

so that

$$1/f = \exp[2.7 \times 10^{-5}(V_0 - V_q)\chi]. \tag{10.3.26}$$

Green concluded from his experiments that Eq. (10.3.26) was valid for all elements, and that, therefore, a set of universal $f(\chi)$ curves could be obtained if, for various values of $(V_0 - V_q)$, the values of f would be plotted as a function of $\log \chi$ (Fig. 10.11). Green thus proposed a function of the type

$$f_p = f(\chi, V_0, V_q).$$

However, as will be seen later, Green's assumption that the mean depth of X-ray production increases linearly with $(V_0 - V_q)$ is inaccurate, and inconsistent with his own experimental results.

Duncumb and Shields [10.17] proposed a modification of the coefficient σ used in Philibert's equation, which introduces in it a dependence upon the critical excitation potential. The proposed model for σ is the following:

$$\sigma = \frac{c}{V_0^{\,n} - V_q^{\,n}} \,. \tag{10.3.27}$$

Theisen [4.39] used for the same purpose the form

$$\sigma = \frac{c}{(V_0 - V_q)^n} \,. \tag{10.3.28}$$

Both models produce convergence of the absorption factor f_p to one when V_0 is equal to V_q. However, the formula proposed by Theisen produces absorption factors which are in serious disagreement with the experiment, and the formula is therefore not used at present. There have been several proposals concerning the numerical values of the parameters c and n, as seen in Fig. 10.18. Those pro-

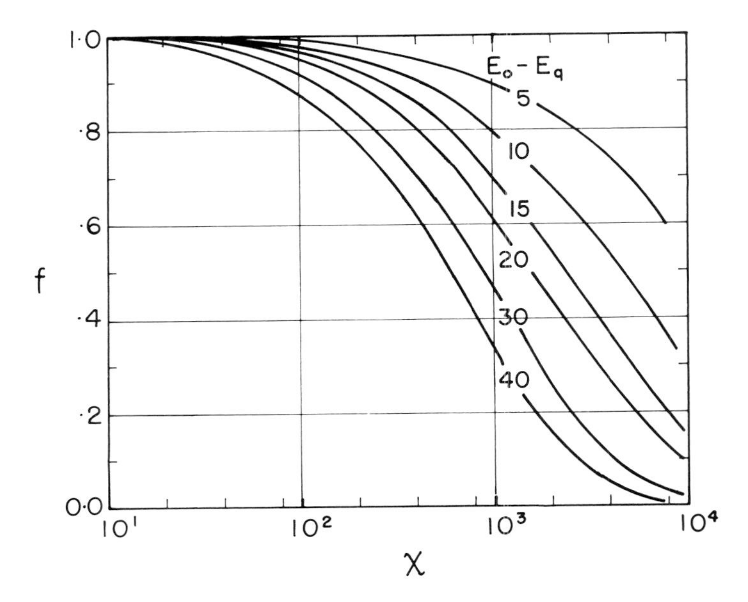

Fig. 10.11. Green's universal absorption correction curves for atomic number 30. The absorption factor is plotted as a function of $(E_0 - E_q)$ and χ [10.10].

posed by Heinrich [10.18] ($c = 4.5 \times 10^5$, $n = 1.65$) were tested and recommended by Duncumb et al. [10.19] as well as by Sweatman and Long [4.25]. The definition

$$\sigma = \frac{4.5 \times 10^5}{V_0^{1.65} - V_q^{1.65}} \tag{10.3.29}$$

is now widely accepted, in combination with Philibert's definition of h ($h = 1.2\ A/Z^2$). There is, of course, an interrelation between σ and h; since both parameters were derived empirically, it would be improper to use them out of the context of Philibert's equation. The formula for σ proposed by Duncumb is clearly a conceptual improvement over previous models for this parameter. It should be noted, however, that there is no theoretical foundation for the particular algorithm used, and that the practical improvement in analysis obtained by introducing V_q in the absorption correction is perhaps less dramatic than generally assumed. When the critical excitation potential is large, the corresponding radiation is of short wavelength, and under proper experimental conditions (high emergence angle, moderate operating potential), the absorption of the radiation in the target is small. On the other hand, when the radiation is soft, and absorption losses are very significant, the critical excitation potential is always small. Calculations with the NBS Monte Carlo program (see Chapter 13) indicate that Eq. (10.3.29) does not provide a very accurate compensation for the effects of the critical excitation potential.

10.3.3 Empirical Treatment of the Absorption Function

In view of the complexity of the physical events in the target, and the difficulty in deriving a theoretical model without substantial simplifications and omissions, a purely empirical treatment of the absorption function, based on the generalization of experimental results on f and on $\phi(z)$ appears to be preferable.

Consider the general form of Philibert's equation, obtainable by multiplying the parentheses in Eq. (10.3.23):

$$\frac{1}{f_p} = 1 + \text{const}_1 \cdot \frac{\chi}{\sigma} + \text{const}_2 \cdot \left(\frac{\chi}{\sigma}\right)^2. \tag{10.3.30}$$

In view of the limited precision of experimental results it does not appear advantageous to add to this equation terms containing powers larger than quadratic. We will also accept Duncumb's proposal that the penetration of the electrons is proportional to a term of the form $V_0^n - V_q^n$, with $(1 < n < 2)$. We will introduce the symbol γ for this parameter which gives the energy dependence of the absorption function (Fig. 10.12):

$$\gamma = V_0^n - V_q^n. \tag{10.3.31}$$

The relation of γ to the function σ [Eq. (10.3.29)] is thus: $\gamma = 4.5 \times 10^5/\sigma$. We can rewrite Eq. (10.3.30) as follows:

$$1/f = 1 + a_1\gamma\chi + a_2\gamma^2\chi^2. \tag{10.3.32}$$

A similar expression can be obtained by expanding the exponential term of Eq. (10.3.26) proposed by Green: Since

$$e^x = 1 + x + \frac{x^2}{2} + \frac{x^3}{6} + \cdots$$

it follows that

$$1/f_p = 1 + a_1(V_0 - V_q)\chi + \frac{1}{2}a_1{}^2(V_0 - V_q)^2\chi^2 + \cdots . \tag{10.3.33}$$

This equation differs from Eq. (10.3.32) by the fact that here the exponent n is assumed to be equal to one.

The constants a_1 and a_2 in Eq. (10.3.32) contain both the numerical constant c which is usually incorporated in the expression for σ Eq. (10.3.27), and the effects of target composition on the depth distribution, which in Philibert's formula were expressed by the parameter h. For instance, for the value for σ given in Eq. (10.3.29), we obtain:

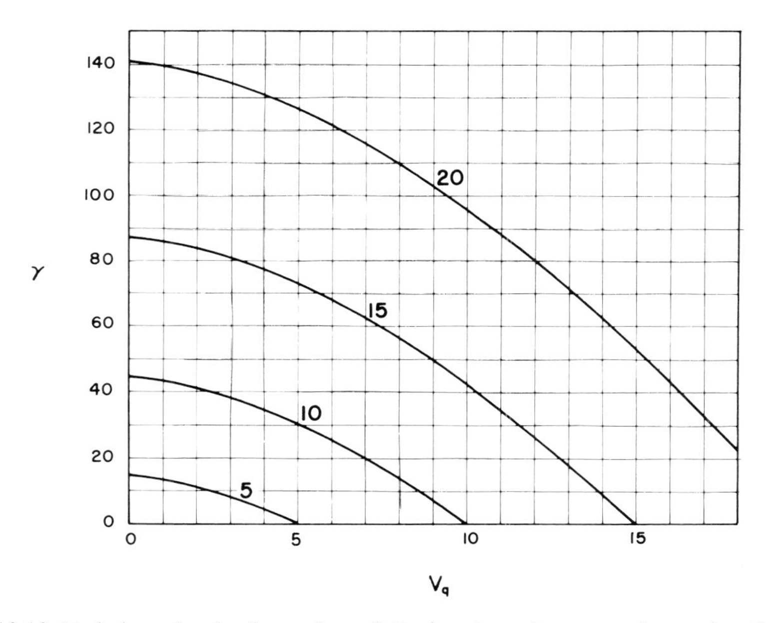

Fig. 10.12. Variation of γ, for four values of V_0 (numbers above curves), as a function of V_q. All potentials are in kilovolts.

$$a_1 = \frac{1}{c} \cdot \frac{1 + 2h}{1 + h} = 2.22 \times 10^{-6} \frac{1 + 2h}{1 + h}$$

$$a_2 = 4.94 \times 10^{-12} \frac{h}{1 + h} .$$

The factor a_2 in Eq. (10.3.32) is less important than a_1, and as a first approximation it can be ignored. Several authors, including Ziebold and Ogilvie [8.1], Belk [10.20], and Borovsky and Rydnik [9.33] have proposed absorption corrections based on a linear equation of this type:

$$1/f_p = 1 + a_1 \gamma \chi. \tag{10.3.34a}$$

This equation is also in agreement with the assumption expressed by Castaing in his thesis [1.1] that at great depth within the target the function $\phi(z)$ approaches an exponential function:

$$\phi(z) \simeq \frac{1}{a_1 \gamma} \exp \left(\frac{-z}{a_1 \gamma} \right) \equiv \sigma \exp(-\sigma z). \tag{10.3.34b}$$

From this approximation it would follow that

$$f_p \simeq \frac{\int_0^\infty \phi(z) \exp(-\chi z) \cdot dz}{\int_0^\infty \phi(z) \cdot dz} = \frac{1}{1 + \chi/\sigma} .$$

In his thesis, Castaing presented results of measurements of f on a graph showing $1/f$ as a function of χ, and he obtained a plot which is closer to linear than the usual plot of f, or $-\log f$, as a function of χ. The plotting of later experimental results on f also shows that the inverse of the absorption function is nearly always close to being a linear function of χ (see Figs. 10.13 to 10.16).

The linear approximation of $1/f_p$ can also be derived from Philibert's equation:

$$\frac{1}{f_p} = \left(1 + \frac{\chi}{\sigma}\right)\left(1 + \frac{h}{1 + h} \frac{\chi}{\sigma}\right) = 1 + \frac{1 + 2h}{1 + h} \frac{\chi}{\sigma} + \frac{h}{1 + h} \left(\frac{\chi}{\sigma}\right)^2 . \tag{10.3.35}$$

As σ is usually larger than χ, and as $(1 + 2h)/(1 + h)$ is larger than $h/(1+ h)$, the quadratic term in the equation is much smaller than the term linear in χ, and can therefore be neglected in the first approximation. This term is important only when the absorption correction is large.

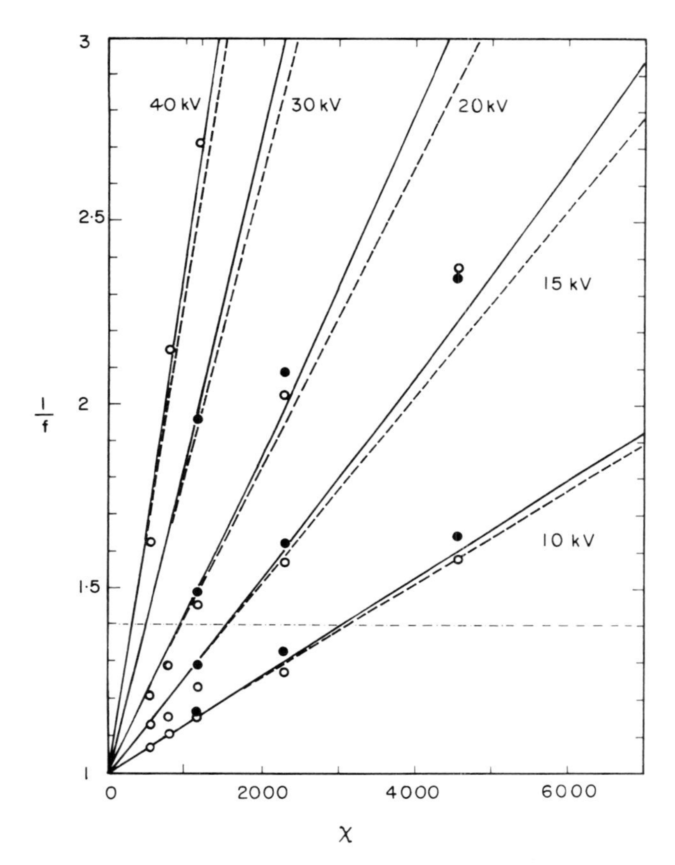

Fig. 10.13. Inverse of the absorption factor for Al Kα radiation in aluminum. Empty circles: Green [10.10]. Full circles: Castaing and Hénoc [10.5]. Solid lines: Eq. (10.3.39). Dashed lines: Eq. (10.3.39) with quadratic term omitted. Above the value of $1/f = 1.4$, the uncertainty in f due to that in the mass absorption coefficient is larger than 1%, according to Eq. (12.3.20).

The values of a_1 and a_2 can be obtained by the following transformation of Eq. (10.3.32):

$$\frac{1/f_{\text{ex}}-1}{\chi} = a_1\gamma + a_2\gamma^2\chi. \tag{10.3.36}$$

If the experimental values of f are used to calculate this expression and to plot it as a function of χ (Fig. 10.17), we obtain, on extrapolation to ($\chi = 0$), the product $a_1\gamma$. Wherever experiments were performed for the same target and line at various operating voltages, a logarithmic plot of $a_1\gamma$ as a function of the

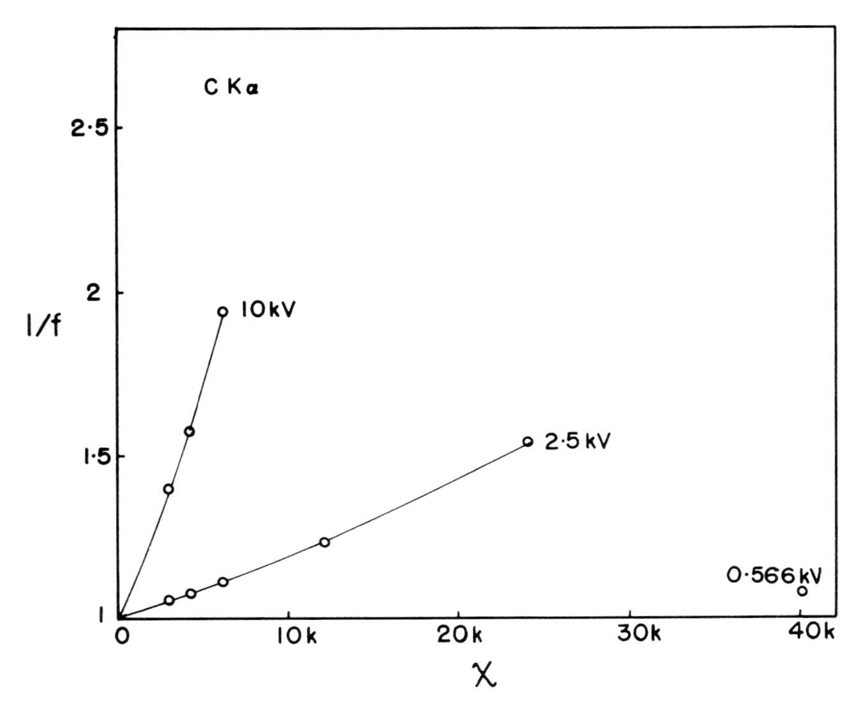

Fig. 10.14. Plot of the inverse absorption function of carbon Kα. Experimental points are from Green [10.10].

operating voltage, V_0, provides information on the variation of γ as a function of V_0, and, specifically, on the appropriate value of the exponent n in Eqs. (10.3.27) and (10.3.31). Duncumb et al. [10.19] showed that reasonably good models for σ can be obtained within a range for n from 1.5 to 1.8, provided that the coefficient c is appropriately adjusted, and several authors have proposed pairs of values of c and n which yield in practice roughly equivalent results (Fig. 10.18). But not all choices are equally valid. For a decision on this matter, the experimental evidence on $a_1\chi$ for Al Kα radiation (Green [10.10]) and Mg Kα radiation (Castaing and Hénoc [10.5]), is particularly useful, since it was obtained by two different methods, covers a wide range of operating voltages, and because the critical excitation potentials of Al Kα and Mg Kα are too low to interfere with the determination of the effect of the operating voltage. Furthermore, a good fit of the absorption function for elements in the atomic number range close to aluminum is particularly important.

A log-log plot of $a_1\gamma$ as a function of V_0 provides a straight line, and its slope corresponds to a value of n close to 1.65. Hence the use of this exponent is justified, at least for elements of atomic number close to 13 (Fig. 10.19). By

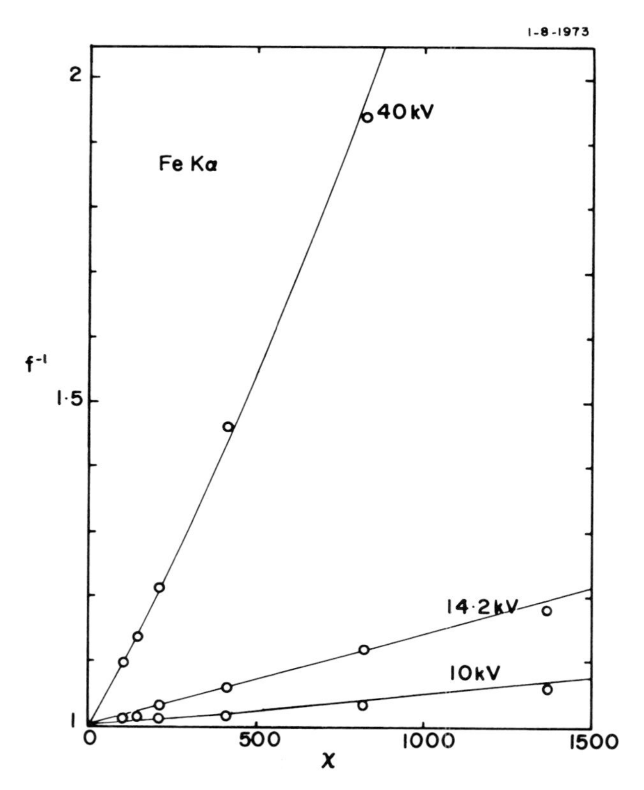

Fig. 10.15. Plot of the inverse absorption function of iron Kα. Experimental points are from Green [10.10].

extrapolation to (V_0 = 1) a value of a_1 from 2.5×10^{-6} to 3×10^{-6} is obtained. To test the behavior of other targets, we can now plot the experimental values of $a_1\gamma$ for all targets and lines as a function of $\gamma = V_0^{1.65} - V_q^{1.65}$. We observe (Fig. 10.20) that in most cases the plotted values of $a_1\gamma$ fall on a line with an inclination of 45° on the log-log plot. The cases in which there is a significant deviation of these lines are mainly those involving lines of short wavelength, such as Ag Kα, Mo Kα, and Au Lα_1, and hence of low absorption losses. The poor fit of such lines is due to the excitation of a significant fraction of the emitted radiation by the continuum (continuum fluorescence). This fluorescent radiation arises at deeper levels than the primary radiation and thus causes the value of the experimental absorption factor to diminish, and that of $a_1\gamma$ to increase. In order to study the absorption of *primary* radiation, a correction for the effect of continuum fluorescence must be applied. Since in the experiments with the aluminum matrix the effects of fluorescence due to characteristic lines are insignificant, the total experimental absorption factor is:

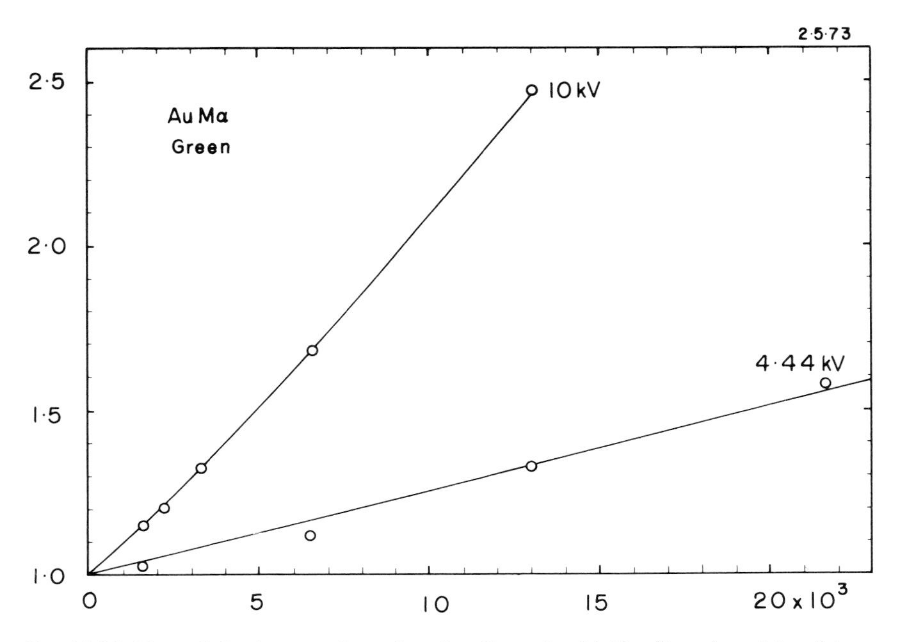

Fig. 10.16. Plot of the inverse absorption function of gold Mα. Experimental points are from Green [10.10];

$$f_{ex} = \frac{I_p f_p + I_c f_c}{I_p + I_c} \tag{10.3.37}$$

so that

$$f_p = f_{ex} + \frac{I_c}{I_p}(f_{ex} - f_c). \tag{10.3.38}$$

As will be discussed in Chapter 11, the values of I_c and f_c can be estimated with sufficient accuracy to obtain a good estimate of f_p for the cases in which continuum fluorescence is significant. If the corrected values are used to produce a plot of $a_1\gamma$ versus γ, the fit of the results to a straight line is greatly improved (Fig. 10.21). This graphical result suggests that the effect of the atomic number of the target on the coefficient a_1 can be neglected without serious loss of accuracy.

We now direct our attention to the parameter a_2 in the quadratic term of Eq. (10.3.32). If this term were equal to zero, as was assumed in Eq. (10.3.34), the values of $(1/f - 1)/\chi$ in Fig. 10.17 would be constant and independent of χ. The experimental results exhibit, however, such a dependence, which can be approximated with good accuracy by postulating that $a_2 = a_1{}^2/4$. Therefore, the absorption function takes the simple form:

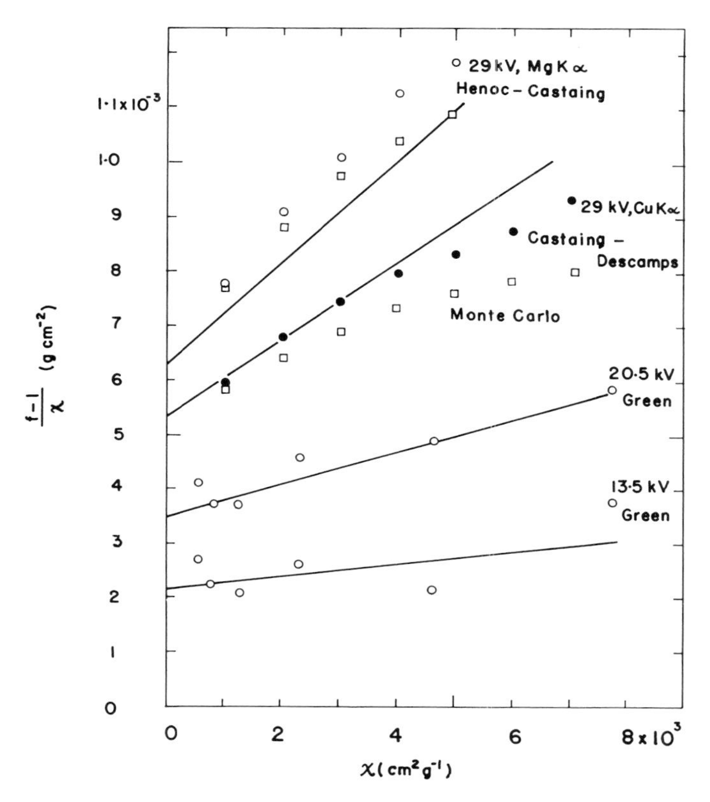

Fig. 10.17. Determination of the coefficient a_2 of the quadratic term of the absorption factor model [Eq. (10.3.32)]. The parameter $(1/f - 1)/\chi$ is plotted against χ [see Eq. (10.3.36)]. The slope of the linear average equals $a_2\gamma^2$, and the cutoff at ($\chi = 0$) is equal to $a_1\gamma$. Solid lines correspond to the equation $1/f_p = (1 + 1.2 \times 10^{-6}\ \gamma\chi)^2$ [see Eq. (10.3.40)]. The upper two sets of data are tracer experiments by Castaing et al. and the lower two sets of data are from Green's experiments with aluminum targets. There is a discrepancy between the two curves at 29 kV which is confirmed by the NBS Monte Carlo procedure and can only be removed if the model for σ (i.e., for γ) by Duncumb and Shields [Eq. (10.3.27)] is abandoned.

$$f = (1 + a\gamma\chi)^{-2} \tag{10.3.39}$$

with $a = a_1/2$. The experimental results shown in Fig. 10.17 suggest a value of $a = 1.2 \times 10^{-6}$, so that we obtain:

$$f = (1 + 1.2 \times 10^{-6}\gamma\chi)^{-2} \tag{10.3.40}$$

as proposed by Heinrich and Yakowitz [10.21], who showed that for a large collection of observations, at moderate absorption losses, on targets of known

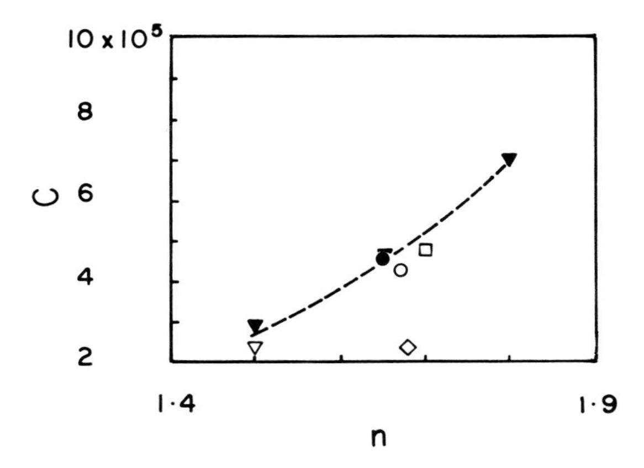

Fig. 10.18. Proposed values of c and n for Duncumb's model for the "Lenard coefficient," σ. ▽: Duncumb and Shields [10.17]; ○: Heinrich [8.6]; ●: Heinrich [10.18]; □: M. L. Barman (2nd Conf., EPASA, 1967, paper 6); ▼ Duncumb et al. [10.19].

composition the residual errors with this model were in average smaller than those obtained with the Philibert-Duncumb-Heinrich model [Eq. (10.3.29)]. For specimens of average atomic number below 15, the factor 1.3×10^{-6} seems preferable.

The model embodied in Eq. (10.3.40) (henceforth called the quadratic model) ignores the expected dependence of the absorption factor on the target composition; the atomic weights and numbers of the target compositions do not enter the function. The fit to experiments is surprisingly good, in view of the dependence of both the stopping power and the backscattering on the target composition. It would be unreasonable to assume that such an effect does not exist; rather, the experimental results indicate that this effect is small, and largely masked by the scatter of experimental values available at present.

For a more detailed discussion of the absorption correction, we must now consider the depth distribution function $\phi(z)$.

10.3.4 Analytical Expressions for the Depth Distribution Function $\phi(z)$

Although the information contained in Fig. 10.21 has been useful in the development of a model for f_p, the spread of the data contained therein is undesirably large. We have used in this figure the values of f derived from tracer experiments; however, in view of the close relationship of the functions f_p and $\phi(z)$, it is useful to test the inverse Laplace transform of a model for f_p, i.e., the corresponding algorithm for $\phi(z)$, against the available experimental evidence.

An analytical expression for $\phi(z)$ is also useful in other ways. As will be shown in Chapter 11, it is needed in the calculation of the intensity of fluo-

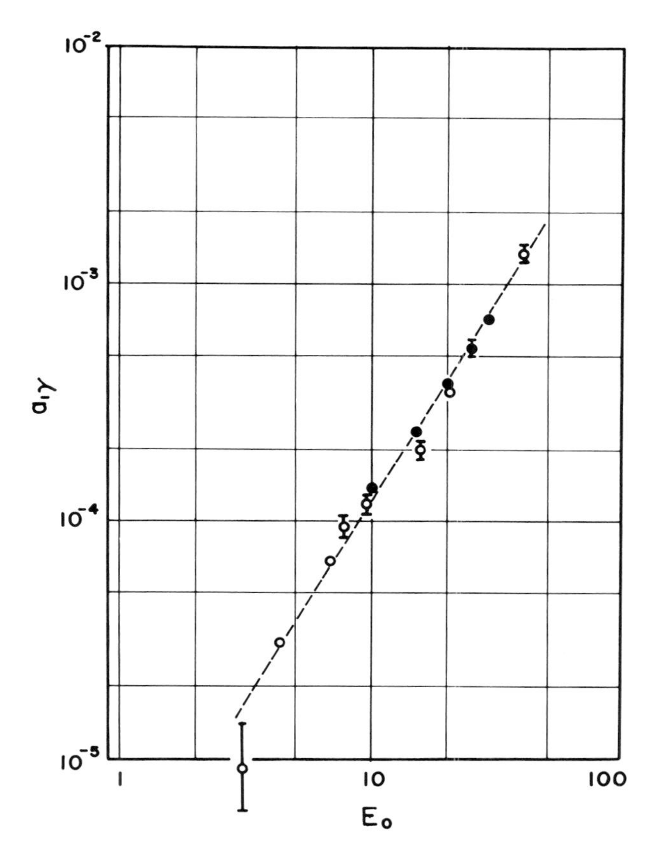

Fig. 10.19. A logarithmic plot of the product $a_1\gamma$, from experimental values by Green (open circles) and Castaing and Hénoc (full circles), for Al $K\alpha$ in aluminum.

rescent radiation; it is also helpful in the definition of the average depth of X-ray emission and of the depth range (see Chapter 12), as well as in the analysis of thin films. The study of the depth distribution function can lead to refinements in the function f_p, which are needed for the analysis with long wavelength X-rays that are strongly absorbed in the target. Since we wish to relate algebraically the functions $\phi(z)$ and f_p, it is desirable that the first of them should have an algebraic Laplace transform [see Eq. (10.1.9)] to represent f_p.

Several authors, e.g., [10.22] have proposed expressions for $\phi(z)$ of the form

$$\phi(z) = \sum_i c_i \cdot \exp(-d_i z). \qquad (10.3.41)$$

The normalization [Eq. (10.1.11)] requires that

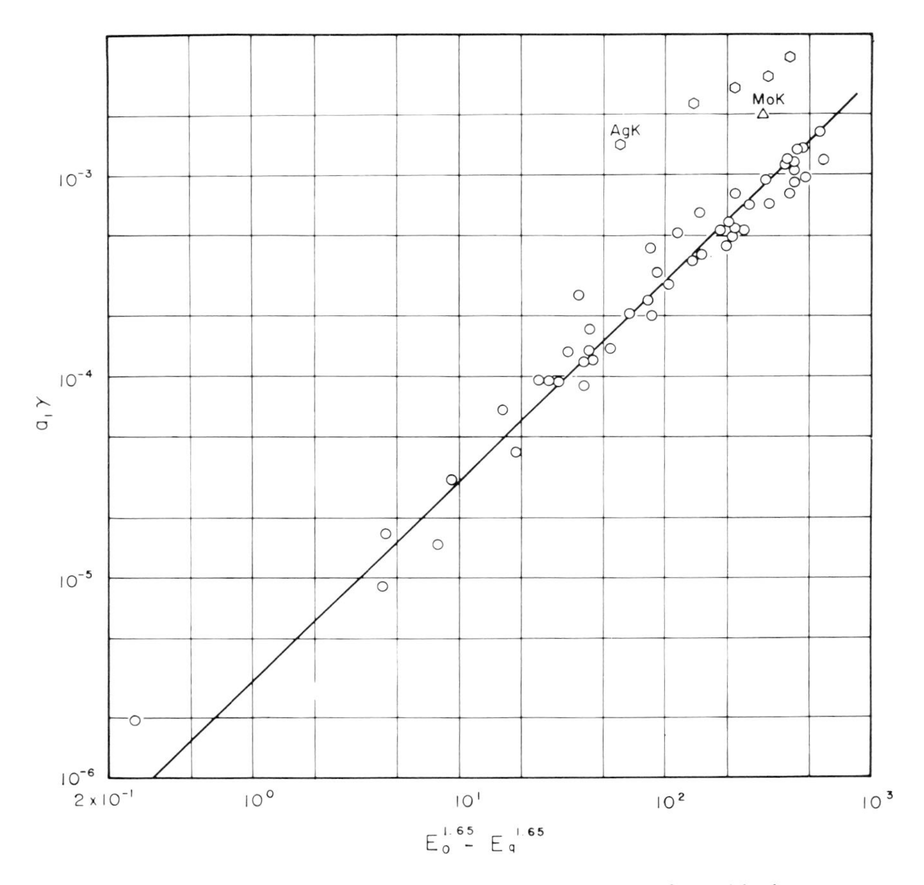

Fig. 10.20. A plot of all available values of $a_1\gamma$ as a function of γ, with the parameter $n = 1.65$.

$$\sum_i (c_i/d_i) = 1. \tag{10.3.42}$$

It was already mentioned that a single-term expression $\phi(z) = \sigma \exp(-\sigma z)$ leads to

$$1/f_p = 1 + \chi/\sigma = 1 + a_1\gamma\chi$$

[see Eq. (10.3.34)] ; however, the experimental evidence on $\phi(z)$ is in disagreement with the assumption of an exponential distribution (see Fig. 12.16). The definition

$$\sigma = 1/a_1\gamma \tag{10.3.43}$$

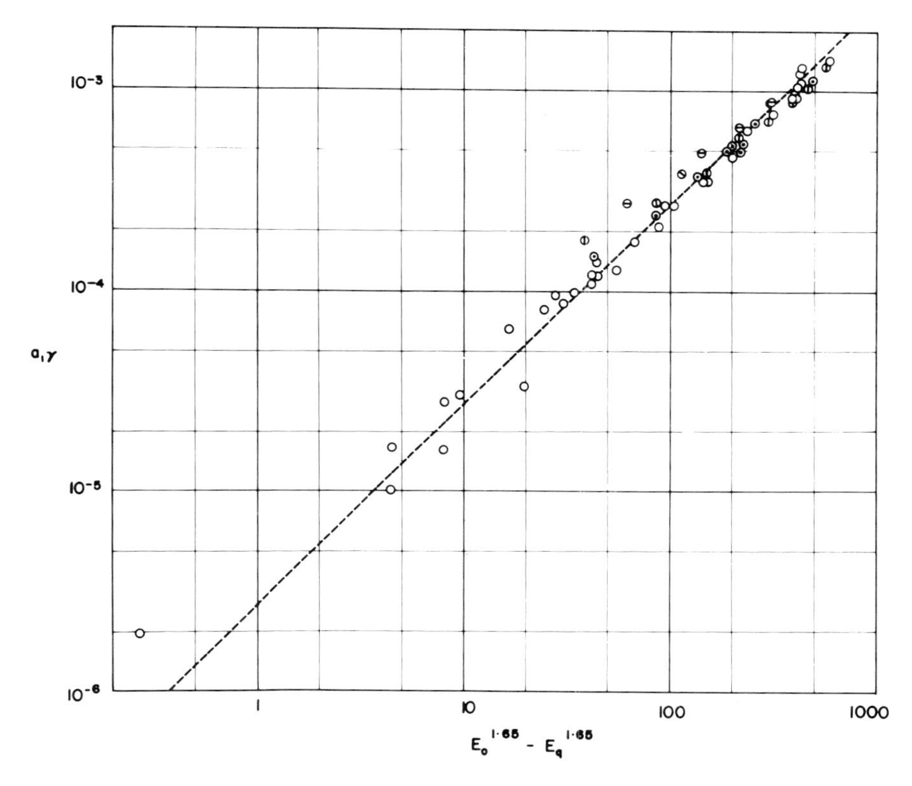

Fig. 10.21. Plot of experimentally obtained values of $a_1\gamma$ as a function of γ, after correction for continuum fluorescence. Dotted circles are tracer experiments and circles with bar are cases of strong continuum fluorescence.

is useful although it incorporates in σ whatever change of a_1 may occur as Z changes. We note that from the value $a = 1.2 \times 10^{-6}$ [Eq. (10.3.40)] we obtain:

$$\sigma = \frac{4 \times 10^5}{V_0^{1.65} - V_q^{1.65}}.$$

in close agreement with Eq. (10.3.29). A special case is that with two exponential terms in which $c_1 = -c_2$. We obtain

$$\phi(z) = \frac{d_1 \cdot d_2}{d_2 - d_1} [\exp(-d_1 z) - \exp(-d_2 z)] \tag{10.3.44}$$

This is the form of Philibert's absorption correction [Eqs. (10.3.21) and (10.3.23)]. The corresponding absorption function is:

$$f_p = \frac{d_1 \cdot d_2}{(d_1 + \chi)(d_2 + \chi)} = \frac{1}{1 + \chi/d_1} \cdot \frac{1}{1 + \chi/d_2}$$

which is equivalent to Eq. (10.3.32), with $a_1\gamma = 1/\mathrm{d}_1 + 1/\mathrm{d}_2, a_2\gamma^2 = (d_1 \cdot d_2)^{-1}$. However, the inverse Laplace transform of the *quadratic model*

$$1/f_p = (1 + a\gamma\chi)^2 = (1 + \chi/2\sigma)^2 \tag{10.3.45}$$

cannot be obtained as a sum of exponential terms. A comparison of Eqs. (10.3.44) and (10.3.45) shows that the quadratic model is a limiting case of Eq. (10.3.44), with $d_1 = d_2$. For this case, the expression (10.3.44) becomes indeterminate. The inverse Laplace transform for the quadratic model is:

$$\phi(z) = \frac{z}{a^2\gamma^2} \exp\left(-\frac{z}{a\gamma}\right) = 4\sigma^2 z \cdot \exp(-2\sigma z). \tag{10.3.46}$$

The maximum of $\phi(z)$ occurs at $z = a\gamma$ and its value is: $\phi(z) = (ea\gamma)^{-1} = 2\sigma/e$. The point of inflection of the curve is at $z = 2a\gamma$. The shape of this function is indeed reasonably close to that of the experimental depth distribution functions (Fig. 10.22).

Equations (10.3.45) and (10.3.46) suggest that a more general function can be obtained if the normalized parameters $Z = z(a\gamma)^{-1}$ and $K = a\gamma\chi$ are used. To conserve unity for the expression $\int_0^\infty \phi(z)dz$ we must also define that $\phi(Z) = a\gamma\phi(z)$. We obtain

$$\phi(Z) = Z \cdot \exp(-Z) \tag{10.3.47}$$

and

$$f_p = (1 + K)^{-2}. \tag{10.3.48}$$

An important shortcoming of the quadratic model for $\phi(z)$ presented in Eqs. (10.3.45) and (10.3.46) is its value at the origin: $\phi(0) = 0$. The implicit assumption that there is no X-ray emission from the target layer close to the surface is clearly contrary to the physical model of electron-target interaction, and to experimental evidence. This objection can also be raised against Philibert's simplified model [Eq. (103.21)], but not against his complete model [Eq. (10.3.17)]. A simple transformation of the quadratic model which removes this imperfection consists of shifting the origin of the function to a point above the specimen surface, by a distance d (g/cm). For a normalized representation, we will stipulate that $\delta = d/a\gamma$ (see Fig. 10.23). The corresponding algorithm for $\phi(z)$, normalized so that $\int_0^\infty \phi(z) \cdot dz = 1$, is

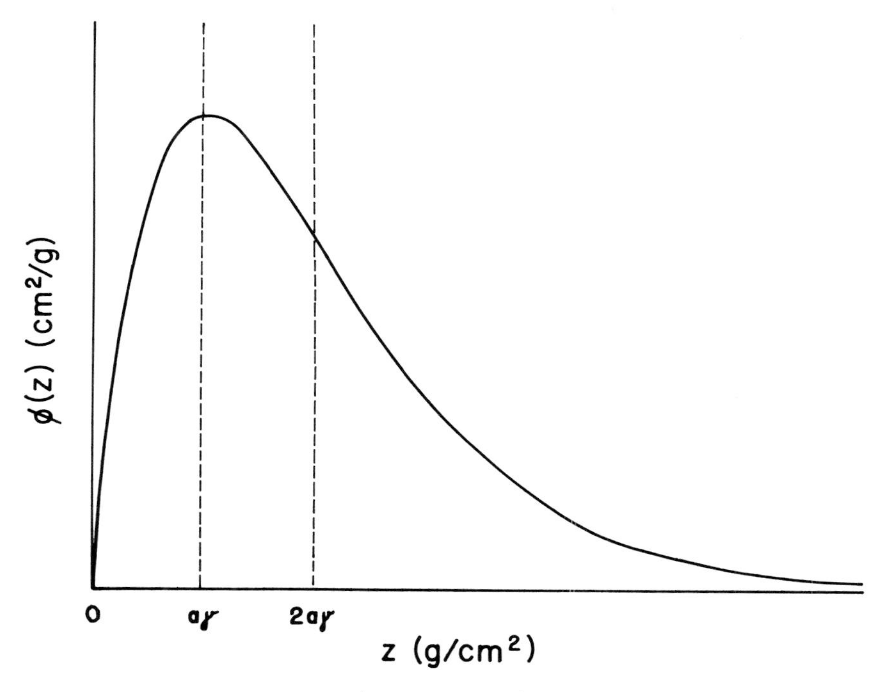

Fig. 10.22. The function $\phi(z) = [z/(a\gamma)] \cdot \exp(-z/a\gamma)$ has the following properties: the area under the curve is equal to one, the maximum occurs at $(z = a\gamma)$; it is equal to $(ea\gamma)^{-1}$, the point of inflection is located at $(z = 2a\gamma)$; the value of $\phi(z)$ at the point of inflection is also equal to $z/(a\gamma e^2)$. The point of inflection is also equal to the mean depth of generation of X-rays as defined in Eq. (13.1.7).

$$\phi(z) = \frac{z \boxed{+d}}{a\gamma(a\gamma \boxed{+d)}} \exp(-z/a\gamma) \tag{10.3.49}$$

and

$$\phi(Z) = \frac{Z + \delta}{1 + \delta} e^{-Z} \tag{10.3.50}$$

hence, $\phi(0) = \delta/(1 + \delta)$. These equations revert to the original model if d or δ become zero. The corresponding expressions for f_p are:

$$f_p = (1 + a\gamma\chi)^{-2}\left(1 + \frac{a\gamma d\chi}{a\gamma + d}\right) = (1 + K)^{-2}\,\frac{1 + \delta(1 + K)}{1 + \delta} \tag{10.3.51}$$

These functions should be an improvement over the quadratic model for very large absorption losses, since in such conditions the emission from shallow layers becomes more important (Fig. 10.23).

Another inaccuracy of the quadratic model and of similar models is their tailing towards great depth, in contrast with the well-defined X-ray range observed in tracer experiments, and similarly in observations concerning electron penetration in targets (see Fig. 2.2 and Figs. 10.3 through 10.5). This tailing can be eliminated by shifting the z axis (the horizontal axis) so that it is intercepted by the function $\phi(z)$ at the X-ray range, z_r (see Fig. 10.24). In other words, the value of $\phi(z)$, according to Eq. (10.3.49), at ($z = z_r$), is subtracted from all values of $\phi(z)$. After normalization to $\int_0^{z_r}(z)dz = 1$ we obtain:

$$\phi(z) = \frac{(z + d)e^{-z/a\gamma} - (z_r + d)e^{-z_r/a\gamma}}{a\gamma(a\gamma + d) - [(a\gamma)^2 + (a\gamma + z_r)(d + z_r)]e^{-z_r/a\gamma}} \quad (10.3.52)$$

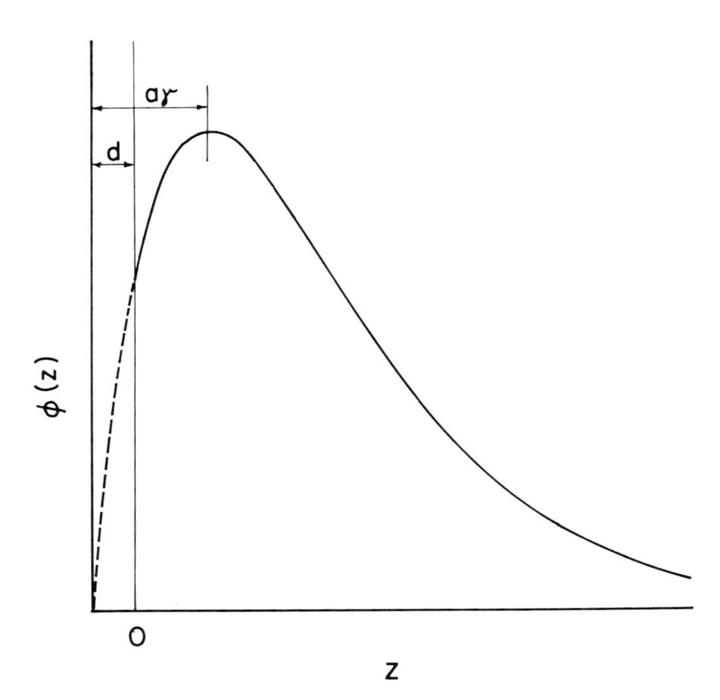

Fig. 10.23. The function represented by Eq. (10.3.49) is obtained by shifting the ordinate (z) by the amount d, and normalizing the area under the curve.

Plate I

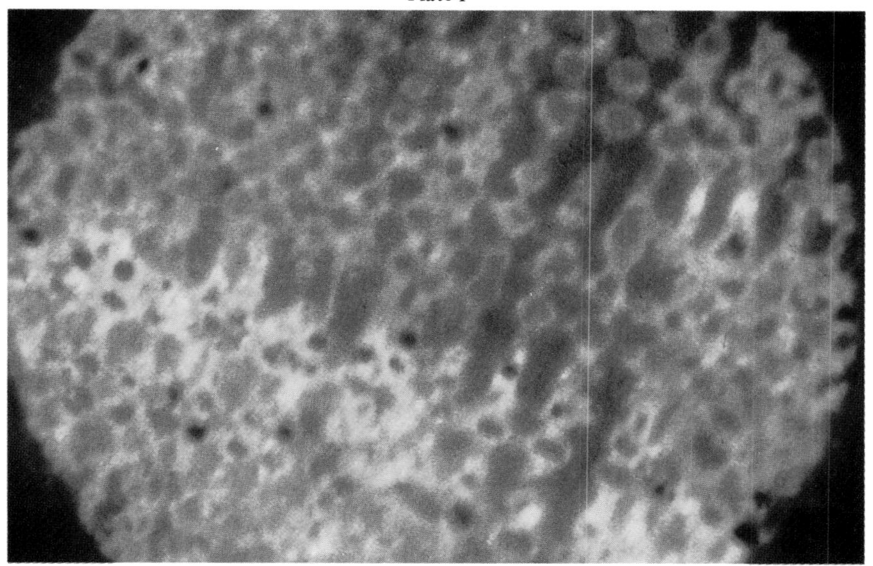

Figure 1. Cross section through stem of fossile crinoid. From ref. /14.25/.

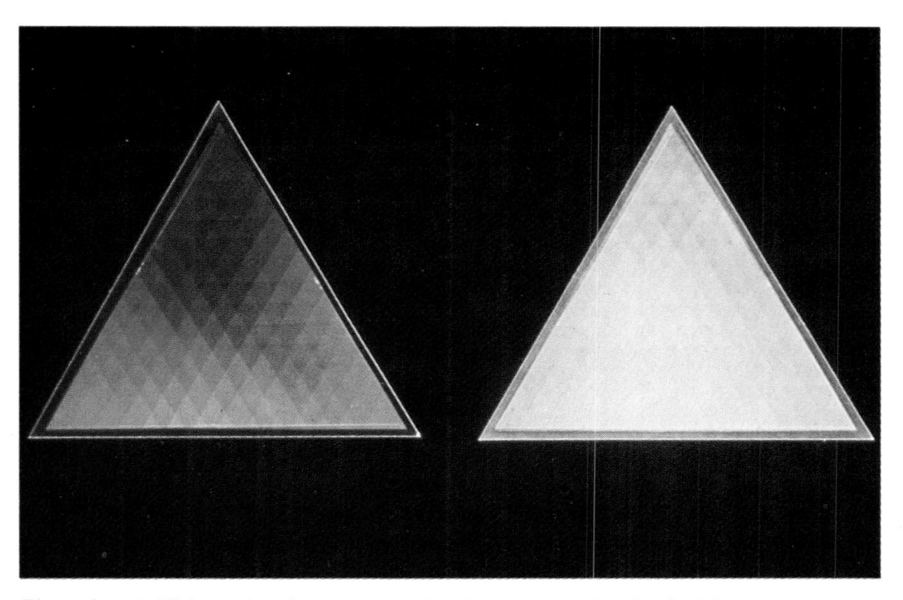

Figure 2. Addition colors from green, red and blue, at two levels of brightness.

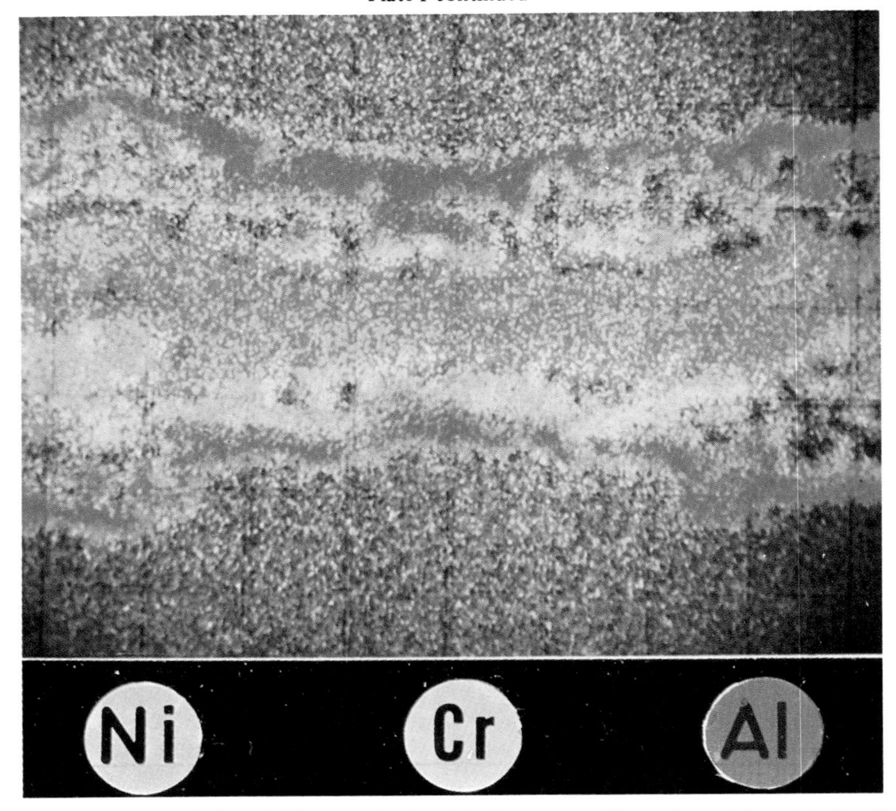

Figure 3. Oxidation within a crack of a nickel-base superalloy. Above: blue: nickel, red: aluminum, green: chromium. No target current signal added. Facing page: red: nickel, light blue: aluminum, green: chromium. Target current added in neutral hue. Observe the improvement in spatial definition and suppression of statistical noise. Partly from reference/15.13/

Plate I continued

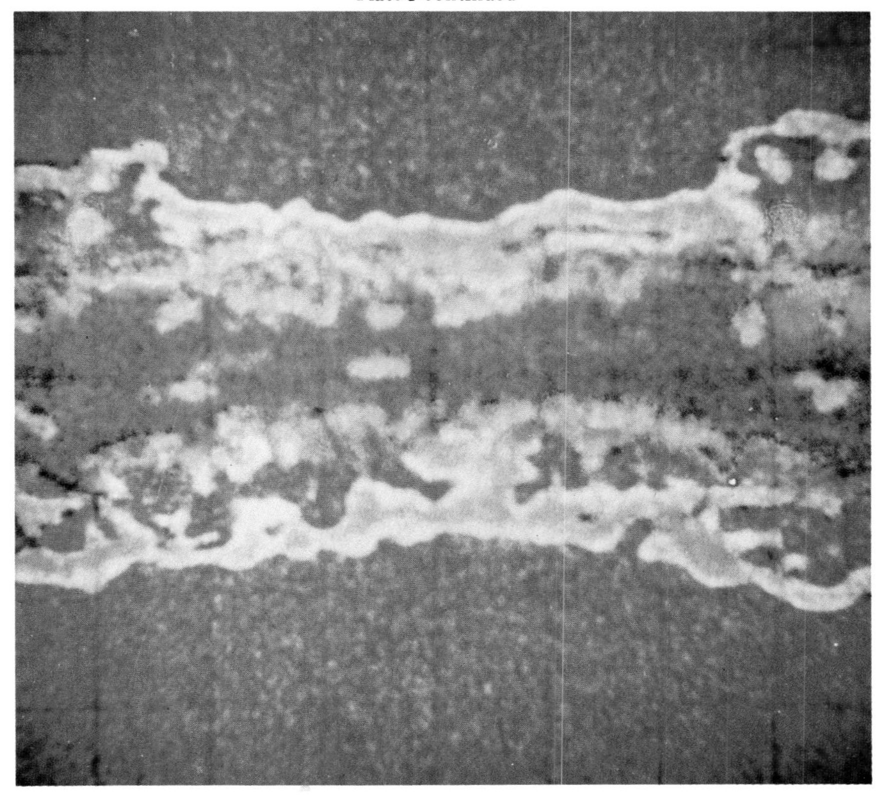

Figure 3. continued

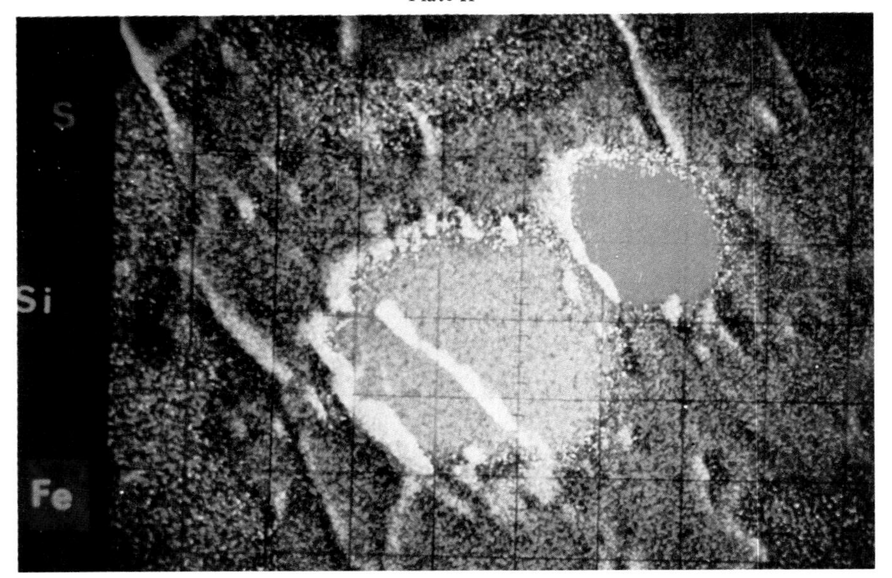

Figure 1. Adjacent spherical inclusions of iron (red) and troilite, FeS (magenta). Red: iron, blue: sulfur, green: silicon. Width of field: 22 μm. E_O = 20 kV. The specimen is a lunar rock from Apollo 11 flight. Target current scan was used to add topographic information.

Figure 2. Silicate rock from Apollo 11 flight. Red: potassium, blue: silica, green: iron. Width of field: 68 μm. E_O = 20 keV. Potassium gathers in the last silicate phase to solidify (magenta). The green phase is ferrous titanate. Target current image added in gray.

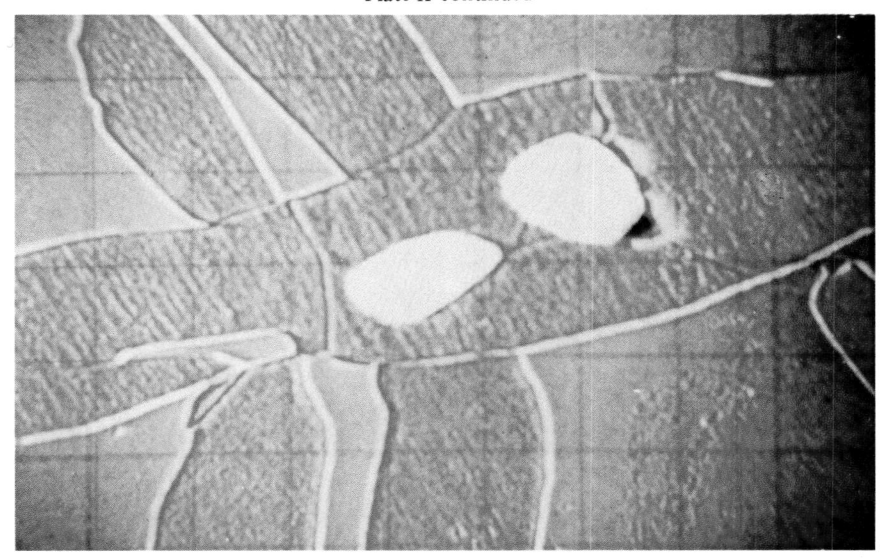

Figure 3. Ferrous meteorite. Red: iron, blue: nickel, green: phosphorus. Width of field: 150 μm. Target current added. The meteorite was slightly etched to enhance surface topography. The yellowish white grains are phosphide. The dark corners are due to the defect of the oscilloscope camera mentioned in Chapter 15.

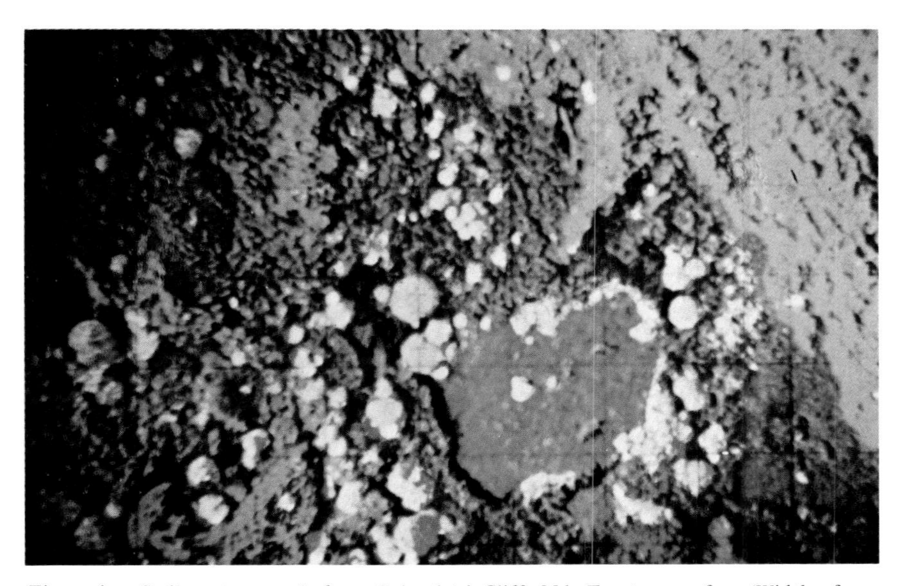

Figure 4. Sedimentary rock from Scientists' Cliff, Md., Fracture surface. Width of scan: 320 μm. Blue: calcium, green: silicon, red: iron. Target current was added, which indicates topographic features, and shows two types of calcium carbonate, of different texture. The reddish spherules are pyrite. The green phase is quartz.

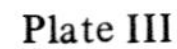

Plate III

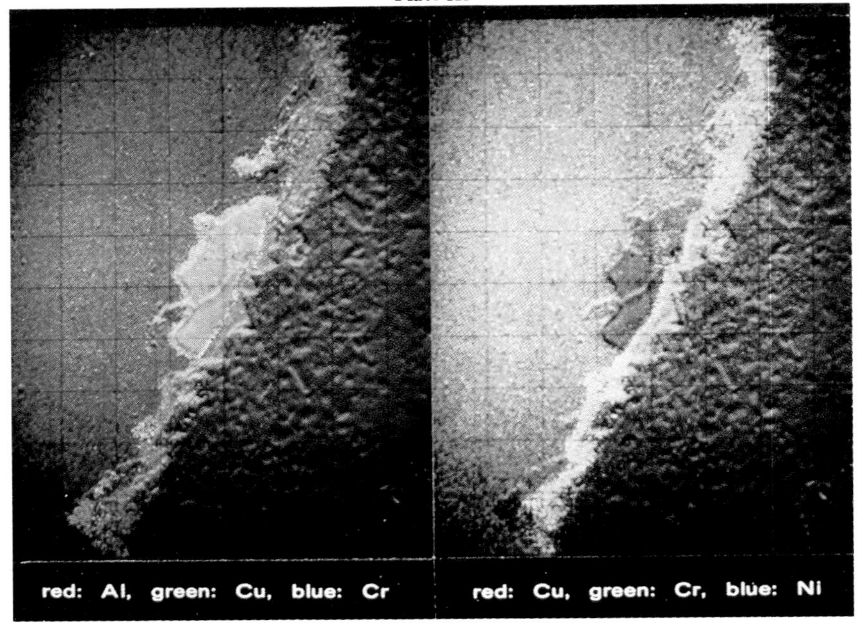

Figure 1. Two-color combinations of area scans of a heat shield after a simulated re-entrance test. Copper has deposited at the surface, and combined with nickel. The red phase, alumina, of the left image, consists of impacted grains from a polishing compound. E_o = 20 keV. Width of scan: 200 μm.

Figure 2. Surface of a heat-treated dental amalgam, with large crystals rich in tin (red) protruding from the surface. Blue: mercury, green: silver. From reference /15.13/.

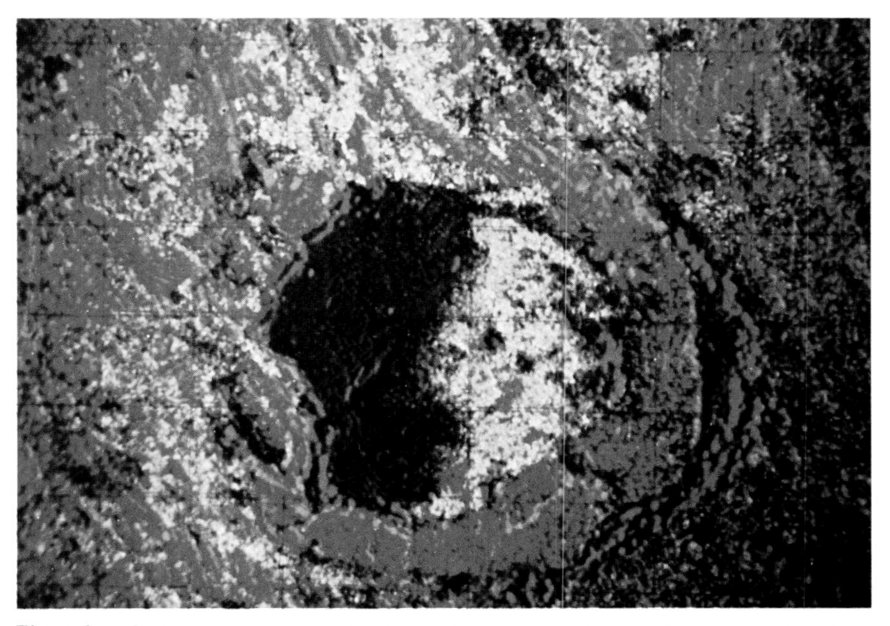

Figure 3. Crater on the surface of a lunar glass spherule. Red: nickel, green: sulfur blue: silicon. Target current image superposed in gray tones. Width of field: 36 μm.

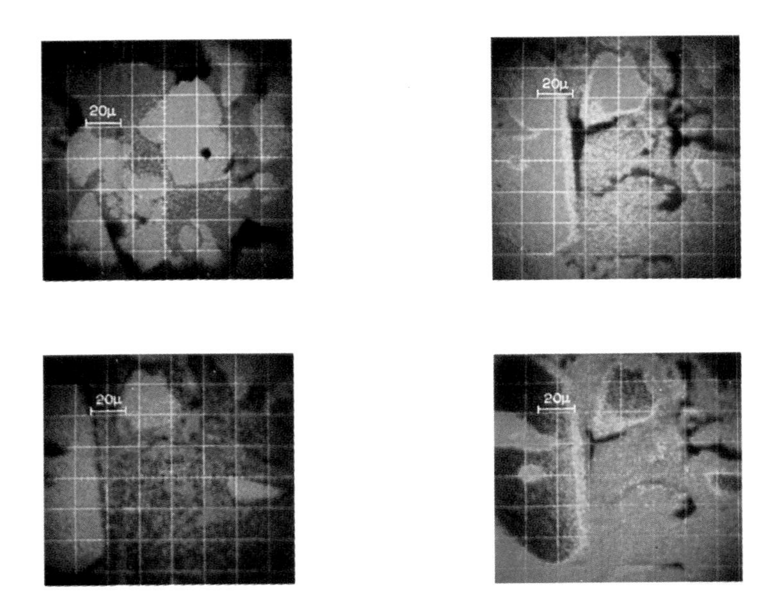

Figure 1. Scanning images from a color CRT. From J. Ficca /15.14/.

Figure 2. Color concentration map, from Schpigler /15.15/.

$$\phi(Z) = \frac{(Z + \delta)e^{-Z} \;\; - (R + \delta)e^{-R}}{1 + \delta \;\; - [1 + (R + 1)(R + \delta)]\, e^{-R}} \tag{10.3.53}$$

where $R = z_r/a\gamma$. The dashed lines indicate how these equations relate to Eqs. (10.3.49) and (10.3.50). The complication of these expressions—which is no problem in data handling by computers—arises from the normalization, and from the fact that all integrations must be limited to the range of z from zero to z_r, in order to avoid negative values for this variable. The corresponding expressions for f_p are:

$$f_p = \frac{1 + \frac{d}{a\gamma}(1 + a\gamma\chi) \;\; - \frac{z_r + d}{a\gamma}\frac{(1 + a\gamma\chi)^2}{a\gamma\chi} e^{-\frac{z_r}{a\gamma}} - \frac{z_r + d}{a\gamma}\frac{1 + a\gamma\chi}{a\gamma\chi} e^{-\frac{z_r}{a\gamma}(1 + a\gamma\chi)}}{(1 + a\gamma\chi)^2 \left\{1 + \frac{d}{a\gamma} \;\; - \left[1 + \left(1 + \frac{z_r}{a\gamma}\right)\left(\frac{z_r + d}{a\gamma}\right)\right] e^{-z_r/a\gamma}\right\}} \tag{10.3.54}$$

and

$$f_p = \frac{1 + \delta(1 + K)}{(1 + K)^2} \frac{\;\; - (R + \delta \;\; \frac{(1 + K)^2}{K} e^{-R} + \left[(R + \delta)\; \frac{1 + K}{K} - 1\right] e^{-R(1 + K)}}{1 + \delta \;\; - [1 + (1 + R)(R + \delta)]\, e^{-R}\}} \tag{10.3.55}$$

As illustrated in Figs. 10.25 and 10.26, good fits to all experimental $\phi(z)$ curves can be obtained by setting the value of the range to $z_r = 6.5 \times 10^{-6}$ g/cm^2, and adjusting the values of a and d to the experimental evidence. However, the advantages of this adjustment are minor, since for high absorption the distribution at large depth is unimportant.

There is clearly a trend with atomic number in the shapes of the $\phi(z)$ function, and it would be tempting to reintroduce an atomic number dependence. I believe, however, that the experimental evidence is still too scanty to guarantee success in such a generalization. Perhaps a carefully adjusted Monte Carlo calculation would provide the needed additional information.

A different generalization of experimental data of the absorption factor was proposed by Andersen and Wittry [10.23], who sought to establish the dependence on Z, A, E_0, and E_q of a parameter called "main depth of ionization," by empirical adjustment of a function originally derived for the Bethe range. By introducing an empirical atomic number dependence, they obtained for most of the range of experimental data on f_{ex} a unique function relating f to their expression for the main depth of ionization. This function is nonlinear, and represented by the authors in a graph drawn on a special scale.

Following a set of assumptions formulated first by Wittry [3.16], Kyser [10.24] developed a model in which $\phi(z)$ has a Gaussian distribution:

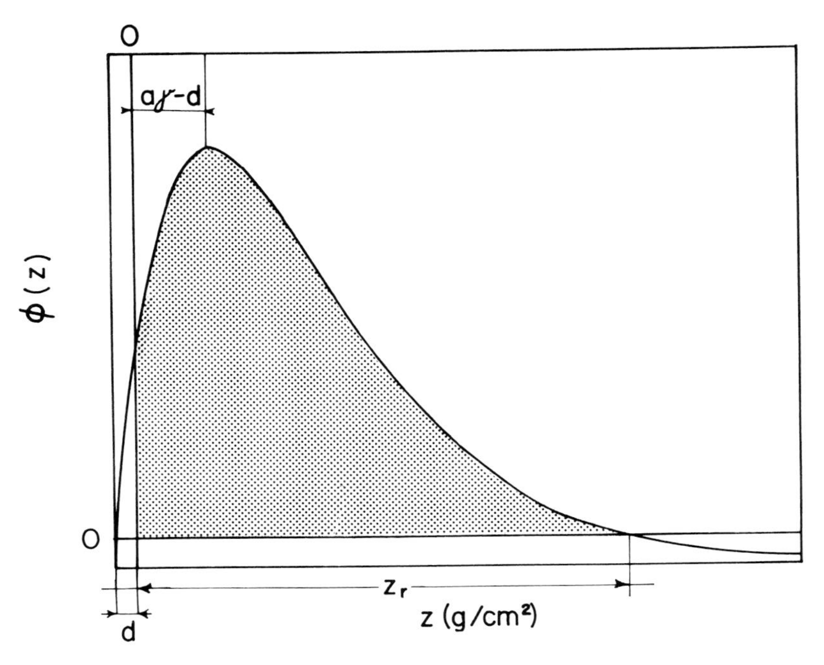

Fig. 10.24. By shifting both coordinates, a nonzero value for $\phi(0)$, and a value of zero at depth z_r is obtained.

$$\phi(z) \propto \exp\left[-\left(\frac{z - z_0}{\Delta z}\right)^2\right]. \tag{10.3.56}$$

In this equation, Δz describes the halfwidth of the Gaussian distribution, and z_0 describes its centroid. The function has a Laplace transform which represents f_p:

$$f_p = \frac{\exp - (z_0/\Delta z)^2 \exp(u^2)[1 - \operatorname{erf}(u)]}{1 + \operatorname{erf}(z_0/\Delta z)} \tag{10.3.57}$$

where $u = \chi\Delta z/z$ and $\operatorname{erf}(u) = 2\pi^{-1/2} \int_0^u \exp(-t^2)dt$. This model is of interest for the analysis of soft X-ray emission because the intercept of $\phi(z)$ at $(z = 0)$ has a nonzero value. However, even with appropriate fitting of the parameters z_0 and Δz, which are scaled to a depth range

$$z_r = 2.56 \times 10^{-3}(E_0/30)^{1.68} \text{ g/cm} \tag{10.3.58}$$

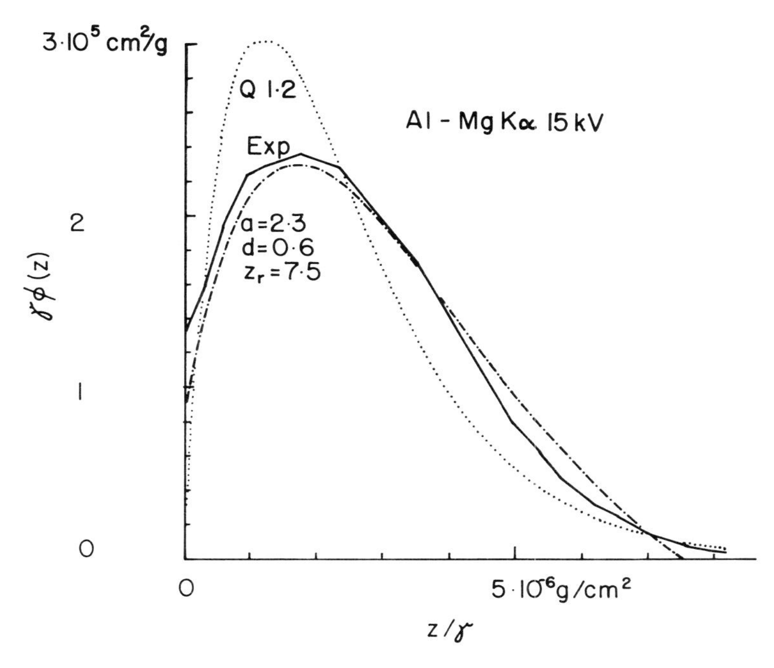

Fig. 10.25. Experimental values, quadratic expression Eq. (10.3.45), and fit of Eq. (10.3.52) to experimental data. Aluminum target with Mg K tracer at 15 kV.

by $z_0 = 0.125z_r$, and $\Delta z = 0.350z_r$, the fit is less than perfect, since the function $\phi(z)$, unlike a Gaussian, is asymmetric. Kyser also proposed an improved function in which an exponential function is added to the Gaussian.

10.4. DISTRIBUTION IN DEPTH OF THE CONTINUOUS RADIATION

Due to the nonspecificity, anisotropy, and polychromaticity of the continuous spectrum, the variable emergence-angle method and the tracer technique cannot be applied to the study of its distribution in depth. Some information can be derived from the change in intensity across absorption edges, but we cannot assume that the depth distribution of continuous X-rays is accurately known. Green and Cosslett [10.25] proposed that the difference in depth of the generation of characteristic and continuous X-rays is negligible, while Birks et al. [10.26] indicated that according to their observations at various emergence angles the continuum is generated, in average, at shallower depth. Reed [6.8] proposed that for the depth distribution of the continuum the Philibert formula [Eq. (10.3.23)] be used, with Heinrich's σ modified by replacing the critical excitation voltage V_q [Eq. (10.3.29)] by the value in kilovolts corresponding to the energy of the continuous photons in question, V_c, and lowering the constant

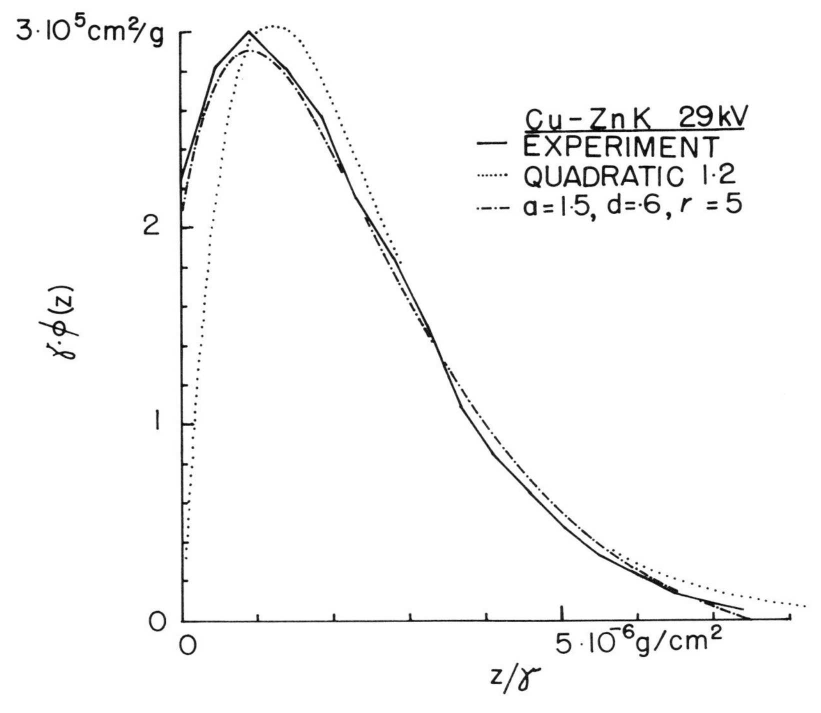

Fig. 10.26. Experimental values, quadratic expression Eq. (10.3.45), and fit of Eq. (10.3.52) to experimental data. Copper target with Zn K tracer at 29 kV.

of σ to 4.0×10^5. There is no question that the cross sections for shell ionization do not vary in the same fashion with the electron energy as the emission efficiency for the continuum does, and hence the absorption corrections must differ somewhat. The difference is not, however, very large, and, until better knowledge is acquired on the subject, it can usually be ignored.

10.5 ANALYSIS WITH INCLINED ELECTRON BEAM

In the preceding discussion of events in the target, we have assumed that the electron beam impinges normally upon the specimen surface. However, in many instruments (Figs. 3.27, 3.28, 3.30, and 3.31), the specimen surface is oriented obliquely with respect to the electron beam. This provides room for X-ray spectrometers at high emergence angles, and for optical microscope objectives of short focal length and high quality. In the scanning electron microscope, the specimen surface is usually inclined with respect to the electron beam. Since many scanning electron microscopes are equipped with X-ray detectors, we will include them in the present discussion.

A moderate beam inclination ($\epsilon \geqslant 45°$) does not affect in principle the quality of quantitative X-ray analysis. However, the behavior of targets bombarded by electrons has usually been studied with normal beam incidence. Therefore, it is necessary to adapt to the inclined beam incidence the principles which were developed for normal incidence.

The increase in backscattering which occurs when the specimen is inclined with respect to the beam (Section 9.3) enhances the backscatter losses in X-ray production; the backscatter correction factor R therefore diminishes. Backscatter corrections for a beam incidence angle of 45° were calculated by Reed [10.27] on the basis of energy distribution of backscattered electrons by Kanter [10.28] and by Kulenkampff and Spyra [9.43]. The ratio of electron backscattering correction factors for 45° and 90°, according to Reed, is seen in Fig. 10.27, which should be compared to Fig. 9.13.

Changes are also required in the expressions for absorption of primary X-rays; these can be predicted fairly rigorously (Fig. 10.28). It can be assumed that the

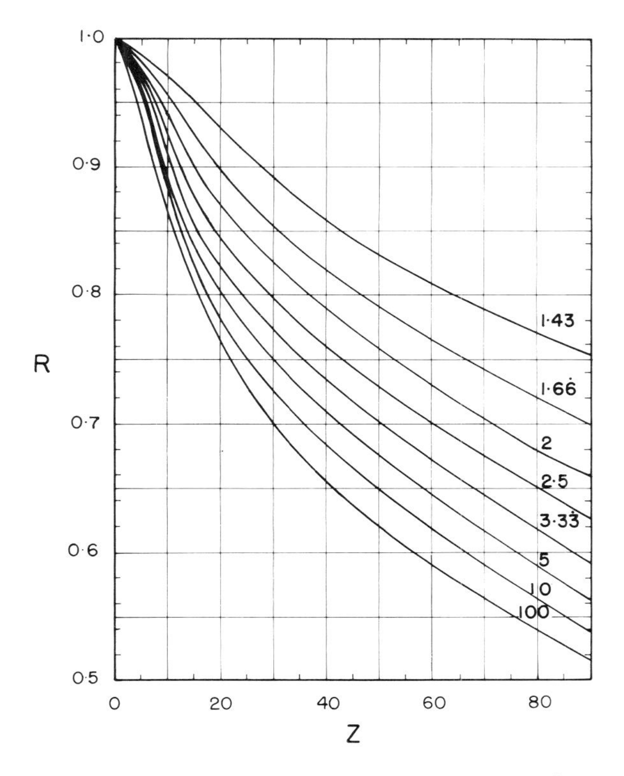

Fig. 10.27. The backscatter factor R for a beam inclination of 45° according to Reed [10.27]. Numbers within the graph are overvoltages, E_0/E_q.

distribution of sites of X-ray generation along the original direction of the electron beam remains virtually unchanged. Therefore, the distribution $\phi(z)$ is about the same as for normal incidence, except that the direction of penetration (z) is not normal to the surface. If the angle between the specimen surface and the electron beam is ϵ, then a photon generated at a depth of penetration z has been produced at the distance $z \cdot \sin \epsilon$ from the specimen surface. If the X-ray emergence angle with respect to the specimen surface is ψ, then the pathlength of the photon within the specimen is equal to

$$z' \cdot \csc \psi = z \cdot \sin \epsilon \csc \psi \qquad (10.5.1)$$

and the intensity generated after the length of penetration z is equal to

$$I'(z) = \frac{\Omega}{4\pi} I(z)dz \cdot \exp(-\mu z \cdot \sin \epsilon \cdot \csc \psi). \qquad (10.5.2)$$

If we compare this equation with Eq. (10.1.2), and follow the ensuing argument concerning the absorption factor, f, we see that for the inclined electron beam case the arguments are equally valid provided that χ is redefined as follows:

$$\chi = \mu \sin \epsilon \csc \psi. \qquad (10.5.3)$$

We neglect here the fact that lateral displacement of the site of X-ray generation, in direction normal to the primary electron beam, also affects the distances of the sites of X-ray generation from the specimen surface. Moreover, the change of surface boundary, with the concomitant change of backscattering, affects the depth distribution of X-ray generation in a complex manner which cannot be described by simple geometrical adjustment. Green [10.29] reported, in fact, in 1964 that Eq. (10.5.1) did not produce a correct absorption correction. Bishop [2.4] suggested, on the basis of his early Monte Carlo calculations that an empirical adjustment by the factor $(1 - 1/2 \cos^2 \epsilon)$ might be more appropriate. Later Monte Carlo calculations by Bishop failed, however, to confirm the earlier results [10.30]. The problem can be solved by means of a greatly refined and thoroughly tested Monte Carlo procedure. Purely empirical adjustments based on the measurements on targets of known composition may instead produce compensations of errors between the backscatter and absorption corrections. A solution based on the use of a Monte Carlo procedure by Curgenven and Duncumb [2.5] was proposed by Love et al. [9.10]. Further tests will indicate if this Monte Carlo procedure is adequate for this particular purpose.

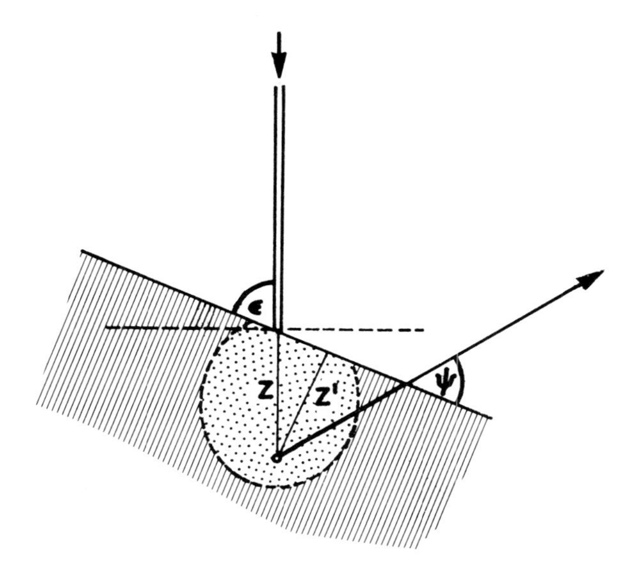

Fig. 10.28. The absorption of primary X-rays in an inclined target.

10. REFERENCES

10.1 Todd, J., In *Handbook of Physics*, Condon, E. U. and Odishaw, H., Eds., McGraw-Hill, New York, 1958, p. I-36.

10.2 Lenard, P., *Ann. Phys. Chem. N.F.* **56**, 255 (1895).

10.3 Williams, E. J., *Proc. Roy. Soc., Ser. A* **130**, 326 (1932).

10.4 Webster, D. L., Hansen, W. W., and Duveneck, F. B., *Phys. Rev.* **43**, 839 (1933).

10.5 Castaing, R. and Hénoc, J., *Proc. 4th Int. Congr. on X-Ray Optics and Microanalysis*, Castaing, R., Deschamps, P., and Philibert, J., Eds., Hermann, Paris, 1966, p. 120.

10.6 Vignes, A. and Dez, G., *Brit. J. Appl. Phys.* **2**, 1309 (1968).

10.7 Brown, J. D. and Parobek, L., *X-Ray Spectrometry* **5**, 36, (1976).

10.8 Brown, J. D. and Parobek, L., *Adv. X-Ray Analysis* **16**, 198 (1973).

10.9 Webster, D. L., *Phys. Rev.* **7**, 599 (1916).

10.10 Green, M., *Proc. 3rd Int. Conf. on X-Ray Optics and Microanalysis*, Pattee, H. H., Cosslett, V. E., and Engström, A., Eds., Academic Press, New York, 1963, p. 361.

10.11 Philibert, J., *Proc. 3rd Int. Conf. on X-Ray Optics and Microanalysis*, Pattee, H. H., Cosslett, V. E., and Engström, A., Eds., Academic Press, New York, 1963, p. 379.

10.12 Archard, G. D., *J. Appl. Phys.* **32**, 1505 (1961).

10.13 Bethe, H. A., Rose, M. E., and Smith, L. P., *Proc. Amer. Phil. Soc.* 78, 573 (1938).

10.14 Bothe, W., *Handbook Phys.* **24**, 18 (1927).

10.15 Reuter, W., *Proc. 6th Int. Conf. on X-Ray Optics and Microanalysis*, Shinoda, G., Kohra, K., and Ichinokawa, T., University of Tokyo Press, Tokyo, Japan, 1972, p. 121.

10.16 Cosslett, V. E. and Thomas, R. N., *Brit. J. Appl. Phys.* **15**, 235, 883, 1283 (1964); also **16**, 779 (1965).

10.17 Duncumb, P. and Shields, P. K., in *The Electron Microprobe*, McKinley, T. D., Heinrich, K. F. J., and Wittry, D. B., Eds., John Wiley & Sons, New York, 1966, p. 284.
10.18 Heinrich, K. F. J., NBS Tech. Note 521, National Bureau of Standards, U. S. Dept. of Commerce, Washington, D.C., 1970.
10.19 Duncumb, P., Shields-Mason, P. K., and da Casa, C., *Proc. 5th Int. Congr. on X-Ray Optics and Microanalysis*, Möllenstedt, G. and Gaukler, K. H., Eds., Springer, Berlin, 1969, p. 146.
10.20 Belk, J. A., *Proc. 4th Int. Congr. on X-Ray Optics and Microanalysis*, Castaing, R., Deschamps, P., and Philibert, J., Eds., Hermann, Paris, 1966, 214.
10.21 Heinrich, K. F. J. and Yakowitz, H., *Anal. Chem.* 47, 2408 (1975).
10.22 Criss, J. and Birks, L. S., in *The Electron Microprobe*, McKinley, T. D., Heinrich, K. F. J., and Wittry, T. D., Eds., John Wiley & Sons, New York, 1966, p. 217.
10.23 Andersen, C. A. and Wittry, D. B., *Brit. J. Appl. Phys. (J. Phys. D) Ser. 2*, **1**, 529 (1968).
10.24 Kyser, D. F., *Proc 6th Int. Conf. on X-Ray Optics and Microanalysis*, Shinoda, G., Kohra, K., and Ichinokawa, T., Eds., University of Tokyo Press, 1972, p. 147.
10.25 Green, M. and Cosslett, V. E., *Brit. J. Appl. Phys. Ser 2*, **1**, 425, (1968).
10.26 Birks, L. S., Seebold, R. E., Grant, B. K., and Grosso, J. S., *J. Appl. Phys.* **36**, 699 (1965).
10.27 Reed, S. J. B., *J. Phys. D, Appl Phys.* **4**, 1910 (1971).
10.28 Kanter, H., *Ann Phys. (Leipzig)* **20**, 144 (1957).
10.29 Green, M., *Proc. Phys. Soc. London* **83**, 435 (1964).
10.30 Bishop H. E., *J. Phys. D, Appl. Phys.* **1**, 673 (1968).

11. Secondary X-Ray Emission

11.1 FLUORESCENCE EXCITED BY CHARACTERISTIC LINES

X-rays are called fluorescent, or secondary, if the emitting orbital levels have been ionized by the absorption of an X-ray photon. If the exciting radiation is a primary X-ray line, then we obtain fluorescence due to characteristic lines. Radiation can also be emitted after ionization of atoms by secondary characteristic radiation; such tertiary emission is not significant, however, in electron probe microanalysis.

To excite a secondary X-ray photon, the primary photon must have an energy exceeding the critical ionization energy E_q of the shell which will subsequently emit the fluorescent photon. But significant intensities of fluorescent radiation are produced only if the energy of the ionizing photon is not much larger than the critical ionization energy. Therefore, the effect is absent, or insignificant, in many cases. In particular, the fluorescence of a line of an element due to emission from another level of the same element (e.g., excitation of L-lines by the K-lines of the same element) is always very small. Fluorescent excitation by characteristic lines is thus negligible in pure elements.

Typically, the penetration of the primary X-rays into the specimen by far exceeds that of electrons. Therefore, the secondary X-ray photons are produced at distances from the point of impact of the electron beam much larger than the range of primary X-ray production. Secondary radiation thus degrades the spatial resolution of electron probe microanalysis, and its absorption factor is lower than that for primary radiation.

The fluorescent radiation generated in addition to the primary emission must be taken into account by means of a fluorescence calculation ("fluorescence correction") in data evaluation procedures used to determine the specimen composition. The computation used for this purpose is theoretically straightforward and can be performed rigorously (Fig. 11.1).

Assume that the target contains the elements a and b, and that the element b emits a primary line of intensity I_{pb} the energy of which exceeds the ionization

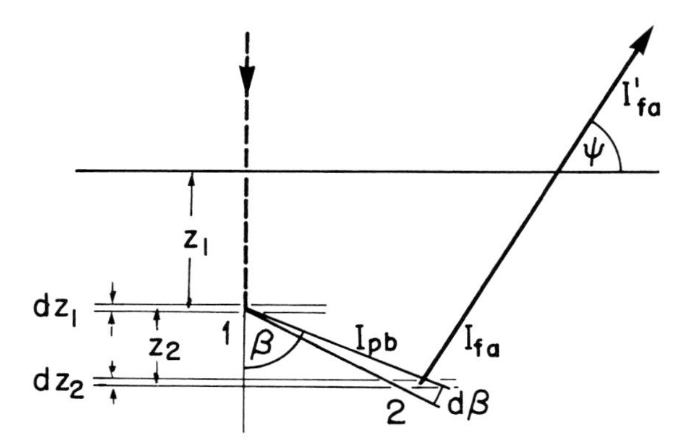

Fig. 11.1. Generation of fluorescent radiation. The meaning of the symbols is explained in the text.

energy for the K-shell of element a. This shell is ionized when the radiation of element b is absorbed in the target, and fluorescent radiation of intensity I'_{fa} is emitted by the specimen, in addition to the primary radiation of the same line from element a. A spectrometer of aperture Ω is aligned to this radiation; the mean angle of emergence of the radiation which is detected is ψ. According to Eq. (9.1.3), the intensity of the emergent fluorescent emission,

$$I'^{*}_{fa} = \frac{\Omega}{4\pi} I^{*}_{fa} f^{*}_{fa}$$

must be calculated.

The distribution in depth of the primary radiation of element b generated within the specimen can be described by the function $\phi^{*}_{b}(z)$ (see Chapter 10). Consider at this point the primary radiation generated at depth z_1, within a thin layer of thickness dz_1 (Fig. 11.1, point 1). The fraction of primary radiation generated between the depth levels z_1 and $z_1 + dz_1$ is, according to Eq. (10.1.13), equal to

$$dI_{pb}(z_1) = I^{*}_{b}(z)dz = I^{*}_{pb}\,\phi_b(z_1)dz_1. \tag{11.1.1}$$

Of this radiation, a fraction, called $d^2I(z_1, \beta)$ is emitted within the angle between β and $\beta + d\beta$, with respect to the electron beam:

$$d^2I_{pb}(z_1, \beta) = dI_p(z_1)\,\frac{2\pi \sin \beta d\beta}{4\pi} = \frac{1}{2}\,dI_{pb}(z_1) \sin \beta d\beta. \tag{11.1.2}$$

Along the trajectory from the depth z_1 to the depth $z_1 + z_2$, the radiation is attenuated, so that, at point 2, its intensity is

$$d^2I'_{pb}(z_1,\beta,z_2) = \frac{1}{2} dI_{pb}(z_1) \sin \beta d\beta \exp - \left[\mu(*b_{qm}) \frac{z_2}{\cos \beta}\right]. \quad (11.1.3)$$

Of the photons absorbed within the target, some cause ionization of the K-shell of the element a, and subsequent emission of the fluorescent radiation of this element. We will now calculate the intensity of the fraction of this fluorescent radiation, which is generated within the layer dz_2, by primary radiation which has propagated within the angle $d\beta$, to the depth $z_1 + z_2$. The length of the path of this primary radiation within the layer dz_2 is $dz_2/\cos \beta$. If the mass fraction of element a in the target is C_a, the fraction of primary radiation absorbed by this element in the layer is

$$d^3I'_{pb}(z_1,\beta,z_2) = -d^2I'_{pb}C_a\mu(a, b_{qm}) \frac{dz_2}{\cos \beta}. \quad (11.1.4)$$

We recall that in these derivations, as elesewhere in the book, the unit of depth is measured in grams per centimeter squared (g/cm^2), and that the symbol $\mu(a, b_{qm})$ denotes the mass absorption coefficient of the line qm emitted from the element b in the absorber a.

We must now determine what fraction of the absorbed radiation ionizes the K-level of the element a. This fraction can be determined from the sudden rise of the X-ray mass absorption coefficient at the absorption edge which corresponds to the energy of the K-level (see Fig. 11.2). Only at wavelengths shorter than the absorption edge is the energy of the X-ray photon large enough to ionize the K-level. Therefore, the mass absorption coefficient increases at the edge, from the value μ_2, of Fig. 11.2, to the value μ_1.
The ratio of these absorption coefficients, μ_1/μ_2, is called the absorption jump ratio of the q-level, r_q. Since the edge shown in Fig. 11.2 is a K-edge, we have: $\mu_1/\mu_2 = r_{kFe}$ (see Section 4.3). The ratio of absorbed photons which ionize this K-level, to those absorbed by all possible mechanisms is equal to

$$\frac{\mu_1 - \mu_2}{\mu_1} = \frac{r_k - 1}{r_k}.$$

We also observe in Fig. 11.2, which is a logarithmic plot of mass absorption coefficients as a function of wavelength, that the slopes of this function on either side of the edge are almost identical. At a wavelength shorter than the absorption edge—e.g., for the absorber Fe, at the wavelength of the Cu Kα line—the ratio of ionizing absorptions to total absorptions is therefore virtually the same as at the absorption edge:

$$\frac{\mu_3 - \mu_4}{\mu_3} \simeq \frac{\mu_1 - \mu_2}{\mu_1} = \frac{r_k - 1}{r_k}. \quad (11.1.5)$$

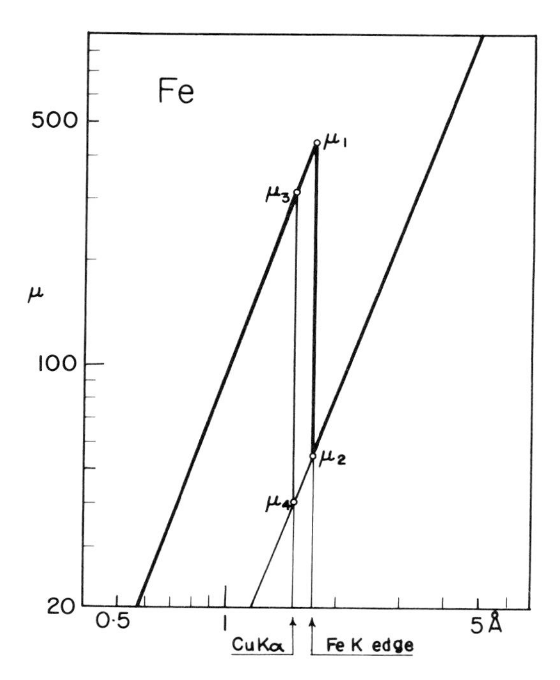

Fig. 11.2. Absorption coefficients of iron, with the position of the Fe K-edge, and the Cu Kα-line.

As we have discussed in Section 4.3, the situation is more complex when the fluorescent radiation emitted by element a is an L-line, since in this case three absorption edges must be taken into consideration (Fig. 4.15). Calling r_1, r_2, and r_3 the absorption jump ratios associated with the edges L_I, L_{II}, and L_{III}, the probabilities of ionization are the following:

a. The exciting line falls between L_{II} and L_{III}. In this case, the levels L_I and L_{II} are not excited, since their critical excitation energies are above the energy of the photon. Therefore, the situation is analogous to that of the K-edge:

$$\left[\frac{r-1}{r'}\right]_{L_{III}} = \frac{r_3 - 1}{r_3}. \tag{11.1.6}$$

b. The exciting line is between L_I and L_{II}. Therefore, the L_I level can not be excited. The probability of excitation of the level L_{II} is

$$\left[\frac{r-1}{r'}\right]_{L_{II}} = \frac{r_2 - 1}{r_2}. \tag{11.1.7}$$

For the ionization of the level L_{III} we must take into account that the fraction of the mass absorption coefficient corresponding to the ionization of the L_{III} level is equal to $(\mu_4 - \mu_5)/\mu_3$. In terms of jump ratios, this ratio is equal to

$$\left[\frac{r-1}{r'}\right]_{L_{III}} = \frac{\mu_4 - \mu_5}{\mu_3} = \frac{\mu_4/\mu_5 - 1}{\mu_3/\mu_5} = \frac{r_3 - 1}{r_2 \cdot r_3}. \tag{11.1.8}$$

c. The wavelength of the exciting line is shorter than L_I. We obtain the following by analogous deduction.
For the level L_I:

$$\frac{r-1}{r'} = \frac{r_1 - 1}{r_1}. \tag{11.1.9}$$

For the level L_{II}:

$$\frac{r-1}{r'} = \frac{\mu_7 - \mu_8}{\mu_6} = \frac{r_2 - 1}{r_1 \cdot r_2}. \tag{11.1.10}$$

For the level L_{III}:

$$\frac{r-1}{r'} = \frac{\mu_8 - \mu_9}{\mu_6} = \frac{r_3 - 1}{r_1 \cdot r_2 \cdot r_3}. \tag{11.1.11}$$

There exist empirical formulas which serve to obtain the jump ratio factors. Springer [11.1] proposed the following:

$$\frac{r_K - 1}{r_K} = 0.924 - 0.00144Z \tag{11.1.12}$$

$$\frac{r_{L_{III}} - 1}{r_{L_{III}} \cdot r_{L_{II}} \cdot r_{L_I}} = 0.548 - 0.00231Z. \tag{11.1.13}$$

Similar relations, but with five absorption edges present, can be deduced for the M-edges. As a generic notation for the probability of ionization we will use the following: $(r_q - 1)/r'_q$. When the notation does not indicate which element is meant, it is understood that the jump ratio is always that of the element which emits the fluorescent radiation.

The fluorescence yield, ω_q, and the relative intensity of the line, or "weight" of the line, p_{qm}, are defined in Section 4.2; they are identical to those used in primary X-ray generation, since the probability of X-ray emission from the ionized state does not depend upon the mechanism of ionization.

The intensity of fluorescent emission of the line qm of element a, excited at depth $z_1 + z_2$ by the primary line of element b generated at depth z_1, and traveling within the hollow cone of semiapex β and thickness $d\beta$, is therefore:

$$d^3 I_{fa}(z_1, \beta, z_2) = -d^3 I'_{pb} \frac{r_{qa} - 1}{r'_{qa}} \omega_{qa} p_{qma} \,. \tag{11.1.14}$$

We recall that the "intensity" is the probability of generation of one photon per incident electron. Curves and formulae for the numerical values of fluorescence yields were given in Section 4.2. As in primary emission, for shells above the K-shell, we must define effective fluorescence yields, in order to take into account the production of Coster-Kronig transitions. A rigorous approach to this problem is required since lines close to the emitting edge, for which effective yields must often be employed, cause strong fluorescent emissions.

If the exciting line falls between the levels L_{II} and L_{III}, only the L_{III} shell is ionized, and no Coster-Kronig transitions are possible. Therefore,

$$\omega_{L\,III\,eff} = \omega_{L\,III} \,. \tag{11.1.15}$$

If the exciting line is located between L_I and L_{II}, only the $L_{II} - L_{III}$ transition is possible. Hence,

$$\omega_{L\,II\,eff} = \omega_{L\,II} \tag{11.1.16}$$

$$\begin{aligned} \omega_{L\,III\,eff} &= \omega_{L\,III} \left[1 + f_{23} \cdot \left(\frac{r_2 - 1}{r_2} \right) \Big/ \left(\frac{r_3 - 1}{r_2 \cdot r_3} \right) \right] \\ &= \omega_{L\,III} \left[1 + f_{23} \cdot \frac{(r_2 - 1) \cdot r_3}{(r_3 - 1)} \right] . \end{aligned} \tag{11.1.17}$$

Finally, if the exciting line is shorter than L_I, then all three Coster-Kronig transformations occur, and

$$\omega_{L\,I\,eff} = \omega_{L\,I} \tag{11.1.18}$$

$$\begin{aligned} \omega_{L\,II\,eff} &= \omega_{L\,II} \left[1 + f_{12} \cdot \left(\frac{r_1 - 1}{r_1} \right) \Big/ \left(\frac{r_2 - 1}{r_1 \cdot r_2} \right) \right] \\ &= \omega_{L\,II} \left[1 + f_{12} \cdot \frac{(r_1 - 1) r_2}{r_2 - 1} \right] \end{aligned} \tag{11.1.19}$$

$$\omega_{\mathrm{L\,III\,eff}} = \omega_{\mathrm{L\,III}} \left[1 + \left(1 + f_{12} \frac{(r_1 - 1)r_2}{r_2 - 1}\right) \cdot f_{23} \cdot \left(\frac{r_2 - 1}{r_1 \cdot r_2}\right) \bigg/ \left(\frac{r_3 - 1}{r_1 \cdot r_2 \cdot r_3}\right) + f_{13} \cdot \frac{r_1 - 1}{r_1} \bigg/ \left(\frac{r_3 - 1}{r_1 \cdot r_2 \cdot r_3}\right)\right] \tag{11.1.20}$$

$$= \omega_{\mathrm{L\,III}} \left[1 + f_{13} \frac{(r_1 - 1) \cdot r_2 \cdot r_3}{r_3 - 1} + f_{23} \left(\frac{r_2 - 1}{r_3 - 1} \cdot r_3 + f_{12} \cdot \frac{r_1 - 1}{r_3 - 1} \cdot r_2 \cdot r_3\right)\right].$$

Similar relations could, in principle, be established for the M-radiation, except that the Coster-Kronig coefficients for this level are practically unknown. However, in the long wavelength region in which M-lines occur, fluorescent emission is negligible, and therefore, the related uncertainties are not of practical interest.

The fluorescent radiation produced at point 2 in Fig. 11.1 is emitted isotropically. A fraction of this radiation, after emerging through the specimen surface, is observed by the X-ray spectrometer. Before emergence, the radiation emitted into the solid angle Ω covered by the spectrometer has been attenuated by the target material. The intensity of the radiation which emerges within the aperture of the spectrometer is therefore

$$d^3 I'_{fa} = \frac{\Omega}{4\pi} d^3 I_{fa} \cdot \exp\left[-\mu(*a) \cdot \frac{z_1 + z_2}{\sin\psi}\right]. \tag{11.1.21}$$

The mass absorption coefficient $\mu(*a)$ is that for the line of element a which is measured, and for the target material. Combining Eqs. (11.1.1), (11.1.3), (11.1.4), (11.1.14), and (11.1.15), we obtain:

$$d^3 I'_{fa} = I^*_{pb} \phi_b(z_1) dz_1 \cdot \frac{1}{2} \sin\beta d\beta \cdot \exp\left[\frac{-\mu(*b) z_2}{\cos\beta}\right] \cdot C_a \mu(a, b) \frac{dz_2}{\cos\beta} \frac{r_q - 1}{r'_q} \omega_{qa} p_a \frac{\Omega}{4\pi} \exp\left[-\mu(*a) \frac{z_1 + z_2}{\sin\psi}\right]. \tag{11.1.22}$$

For simplicity, we have dropped the subindices qm; we also recall that the jump ratio r_q is that of element a. If we denote the conversion ratio from primary to secondary intensities by

$$G = C_a \mu(ab) \frac{r_q - 1}{r'_q} \omega_{qa} p_a \tag{11.1.23}$$

we can transform Eq. (11.1.22) to

$$d^3I'_{fa} = \frac{\Omega}{8\pi} I^*_{pb}\phi_b(z_1)G \cdot tg\beta \cdot \exp\left[-\frac{\mu(*a)z_1}{\sin\psi}\right] \tag{11.1.24}$$
$$\cdot \exp\left[-\left(\frac{\mu(*b)}{\cos\beta} + \frac{\mu(*a)}{\sin\psi}\right) z_2\right] dz_1 \cdot dz_2 \cdot d\beta.$$

This is the differential expression for fluorescence due to characteristic lines. To get the total fluorescent radiation of the line qm of element a excited by one line of element b, and emitted from the target into the aperture of the detector, we must integrate this expression over the complete range of depth of primary X-ray excitation, over all angles of emission of the primary radiation, and over the complete range of depth of the emission of fluorescent radiation:

$$I'_{fa} = \frac{\Omega}{8\pi} GI^*_{pb} \int_0^\infty \phi_b(z_1) \exp\left(\frac{-\mu(*a)}{\sin\psi} \cdot z_1\right) \tag{11.1.25}$$
$$\cdot \int_0^\pi \int_{-z_1}^\infty tg\,\beta \cdot \exp\left[-\left(\frac{\mu(*b)}{\cos\beta} + \frac{\mu(*a)}{\sin\psi}\right) z_2\right] dz_2 \cdot d\beta \cdot dz_1.$$

The integration of this equation over z_2 is performed in two parts [1.1]. We consider separately the fluorescent radiation produced at depths below z_1 (downward fluorescence), and that generated above z_1 (upward fluorescence).

For secondary photons generated below the depth z_1 we obtain the intensity $I'\downarrow_{fa}$ as follows:

$$I'\downarrow_{fa} = \frac{\Omega}{8\pi} GI^*_{pb} \int_0^\infty \phi_b(z_1) \exp\left(-\frac{\mu(*a)}{\sin\psi} z_1\right) \tag{11.1.26}$$
$$\cdot \int_0^{\pi/2} \int_0^\infty tg\,\beta \exp\left[-\left(\frac{\mu(*b)}{\cos\beta} + \frac{\mu(*a)}{\sin\psi}\right) z_2\right] dz_2 \cdot d\beta \cdot dz_1.$$

We abbreviate by defining

$$g = -\left[\frac{\mu(*b)}{\cos\beta} + \frac{\mu(*a)}{\sin\psi}\right]$$

and we resolve the double integral at the end of Eq. (11.12.26) as follows:

$$\int_0^{\pi/2}\int_0^{\infty} tg\,\beta \exp(gz_2)dz_2 d\beta = \int_0^{\pi/2}\int_0^{-\infty} \frac{tg\,\beta}{g} \exp(gz_2)d(gz_2)d\beta$$

$$= \int_0^{\pi/2} \frac{tg\,\beta}{g} \left| \exp(gz_2) \right|_0^{-\infty} d\beta = \int_0^{\pi/2} \frac{tg\,\beta}{g} d\beta = \int_0^{\pi/2} \frac{\sin\beta d\beta}{\mu(*b) + \chi(*a)\cos\beta}$$

$$= \frac{1}{\chi(*a)} \left| \ln[\mu(*b) + \chi(*a)\cos\beta] \right|_{\pi/2}^{0} = \frac{\ln\left[1 + \frac{\chi(*a)}{\mu(*b)}\right]}{\chi(*a)} = \frac{1}{\mu(*b)} \cdot \frac{\ln(1+\mathrm{u})}{\mathrm{u}}$$

$$(11.1.27)$$

with $u = \chi(*a)/\mu(*b)$. We therefore obtain:

$$I'\downarrow_{fa} = \frac{\Omega}{8\pi} \frac{GI'^*_{pb}}{\mu(*b)} \int_0^{\infty} \phi_b(z_1) \exp\left(\frac{-\mu(*a)z_1}{\sin\psi}\right) \frac{\ln(1+\mathrm{u})}{\mathrm{u}} dz_1. \quad (11.1.28)$$

Since

$$\frac{\mu(*a)}{\sin\psi} = \chi(*a)$$

and

$$f_{pb} = f[V_{qb}, \chi(*b)] = \int_0^{\infty} \varphi_b(z) \exp[-\chi(*b)z]\,dz_1,$$

we can write the downwards fluorescence as follows:

$$I'\downarrow_{fa} = \frac{\Omega}{8\pi} \frac{GI^*_{pb}}{\mu(*b)} f_p[V_{qb}\chi(*a)] \frac{\ln(1+\mathrm{u})}{\mathrm{u}}. \quad (11.1.29)$$

In this equation, $f_p\,[V_{qb}, \chi(*a)]$ has the form of the absorption factor f_p, except that the critical excitation potential V_{qb} is that of element b, while the parameter $\chi(*a) = \mu(*a)\csc\psi$ contains the mass absorption coefficient of the target for the secondary radiation emitted by element a. In practice the difference between $f_p\,[V_{qb}, \chi(*a)]$ and the usual absorption factor, $f_p\,[V_{qa}, \chi(*a)]$, is not important, since the fluorescent emission is strong only when the wavelengths of the primary and secondary radiations do not differ much; in typical cases of strong fluorescence, both absorption factors will be close to unity.

For the upward fluorescence ($-z_1 < z_2 < 0$), we obtain similarly:

$$I'\uparrow_{fa} = \frac{\Omega}{8\pi} GI^*_{pb} \int_0^\infty \phi_b(z_1) \exp[-\chi(*a)z_1]\, dz_1 \cdot \int_{\pi/2}^{\pi} \int_0^{-z_1} tg\,\beta \exp(gz_2) dz_2 d\beta. \qquad (11.1.30)$$

We first integrate over z_2:

$$\int_0^{-z_1} \exp(gz_2) dz_2 = \frac{1}{g}[\exp(-gz_1) - 1] \qquad (11.1.31)$$

and

$$I'\uparrow_{fa} = \frac{-\Omega}{8\pi} GI^*_{pb} \int_0^\infty \phi_b(z_1) \exp[-\chi(*a)z_1]\, dz_1 \cdot \int_{\pi/2}^{\pi} \frac{tg\,\beta}{\frac{\mu(*b)}{\cos\beta} + \chi(*a)} \left\{ \exp\left[\frac{\mu(*b)}{\cos\beta} + \chi(*a) \right] z_1 - 1 \right\} d\beta. \qquad (11.1.32)$$

At this point, an analytical expression for $\phi(z_1)$ is required. A single exponential function, such as used by Castaing [1.1] suffices since most fluorescent photons are produced at a depth below that of the production of primaries: $\phi(z) = d_1 \exp(-d_1 z)$. We obtain:

$$I'\uparrow_{fa} = -\frac{\Omega}{8\pi} GI^*_{pb}\, d_1 \int_0^\infty \int_{\pi/2}^{\pi} \frac{\sin\beta}{\mu(*b) + \chi(*a)\cos\beta} \cdot \left\{ \exp\left[-\left(d_1 - \frac{\mu(*b)}{\cos\beta} \right) z_1 \right] - \exp[-(d_1 + \chi(*a))] \right\} d\beta dz_1$$

$$= -\frac{\Omega}{8\pi} GI^*_{pb} d_1 \int_{\pi/2}^{\pi} \frac{\sin\beta}{\mu(*b) + \chi(*a)\cos\beta} \left[\frac{-1}{d_1 - \frac{\mu(*b)}{\cos\beta}} - \frac{-1}{d_1 + \chi(*a)} \right] d\beta$$

$$= -\frac{\Omega}{8\pi} GI^*_{pb} \frac{d_1}{d_1 + \chi(*a)} \int_{\pi/2}^{\pi} \frac{\sin\beta}{\mu(*b) + \chi(*a)\cos\beta} \left[\frac{\mu(*b) + \chi(*a)\cos\beta}{\mu(*b) - d_1\cos\beta} \right] d\beta$$

$$= -\frac{\Omega}{8\pi} GI^*_{pb} \frac{d_1}{d_1 + \chi(*a)} \int_{\pi/2}^{\pi} \frac{\sin\beta d\beta}{\mu(*b) - d_1\cos\beta} \qquad (11.1.33)$$

$$= \frac{\Omega}{8\pi} GI^*_{pb} \frac{d_1}{d_1 + \chi(*a)} \frac{\ln\left(1 + \frac{d_1}{\mu(*b)}\right)}{d_1}$$

$$= \frac{\Omega}{8\pi} GI^*_{pb} \frac{d_1}{d_1 + \chi(*a)} \cdot \frac{1}{\mu(*b)} \cdot \frac{\ln(1 + \nu)}{\nu}$$

with

$$\nu = \frac{d_1}{\mu(*b)}$$

and since

$$f_p = \frac{d_1}{d_1 + \chi(*a)}$$

we obtain, finally

$$I'\uparrow^*_{fa} = \frac{\Omega}{8\pi} GI^*_{pb} f_p [V_{qb}, \chi(*a)] \frac{1}{\mu(*b)} \frac{\ln(1 + \nu)}{\nu}. \qquad (11.1.34)$$

The value of ν depends on the choice of function for $\phi(z)$; the corresponding values of d_1 are listed in Table 11.1. If the analyst prefers the use of a model having more than one exponential term [Eq. (10.3.41) with $i > 1$], we obtain, analogously,

$$I'\uparrow^*_{fa} = \frac{\Omega}{8\pi} GI^*_{pb} f_p [V_b, \chi(*a)] \frac{1}{\mu(*b)} \sum_i \frac{c_i}{d_i} \frac{\ln(1 + \nu_i)}{\nu_i} \qquad (11.1.35)$$

with $\nu_i = d_i/\mu(*b)$. Criss and Birks [10.22] have developed an alternative formulation which uses for the calculation of I'_{fa} the factor f_p rather than $\phi(z)$; this factor, however, must be algebraically defined. Because the most common expressions for f_p are related to algorithms for $\phi(z)$ of the form shown in Eq. (10.3.41), Criss' approach is not commonly used in practice. Criss has also explicitly derived the ν_i terms corresponding to Philibert's equation, and these were subsequently used in the program COR [9.14], which will be discussed later.

By adding Eqs. (11.1.29) and (11.1.35) we obtain the total fluorescent intensity produced by one line of element b which excites the line of interest of element a:

$$I'^*_{fa} = I'\downarrow^*_{fa} + I'\uparrow_{fa} = \frac{\Omega}{8\pi}\frac{I^*_{pb}G}{\mu(*b)} f[V_{qb}, \chi(*a)]$$

$$\cdot \left[\frac{\ln(1+u)}{u} + \sum_i \frac{c_i}{d_i} \frac{\ln(1+v_i)}{v_i} \right] \quad (11.1.36)$$

$$= \frac{\Omega}{8\pi} I_{pb} C_a \frac{\mu(ab)}{\mu(*b)} \frac{r_a - 1}{r'_a} \omega_{qa} p_a f[V_{qb}, \chi(*a)]$$

$$\cdot \left[\frac{\ln(1+u)}{u} + \sum_i \frac{c_i}{d_i} \frac{\ln(1+v_i)}{v_i} \right]$$

In most cases of fluorescent excitation more than one line is capable of ionizing the shell which emits the secondary radiation. In a rigorous procedure [9.14] the relative positions of lines and absorption edges must be scanned in order to determine all cases of excitation present in a particular specimen and the calculations summarized in Eq. (11.1.36) must be repeated for each exciting line. The equation is equally valid for all cases of secondary excitation, regardless of the series of exciting lines and levels of ionized shells. The relations between primary X-ray lines and shells of elements which may produce significant fluorescence emission are summarized in Table 11.2. This table enables the analyst to decide which interactions must be calculated, although an automated search program—such as the one incorporated in the programs COR and FRAME is preferable.

Table 11.1. Single-Exponential Models for $\phi(z) = d_1 \exp(-d_i z)$.

AUTHOR	REFERENCES	d_1
Castaing	[1.1]	$d_1 = \sigma$
Ziebold and Ogilvie	[8.1]	$d_1 = \sigma$
Belk	[10.20]	$d_1 = \frac{1+h}{1+2h}\sigma$
Borovsky	[9.33]	$d_1 = \frac{3 \times 10^6}{V_0^n - V_q^n}$ $(n = 1.5\text{-}1.7)$
Heinrich and Yakowitz	Eq. (10.3.34)	$d_1 = \frac{1}{a_1\gamma} = \frac{3.33 \times 10^6}{V_0^n - V_q^n}$ $(n = 1.65)$

Table 11.2. Excitation of Fluorescent X-ray Emission.

ATOMIC NUMBER	K-SHELL EXCITED BY				L_{III}-SHELL EXCITED BY			
	Kα OF	Kβ OF	Lα OF	$L\beta_1$ OF	Kα OF	Kβ OF	Lα OF	$L\beta_1$ OF
10	11	11	29	28				
11	12	12	31	31				
12	13	13	34	33				
13	14	14	36	36				
14	15	15	39	38				
15	16	16	41	41				
16	17	17	44	43				
17	18	18	46	45				
18	19	19	49	48				
19	20	20	52	50				
20	21	21	54	53				
21	22	22	57	55				
22	24	23	59	57				
23	25	24	62	59				
24	26	25	64	62				
25	27	26	67	64	9		26	25
26	28	27	69	66	10		27	26
27	29	28	72	68	10		28	27
28	30	29	74	70	11		29	28
29	31	30	77	72	11		30	29
30	32	31	79	74	11	11	31	30
31	33	32	82	77	12	12	32	31
32	34	33	84[a]	79	12	12	33	33
33	35	34	87	81	13	13	34	34
34	37	35	90	83	13	13	35	35
35	38	36	92	85	14	13	36	36
36	38	37		87	14	14	37	37
37	39	38		88	15	14	39	38
38	40	39		90	15	15	40	39
39	41	40		92	16	15	41	40
40					16	16	42	41
41					17	16	43	42
42					17	17	44	43
43					18	17	45	44
44					18	18	46[a]	46
45					19	18	48	47
46					19	18	49	48
47					20	19	50	49
48					20	19	51	50
49					21	20	52	51

[a] $L\alpha_1$ only.

Table 11.2. (Continued)

ATOMIC NUMBER	K-SHELL EXCITED BY Kα OF	Kβ OF	Lα OF	$L\beta_1$ OF	L_{III}-SHELL EXCITED BY Kα OF	Kβ OF	Lα OF	$L\beta_1$ OF
50					21	20	53	52
51					22	21	55	53
52					22	21	56	54
53					23	22	57	55
54					23	22	58	56
55					24	23	59	57
56					24	23	61	58
57					25	24	62	59
58					25	24	63	61
59					26	25	64	62
60					26	25	65	63
61					27	25	66	64
62					27	26	67	65
63					28	26	69	66
64					28	27	70	67
65					29	27	71	67
66					29	28	72	68
67					30	28	73	69
68					30	29	74[a]	70
69					31	29	75[a]	71
70					31	30	77	72
71					31	30	78	73
72					32	30	79	74
73					32	31	80	75
74					33	31	81[a]	76
75					33	32	82[a]	77
76					34	32	84	78
77					34	33	85	79
78					35	33	86	80
79					35[b]	34	87[a]	81
80					36	34	88[a]	82
81					37	35	90	83
82					37	35	91	84
83					38	36	92	84
84					38	36		85
85					39	37		86
86					39	37		87
87					40	38		88
88					40	38		89
89					41	39		90

[a] $L\alpha_1$ only.
[b] $K\alpha_1$ only.

11.1.1 SPATIAL DISTRIBUTION OF FLUORESCENT RADIATION

The distribution of generated fluorescent X-ray photons, in depth as well as in the lateral dimensions, depends on how far the primary radiation travels within the specimen before exciting secondary radiation. The path of the primary radiation is thus added to that of the originating electrons; in average, the point of secondary photon generation is more distant from the point of beam impact than that of the primary photon. Moreover, under the conditions which produce significant secondary emission, the electron absorption and deceleration is stronger than the attenuation of primary X-rays. Therefore, the secondary X-rays are produced within a much larger, and less well-defined volume, than primary radiation.

The wide spatial distribution of secondary emission is detrimental to the spatial resolution of electron probe microanalysis and can cause errors in quantitation. The fluorescence corrections which aim to compensate for the secondary emission may be inaccurate if the region around the volume which is primarily excited does not contain the element of interest, or contains it at a different concentration. It is especially difficult or impossible to estimate low levels of an element in small volumes surrounded by a material which is rich in this element, if secondary excitation is present. This is, for instance, the case in the determination of chromium in an inclusion embedded in a steel which itself is rich in chromium.

Birks et al. [11.2] used a Monte Carlo calculation to determine the fluorescent depth distribution function, ϕ_f. The authors proposed that a graph representing a general depth distribution function can be obtained. An analytic solution for ϕ_f was obtained by Brown [11.3], who also investigated this function by means of tracer experiments.

The attenuation factor for secondary radiation, f_f, is of interest since it appears formally in Eqs. (9.1.2) and (9.1.3). It is implicit in Eq. (11.1.36), and can be obtained explicitly from it, considering that when the corresponding mass attenuation factor, $\mu(*a)$ vanishes, the condition $(u \to 0)$ holds. Since $f[V_b, \chi(*a)]$ then becomes one, and

$$\frac{\ln(1+u)}{u} \to \ln e = 1,$$

(see Fig. 11.3) while the rest of Eq. (11.1.36) is unchanged, we obtain:

$$f_f^* = \frac{f[V_b, \chi(*a)] \left[\frac{\ln(1+u)}{u} + \frac{\ln(1+v)}{v}\right]}{1 + \frac{\ln(1+v)}{v}}. \tag{11.1.37}$$

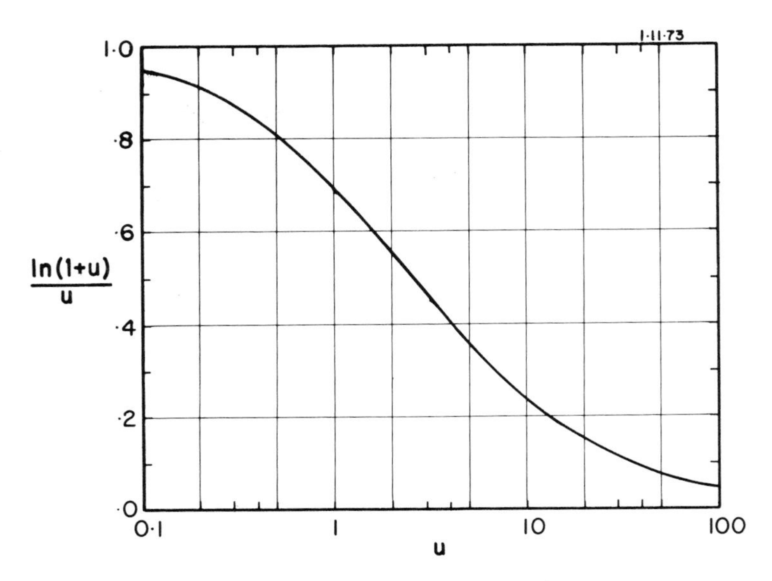

Fig. 11.3. The function ln $(1 + u)/u$.

The term ln $(1 + v)/v$ is typically smaller than ln $(1 + u)/u$, and, in cases of strong fluorescence, $f[V_b, \chi(*a)]$ is close to one; hence, the following approximation is usually adequate:

$$f_f^* \simeq \frac{\ln(1+u)}{u}, \qquad u = \chi(*a)/\mu(*b). \tag{11.1.38}$$

We will demonstrate this approximation with an example. Consider the excitation of Fe Kα by Ni Kα, at 20 kV, with an X-ray emergence angle of 20°. The full expression, Eq. (11.1.37), gives a value of $f_f = 0.391$, while the approximation gives 0.380. For the same system, at 30 kV, and with an emergence angle of 60°, the full expression yields 0.585, while the approximation gives a value of 0.576.

11.1.2 SIGNIFICANCE OF THE TERMS OF THE FLUORESCENCE EQUATION

As already mentioned, fluorescence due to characteristic lines is significant only under certain conditions and can be neglected frequently even when a line of one element in the specimen has an energy above that of the characteristic ionization potential of the line of another element present. In order to avoid unnecessary complexity in the data reduction scheme, we must know which situations require the fluorescence calculation. To establish guidelines to that effect, we shall consider in more detail the individual terms of the fluorescence equation [Eq. (11.1.36)].

Equation (9.1.3) indicates that the emitted fluorescent radiation must be added to the emitted primary radiation to obtain the total characteristic radiation. However, the intensity of primary radiation, even for a pure element, can vary over a wide range (Fig. 11.4). Therefore, the relative importance of the emitted fluorescent intensity, I'_{fa}, of element a, is best appreciated if Eq. (11.1.36) is modified so as to give the ratio of emitted fluorescent and primary radiations:

$$\frac{I'^*_{fa}}{I'^*_{pa}} = \frac{1}{2}\frac{I^*_{pb}}{I^*_{pa}} C_a \frac{\mu(ab)}{\mu(*b)} \cdot \frac{r_q - 1}{r'_q} \omega_a p_a \cdot \frac{f[V_b \chi(*a)]}{f[V_a \chi(*a)]}$$

$$\cdot \left[\frac{\ln(1+u)}{u} + \Sigma \frac{c_i}{d_i} \frac{\ln(1+v_i)}{v_i}\right]. \qquad (11.1.39)$$

We may introduce here several simplifications which do not invalidate the evaluation of the important terms. We will choose a simple exponential model for $\phi(z)$, so that only one term appears in the summation within brackets. We also assume that the ratio $f[V_b\, \chi(*a)]/f[V_a\, \chi(*a)]$ is virtually equal to one and can thus be omitted. Furthermore, we will write the above expression for a very dilute solid solution of element a in element b, so that $\mu(*b)$ becomes $\mu(bb)$. Finally, we assume that Castaing's first approximation [Eq. (9.1.4)] holds, so that

$$\frac{I^*_{pb}}{I^*_{pa}} = \frac{C_b}{C_a} \cdot \frac{I_p(b)}{I_p(a)} \cdot \qquad (11.1.40)$$

With these assumptions, we obtain: (11.1.41)

$$\frac{I'^*_{fa}}{I'^*_{pa}} = \underbrace{\frac{C_b}{2} \cdot \frac{I_p(b)}{I_p(a)}}_{F(p)} \cdot \underbrace{\frac{\mu(ab)}{\mu(bb)} \cdot \frac{r_{qa} - 1}{r'_{qa}}}_{F(\mu)} \cdot \underbrace{\omega_a p_a}_{F(a)} \cdot \left[\underbrace{\frac{\ln(1+u)}{u}}_{F(\psi)} + \underbrace{\frac{\ln(1+v)}{v}}_{F(V)}\right].$$

We will now consider separately the groups of terms indicated by the brackets under the equation.

Consider first the wavelength-dependent factor, $F(\mu)$. The ratio $\mu(ab)/\mu(bb)$ relates the absorption of the primary X-ray line of element b in the target b to that of the same radiation which has been absorbed by the element a present in

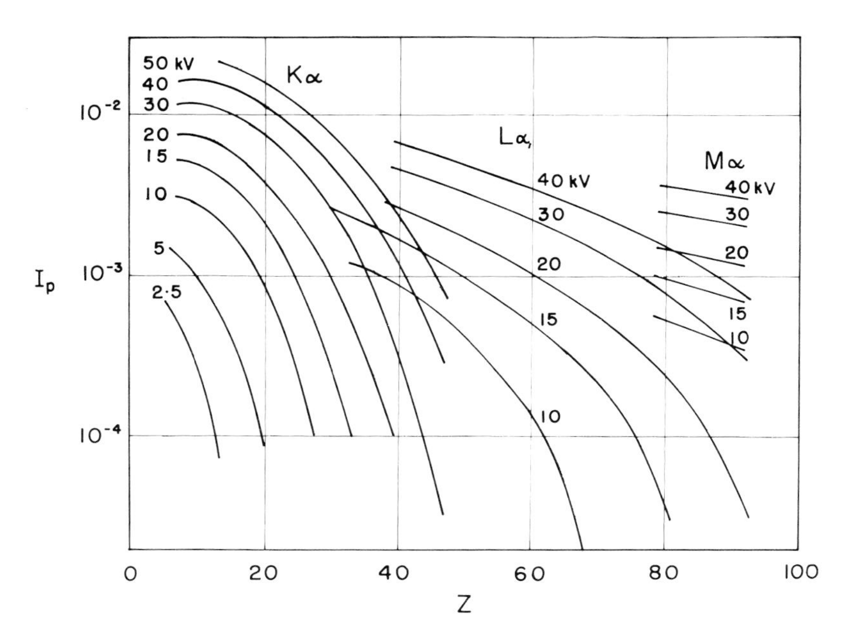

Fig. 11.4. Generated primary X-ray intensities for pure elements as a function of Z and E_0, according to Eq. (9.2.26).

the target [Eqs. (11.1.3) and (11.1.4)]. The concentration C_a which would also belong here, has canceled in Eq. (11.1.41) in which the secondary emission is given as a fraction of primary emission when Castaing's first approximation, Eq. (11.1.40) was introduced.

The term $(r_{qa} - 1)/r'_{qa}$ indicates the fraction of the absorbed primary radiation that is used in ionization of the shell which emits the fluorescent radiation of interest. Equations (11.1.5) through (11.1.13) deal with the value of this term. Whenever an X-ray line has an energy level below that required to ionize the shell or subshell of interest, ionization does not take place and hence this term is equal to zero. Therefore, to determine in the first place which lines excite shell levels of other lines, the energies—or wavelengths—of all emitted primary lines from a target must be compared with the energy levels of all shells of all elements present in the target. This comparison is not a trivial task, particularly when lines and shells above the K-level are involved. In practice, the sorting of lines and shells is best performed automatically, by a computer.

The simplest case of fluorescence excitation is that by a homologous line (e.g., $b\mathrm{K}\alpha \to a\mathrm{K}\alpha$, or $b\mathrm{L}\alpha_1 \to a\mathrm{L}\alpha_1$). In such a case, according to Table 11.2, the atomic number of the exciting primary element must be higher, at least by one or two, than that of the excited element. Consider now the changes in fluorescent intensity that occur if the excited element, a, is maintained unchanged, while the atomic number of the exciter, b, is raised. Figure 11.5 shows the changes produced both in the ratio $\mu(ab)/\mu(bb)$ and in the function $F(\mu)$, for K-lines as well as for L-lines. In both cases the function $F(\mu)$ diminishes rapidly with the increasing atomic number of the exciting element b. The same effect is observed in the function $F(\mu)$ for the cases of fluorescence of L-lines by K-lines and vice versa. Similar, though less dramatic, effects occur in $F(p)$ and $F(\psi)$. Therefore, *the fluorescent emission is strongest when the energy of the exciting line is just above that needed to ionize the level of interest.* It should be noted that the apparent complications due to the multiple absorption jumps for element b in Fig. 11.5 cancel in $F(\mu)$ through the multiplication with $(r_{qa} - 1)/r'_{qa}$, due to the use of the effective jump ratios [Eqs. (11.1.8) through (11.1.13)].

Consider next the factor $F(p)$, i.e., the ratio of primary emissions from the pure elements a and b. As Fig. 11.4 indicates, at any voltage, the intensity of Kα

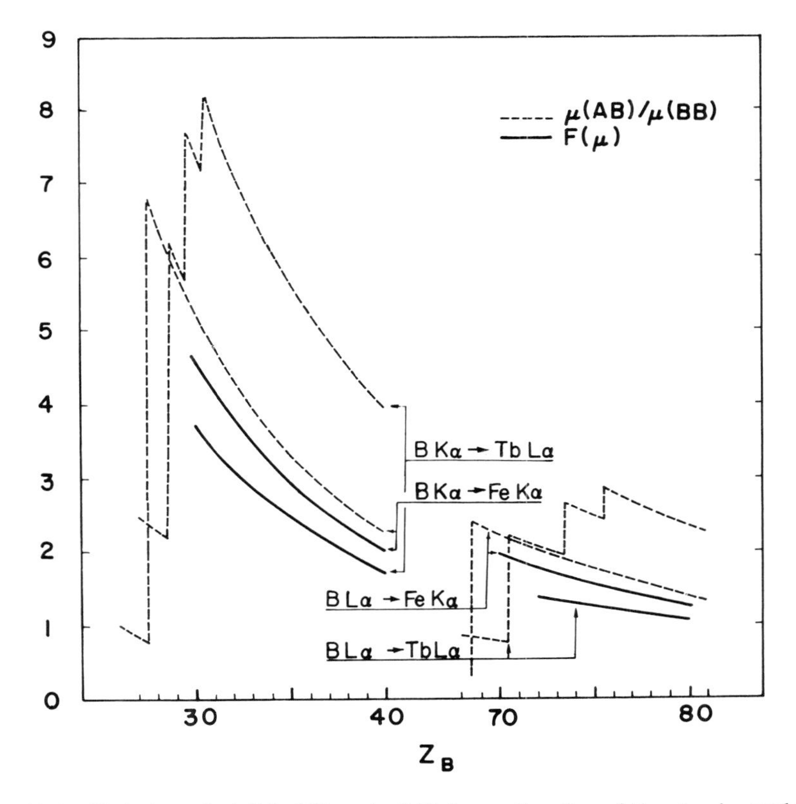

Fig. 11.5. Variation of $\mu(ab)/\mu(bb)$, and of $F(\mu)$, as a function of the atomic number of the exciting element, b.

emission diminishes with increasing atomic number. This may be surprising in view of the sharp increase of the fluorescence yield with the atomic number (Fig. 4.7) but it is related to the respective values of the mean excitation potential (Fig. 9.4) and the effects of overvoltage. Therefore, for any given fluorescent emitter element, the factor $F(p)$ decreases with increasing atomic number of the exciting element, and in consequence the trend observed in $F(\mu)$ is reinforced.

The factor $F(a)$ is the probability that ionization of the corresponding level in element a will produce emission of the qm line of element a. It is the product of the fluorescent yield of a and of the weight of the fluorescent line. The main effect on $F(a)$ of increasing the atomic number of element a is a rapid increase of the fluorescent yield, and hence of the product $\omega_a p_a$ (Fig. 11.6). Therefore, other conditions being the same, *the fluorescent excitation increases with the atomic number of the element which emits it.*

The term $F(\psi)$ is the only one depending on the X-ray emergence angle, ψ, and on the X-ray absorption coefficient for the fluorescent radiation. As we have seen [Eq. (11.1.38)], it is virtually equal to the secondary absorption factor, f_f. This term was obtained in the computation of fluorescent radiation generated below the primary excitation; if the fluorescent radiation were not partially absorbed, $F(\psi)$ would be equal to one. The function $F(\psi)$, unlike f_p, is independent of the energy of the primary electrons. (Rather, this energy affects the value of $f[V_b, \chi(*a)]$.) It increases, however, with increasing X-ray emergence angle (Fig. 11.7). Notwithstanding, it would be impractical to reduce the fluores-

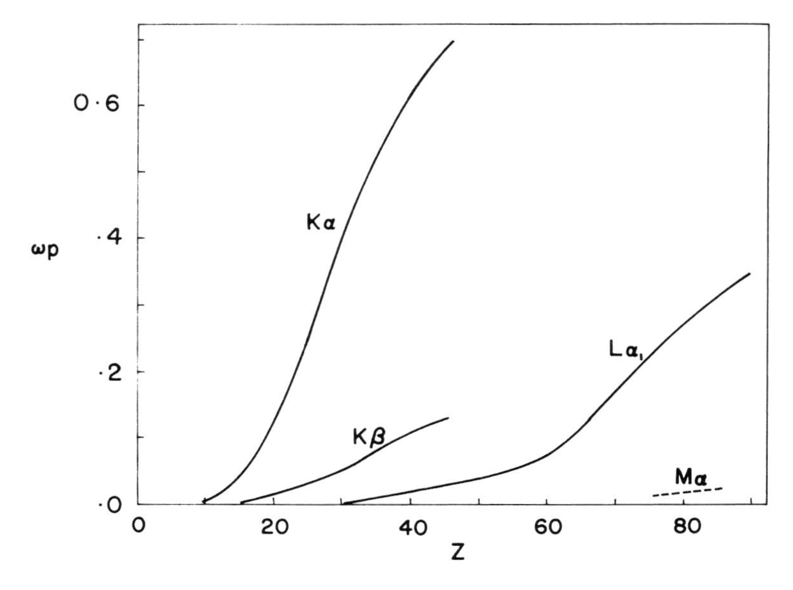

Fig. 11.6. Variation of $F(a) = \omega_a p_a$ with the atomic number of element a.

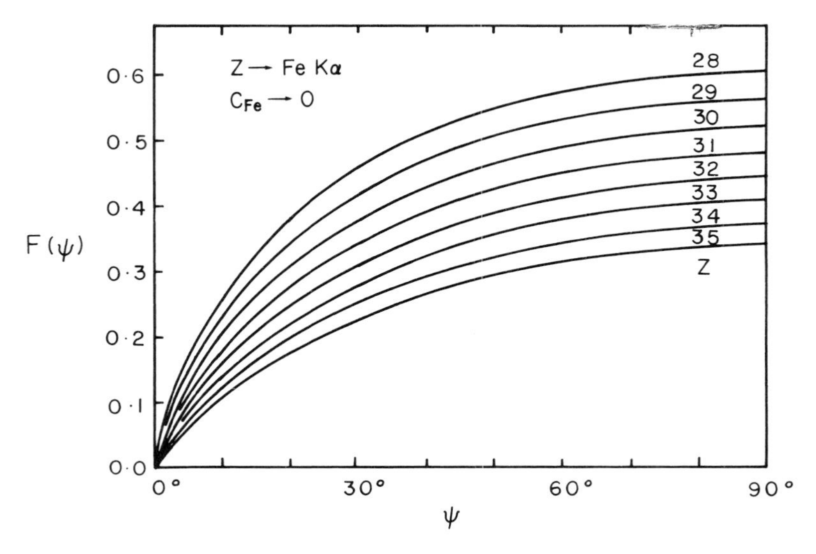

Fig. 11.7. For a given excited element and line the factor $F(\psi)$ increases with the X-ray emergence angle, ψ, and with decreasing atomic number of the exciting element. In this figure, only the interaction Z Kα → Fe Kα is considered.

cence emission, and hence improve the spatial resolution of the technique, by reducing the X-ray emergence angle, since the fluorescence emission cannot be significantly diminished at X-ray emergence angles above 10°. Furthermore, the term $F(\psi)$ must be added in Eq. (11.1.41) to the term $F(V)$, which is not affected by changes in the emergence angle.

The term $F(V)$ is related to emission of fluorescent radiation generated above the corresponding primary emission [Eq. (11.1.34)]. As the parameter $\nu = d_1/\mu(*b)$ indicates, $F(V)$ depends on the depth distribution function $\phi(z)$, and hence on the operating voltage. At low operating voltages, the primary radiation is produced virtually at the specimen surface, and all secondary emission occurs at deeper levels below the primary emission; hence, $F(V)$ is negligibly small. At very high operating voltages, however, most of the upward primary radiation is absorbed by the specimen. Therefore, $F(V)$ increases (Fig. 11.8). In the complete fluorescence Eq. (11.1.36) the absorption of secondary radiation of element a would be taken into account by $f[V_b, \chi(*a)]$, which, at high voltages, is smaller than one. Because $F(V)$ is typically smaller than $F(\psi)$, the *magnitude of the relative fluorescent emission,* I'^*_{fa}/I'^*_{pa}, *does not vary considerably with the operating voltage.* Furthermore, the sum $F(\psi) + F(V)$ decreases, for any given emitting element, with increasing atomic number of the fluorescer. This further enhances the effect discussed for $F(\mu)$ (Fig. 11.9).

Finally we will consider the effects of composition changes within a binary

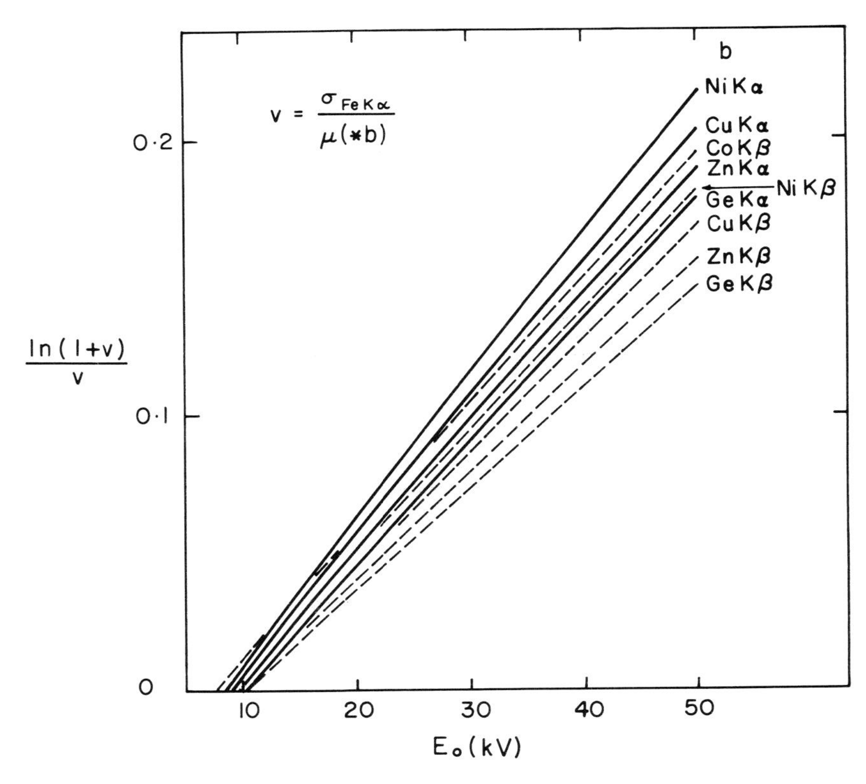

Fig. 11.8. The term ln $(1 + \nu)/\nu$ increases almost linearly with the overvoltage. Secondary emission of Fe Kα is caused by the lines of element b listed in the upper right-hand corner.

system on the fluorescent intensity. To apply Eq. (11.1.41) to the entire binary composition range, we merely replace $\mu(bb)$ by $\mu(*b)$. We see that concentration changes only affect the factors C_b, $\mu(ab)/\mu(*b)$, and the last term within brackets, of the equation representing I'_f/I'_p. The change of the term within brackets, $[\ln(1 + u)/u + \ln(1 + \nu)/\nu]$, is slight (Fig. 11.10). As the concentration of element b increases, both the ratio of mass absorption coefficients and of the concentration C_b contribute to a rapid increase of I'_f/I'_p. In fact, the increase is too sharp for low concentrations of a to be adequately represented by a hyperbolic approximation (Fig. 11.11). For this reason, the empirical approach must be used with caution when strong secondary emission is produced.

11.1.3 Approximate Solutions to Fluorescence Excitation

If computers of sufficient size are available, it is best to follow for the fluorescence calculation the rigorous scheme summarized in Eq. (11.1.36), including the full calculation of the exciting primary intensity. Such facilities were not

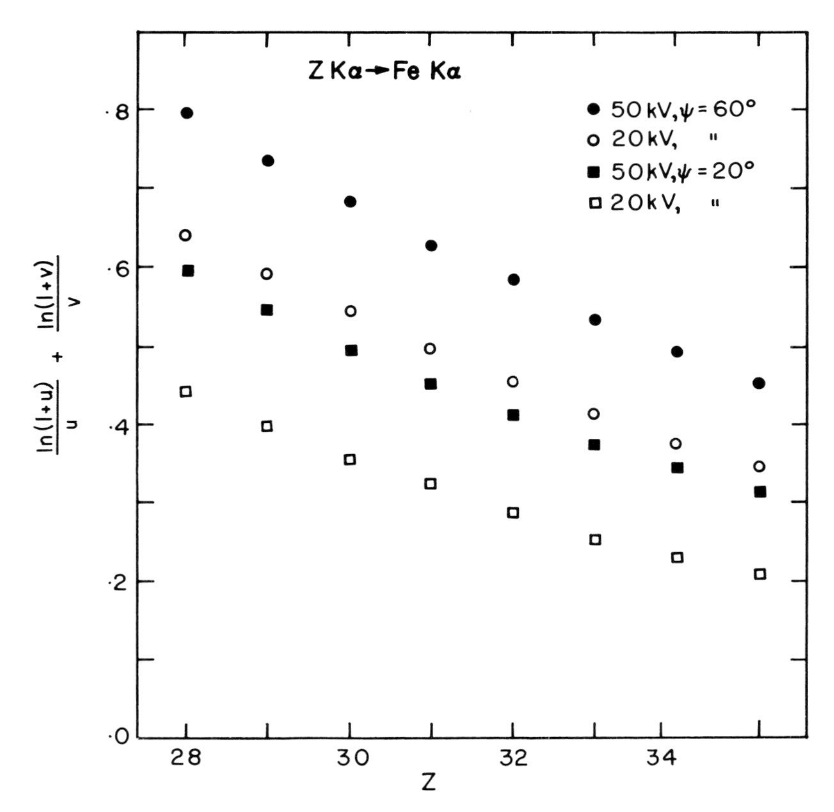

Fig. 11.9. Variation of $F(\psi) + F(V)$ with the atomic number of the element which emits the exciting radiation, Z. The sum increases with operating voltage and with increasing X-ray emergence angle. It decreases with increasing atomic number of the exciting element.

common in microprobe laboratories until recent times. Even now, the on-line use of small computers for signal evaluation may force the analyst to use simpler procedures.

The traditional *ZAF* scheme [Eq. (9.1.10)] requires the ratio of secondary to primary emission, given by Eq. (11.1.39). Castaing [1.1] proposed an approximation for the K → K excitation in which the primary generated intensities are considered to depend upon the respective excitation voltages according to Eq. (9.2.33):

$$\frac{I(b)}{I(a)} = \frac{E_{qa}}{E_{qb}} \frac{A_a}{A_b} \frac{\omega_b}{\omega_a} .$$

Castaing also observed that in most cases, both the Kα and the Kβ-line of the exciting element b ionize the K-shell of the emitting element a. He suggested

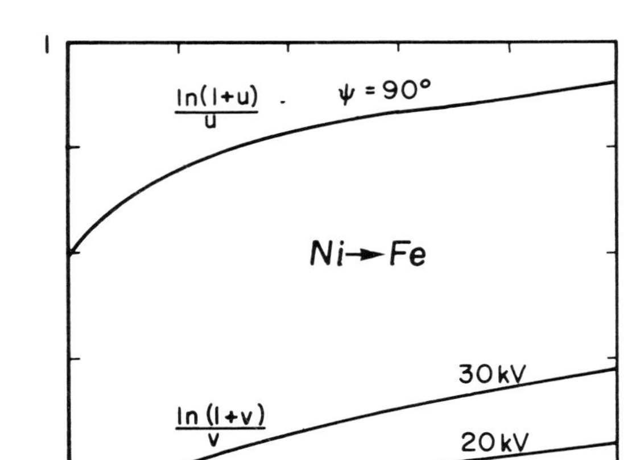

Fig. 11.10. The values of ln $(1 + u)/u$ and ln $(1 + v)/v$ vary moderately with the composition of the binary alloy. ln $(1 + v)/v$ increases with increasing operating voltage.

that the added effect of excitation by the b Kβ-line could be approximated by setting equal to one the weight of the Kα-line, $p_{K\alpha}$, in Eq. (11.1.17). In this way, the separate calculation of the excitation by Kβ-lines can be omitted, and we obtain for the combined effect of both b K-lines on the a Kα-line:

$$\frac{I'_{fa}}{I'_{pa}} = \frac{C_b}{2} \cdot \frac{A_a}{A_b} \cdot \frac{E_{ka}}{E_{kb}} \cdot \omega_b \frac{\mu(ab)}{\mu(*b)} \frac{r_{Ka} - 1}{r_{Ka}} \cdot \left[\frac{\ln(1 + u)}{u} + \frac{\ln(1 + v)}{v}\right] \cdot \tag{11.1.42}$$

For the primary depth distribution Castaing chose a simple exponential approximation, so that $v = \sigma/\mu(*b)$; the parameter u has the usual significance: $u = \chi(*a)/\mu(*b)$.

Reed and Long [11.4] pointed out that a better approximation for $I(b)/I(a)$ can be derived from the equation proposed by Duncumb and Cosslett [Eq. (9.2.34)]:

$$\frac{I(b)}{I(a)} = \frac{\omega_b}{\omega_a} \cdot \frac{A_a}{A_b} \left(\frac{U_b - 1}{U_a - 1}\right)^{1.67} \cdot \frac{p(b)}{p(a)}.$$

The effect of this substitution is to reduce the magnitude of the fluorescence correction (Fig. 11.11).

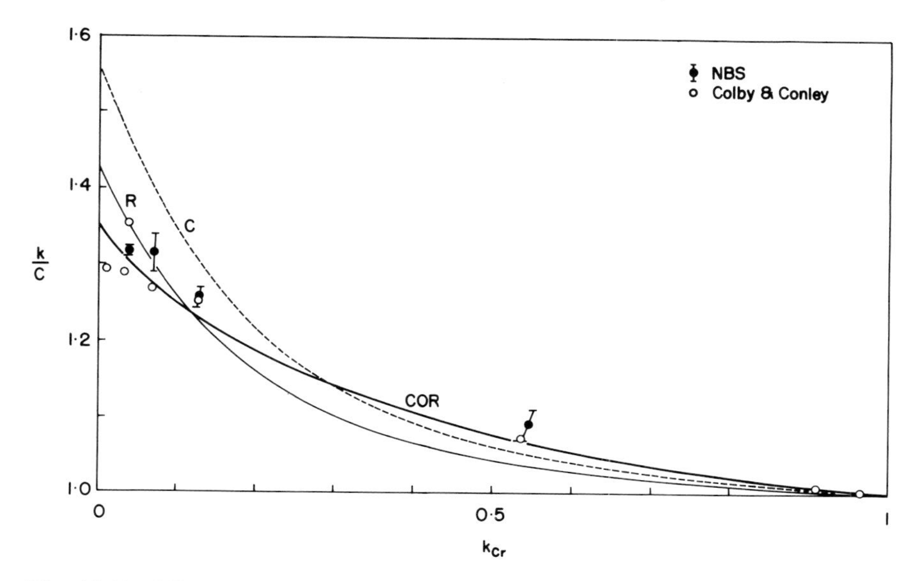

Fig. 11.11. Effects of fluorescence in the binary system chromium-iron. C: Castaing [Eq. (11.1.42)]; R: Reed [Eq. (11.143)]; COR [Eq. (11.1.39)]. X-ray emergence angle: 52.5°. Circles mark the experimental results. Ranges marked are attributed mainly to inhomogeneity of the standard specimens. Operating potential: 20 kV.

Reed also considered fluorescent excitation other than K-K. For the L-L excitation, the omission of the weight of lines compensates less well for the multiple excitation of the Lα-line, but the results are still acceptable, since the L-fluorescence is not as strong as the K-fluorescence. The complexities arising from effective yields and jump ratios are also ignored. The main uncertainty in the calculation is that arising from the ratio of primary intensities of different shells of the two elements. To take into account this ratio, Reed introduced, for the L-K fluorescence, the factor 4.2, and 0.24 for K-L fluorescence. We thus obtain:

$$\frac{I'_{fa}}{I'_{pa}} = \frac{C_b}{2} P_{ij} \frac{A_a}{A_b} \left(\frac{U_b - 1}{U_a - 1} \right)^{1.67} \omega_b \frac{r_{qa} - 1}{r_{qa}} \frac{\mu(ab)}{\mu(*b)} \left[\frac{\ln(1+u)}{u} + \frac{\ln(1+v)}{v} \right] \tag{11.1.43}$$

with $P_{KL} = 0.24$ and $P_{LK} = 4.2$.

The K-K fluorescence is most significant for elements of atomic number larger than 20. For these elements, the absorption jumps do not change drastically with atomic weight. Therefore, Reed and Long could introduce the approximate values:

$$\frac{r_K - 1}{r_K} = 0.88 \qquad \frac{r_L - 1}{r_L} = 0.75\,.$$

Several applying parameters were combined and given in tabular form. Numerical values suggested for the "Lenard coefficient" σ were also given.

The method of Reed is incorporated in several data reduction schemes, such as the programs FRAME [4.42] and MAGIC [11.5].

The complete rigorous correction technique [Eq. (11.1.36)], separately calculated for each exciting line, was used in COR [9.14]. Other approaches to simplified corrections developed by Birks [11.6] and by Wittry [11.7] were evaluated by Duncumb and Shields [11.8], and are of historical interest only.

11.2 FLUORESCENCE EXCITED BY THE CONTINUUM

The mechanisms of fluorescent excitation by the continuum and by characteristic lines are identical. Therefore, Fig. 11.1 can be used to describe continuum fluorescence, and Eq. (11.1.36) also applies to continuum fluorescence. However, the differences in the generation of the respective primary radiations must be taken into account since they require significant modifications of the calculation procedure, and limit the accuracy of the continuum fluorescence correction.

The most important differences between the continuum fluorescence and the fluorescence by characteristic lines are:

a. The continuum emission is polychromatic; hence, Eq. (11.1.36) must be integrated over the wavelength range of the continuum which can excite a given shell.
b. All targets, including pure elements, emit continuous radiation.
c. The laws of continuum production, such as the distribution in depth of the continuum and the cross sections for its generation, are not very accurately known.

   With respect to the relative intensity of continuum fluorescence, we cannot expect that any compensation may occur between errors in the calculation of the exciting continuous radiation, and the primary characteristic radiation of the element which is also secondarily excited. The situation is similar, but worse, than for characteristic fluorescence involving different shells.
d. The generation of continuous radiation is anisotropic and the generated X-rays are polarized.

The anisotropy of continuum generation is strong in thin targets, but attenuated in electron-opaque targets due to the changes in direction of electrons after scattering acts. The anisotropy is generally neglected in the treatment of fluorescence by the continuum. This is plausible since fluorescence by the continuum is a second-order effect, difficult to observe separately from the much stronger primary emission. The assumption may not hold, however, when continuum emission from thin targets is considered.

From Kramers' law of continuum generation [Eq. (6.3.7)], we obtain, by integrating over all variables which depend on the primary wavelength or energy, the following transformation of Eq. (11.1.3):

$$I'^{*}_{ca} = \frac{\Omega}{8\pi} C_a \omega_a p_a \cdot k \cdot Z^*$$

$$\times \int_{E_{qa}}^{E_0} f[E_c \chi(^*a)] \frac{\mu(aE)}{\mu(^*E)} \frac{r_a - 1}{r'_a} \left(\frac{E_0}{E} - 1\right) \left[\frac{\ln(1+u)}{u} + \frac{\ln(1+v)}{v}\right] dE \tag{11.2.1}$$

where

$$u = \frac{\chi(^*a)}{\mu(^*E)} \quad \text{and} \quad v = \frac{d_1}{\mu(^*E)}.$$

The term $(r_a - 1)/r'_a$ changes at the absorption edges and must therefore be included in the integral. The above integration can also be performed in terms of wavelengths, as preferred by Hénoc et al. [9.14].

It is typical for continuum fluorescence that it can be excited efficiently by radiation of much higher energy (shorter wavelength) than the secondary emission, as can be verified by numerical integration in small steps of Eq. (11.2.1). In this respect, continuum fluorescence differs markedly from fluorescence by characteristic lines. Although we do not know with high accuracy the distribution in depth of continuous radiation, we can assume that it is similar to that of characteristic radiation. The distribution in depth of hard radiation–which is efficient for continuum excitation–is relatively shallow. Models are therefore used in which the continuum is produced at the specimen surface (Hénoc [11.9]).

If one assumes that the depth distribution of the continuum is equal to that of characteristic rays, the results differ but little (Table 11.3). If $f(V) = \ln(1+v)/v$ and $f[E_c, \chi(a)]$ are neglected, we can calculate the mean energy of the continuum which *generates* fluorescence by the continuum:

$$\overline{E} = \frac{\int_{E_q}^{E_0} (E_0 - E)\, dE}{\int_{E_q}^{E_0} (E_0/E - 1)\, dE} = \frac{\frac{1}{2}(U_0 - 1)^2 \cdot E_q}{U_0 (\ln U_0 - 1) + 1} \tag{11.2.2}$$

and, with Eq. (9.2.33),

$$\overline{E} \simeq 1.37\,(U_0 - 1)^{0.33} \cdot E_q. \qquad (11.2.3)$$

The equation yields, for excitation of Fe Kα at 20 kV, the value $\overline{E} \simeq 1.5E_q$. The mean energy of the continuum which produces *emitted* fluorescence by the continuum is lower, since the term $\ln(1 + u)/u$ decreases with increasing energy. Even so, this energy is considerably higher than that of a line which is an efficient exciter for fluorescence. This trend is related to the changes with energy of the respective primary exciting radiations. Because fluorescent emission comes, in average, from greater depth than primary radiation, the continuum fluorescence absorption factor f_c is smaller than the corresponding primary absorption factor, f_p. Therefore, the relative emergent fluorescent intensity I'_c/I'_p is smaller than the relative generated fluorescent intensity I_c/I_p and the difference increases with the operating voltage (Fig. 11.12). The ratio I'_c/I'_p is fairly insensitive to the X-ray emergence angle (Fig. 11.13). It is also noted that I_c/I_p increases slightly with the operating voltage.

The excitation of continuum fluorescence can be conveniently discussed for the simple case of self-excitation of an element. In this case, $\mu(*E)$ is equal to $\mu(aE)$, C_a is equal to one, Z^* is equal to Z_a, $\chi(*a)$ is equal to $\chi(aa)$, and the

Table 11.3. Effect of Assumed Depth Distribution and Kramers' Constant on I'_c/I'_p (%).

Line	System	Composition	Kramers' Variable	Kramers' Variable	Kramers' Constant
			↓	↓ ↑	↓ ↑
AgLα		Ag	0.497	0.623	0.990
2.984 keV	AgAu	20% Ag	0.752	0.765	1.085
AuLα		Au	4.69	5.18	7.23
9.713 keV	AuCu	20% Au	3.44	3.75	4.58
	AuAg	20% Au	5.31	5.68	8.64
CuKα		Cu	3.52	3.96	4.45
8.048 keV	CuAu	20% Cu	5.20	5.84	8.01

Note: The ratio of emitted continuum fluorescence and primary intensities (in percent) is shown for several binary alloys. Numbers beneath the symbol of the line are photon energies (keV). The operating voltage was 20 keV; the X-ray emergence angle, 52.5°. In the first column of ratios, secondary X-ray emission above the point of generation of the primary radiation is neglected [Fig. 11.16(C)]. In the second and third columns of ratios, both upwards and downwards fluorescence are included [Fig. 11.16(A)]. The choices for constant and variable values for Kramers' "constant" are those indicated in Fig. 6.1. It can be appreciated that the choice of Kramers' constant affects the results more than the respective assumptions on the secondary emission.

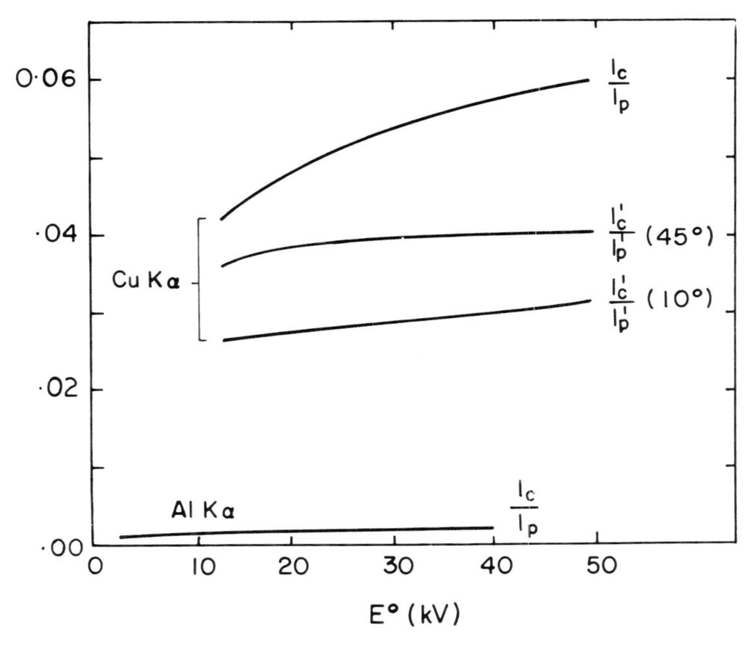

Fig. 11.12. Ratios of emitted and observed intensities from fluorescence due to the continuum to primary radiation. Emissions are from pure elements.

ratio $\mu(aE)/\mu(^*E)$ is one, for all energies. Therefore, with the assumption of generation at the surface of the continuum, we obtain:

$$I'_{ca} = \frac{\Omega}{8\pi} k\omega_a p_a Z_a \int_{E_{qa}}^{E_0} \frac{r_a - 1}{r'_a}\left(\frac{E_0}{E} - 1\right) \frac{\ln(1+u)}{u}\, dE. \tag{11.2.4}$$

From a practical point of view it is interesting to know the ratio of fluorescent to primary intensities. But a simple expression for the integral in Eq. (11.2.4) cannot be obtained. It is simpler to compare, instead of emitted intensities, the generated intensities, and to keep in mind that, due to the larger depth of generation of secondary radiation, the ratio of emitted intensities is somewhat smaller, and that the difference diminishes slightly with increasing operating voltage. With the use of Eq. (9.2.32), we obtain the following approximation:

$$\frac{I_{ca}}{I_{pa}} = \text{constant} \frac{Z_a \dfrac{r_a - 1}{r'_a} \displaystyle\int_{E_q}^{E_0} \left(\frac{E_0}{E} - 1\right) dE}{(1/A_a)(U_0 \ln U_0 - U_0 + 1)} \simeq \text{constant } Z_a^2 \frac{r_a - 1}{r'_a} E_q. \tag{11.2.5}$$

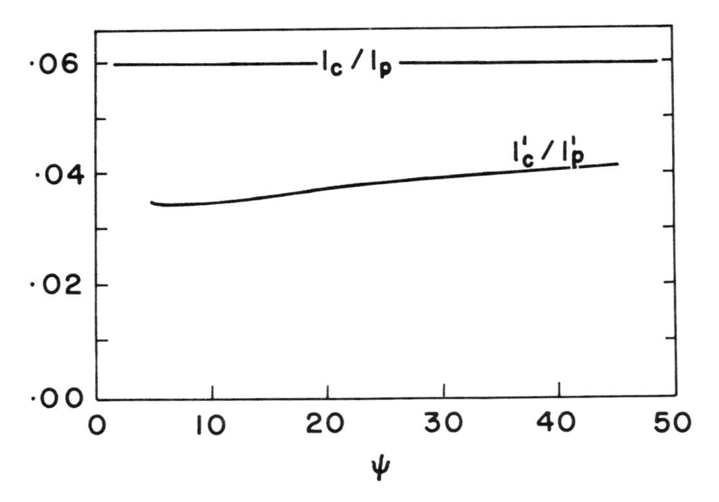

Fig. 11.13. Variation of the relative intensity of emergent continuum fluorescence as a function of the X-ray emergence angle. Cu Kα radiation in copper. Operating voltage: 49.4 kV.

In reasonable agreement with this approximation, the relative fluorescent intensities, within each series of lines, increase rapidly with atomic number, and vary little with operating voltages (Fig. 11.14).

In fact, we find that, regardless of the respective X-ray series, the ratios I_c/I_p for all elementary targets can be approximated by a monotonically increasing function of the critical excitation energy, E_q (Fig. 11.15).

We can now consider a binary target. Its continuum fluorescence emission, relative to that of the pure element, is

$$\frac{I'^{*}_{ca}}{I'^{0}_{ca}} = \frac{C_a \cdot Z^* \displaystyle\int f\,[E_c\chi(^*a)]\,\frac{\mu(aE)}{\mu(^*E)}\left(\frac{E_0}{E} - 1\right)\frac{\ln(1+u)}{u}\,dE}{Z_a \displaystyle\int f\,[E_c\chi(aa)]\left(\frac{E_0}{E} - 1\right)\frac{\ln(1+u_0)}{u_0}\,dE} \tag{11.2.6}$$

with $u = \chi(^*a)/\mu(^*E)$ and $u_0 = \chi(aa)/\mu(aE)$. The most significant factors in Eq. (11.2.6) are $C_a, Z^*/Z_a$, and, within the upper integral, $\mu(aE)/\mu(^*E)$. If other parameters remain unchanged, the fluorescent intensity, I_c is proportional to the concentration C_a. Since, in first approximation, the primary intensity, I_p, is also proportional to C_a, the variations of the ratio I_c/I_p with composition depend mainly on the ratio of atomic numbers, Z^*/Z_a, and of absorption coefficients, $\mu(aE)/\mu(^*E)$. Contrary to what happens in characteristic fluorescence, the relative intensity of the continuum fluorescence may either diminish or increase with the concentration of the excited element. A high level of continuum

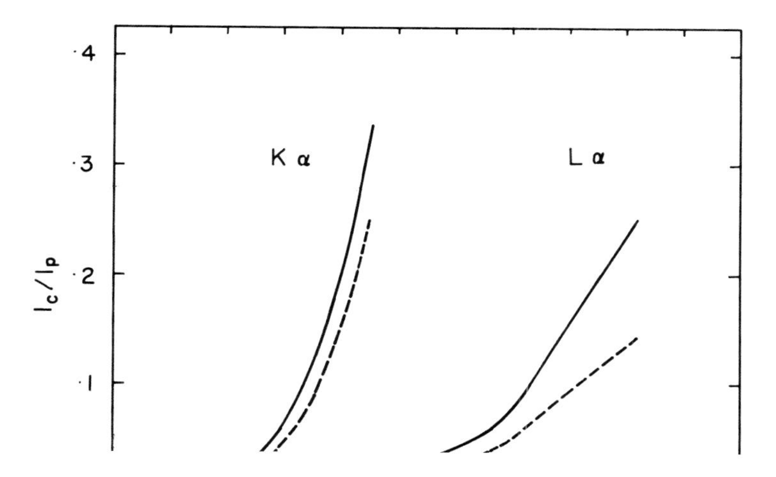

Fig. 11.14. Relative generated (full curves) and emitted (broken curves) intensities of fluorescence due to the continuum for elements of atomic number Z, excited at an operating potential of 40 kV, and observed at an X-ray emergence angle of 30°.

fluorescence with diminishing concentration of C_a occurs when the excited radiation is of short wavelength, and the matrix is of very low atomic number (see Table 11.4) due to the variation of the respective absorption coefficients.

For analogous regions between absorption edges (e.g., at energies above the K-edge), the following relation holds, according to Eq. (4.3.10):

$$\frac{\mu(aE)}{\mu(^*E)} \simeq \left(\frac{Z_a}{Z^*}\right)^3 \quad \text{so that} \quad \frac{I'^*_{ca}}{I'^0_{ca}} \propto \frac{\mu(aE)}{\mu(^*E)} \cdot \frac{Z^*}{Z_a} \simeq \left(\frac{Z_a}{Z^*}\right)^2 . \tag{11.2.7}$$

In general, therefore, *fluorescence due to the continuum increases rapidly with decreasing mean atomic number of the target.* The trend reverses when absorption edges are present at energies above the critical excitation potential of the line of interest. At the absorption edges, the ratio of absorption coefficients changes suddenly. Therefore, in the presence of absorption edges, *the integration must be performed separately for each region between consecutive absorption edges of all elements in the specimen* from the Duane-Hunt limit of continuum emission to the critical excitation potential of the element and line of interest. For accurate calculations the deviation from Kramers' law, which Green expressed as variations of the factor ½k (Section 6.3, Fig. 6.1), cannot be neglected. Hence, Eq. (11.2.6) must be written as follows:

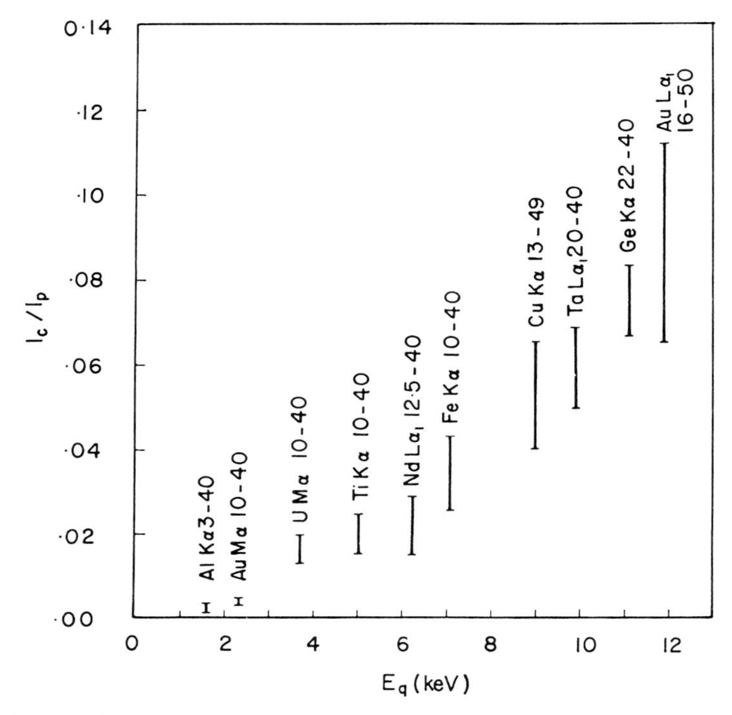

Fig. 11.15. Relative generated intensity of continuum fluorescence for various elements, lines, and electron energies, indicated in kilo-electron volts (keV) after the line notation. Bars indicate the range, from the lowest (bottom) to the highest (top) electron energy. Horizontal scale gives the critical excitation potential of the lines.

$$\frac{I'^{*}_{ca}}{I'^{0}_{ca}} = \frac{C_a Z^* k^* \int f[E_c \chi(^*a)] \dfrac{r_a - 1}{r'_a} \dfrac{\mu(aE)}{\mu(^*E)} \left(\dfrac{E_0}{E} - 1\right) \dfrac{\ln(1+u)}{u} \, dE}{Z_a k_a \int f[E_c \chi(aa)] \dfrac{r_a - 1}{r'_a} \left(\dfrac{E_0}{E} - 1\right) \dfrac{\ln(1+u_0)}{u_0} \, dE}. \tag{11.2.8}$$

Hénoc [11.9] developed a method for the formal integration of the expression (11.2.1). A similar procedure was later proposed by Springer [11.10]-[11.12]. Hénoc assumed that the continuous radiation is generated at the specimen surface, while Springer postulated that the depth distribution of the continuum is equal to that of the characteristic radiation. But, since Springer also neglected the generation of secondary radiation above the corresponding site of primary radiation, his model is not more realistic than Hénoc's (see Fig. 11.16).

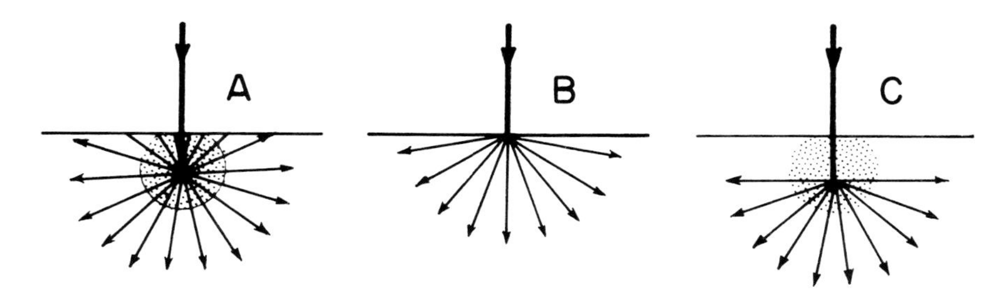

Fig. 11.16. Models for the generation of continuous radiation. (A) Real conditions: continuous radiation is emitted in all directions, from any point within the excited region (dotted). (B) Hénoc model: the continuum radiates isotropically from a point at the specimen surface. (C) Springer model: continuous radiation is emitted in downward directions only, from any point within the excited region.

The formal integration of Eq. (11.2.1) requires a simple law for the change of the mass absorption coefficient with the photon energy E : $\mu \propto E^{-3} \propto \lambda^3$ (from Eq. (4.3.10)]. But this exponent always overestimates the change of μ. Since mass absorption coefficients must be calculated for wide energy ranges, the cubic power law may introduce significant error. More importantly, the presence of absorption edges makes it impracticable, for all but the simplest specimens, to calculate the effects of continuum fluorescence without a computer program that seeks and arranges in the order of energies the absorption edges of all elements in the specimen and standard. Such a searching and sorting routine is contained in the program COR for electron probe data reduction, by Hénoc et al. [9.14]. In view of the complexity of such routines, the advantages in simplicity of formal integration schemes become insignificant. It is therefore preferable to numerically perform the integration according to Eq. (11.2.8), over all intervals between absorption edges, as is done in COR. In the numerical integration, the correct noninteger exponents for Eq. (4.3.10) can be used.

Effects of Continuum Fluorescence. In spite of the considerable effects we have described, the treatment of continuum fluorescence in the analytical practice is still, in general, neglected. The correction is frequently ignored [9.3], or performed by means of oversimplified models, and recently acquired knowledge concerning the continuum emission is not being incorporated in correction procedures. It is true that in most cases the omission of the continuum correction does not produce serious errors in quantitation. Significant effects, however, can arise in the emission of hard lines from matrices of low atomic number. Table 11.4 illustrates this point. A series of elements is assumed to be dissolved, at a mass fraction of 0.01, in sodium tetraborate. By means of the program COR [9.14], the following results are calculated: the generated primary intensity divided by the mass fraction (I_p/C), the primary absorption factor (f_p), that for continuum fluorescence (f_c), the average absorption factor (f), the ratio of generated continuum fluorescence over primary emission (I_c/I_p), the ratio of the emitted intensities (I'_c/I'_p), the relative intensity that would be generated when the standard is a pure element and the fluorescence from the continuum were neglected (k_{noc}), and the relative intensity obtained when the continuum fluorescence is included (k_c). It can be seen that with the increasing atomic number of the emitter the effects of the continuum increase drastically. With Zr Kα emission, 87% of the X-ray emission from the borax solution is due to continuum fluorescence, and the omission of the corresponding correction produces an error of −45% in the prediction of the relative X-ray intensity. Since borax has a mean atomic number of the same order as biological tissue, the implications for the analysis of thick biological tissue samples are obvious.

11.3 EXPERIMENTAL INVESTIGATION OF SECONDARY EMISSION

The observation of fluorescent excitation is hindered by the much stronger primary emission always present when secondary radiation is produced. Although the effects of fluorescence can be shown by careful measurement of the emission from well-characterized standard specimens (see Fig. 11.17), such a procedure is affected by considerable uncertainties, particularly in the case of continuum fluorescence, which is frequently of interest when a large atomic number effect is also present. In the cases of characteristic L-K and K-L excitation, significant fluorescent excitation has sometimes been ignored [8.6]. The most effective experimental approaches to fluorescence, due either to characteristic or continuous radiation, are based on secondary excitation at distances larger than the electron range from the point of electron beam impact. Such experiments include the analysis of coated targets ("aluminum method," [11.9]), tracer targets such as those discussed in Chapter 10, and undiffused couples.

Table 11.4. Effects of Continuum Fluorescence in Borax Glasses.

	Line and Z	kV	I_p/C	f_p	f_c	$\overline{f}$	I_c/I_p	I'_c/I'_p	k_{noc}	k_c
P	NiKα	20	1.33×10^{-3}	0.979	0.797	0.972	0.0420	0.0342	0.0089	0.0098
*	28		1.18×10^{-3}	0.993	0.443	0.868	0.2957	0.1320		
P	ZnKα	20	9.14×10^{-4}	0.984	0.816	0.975	0.0550	0.0456	0.0085	0.0098
*	30		7.64×10^{-4}	0.996	0.479	0.838	0.4405	0.2117		
P	GeKα	30	1.98×10^{-3}	0.970	0.763	0.955	0.0795	0.0626	0.0085	0.0104
*	32		1.64×10^{-3}	0.994	0.430	0.763	0.6917	0.2992		
P	ZrKα	40	1.17×10^{-3}	0.977	0.787	0.944	0.2096	0.1688	0.0084	0.0152
*	40		9.56×10^{-4}	0.997	0.496	0.650	2.246	1.1168		

P: pure element.
*: 1% in $Na_2B_4O_7$.

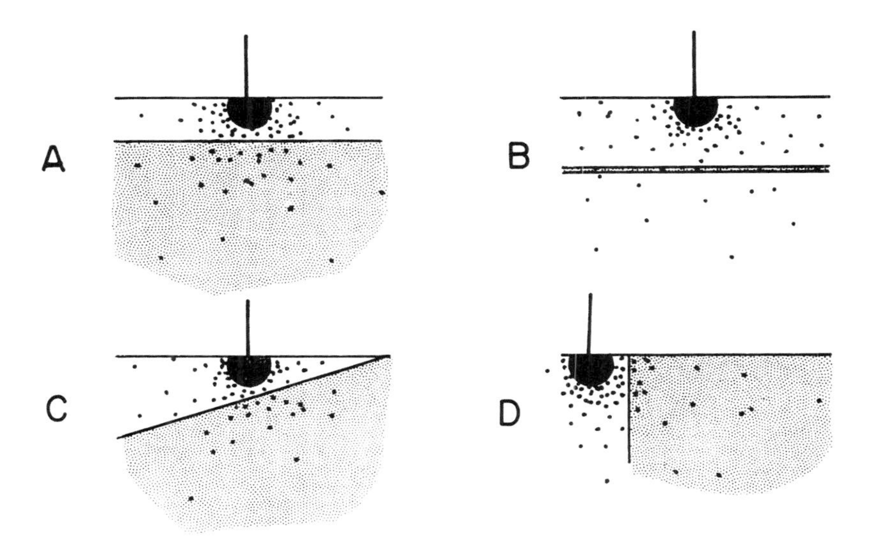

Fig. 11.17. Target configurations for the investigation of indirect radiation. (A) "aluminum method." (B) Tracer technique. (C) Wedge configuration. (D) Undiffused couple. The black region is that of primary X-ray generation; dots are possible sites of secondary X-ray generation. The gray phase contains the element which emits fluorescent radiation.

In the aluminum method, the primary excitation is contained within a surface layer, which is applied onto the target which is secondarily excited. The electrons cannot reach this element. To determine the fluorescence effect, corrections are applied for geometrical effects on the secondary target, and for absorption of the emergent radiation by the surface layer.

A wedge target, similar in principle to the aluminum-covered target, has been proposed as a means for investigating the depth distribution of primary radiation [11.13] and could also be used in studying fluorescent excitation.

The use of the undiffused couple [11.4], [11.14], [11.15] is based on the lateral diffusion of the secondary X-rays. By extrapolating the fluorescent signal to the interface of the primary and secondary target materials, one obtains an intensity which is equivalent to full excitation of the primary target material, and half of the full excitation by fluorescence of the secondary target material. There are, however, several potential sources of error, such as smearing and other artifacts in the preparation of the target, and effects due to the peripheral parts of a poorly defined electron beam, or of scattered high-energy electrons outside the beam. The mathematical aspects of the technique were explored by Maurice et al. [11.15].

Although experiments can be used to evaluate approximate calculations of secondary excitation, a rigorous calculation of the effects of approximations is preferable. The experimental approaches should be used to investigate the uncertainties in the ratio of intensities of the two primary excitations which are involved. This is of particular importance for the continuum fluorescence, since the estimate of the generated continuum intensity is still subject to revisions.

11. REFERENCES

11.1 Springer, G., *Fortschr. Miner.* **45,** 103 (1967).

11.2 Birks, L. S., Ellis, D. J., Grant, B. K., Frish, A. S., and Hickmann, R. B., in *The Electron Microprobe,* McKinley, T. D., Heinrich, K. F. J., and Wittry, D. B., Eds., John Wiley & Sons, New York, 1966, p. 199.

11.3 Brown, J. D., Ph.D. Thesis, Univ. of Maryland, College Park, Md., 1966.

11.4 Reed, S. J. B. and Long, J. V. P., *Proc. 3rd Int. Conf. on X-Ray Optics and Microanalysis,* Pattee, H. H., Cosslett, V. E., and Engström, A., Eds., Academic Press, New York, 1963, p. 317.

11.5 Colby, J. W., *Adv. X-Ray Analysis* **11,** 287 (1968).

11.6 Birks, L. S., *J. Appl. Phys.* **32,** 387 (1961).

11.7 Wittry, D. B., Ph.D. Thesis, California Institute of Technology, Pasadena, Calif., 1957; also ASTM Special Publ. 349, American Society for Testing and Materials, Philadelphia, Pa., 1964, p. 128.

11.8 Duncumb, P. and Shields, P. K., *Proc. 3rd Int. Conf. on X-Ray Optics and Microanalysis,* Pattee, H. H., Cosslett, V. E., and Engström, A., Eds., Academic Press, New York, 1963, p. 329.

11.9 Hénoc, J., in *Quantitative Electron Probe Microanalysis,* Heinrich, K. F. J., Ed., NBS Special Publ. 298, National Bureau of Standards, U.S. Government Printing Office, Washington, D.C., 1968, p. 197.

11.10 Springer, G., *Neues Jahrbuch Fuer Mineralogie, Abhandlungen.* **106,** 241 (1967).

11.11 Springer, G. and Rosner, B., *Proc. 5th Int. Conf. on X-Ray Optics and Microanalysis,* Möllenstedt, G. and Gaukler, K. H., Eds., Springer, Berlin, 1969, p. 170.

11.12 Springer, G., *Proc. 6th Int. Conf. on X-Ray Optics and Microanalysis,* Shinoda G., Kohra, K., and Ichinokawa, T., Eds., University of Tokyo Press, Tokyo, Japan, 1972, p. 141.

11.13 Schmitz, U., Ryder, P. L., and Pitsch, W., *Proc. 5th Int. Conf. on X-Ray Optics and Microanalysis,* Möllenstedt, G. and Gaukler, K. H., Eds., Springer, Berlin, 1969, p. 104.

11.14 Dils, R. R., Zeitz, L., and Huggins, R. A., *Proc. 3rd Int. Conf. on X-Ray Optics and Microanalysis,* Pattee, H. H., Cosslett, V. E., and Engström, A., Eds., Academic Press, New York, 1963, p. 341.

11.15 Maurice, F., Seguin, R., and Hénoc, J., *Proc. 4th Int. Conf. on X-Ray Optics and Microanalysis,* Castaing, R., Deschamps, P., and Philibert, J., Eds., Hermann, Paris, 1966, p. 357.

12. Practice of Quantitative Electron Probe Microanalysis

12.1 THE ITERATION PROCEDURE

The relative X-ray intensity emitted from a target of known composition can be calculated by Eq. (9.1.3), with the aid of the expressions for its terms developed in Chapters 9 through 11. The analytical practice, however, requires the inverse process; namely, to determine the composition of an unknown specimen, or specimen region, on the basis of the X-ray intensities emitted by it. Because graphic solutions to the problem are impractical when the number of component elements is larger than two, mathematical treatments are required. For a specimen containing n elements, all of which emit at least one measurable X-ray line, the theory summarized in Eq. (9.1.3) yields the following system of equations:

$$\begin{aligned} k_a &= F_a(\underline{C_a}, C_b, \ldots, C_n) \\ k_b &= F_b(C_a, \underline{C_b}, \ldots, C_n) \\ &\cdots\cdots\cdots\cdots\cdots \\ k_n &= F_n(C_a, C_b, \ldots, \underline{C_n}) \\ \Sigma C_i &= 1. \end{aligned} \tag{12.1.1}$$

The underlined unknowns are most significant in determining the relative intensity of the radiation from the element corresponding to their subscript. The resolution of the analytical problem, however, requires the following set of equations:

$$
\begin{aligned}
C_a &= F'_a(\underline{k_a}, k_b, \ldots, k_n) \\
C_b &= F'_b(k_a, \underline{k_b}, \ldots, k_n) \\
&\ldots\ldots\ldots\ldots\ldots\ldots \\
C_n &= F'_n(k_a, k_b, \ldots, \underline{k_n}).
\end{aligned}
\tag{12.1.2}
$$

In view of the complexity of the analytical function it is impossible to arithmetically transform the matrix Eqs. (12.1.1) into Eqs. (12.1.2). This can, however, be achieved by a method of successive approximations called the iteration procedure. The principle of such a procedure is outlined in Fig. 12.1. The technique illustrated here (forward iteration) is based on estimating the concentration of each element by means of a form of the analytical function which gives the concentration of the element as a function of the relative intensity:

$$C_a = F'^*_a(k_a). \tag{12.1.3}$$

The asterisk indicates the dependence of F'_a upon the concentrations of all elements besides a, as shown implicitly in Eq. (12.1.2). An equation of this type is the *ZAF* equation [Eq. (9.1.10)] . To the contrary, Eq. (9.1.3) cannot be used in this context since it does not explicitly contain the concentration of the element to be determined. The sequence of the procedure is as follows. In the first step, the relative intensities of all elements, k_i, are either entered in the program or calculated from the experimental data. In the next step, a first estimate of the concentrations of all elements is made. Usually it is postulated at this point that C_i, for each element i, is equal to k_i. (Variables with subscript i must be calculated for all n elements present.) With the aid of these first estimates C_{oi}, we now obtain an initial estimate of the functions F'^*_i (first step within the iteration loop). Then, applying Eq. (12.1.3), we calculate the next estimate of C_i, called C_{i1}.

If the initial estimates of C_i had been perfect, no further calculation would be required. To test this possibility, we proceed with a stopping test: if the difference between each C_{i1} and the corresponding value of C_{i0} is smaller than a predetermined value Δ, (stopping criterion), it is deemed that further calculation would not produce significant improvement in subsequent iterations; the results are therefore directly transmitted to the output of the program (e.g., they can be printed). If, however, the stopping criterion is not met, the estimates C_{i1} are used to calculate a better approximation of F'^*_i, and all steps up to the stopping test are repeated. Only when, in the mth iteration, the difference between C_{im} and $C_{i(m-1)}$ is smaller than Δ for each element, is the iteration process terminated, and the last set of values of C_i issued.

The gradual approaching of the final values of C_i is called *convergence*. The

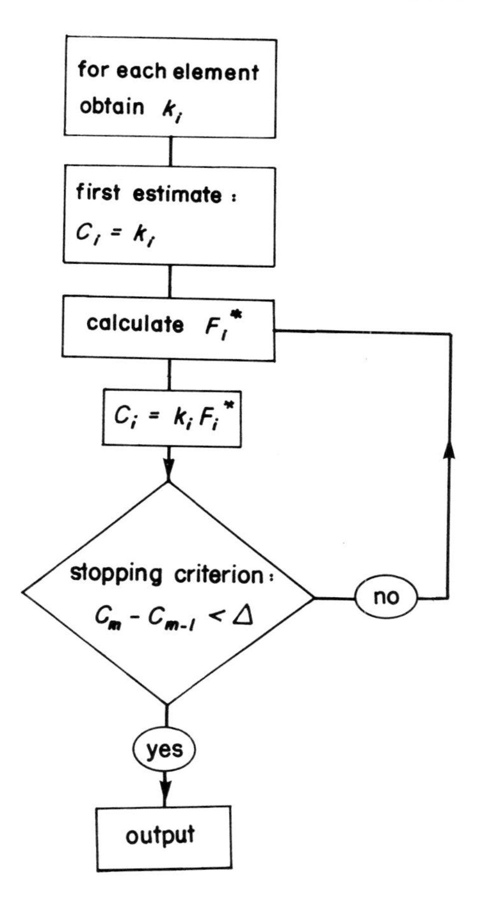

Fig. 12.1. Basic iteration loop.

iteration procedure, to be successful, must converge, and moreover, it must converge to the "right endpoint." In other words, the result of the calculation must be the same as that which would be obtained if known concentrations had been fed into the determination of the function $F_i'^*$. This is usually the case if convergence is obtained. However, failure of convergence is often observed with the iteration scheme we have described. This occurs when the errors in the estimates of concentration cause the estimate of $F_i'^*$ to be so poor that the next set of estimates is even less accurate than the previous one. The inaccuracies discussed here are caused by the variation of $F_i'^*$ with composition, and are not related to, or including, the systematic errors of the model on which $F_i'^*$ is based. Nonconvergence is, however, more frequent when the corrections implied in $F_i'^*$ are large (particularly if the primary absorption factor, f_p, is very low), since in such a case $F_i'^*$ is highly concentration dependent. Lack of convergence frequently takes the form of a divergent oscillation; in alternate iterations, the error changes from positive to negative and vice versa, while its magnitude increases with the number of iterations.

The chances for convergence in the iteration can be increased through the following steps (Fig. 12.2):

1. The concentrations of all elements are normalized at the start of each loop by forcing the sum of weight fractions to one:

$$C_i' = C_i \Big/ \sum_{j=1}^{n} C_j .$$

   This normalization affects the values of the correction factors which determine the analytical function, but the sum of the final concentrations is not forced to one, unless another normalization is performed at the end of the last loop. Performing such a final normalization would be of doubtful value, since, in the absence of this step, the value of the sum of element concentrations may reveal the existence of a gross error, or the omission of an element. The analyst should be cautioned, however, that obtaining a sum of element concentrations close to unity is not proof of accuracy of the analysis, since with faulty models for the analytical function, compensation of errors frequently drives the sum of results close to unity, although the individual values may be grossly incorrect. It is true, though, that, with a proven data reduction procedure, failure to obtain a sum of concentrations close to unity (e.g., between 0.98 and 1.02) indicates gross errors of measurement, omission of one or more elements, or other mistakes in the analytical procedure.
2. In order to avoid diverging oscillation of successive values, a damping step is introduced, in which the change of concentrations obtained in successive iterations is diminished. A simple procedure is to halve the change of concentration produced in each iteration: at the mth iteration,

$$C_{im}' = \frac{1}{2}(C_{im} + C_{i(m-1)}).$$

   More complicated damping procedures, such as the Wegstein iteration scheme [12.1], may extend the damping process over more than one loop. Even with these precautions, convergence may not always be achieved in this procedure. Therefore, to avoid needless calculations, computer programs sometimes incorporate an abort procedure, in which the attempt at solution is stopped if convergence has not been achieved after a preset number of iterations, and a message indicating failure of the calculation is issued.
3. Since failure of convergence is most probable when the correction is large, usually due to excessively low primary absorption factors, it is a good practice to choose operating conditions which will produce little X-ray absorption in the target.

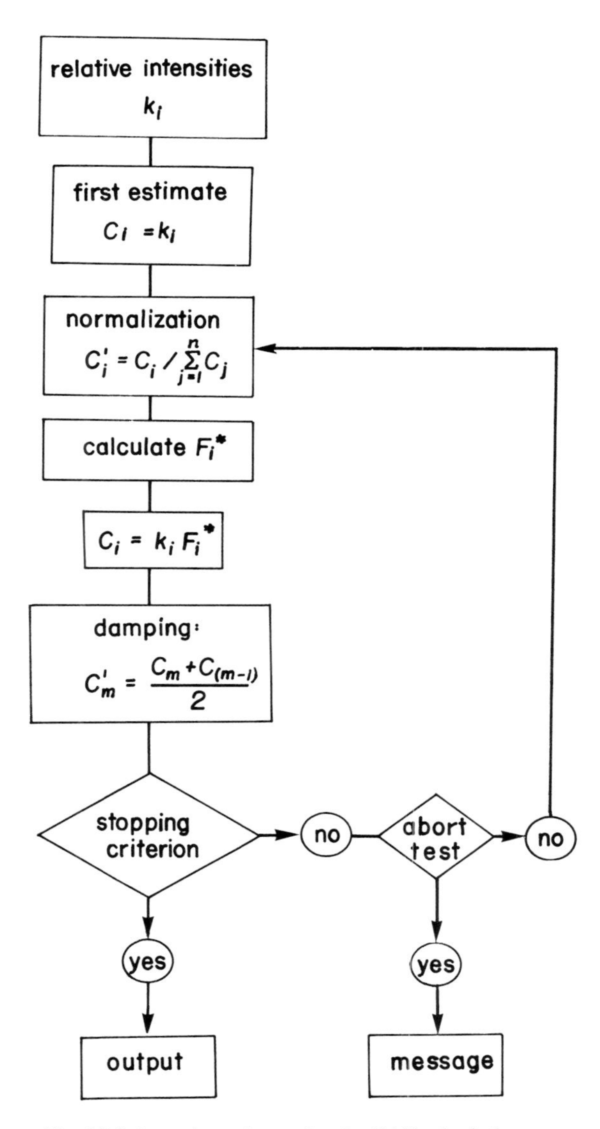

Fig. 12.2. Iteration scheme for the ZAF calculation.

Hyperbolic Iteration Procedure of Criss and Birks. A variation of the iteration procedure proposed by Criss and Birks [10.22] has virtually eliminated the chances of failure to converge, and, at the same time, permits the application of the rigorous procedure based on Eq. (9.1.3), although this equation does not explicitly contain the concentration of the element in question.

The procedure follows the scheme shown in Fig. 12.1. However, the analytical function F'^* used in this case is not the ZAF equation, but the hyperbolic approximation in the form of Eq. (8.2.7). This approximation permits calculating the concentration, starting from the relative intensity, since the coefficient α in Eq. (8.2.7) can be determined if the pair of parameters (C,k) is available for any point along the hyperbola. This pair is obtained in the procedure by means of the Eq. (9.1.3), applied to the best estimate of composition available at the start of the iteration loop.

The mechanism of the hyperbolic iteration is best followed by comparing the flow diagram of Fig. 12.3 with the representation of the analytical curve on Fig. 12.4, which was described in Section 8.2. We have chosen, for clarity, an analytical function which deviates far more from linearity than would be desirable in a practical case. We commence with a first estimate of the specimen composition, by assuming a linear calibration curve ($C_i = k_i$). The concentrations of all elements thus obtained will represent a composition fairly close to that of the specimen, which can be used as a fictitious standard to calculate the α coefficients for the hyperbolic equation. For this purpose, we use the theoretical equations incorporated in Eq. (9.1.3), to predict the relative intensities to be expected, under the conditions of analysis, from this fictitious standard. For each element, we thus have the two parameters, C_o and k, of the fictitious standard, which permit calculating the corresponding α coefficient:

$$\alpha = \frac{1-k}{k} \cdot \frac{C_0}{1-C_0} \qquad (C_0 = k_{\text{ex}})$$

[see Eq. (8.2.4)]. Assuming that the same α is also valid for the composition of the unknown specimen, we obtain the next estimate of composition:

$$C_1 = \frac{\alpha k_{\text{ex}}}{1 + k_{\text{ex}}(\alpha - 1)} \, .$$

In terms of the graph showing the analytical function, the first compositional estimate is represented by the point B, the fictitious standard by the point C, and the specimen composition by the point A. As the graph in Fig. 12.4 indicates, in a binary system which strictly obeys the hyperbolic approximation, the true composition of the specimen would be obtained in the first iteration. More than one iteration is required, however, if the binary system deviates from the hyperbolic approximation (Figs. 12.5 and 12.6), or in a system containing more than two elements. In the latter case, the analytical curve for each system changes slightly as successive adjustments are made in the concentrations of the other components, so that the graphic representation of the iteration is as shown in Fig. 12.7.

As in the previous technique, it is advantageous to insert a normalization step

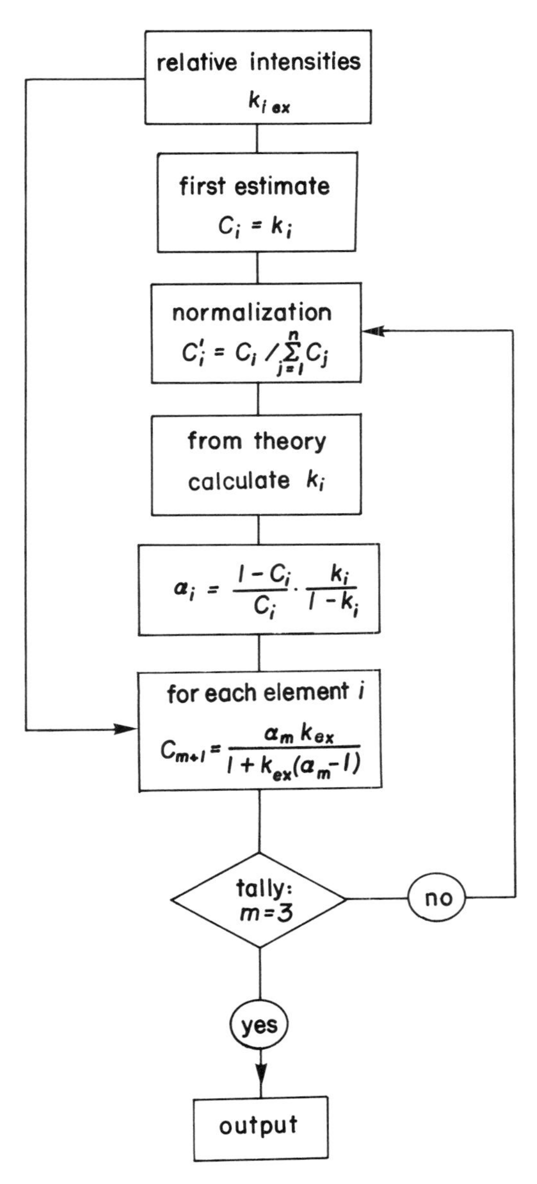

Fig. 12.3. Flow diagram of the iteration after Criss and Birks [10.22].

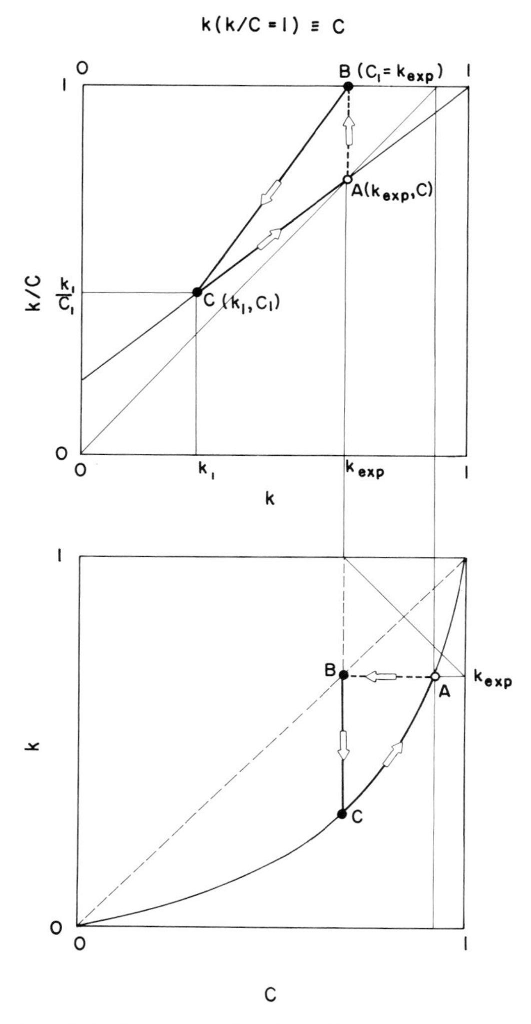

Fig. 12.4. Illustration of the hyperbolic iteration. The system is binary and the hyperbolic approximation is assumed to be accurate. Upper graph: k/C versus k. Lower graph: conventional analytical curve. A: true coordinates; B: first approximation; C: fictitious standard.

for the sum of mass fractions before each iterative loop. A damping step, however, is not required and would in fact prolong the calculation. The system we have described has been tested in hundreds of cases, including those which failed to yield convergence with other iteration techniques, and in each case a rapid convergence was observed. Therefore, an abort test is unnecessary, and the iteration can be automatically terminated after four loops, without a convergence test, as is done in the computer program FRAME [9.14].

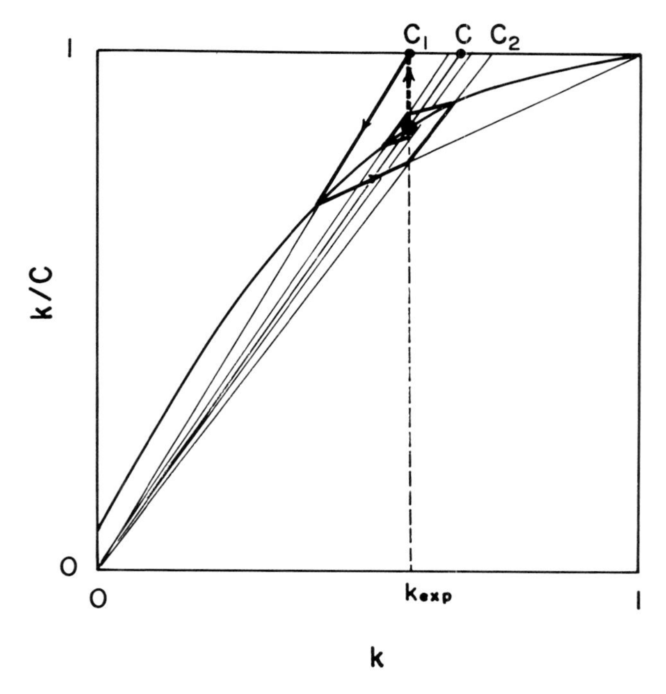

Fig. 12.5. Hyperbolic iteration; binary system with imperfect fit of the hyperbolic approximation which is shown as a curvature of the analytical curve. Oscillating convergence.

12.1.1 Variations in the Iteration Scheme

With trivial changes, the results of the iteration can be expressed in mole fractions instead of weight fractions. More basic changes are the following:

a. *Calculation of Relative Intensities from a Target of Known Composition.* These are useful in theoretical studies and also to determine the α factors to be applied in the semiempirical approach of Bence and Albee [8.4]. The programming is easy, particularly in the hyperbolic iteration scheme, in which this calculation [Eq. (9.1.3)] is normally a part of the iteration loop.
b. *Use of Multielement Standards.* If a multielement standard rather than a pure element is used, the intensity observed from the standard, I'^s, can be used to calculate the intensity that would be emitted from a hypothetical elementary standard, $I'(a)$:

$$I'(a) = I'^s/k^s.$$

The value of the relative intensity of the standard with respect to the element, k^s, is obtained as in case a, and the intensity emitted by the specimen, relative to that from the element, is

$$k^* = \frac{I'^*}{I'^s} \cdot k^s.$$

The variations which follow apply, singly or in combination, when some of the elements in the specimen are not determined, either because they do not emit appropriate lines, or because the number of elements present exceeds that of available spectrometers.

c. *Fixed Composition for Major Components*. If the analyst determines trace components in a matrix in which the composition is known, then provisions can be made, at the start of the iteration loop, to input these known concentrations, which are maintained unchanged during the iteration procedure.

d. *Stoichiometric Relations*. These are mainly used in the analysis of specimens in which the elements are combined with the stoichiometric equivalent of oxygen. Therefore, oxygen can be determined indirectly. Problems arise, however, with elements such as iron which may be present in more than one

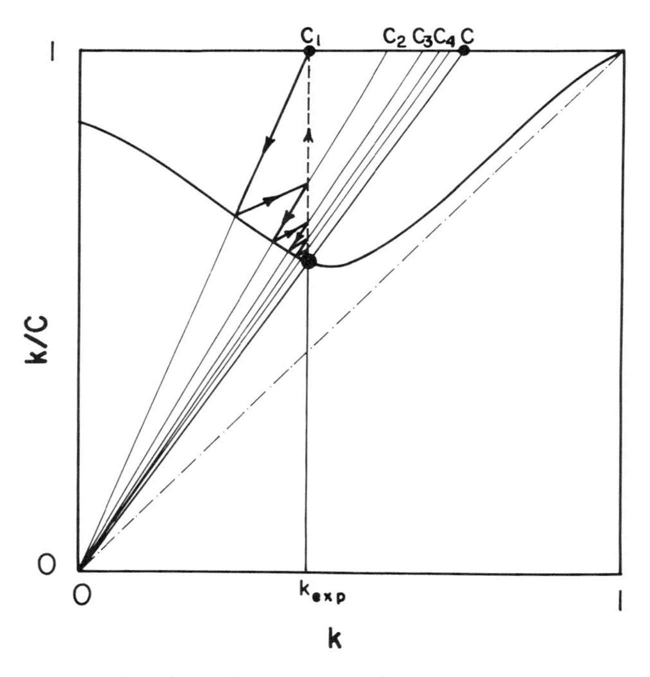

Fig. 12.6. Hyperbolic iteration; binary system with poor fit of the hyperbolic approximation. Unidirectional convergence.

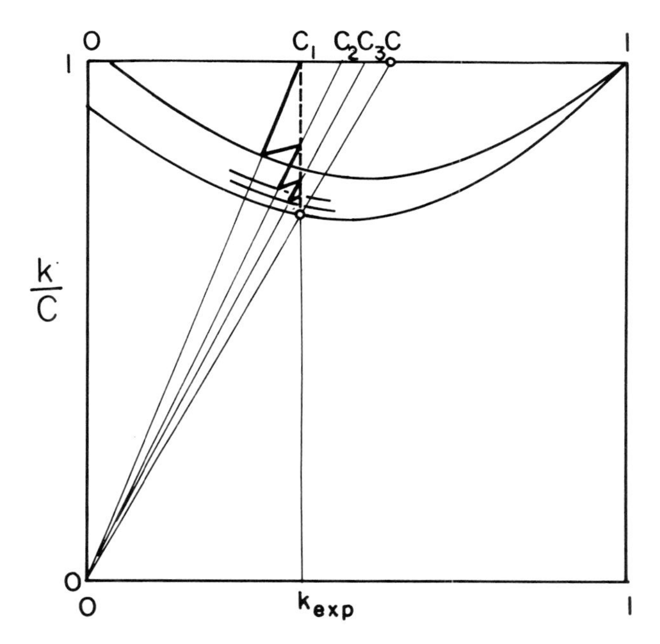

Fig. 12.7. Hyperbolic iteration; multielement system with poor fit to the hyperbolic approximation. Due to changes in the assumed matrix composition, the calibration curve changes in each iterative step.

valence state. In the presence of such elements, unique solutions cannot be given unless an a priori decision concerning their valence state is made. When the element in question is present at high concentration, failure of the sum of elements to approach unity may be an indication of an error in this decision.

e. *Element by Difference.* Since the matrix Eq. (12.1.1) is overdetermined, one element (or a stoichiometric combination of elements, such as CO_2) can be determined by difference. This recourse may be useful in simple cases such as the analysis of binary specimens. The estimate of the element by difference is made at the start of each loop and at the end of the calculation. The procedure has several disadvantages over the preceding techniques. The accuracy in the determination by difference is poor when the element in question is present at low concentration. This is frequently the case if the element is of low atomic number. Furthermore, in this procedure the check normally provided by the sum of elements is lost. Therefore, elements may be missed and other mistakes may pass unobserved. Finally, in the less effective iteration schemes—but not in the Criss procedure—the probability of convergence diminishes. Therefore, the procedure should be used as a last recourse only.

12.2 STANDARDS

Both the theoretical and empirical procedures we have discussed previously are based on the comparison of the X-ray emission from the unknown specimen with that of one or more standard reference materials of known composition. In virtually all cases, electron probe microanalysis is based on comparative measurements; therefore, the choice and use of standards is an integral part of the analytical procedure. Although the range of materials used for this purpose is very wide, standards fall basically into three groups:

1. *Elements and Simple Compounds*. Elements and simple compounds can be used in conjunction with the theoretical procedure we have discussed in Chapters 9 through 11. It is advantageous that with such standards a strong signal is obtained from the line to be used in the analysis. The differences in composition between the standard and the specimen are considered of less importance, provided that the intensity dependence upon composition can be handled efficiently by the theoretical corrections, or by the hyperbolic approximation, if used, and that the spectrometer response is not affected by line shifts.
2. *Matching Standards for Empirical Procedures*. This class of standards must fit the specimen in composition as closely as possible, so that errors due to matrix effects and line shifts are insignificant. With the use of such standards, the analysis is essentially reduced to a comparison of emitted intensities, and reliance on theoretical considerations is minimized.
3. *Standards Used to Test Theoretical Models*. In the use of such standards there is usually a latitude of allowable compositions, as long as the effects of the model which are to be tested are significant. The requirements for accurate knowledge of composition and for homogeneity are stringent.

From a compositional point of view, standards may be elements, metallic or nonmetallic compounds, single or multiphase alloys, minerals or glasses, etc. In addition, special standards are used in situations in which spatial resolution is important, such as resolution standards, thin films or coatings, particles, tracers, and undiffused couples.

Pure Elements. Most elements are easy to obtain, and, unless they are used to measure the background for trace determinations of other elements, the purity requirements need not exceed the precision of the line intensity measurements; hence, a purity of 99.9% is sufficient. Their use is limited by the fact that many elements are not solid (halogens, sulfur, selenium, mercury), have low melting points (gallium), or are reactive (alkalies, calcium, strontium, barium). With emitters of lines of long wavelength serious errors may also arise due to chemical effects such as wavelength shifts; for instance, in the use of magnesium and aluminum as standards for these elements in oxidic specimens. The extent of such effects depends on the resolution of the spectrometers used in the measurement, which may vary widely for different crystals (see Fig. 5.20 and Table 12.1). For this reason, high resolution of X-ray spectrometers is not always advantageous.

Chemical Compounds. When pure elements cannot be used as standards, the problem of accurate knowledge of the standard composition becomes very important. One solution to this problem is to use stoichiometric compounds. The problems which may arise from lack of thermal and electric conductivity are discussed in detail in Chapter 16. But many compounds, such as sulfides, phosphides, silicides, and intermetallic compounds may not be stoichiometric. In natural products, such as silicate minerals, substitution of one element by another is frequent. Unless stoichiometry can be considered proven, the composition of the standard must be established, by careful synthesis, by chemical analysis, or both. It should also be recognized that the published phase equilibrium diagrams of intermetallic compounds frequently indicate stoichiometry when the corresponding compound may either be nonstoichiometric or have a significant range of composition.

Solid Solutions. When solid solutions are used as standards, both the average composition and the degree of homogeneity must be determined. A liquid solution and the solid in equilibrium with it generally differ in composition. Therefore, coring always occurs to some degree during solidification. This effect can be reduced by rapidly quenching small portions of the alloy from a single-phase melt [12.2]. The measurements of Goldstein et al. [8.3] show that usable standards can be obtained by quenching a molten alloy outside the composition range of solubility for the solid state in the equilibrium phase diagram. If the grains formed are much smaller than the region excited by the electron beam, the effects of inhomogeneity on the emitted intensity are negligible. However, some materials obtained in this manner recrystallize spontaneously at room

Table 12.1. Effects of Line Shifts in Quantitation.

CRYSTAL USED	LINE FOCUSED AT RADIATION FROM	TRUE ALUMINUM CONCENTRATION IN Al_2O_3	OBTAINED	SPECTROMETER RESOLUTION (eV)
ADP	Al	0.5292	0.4779	1
	Al_2O_3		0.5977	
RAP	Al		0.5265	28
	Al_2O_3		0.5345	

Note: Analysis of Al_2O_3, with aluminum used as standard. Conditions: operating voltage is 15 kV, ψ = 52.5°. Line Observed: Al Kα. The specimens were coated with carbon. The better agreement obtained with the RAP crystal is due to its poor resolution which suppresses the effects of the line shift.

temperature. For instance, Goldstein et al. observed that quenched magnesium-aluminum alloys became inhomogeneous, on a micrometer scale, within a few days.

When metallic solid solutions are prepared for the use of microanalytical standards, precautions must be taken to avoid inhomogeneity on both the macroscopic and microscopic scale. The material may be mixed mechanically (e.g., by repeated extrusion), or by prolonged heat treatment. The final product must be chemically analyzed (if possible, by more than one laboratory and technique), and tested for uniformity of composition. The steps taken in the preparation and testing of such standards, in the binary systems of silver-gold and copper-gold, are described in detail in a publication of the National Bureau of Standards [12.3]. The testing for homogeneity of microanalytical standards is further discussed in [12.4]. (See also Fig. 6.16.)

Oxidic Standards. Mineralogists naturally tend toward the use of oxidic standards, because many of the elements of interest to them cannot be used as standards in their elementary form, or emit soft lines subject to chemical shifts [4.23], [7.15], [9.9], [12.5]. Excellent synthetic minerals can be prepared for standardization of silicates [12.5]; however, their manufacture is tedious. Hence, the use of natural standards is prevalent among mineralogists. Since this choice permits, at least in principle, the use of materials close in composition to the specimen, a simple data reduction approach such as the hyperbolic approximation can be used to advantage [8.4]. Such a strategy however, is not exempt of risks. Natural minerals are frequently inhomogeneous; in some silicates, even intragranular composition ranges can be substantial. Well-analyzed samples of hand-selected mineral grains are difficult to obtain, and the reliability of chemical analysis should not be overestimated [12.6]-[12.8], particularly if the analysis has not been performed recently, with inclusion of spectrographic tests for the presence of the rarer elements. The stability of minerals under the impact of intense electron beams must also be proven before they can be used or recommended as standards. Minerals may decompose, losing water or carbon dioxide, or other volatile compounds. The mobility within the specimen or evaporation of metals such as sodium and potassium must also be carefully considered, particularly in glassy silicates. More about these difficulties will be said in the discussion of the specimen (Chapter 16).

Standardization with Large Beams. Because compositions desirable for standard reference materials may fall outside the ranges of single-phase composition, and even single-phase standards may have some degree of segregation, it is occasionally proposed that inhomogeneous standards can be used provided that the exciting electron beam is of sufficient width to average the effects of local concentration differences. In order to estimate the errors that may be committed in this approach, we must consider the analytical calibration curve for an element in a binary system (Fig. 12.8). Assume that the analytical curve for homogeneous targets is that which passes through the points B, E, C, F, and A. If

C^* is the mass fraction of the measured element in the specimen, the relative emitted X-ray intensity will be k_h (point C).

The largest deviation from the calibration curve will occur if the standard is a mixture of grains consisting of the pure elements, a and b. If the dimensions of the electron trajectories within the specimen are negligibly small compared with those of the grains, virtually every electron trajectory will be contained entirely in one of the two species of grains. If the component elements a and b have the densities ρ_a and ρ_b, the probability of an electron trajectory being within the element a is therefore equal to $(C_a/\rho_a)/(C_a/\rho_a + C_b/\rho_b)$. The analytical calibration curve for the totally segregated case is thus a hyperbola, which, on Fig. 12.8, is assumed to pass through points A, D, and B. If the two densities were equal, the curve would be a straight line from point A to point B. The same curve would also be valid for integrated results of line or area scans with a focused beam. As long as all target events, from the point of electron impact to that of X-ray photon emergence, occur within a single element, no interelement effect can occur. Therefore, a standard of this nature, of average concentration C^* of the element being measured, would produce a relative X-ray intensity equal to k_i.

A more favorable situation occurs if the large grains, instead of being pure elements, have the concentrations C_1 and C_2 of the element in question. We obtain, for full segregation, a calibration curve through points E, G, and F. If the grains are so small that a significant fraction of electron trajectories passes through both phases, the possible range of relative X-ray intensities spans, for

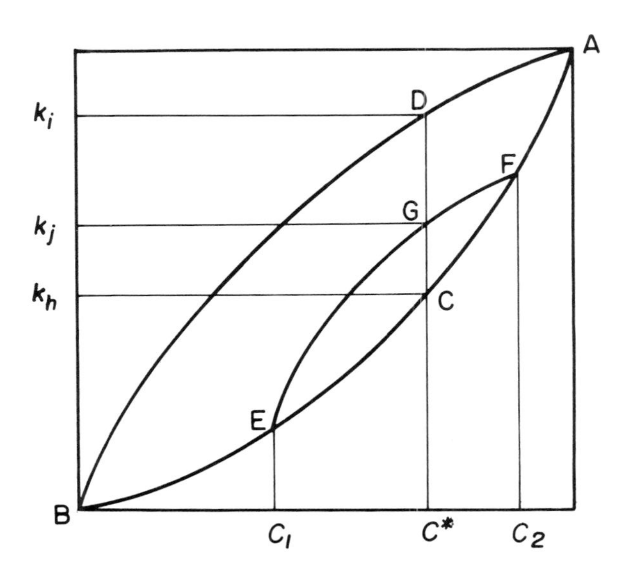

Fig. 12.8. Plot of concentration versus relative X-ray intensity for a two-phase standard analyzed with a wide beam.

pure element grains from k_i to k_h, and for grains of concentrations C_1 and C_2, from k_j to k_h. The range of uncertainty is much narrower in the second case, although the compositions of the two phases may differ considerably from the average composition.

Analogous considerations hold for inhomogeneous standards within a one-phase region. The effects of inhomogeneity may vary from insignificant to serious depending on the divergence of the two limiting analytical curves, the range of composition within the standard, and the average dimensions of compositional domains. It should be noted, however, that the successful use of such standards also depends on the premise, which may not always be correct, that the average composition of the specimen and area used in the microprobe measurement is the same as that of the portions of the standard which were analyzed chemically. It is therefore preferable not to use such standards if suitable homogeneous single-phase standards can be substituted.

Another reason for using large electron beams, both on specimens and standards, is to reduce the effects of contamination, charging, and compositional changes due to the electron beam in some types of materials such as glasses [12.9]. The accuracy of the technique may be limited in this case by the defocusing of the crystal spectrometers.

Standards Used for the Testing of Correction Procedures. The observations we made concerning standards for analysis are also valid for those specimens of known composition which are used to test the validity of correction procedures. For such tests, there is usually a greater latitude in useful compositions; on the other hand, the requirements for accurate knowledge of composition and homogeneity should be very strict, and the standard materials to be used should be screened *before* the testing experiments. If several compositions within a binary system have been prepared, consistency tests can be performed by means of the plots discussed in connection with the hyperbolic approximation [8.6] (Figs. 8.5 through 8.7 and Fig. 12.9). Inconsistency in the postulated compositions can be demonstrated, particularly if measurements at various operating voltages show similar deviations, or if the measurements on both components of a binary specimen show compensating deviations. If more than one standard fails to fit a clearly defined trend (k/C must be a smooth, though not necessarily linear, function of k), the whole set of standards should be rejected.

Mounted standards should be conserved and stored carefully, to avoid contamination. Some metallic standards (e.g., lead and uranium) oxidize at room temperature and must be polished prior to each use.

In an extensive research program at the U.S. National Bureau of Standards the usefulness of glasses as standards, both for analysis and for the testing of correction procedures, was investigated. Tables 12.2 and 12.3 show the compositions of some such glasses. It can be observed that an extremely wide range of elements can be incorporated in glasses, and that a very wide range of mean

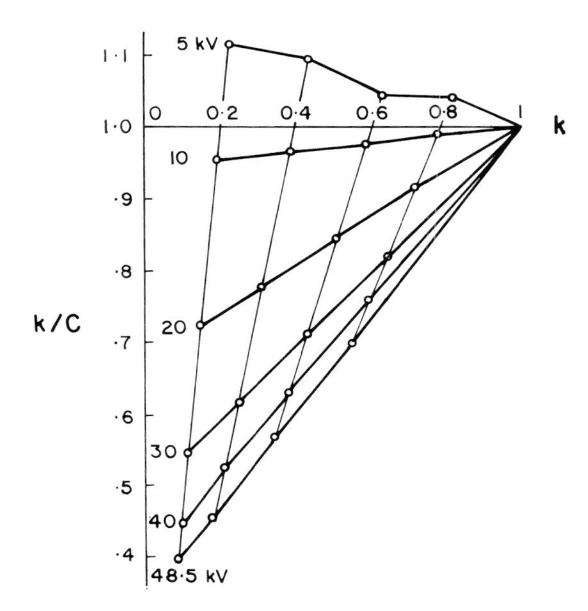

Fig. 12.9. Measurements of relative X-ray intensity with the Ag Lα_1 line on a series of gold-silver alloys (NBS Standard Reference Material 481[12.3]). Bars indicate $\pm 1\sigma$, where repetitions were performed. Measuıements at 5 kV are less self-consistent than those at higher operating potentials.

atomic numbers can be covered, particularly if materials other than silicates are included. It should also be noted that the problems of mobility of alkali ions can be minimized or eliminated if glasses with a high concentration of these elements are avoided. Figure 12.10 shows the elements which are presently incorporated in NBS glasses. The largest expense in the preparation of such standard glasses is usually the chemical characterization of their composition.

12.3 THE ACCURACY OF QUANTITATIVE ELECTRON PROBE MICROANALYSIS

Poole and Thomas [12.10] tested the efficacy of several data reduction procedures by applying them to a large number of analyses of known specimens by several laboratories, and presenting the respective error histograms. In a second study of this type Poole [12.11] included no less than 229 analyses. The main concern of these authors was with the atomic number correction, or generation calculation, which they thought to be the main responsible for the inaccuracies in the analyses. They demonstrated that all data handling schemes which were tested by them produced an improvement over the uncorrected relative intensities used as estimates of concentration, and that some procedures were more

Table 12.2. Composition of Typical Glass Standards for Microanalysis Produced at NBS.

	K249	K252	K309	K453	K456	K458	K408	K508	K521
Li_2O									0.07
B_2O_3								0.298	0.68
Al_2O_3	0.05		0.15						
SiO_2	0.30	0.40	0.40		0.287	0.493	0.30	0.258	
CaO			0.15						
MnO		0.05							
Fe_2O_3			0.15						
CoO		0.05							
CuO		0.05							
ZnO		0.10				0.038			
GeO_2				0.412					
SrO								0.444	
BaO	0.10	0.35	0.15			0.468			
CeO_2									0.25
Eu_2O_3							0.05		
Ta_2O_5	0.10								
PbO	0.425			0.587	0.712		0.65		
Bi_2O_3	0.025								

Note: All concentrations are weight fractions.

effective than others. However, regardless of the correction procedure, there remained a distressingly wide error distribution, and some analyses fell very wide from the goal (more than 10%) in a pattern that was difficult to explain by any statistical distribution. Heinrich subjected the results of the first publication by Poole and Thomas to a critical analysis, and presented strong evidence that the most serious errors in the analyses were not due to failure of the atomic number correction, but to poor characterization or homogeneity of specimens, to excessive absorption losses of the analyzed X-rays, and to the omission of fluorescence corrections [8.6].

There have been frequent attempts to improve the accuracy of microprobe analysis by adjustment of a single correction model, e.g., for stopping power, or by using a new set of mass absorption coefficients. It is, however, a characteristic of microprobe analysis that its accuracy is limited by a fairly large number of factors, ranging from instrumental parameters to theoretical considerations. Progress in the last ten years in achievable accuracy is mainly due to a systematic study of all sources of errors, and to the application of error propagation formula to various parameters of importance in electron probe microanalysis.

Classification of Errors in Electron Probe Analysis. If a measurement on a specimen, after application of data reduction procedures, yields an unexpected

Table 12.3. Analysis of Mineral Glasses.

K-411	NOMINAL COMPOSITION	WET CHEMICAL LAB A	WET CHEMICAL LAB B	ELECTRON MICROPROBE (1)	ELECTRON MICROPROBE (2)	ELECTRON MICROPROBE (3)
SiO_2	55	54.28	54.36	54.6[c]	55.6[a]	54.5[b]
FeO	15	14.49	14.34	14.7[b]	14.4[b]	14.4[b]
MgO	15	14.64	14.69	15.1[a]	15.0[a]	15.3[c]
CaO	15	15.53	15.41	15.5[a]	15.4[a]	15.6[b]
K-412						
SiO_2	45	45.38	45.32	45.2[c]	46.0[a]	45.2[b]
FeO	10	10.10	9.82	10.1[b]	9.88[b]	9.88[b]
MgO	20	19.33	19.32	19.6[a]	19.6[a]	19.9[c]
CaO	15	15.29	15.21	15.5[a]	15.3[a]	15.5[b]
Al_2O_3	10	9.26	9.28	9.26[c]	9.24[c]	9.54[b]

Standards: [a]Diopside; [b]Juan de Fuca, basaltic glass; [c]Enal 10, synthetic glass ($MgSiO_3$ + 10% Al_2O_3) ($CaMgSi_2O_6$).

Concentrations are weight percents of oxides. The nominal compositions were obtained from weighing the components before fusion. All electron probe analyses were made by the same operator at 15 kV; the X-ray emergence angle was 52.5° and the data reduction program was FRAME. Standards were used as indicated by superscripts.

result, the analyst should first check if a *mistake* has not been committed in one of the many operations needed to perform an analysis. A mistake is a faulty procedure such as the setting of the spectrometer on a wrong wavelength, the pushing of a wrong switch, use of faulty computer input cards, omission of an important correction, etc. Mistakes should not be considered in the evaluation of the accuracy of a procedure, although, as will be seen in the next chapter, pro-

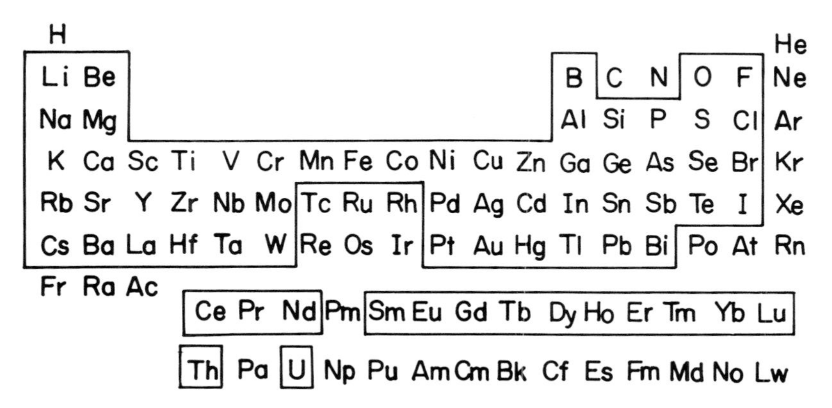

Fig. 12.10. Elements which were incorporated in experimental glasses for electron probe microanalysis and ion probe microanalysis at the National Bureau of Standards.

grams for data evaluation should be arranged so as to minimize the possibility of mistakes, and to produce warnings when certain mistakes have been committed.

If a mistake has not been found, the experiment should be repeated under unchanged conditions. The result of the repetition will not, in general, be exactly the same as in the first experiment. After repetition or repetitions, the *random error* of the process can be estimated, and expressed by *statistics* such as the standard deviation (see the Appendix). If the inaccuracy of the analysis is entirely due to statistical variations, these should be compared with the prediction of variation due to Poissonian counting statistics (Section 6.2).

An error which persists on repetition, or changes in an orderly fashion as a function of time, is called *systematic*. In particular, errors which change in a recognizable manner with time (either periodically or progressively), are attributed to *drift* of some parameter or parameters related to the measurement (Section 6.6).

Random errors and drifts are usually attributable to the instrument. If the level of random error is that expected from Poisson statistics, then the corresponding errors cannot be reduced except by increasing the number of pulses collected per determination. Unless a stronger X-ray line is available, this requires increasing the counting time per measurement, the beam current, or the operating potential. The first solution may increase drift effects, and be uneconomical if many measurements must be performed. The second solution may increase contamination, heating, decomposition, or electrostatic charging of the target, as well as produce a beam of larger diameter. The third alternative, increase in operating potential, reduces the spatial resolution and may cause excessive error due to X-ray absorption within the target. If the random error significantly exceeds that expected from counting statistics, several sources of instability may be suspected. Detector noise can be observed on an oscilloscope representation of the amplified detector output, or confirmed by the observation of signal bursts on a ratemeter output of low time constant. Fluctuations of the beam intensity can be recognized by simultaneous changes in X-ray signals from different detectors, as well as in the target current. Instability due to electrostatic charging of the specimen or of some object in the column frequently produces a characteristic signal pattern of the target current, which gradually diminishes until after a discharge it suddenly returns to the normal level (Fig. 12.11). It is recommended that the operator produce ratemeter tracings of X-ray signals and of target current, during routine analyses, and that the amplified pulses of the X-ray detector be displayed during operation; this will make an early detection of such problems more probable. Interfering noise patterns can be produced from a variety of external sources, including elevators, emission sparks from light-optical spectrometers, and induction furnaces. The more sporadic the noise source, the more exasperating is its investigation. Care in grounding procedures, filtering the incoming power lines, and shielding of sensitive parts may assist in eliminating external sources of noise.

12.3.1 Propagation of Errors [12.12]

The significance of errors, random or systematic, in parameters which affect the analytical accuracy [12.13], depends on the degree to which they propagate into the final result. We must therefore, in each case, study the propagation of errors through analytical models [12.14]. Since this task only requires an approximate estimate of the effect of an error or uncertainty in a parameter, the simplification of correction models is, for this purpose, admissible and even desirable.

In the case of systematic errors, the propagation of errors is estimated by the equation:

$$\Delta f(x) \simeq \Delta x \cdot \frac{df(x)}{dx} \tag{12.3.1}$$

in which Δx and $\Delta f(x)$ are small changes in the variable x and in the function $f(x)$, and $df(x)/dx$ is the first derivative of the function with respect to x. Consider, for instance, the effect of an error in the composition of the standard, ΔC_s. Assuming a linear calibration curve, and neglecting the effects of background, we obtain the function

Fig. 12.11. Periodic shrinkage of X-ray production due to interruption of path of the target current to ground, starting at point A, and sudden increases whenever the specimen discharges by arcing.

$$C_* = C_s \cdot \frac{Y_*}{Y_s}$$

in which Y_* and Y_s are the count rates for specimen and standard, and C_* and C_s are the corresponding concentrations of the element of interest. If there is an error in C_s, called ΔC_s, we obtain:

$$\Delta C_* = \Delta C_s \cdot \frac{dC_*}{dC_s} = \Delta C_s \cdot \frac{Y_*}{Y_s} \quad \text{or} \quad \frac{\Delta C_*}{C_*} = \frac{\Delta C_s}{C_s}. \tag{12.3.2}$$

In analogous fashion, we obtain the effect of an error in the deadtime, τ, upon the measured count rate, Y, according to Eq. (5.3.18):

$$Y = Y'(1 - Y'\tau)^{-1}$$

$$\Delta Y = Y'(1 - Y'\tau)^{-2}(-Y')\Delta\tau \quad \text{or} \quad \frac{\Delta Y}{Y} = -Y\Delta\tau. \tag{12.3.3}$$

If one or more independent parameters of a function have a known statistical distribution, the variance of the function can be obtained by Eq. (A.14)

$$\sigma_f^2 = \left(\frac{\partial f}{\partial x}\right)^2 \sigma_x^2 + \left(\frac{\partial f}{\partial y}\right)^2 \sigma_y^2 + \cdots$$

(see the Appendix). If, for instance, the ratio of two X-ray intensities, $k = Y_*/Y_s$, is determined, the variance of this ratio depends on the variances of the count rates Y_* and Y_s as follows:

$$\sigma_k^2 = Y_s^{-2}\sigma_{Y_*}^2 + Y_*^2\sigma_{Y_s}^2$$

or

$$\frac{\sigma_k^2}{k^2} = \frac{\sigma_{Y*}^2}{Y_*^2} + \frac{\sigma_{Y_s}^2}{Y_s^2}. \tag{12.3.4}$$

If Y_* and Y_s were determined by the respective pulse counts of N_* and N_s photons, obtained in the counting time t, the theoretical counting error due to Poisson statistics (see the Appendix) can be estimated: since $t \cdot Y_* = N_*$, and $t \cdot Y_s = N_s$, $\sigma_{Y_*} = N_*^{1/2}/t$ and $\sigma_{Y_s} = N_s^{1/2}/t$ so that

$$\frac{\sigma_k^2}{k^2} = \frac{1}{N_*} + \frac{1}{N_s}, \qquad \sigma_k = k\sqrt{1/N_* + 1/N_s}.$$

The effect of errors in τ upon the intensity ratio k is obtained similarly:

$$k = \frac{Y_*}{Y_s} = \frac{Y'_*(1 - Y'_s\tau)}{Y'_s(1 - Y'_*\tau)} \simeq \frac{Y'_*}{Y'_s}\,[1 + (Y'_* - Y'_s)\tau]$$

$$\frac{\Delta k}{k} \simeq (Y_* - Y_s)\Delta\tau. \tag{12.3.5}$$

We notice that the effect of an error in the parameter τ vanishes as the standard becomes identical to the specimen, so that Y_* is equal to Y_s, regardless of the value of τ.

In trace analysis, the accuracy of the result is determined mainly by the measurement of the trace signal and of the respective background signal. The error in the standard (assuming that a pure element was used) can be neglected. If the factor α of the hyperbolic equation, Eq. (8.2.3), is known, we obtain:

$$k = \frac{Y_* - Y_{B*}}{Y_s - Y_{Bs}} \simeq \frac{N_* - N_B}{N_s}\,; \quad \sigma_k^2 = \frac{N_* + N_B}{N_s^2}$$

and as $C \to 0$, $k \to 0$ hence, from Eq. (8.2.4), $C \to \alpha k$ so that

$$\frac{\sigma_C}{C} \simeq \frac{\sigma_k}{k} = \frac{\sqrt{N_* + N_B}}{N_* - N_B}\,. \tag{12.3.6}$$

We will obtain similar estimates as we discuss other sources of error in electron probe microanalysis. Equation (A.14), which provides a means of simultaneously considering the effects of errors in several parameters, cannot be used for systematic errors, and an equivalent expression cannot be given. It is best to consider each error individually. The total combined error must be equal to or smaller than the sum of the absolute values of the individual errors [12.15].

12.3.2 Sources of Errors

A variety of possible sources of drift were discussed in Section 6.6. Drift due to alteration or contamination of the specimen can be identified by the return to the initial signal level when the specimen or the beam is moved to a previously unexposed spot of the same composition. Contamination is particularly important in the analysis with X-ray lines of long wavelength. When problems of specimen decomposition or contamination become significant, it is advisable to make the irradiated spot as large as possible, within the restrictions due to loss of spatial resolution and spectrometer defocusing. Sometimes it may become necessary to repeatedly change the analyzed location on the specimen, and to extrapolate the ratemeter output to the initial point of irradiation, or to move the specimen, slowly but continuously, under the beam (e.g., in the analysis of glasses).

If the measurements are stable at each point of a standard, but vary from point to point, the standard is inhomogeneous. This defect is often a significant source of error in natural mineral standards. If present, it cannot be safely eliminated by averaging, since there is always a suspicion that the used region may not be representative of the material which was originally analyzed chemically.

If virtually identical signals are obtained from various points of the specimen or standard, the principal remaining sources of error in the measurement are coincidence losses in the detector system, and faulty background corrections. The first possibility must be considered at high count rates. Although the analyst should not exceed the count rates for which effective dead-time corrections can be made, on the basis of previous calibrations, he must also consider that erratic behavior at high count rates may occur if the mean height of the amplified pulses is incorrect, with respect to the discriminator level or the single-channel analyzer window settings.

The errors related to background determination have been discussed, in a general way, in Section 6.4. When crystal spectrometers are used, the problems which may arise are seldom serious, except for trace determinations (i.e., in electron probe microanalysis, for concentrations less than 0.1%). Particularly in the case of minerals the argument is valid that the background determination is simple if standards of composition close to that of the specimen, and not containing a measurable concentration of the element in question, are available. The absence in the background standard of the element to be measured must be proven, particularly if the standard is a natural product. The background problems are much more severe when solid-state detectors are used in the energy-dispersive mode. This situation will be discussed in detail in Section 12.6.

Errors Due to Chemical Shift. As mentioned in Section 4.2.2, the positions and shapes of lines at wavelengths of 4Å or longer can change observably as a consequence of differences in chemical binding and coordination. The significance of these shifts is related to the resolution which curved-crystal spectrometers have for this spectral region, and chemical shifts are not observed, of course, when energy-dispersive detector devices are used. As can be seen in Fig. 5.20, the resolution of crystal spectrometers is often much narrower than the width of the emission bands which in this region replace the lines. Therefore, the spectrometer never receives the full width of the emitted band, and the definition of the relative intensity becomes dependent on the bandwidth of the spectrometer. Integral measurement of the bands, instead of measurement at the fixed wavelength emitting the highest intensity, has been proposed; however, even with integral measurements, large chemical effects may still affect the accuracy of the determination of relative intensities. The consequences of chemical shift were illustrated in Table 12.1, which contains the results of analyses of aluminum oxide, with the use of metallic aluminum as standard. Two different crystals were used, and in each case the spectrometer was aligned for maximum signal, once on the metal and then on the oxide. The effects of chemical shift enhance in each case the measurement on the target on which the spectrometer was aligned. However, the effect is much stronger when the high-resolution crystal (ammonium dihydrogen phosphate) is used, than with the rubidium acid phthalate crystal, of poor resolution for the aluminum Kα-line. The data reduction procedure was by the program FRAME (see Section 12.5). The surest way to minimize such chemical shifts is to use standards close in composition to the specimen when lines of long wavelength are measured.

Spectrometer Defocusing. Serious errors can be committed in the use of crystal spectrometers if the emitting region is displaced accidentally from the focusing position. For this reason, beam deflection should normally not be used in quantitative applications except with nonfocusing energy-dispersive spectrometers.

There is also, in instruments with crystal spectrometers, the possibility of defocusing due to accidental changes in elevation of the specimen, particularly if the spectrometer is poorly aligned. Such a possibility may be tested for by observing the signal intensity after the specimen has been changed in elevation and carefully repositioned, with the aid of the optical microscope. Well-polished specimens may show very little detail in the light microscope, so that errors in the focusing of the optical image are possible.

12.3.3 The Accuracy of the Empirical Method

Next we will discuss errors in the evaluation of relative X-ray intensities. We must separately consider the empirical and the theoretical procedures.

In the empirical procedures (by calibration curves or by the hyperbolic approximation), the calibration is usually provided by means of multielement standards, the composition of which is supposed to be known. The difficulties arising in the manufacture and characterization of such standards were discussed in the preceding section.

The hyperbolic approximation method requires the determination of the parameter α, for a binary system, by means of the measurement of the relative X-ray intensity obtained from at least one known standard. If both the concentration, C_s, and the relative intensity, k_s, have been obtained, the parameter α is calculated from Eq. (8.2.4). If the presumed composition of the standard is in error by ΔC_s, the corresponding relative error in α, according to Eq. (12.3.1), is

$$\frac{\Delta\alpha}{\alpha} = \frac{1}{1 - C_s} \circ \frac{\Delta C_s}{C_s} . \tag{12.3.7}$$

Conversely, if α is affected by an error, $\Delta\alpha$, the corresponding error in the mass fraction C of the element in question in the specimen, is

$$\frac{\Delta C}{C} = (1 - C)\frac{\Delta\alpha}{\alpha}. \tag{12.3.8}$$

For low concentrations, this reduces to

$$\frac{\Delta C}{C} = \frac{\Delta\alpha}{\alpha}.$$

Combining Eqs. (12.3.7) and (12.3.8), we obtain

$$\frac{\Delta C}{C} = \frac{1 - C}{1 - C_s} \cdot \frac{\Delta C_s}{C_s}. \tag{12.3.9}$$

Equations (12.3.7) and (12.3.9) indicate that for a good definition of the calibration curve, C_s should not be too large; however, very small values for C_s are also undesirable, since they are conducive to large errors in the ratio $\Delta C_s/C_s$. If the standard is of composition nearly equal to that of the specimen, the relative errors in the analysis are proportional to those made in determining C_s.

If the errors in the characterization of the standard are negligible, the main source of error to consider is in the fit of the hyperbolic approximation, which, as discussed in Section 8.1, is imperfect in the presence of large atomic number differences and fluorescent contributions (Fig. 12.12). The resulting error increases with the composition difference between specimen and standard. In most systems, however, the fit is within the limits of precision of routine measurements. The results of a large number of measurements on standard reference materials in the systems gold-silver and gold-copper performed at the National Bureau of Standards [12.3] fail to indicate a single case in which a significant deviation from the hyperbolic model is observed. (See, however, Fig. 11.11, for failure of the model in case of strong fluorescent excitation.)

When the coefficient α for each element is determined, not by measurement of the emission of a real standard, but by calculating from theory the intensity which would be emitted from a hypothetical standard, then the error in C_s is zero, but the errors and uncertainties of the theoretical procedure enter in determining the error of the coefficient α. In addition, if the hypothetical standard differs significantly in composition from the specimen, deviations from the hyperbolic approximation also codetermine the final error of the analysis. Therefore, if the necessary computation devices are available, it is preferable to use the theoretical procedure, with iterations as described in Section 12.1.

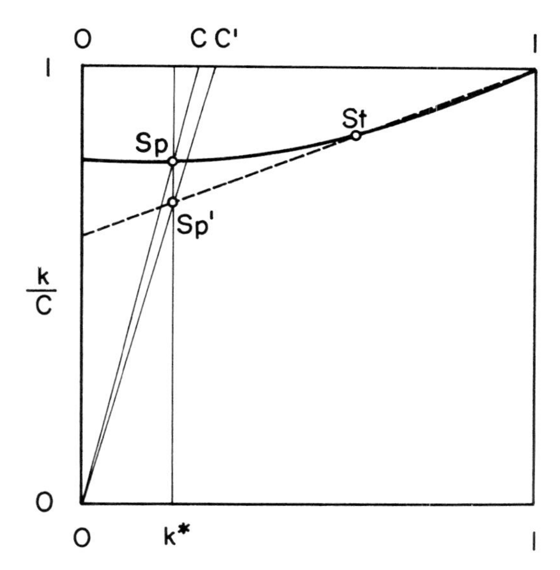

Fig. 12.12. According to the hyperbolic approximation, the measurement of the standard (St) generates an estimate of the analytical curve (hatched line). If the true analytical curve is instead the continuous line, for a specimen having the relative X-ray intensity k^*, the calculated intensity, C', will differ from the true value, C. The difference increases with the distance from the point Sp, representing the specimen, to St, which represents the standard.

12.3.4 The Accuracy of the Theoretical Correction Method

Equation (9.1.3) indicates that the observed intensities from specimen and standard are the sums of several components, each produced by a different mechanism of excitation. For each of these components, propagation of error indicates that

$$\frac{\Delta k}{k} = \frac{\Delta I'_n}{I'_n + \Sigma I'_i} \; . \tag{12.3.10}$$

In this equation, I'_n represents one of the components (after attenuation on emerging from the specimen) and $\Sigma I'_i$ is the sum of all other emerging components. The emitted primary radiation is much more intense than the other components, so that relative errors in the estimate of the primary radiation cause relative errors of almost the same magnitude in the estimation of the relative X-ray intensity, k. The situation is different with fluorescent excitation; the fluorescent intensity is not larger than 5% of the primary radiation so that the relative error caused by it in the value of k will be 20 or more times smaller than the error in the estimate of fluorescence emission which has caused it. For this reason, the accuracy requirements for the estimate of primary radiation are particularly stringent.

Primary X-Ray Generation [9.25]. Insofar as primary X-ray generation is concerned, parameters which are not composition dependent, such as ω_q, p_{qm}, b, and z_q [Eq. (9.2.25)] appear in the calculation of intensities from both the specimen and the standard and therefore cancel in the intensity ratio. The uncertainties in the primary generation are thus mainly related to the two parameters R^* and J—the backscatter correction factor and the mean excitation energy—which vary strongly with the target composition.

The uncertainties in R^* are quite significant, as shown in Fig. 9.15. According to Eq. (9.2.27), the primary generated intensity is proportional to this parameter. However, the backscatter correction factor varies monotonically with the atomic weight of the target; therefore, the uncertainties in R will compensate if the specimen and the target are of similar atomic number. Whenever this is not the case, the uncertainty in R is a significant limitation to the accuracy of electron probe microanalysis. To judge from the graph of Fig. 9.13, this uncertainty may produce errors of 2 to 3% in the analysis when the difference of atomic number between standard and specimen is large. From Eqs. (9.2.38) and (12.3.1) we can derive the propagation of errors in the factor R:

$$\frac{\Delta C}{C} = \frac{\Delta R^s}{R^s} - \frac{\Delta R^*}{R^*} . \tag{12.3.11}$$

It is, however difficult to separate the effect of R from that of the mean ionization energy, and sometimes also of the absorption and fluorescent corrections. Further measurements of the energy distribution of backscattered electrons will help to reduce this uncertainty.

According to Eq. (9.2.26) through (9.2.29), the uncertainty in the primary ionization term resides mainly in the assigned values of J_i; other variables which may be affected with uncertainties cancel between the specimen and the standard. The considerable uncertainties in J_i, and the controversial proposal of Duncumb and DaCasa, were discussed in Subsection 9.2.2. It should be noted that the considerable variation among proposed values of J_i extends over the entire range of atomic number. Furthermore, the meaning of J_i in Eq. (9.2.25) and following is somewhat clouded because the equations are based on the acceptance of Eqs. (9.2.2) and (9.2.11). There is considerable disagreement among experimental values for the ionization cross section, and it is doubtful that its variation with respect to atomic number is correctly represented by any of the equations in Subsection 9.2.3. It would be impossible to separate the effects of variations in Q_q from those of J, as far as the atomic number is concerned. Although, for each element, the effects of the ionization cross section largely cancel, consequences of the cross-section uncertainty may cause apparent variations in J. Therefore, the cross-section uncertainty deserves further investigation.

The study of the propagation of uncertainties in J_i is best started with an infinitely dilute solution of element a in element b. For the determination of the mass fraction of a, we can use a simplified expression [Eq. (9.2.4) and Eq. (9.2.39)]:

$$k_a = C_a \frac{R_b}{R_a} \cdot \frac{Z_a/A_a}{Z_b/A_b} \frac{\ln\left(\frac{\epsilon \bar{E}}{J_a}\right)}{\ln\left(\frac{\epsilon \bar{E}}{J_b}\right)} .$$

The effects of errors in J_a and J_b are:

$$\frac{\Delta C_a}{C_a} = \frac{\frac{\Delta J_a}{J_a}}{\ln\left(\frac{\epsilon \bar{E}}{J_a}\right)} \quad \text{and} \quad \frac{\Delta C_a}{C_a} = \frac{-\frac{\Delta J_b}{J_b}}{\ln\left(\frac{\epsilon \bar{E}}{J_b}\right)} . \tag{12.3.12}$$

The effects of the uncertainty in J_i are thus most significant when the atomic numbers of the elements in the target—and hence the differences in J_i—are largest, and when the mean energy of the electrons, $\bar{E}$, is smallest, i.e., at low operating voltages.

If we accept the hyperbolic approximation for the atomic number effect, we can extend the expressions for errors due to J_i to intermediate concentrations: since $\alpha = (C/k)_{C(0)}$ and $\Delta C/C = (1 - C)(\Delta\alpha/\alpha)$ we obtain

$$\frac{\Delta C}{C} = \frac{1 - C}{\ln\left(\frac{\epsilon \bar{E}}{J_a}\right)} \frac{\Delta J_a}{J_a}$$

$$\frac{\Delta C}{C} = \frac{-(1 - C)}{\ln\left(\frac{\epsilon \bar{E}}{J_b}\right)} \cdot \frac{\Delta J_b}{J_b} . \tag{12.3.13}$$

The consequences of using the simplification implied in Eq. (9.2.36), rather than the full integration over the electron energy, as in Eq. (9.2.16) and following, have been analyzed in [9.25], where it is shown that except for extreme cases, such as the determination of a trace of aluminum in gold, the errors caused by the simplification do not exceed 1%.

Further errors in the generation calculation could be caused by flaws in the proposed addition rules for stopping power and for the backscatter correction factor R [Eq. (9.3.7)], as well as by the failure of Bethe's law at low electron energies.

The Accuracy of the Primary Absorption Calculation [12.16]. By choosing the instrumental and operating conditions, the analyst can influence to a large degree the absorption of primary X-rays in the specimen. In the first place, the parameter $\chi = \mu \csc \psi$ depends on the X-ray emergence angle ψ. Although this angle is, in general, fixed for a given instrument, values chosen by builders of electron probe microanalyzers have, in the past, varied from 6 to 60°. For a given specimen, X-ray line and X-ray emergence angle, the value of the absorption factor f depends on that of γ, and hence of the operating voltage, V_0. In the first instruments built, the existing designs for objective lenses of short focal length dictated low values for ψ. Furthermore, the low efficiency of early X-ray spectrometers induced the operators to raise the operating voltages as high as the requirements of spatial resolution, or the instrument, would permit (typically 30 kV). However, as shown by Eq. (10.3.32), the losses through absorption in the target grow quickly with increasing operating voltage. Therefore, absorption losses were often large, and caused serious analytical errors. In modern instruments, X-ray emergence angles below 30° are no longer used. Errors in the analytical result can arise from the absorption calculations, either because of failure of the model for the absorption function, or because of errors in the values of the parameters used in the function, i.e., errors in μ, ψ, or V_0.

Assume, for simplicity, that the atomic number correction is negligible (Castaing's first approximation). For an elementary standard, we have:

$$C^* = \frac{I_p{}^*}{I_p{}^s}.$$

We can also neglect secondary radiation and we obtain

$$k = C^* \frac{f_p{}^*}{f_p{}^s} \quad \text{or} \quad C^* = k \frac{f_p{}^s}{f_p{}^*}.$$

It follows that the uncertainties in C^* are related to those in $f_p{}^*$ and $f_p{}^s$ by

$$\Delta C^* = \frac{-k f_p{}^s}{(f_p{}^*)^2} \Delta f_p{}^* \quad \text{or} \quad \frac{\Delta C^*}{C^*} = -\frac{\Delta f_p{}^*}{f_p{}^*} \tag{12.3.14}$$

and

$$\Delta C^* = \frac{k}{f_p{}^*} \Delta f_p{}^s \quad \text{or} \quad \frac{\Delta C^*}{C^*} = \frac{\Delta f_p{}^s}{f_p{}^s}. \tag{12.3.15}$$

Relative errors in $f_p{}^*$ or $f_p{}^s$ thus produce a relative error of the same magnitude in the estimate of the concentration C^*.

If for the absorption function we use Eq. (10.3.34):

$$f = (1 + a_1\gamma\chi)^{-1} = [1 + a_1(V_o{}^n - V_q{}^n)\mu \csc \psi]^{-1} \tag{12.3.16}$$

we obtain, for the dependence of f_p upon uncertainties in μ, ψ, and V_0, the following:

$$\frac{\Delta f}{f} = (1 - f)\frac{\Delta \mu}{\mu}, \tag{12.3.17}$$

$$\frac{\Delta f}{f} = -(1 - f) \operatorname{ctg} \psi \Delta \psi , \tag{12.3.18}$$

and

$$\frac{\Delta f}{f} = (1 - f)^n \frac{U^n}{U^n - 1} \cdot \frac{\Delta V_0}{V_0} . \tag{12.3.19}$$

Of these relations, the first is the most important because the mass absorption coefficients are subject to large uncertainties, particularly for long wavelengths, at which the absorption is significant. If one assumes that errors in μ can be as large as 5%, and if one wishes to determine f_p with an uncertainty not exceeding 1%, one obtains:

$$0.01 \geqslant 0.05(1 - f), \text{ so that } f \geqslant 0.8. \tag{12.3.20}$$

The requirement that the absorption factor f_p should be equal to or larger than 0.8 seems somewhat severe, since partial compensation usually occurs between the errors in μ^* and μ^s, which have effects of opposite sign. This compensation increases with the similarity in composition of specimen and standard. From this viewpoint, the use of elements as standards is undesirable when the absorption is strong, since the absorption coefficients for the emitting elements are usually much lower than for the multielement specimen. In spite of possible compensation, it is good practice to set a limit $f \geqslant 0.6$ to assure accuracy in quantitation.

To reduce the absorption in the target of soft X-rays, it is advisable to use instruments with high X-ray emergence angles, and to operate at low accelerating voltages. Figure 12.13 illustrates the relation between μ, ψ, and χ, and indicates the highest operating voltages at which reasonably high values ($\geqslant 0.6$) of f_p are obtained. However, according to Eq. (9.2.32) the intensity of the generation of X-ray lines increases with the operating voltage (see Fig. 6.19). To achieve adequate signal intensities for soft lines (e.g., in the analysis of silicate minerals), one must therefore use high X-ray emergence angles and spectrometers of high efficiency.

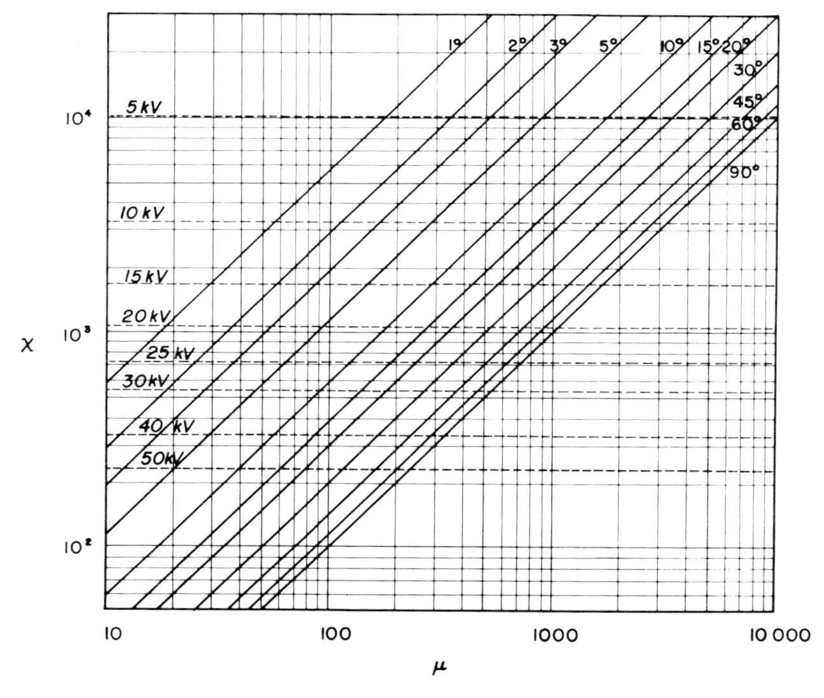

Fig. 12.13. Relationship of the mass absorption coefficient, μ, the absorption parameter, χ, and the X-ray emergence angle, ψ (indicated in degrees). The horizontal dashed lines indicate the maximum operating voltage at which the corresponding value of χ will produce an absorption factor larger than 0.7, according to the equation

$$1/f = 1 + 3 \times 10^{-6} V_0^{1.65} \chi.$$

For simplicity, the effect of the critical excitation potential is neglected here.

Errors in the estimate of the X-ray emergence angle can be caused by misalignment of the spectrometer or the specimen stage, or by local surface irregularities. We recall that a real spectrometer extends over a significant range of emergency angles; in crystal spectrometers not all parts of the crystal may diffract with equal efficiency, and therefore the assumed mean X-ray emergence angle may be in error. As Eq. (12.3.18) indicates, the effects of errors in ψ increase very rapidly with decreasing ψ. This effect is another good reason why instruments used for the measurement of X-rays of long wavelength should have a high ($\geqslant 30°$) X-ray emergence angle.

Errors due to the alignment of the spectrometer and the specimen stage may compensate at least partially, between specimen and standard, while those due to lack of flatness of the specimen are not compensated. In the analysis of very small regions, such as inclusions (Fig. 12.14), the disadvantages of low X-ray emergence angles become particularly significant. In such specimens, it is difficult to obtain flatness in the region of interest, due to differences in hardness of the phases. But it is also possible, with low emergence angels, that part of the radiation excited within an inclusion may exit through a region of different composition. This could cause errors in the absorption calculation.

Errors in the operating potential are, in general, less important. They may arise from faulty calibration of the voltmeter, or from failure to take into account the potential drop through the gun bias resistor [Eq. (3.2.12)]. Further errors may arise through passage of the electron beam through conductive surface coatings applied to nonconductive specimens. These are largely compensated if the specimen and the standard were coated simultaneously. Equation (10.3.31) would seem to indicate that the consequences of such errors are worst at low overvoltages; however, under such conditions the absorption is quite small. The overall effect of errors in V_o is almost independent of the overvoltage [12.16], but it increases with decreasing X-ray emergence angle.

Besides the uncertainties in input parameters, the accuracy of the absorption calculation is also affected by uncertainties in the model itself, as was discussed in Chapter 10. Diverse models of absorption calculation often lead to significant divergence in the result. The effect of the model is mainly represented, in Eq. (10.3.34), by the coefficient a_1 (which, in Eq. (10.3.39), is equal to $2a$). The effect of errors can be obtained by the equation

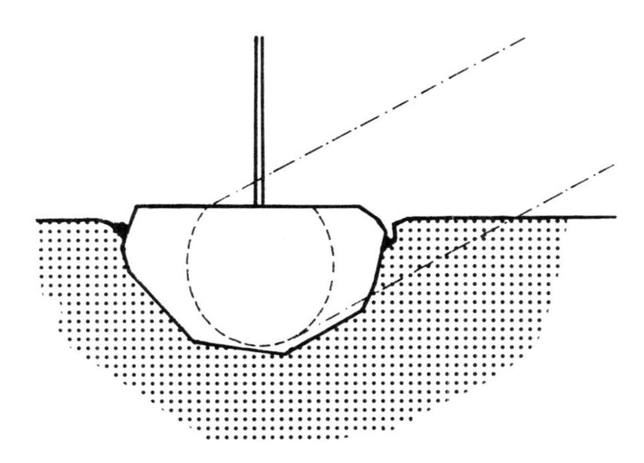

Fig. 12.14. If the phase to be analyzed is small, and if the X-ray emergence angle is small, the emergent X-rays may cross regions of composition different from that of the phase of interest. The surface flatness may also be affected and polishing artifacts may accumulate in holes. The accuracy of the absorption correction may suffer under such circumstances.

$$\frac{\Delta f}{f} = (1 - f)\frac{\Delta a_1}{a_1} . \qquad (12.3.21)$$

The uncertainties can again be minimized by using low operating potentials and high X-ray emergence angles. The most effective reduction of uncertainties is obtained, however, if well-characterized standards similar in composition to the specimen can be used.

The absorption of primary X-rays is the most serious potential source of errors in quantitative electron probe microanalysis, and it is one that is under the control of the discerning operator. The absorption calculation should therefore always be considered with the greatest care, particularly if the analysis involves elements which emit X-ray lines of long wavelength only ($Z \leqslant 25$).

Interesting conclusions can be derived from the consideration of the effect of errors in the factor a of the absorption function [Eq. (10.3.39)]. From

$$k' = \frac{I_{p*} f_{p*}}{I_{ps} f_{ps}} = \frac{I_{p*}}{I_{ps}} \cdot \frac{(1 + a\gamma\chi_*)^2}{(1 + a\gamma\chi_s)^2}$$

we obtain:

$$\frac{\Delta k'}{k'} = 2\left[\frac{\gamma\chi_*}{1 + a\gamma\chi_*} - \frac{\gamma\chi_s}{1 + a\gamma\chi_s}\right]\Delta a \simeq 2\gamma(\chi_* - \chi_s)\Delta a. \qquad (12.3.22)$$

The same result is obtained more easily if the simplified model

$$f_p = (1 + 2a\gamma\chi)^{-1}$$

is used. The consequences of error in the model can be reduced in three ways: by reducing the error of the coefficient $2a$, by reducing the factors γ and χ, and by reducing the difference $\chi_* - \chi_s$. The first approach is equivalent to improving the model, the second approach is equivalent to working at low operating voltages and high X-ray emergence angles, and to using lines of short wavelength, and the third approach means using a standard as similar as possible to the specimen. Analogous solutions are found for all systematic but unknown errors, and the three approaches mentioned above, or a combination thereof, improve the accuracy in most cases.

Fluorescence Due to Characteristic Lines. Since the relative fluorescent emission due to a characteristic line, $I'_f/I'_p = r_f$, is rarely larger than 5%, fairly large errors in I'_f can be tolerated. The error propagation is as follows:

$$\frac{\Delta C}{C} \simeq \frac{\Delta k}{k} = \frac{\Delta I'_f}{\Sigma I'} = \Delta r_f \cdot \frac{I'_p}{\Sigma I'} \simeq \Delta r_f. \qquad (12.3.23)$$

To observe the effect of the more important parameters on the accuracy of the fluorescence calculation, it is best to use a simplified model such as Reed's [Eq. (11.1.43)] or Castaing's [Eq. (11.1.42)]. The subject is treated in detail in [12.17]. We obtain:

$$\frac{\Delta C_a}{C_a} = r_f \frac{\Delta \omega}{\omega} . \qquad (12.3.24)$$

In the rigorous version of the calculation, [Eq. (11.1.36)], the fluorescent yield is that of the element a, which emits the secondary radiation, while in the versions of Reed and Castaing, the fluorescent yield of the exciting element, b, is used.

$$\frac{\Delta C_a}{C_a} = r_f \frac{\Delta C_b}{C_b}, \quad \frac{\Delta C_a}{C_a} = r_f \frac{\Delta r_{qa}}{r_{qa}^2}, \quad \frac{\Delta C_a}{C_a} = \frac{\Delta p_{ij}}{p_{ij}} . \qquad (12.3.25)$$

The effects of other parameters, such as the mass absorption coefficients, the emergence angle, and voltage are of less importance. Their role is qualitatively discussed in Section 11.1, and their effects on the analytical error are shown graphically in [12.17]. Errors up to 20% may arise in some cases from omission of the fluorescence correction [8.6]; the next largest source of error is probably the simplification of the calculation (Fig. 11.11). The choice of primary depth distribution in the factor ν of the absorption part of the correction has been the subject of some discussion, but it has little effect on the accuracy of analysis. It is not, however, admissible in accurate analysis to ignore the fluorescence above the depth of primary excitation [i.e., the term $F(V)$ of Eq. (11.1.41)]. The consequences of omitting the fluorescence due to the continuum are discussed in Section 12.5 (see Fig. 11.15 and Table 11.4).

Selection of the Accelerating Potential. Of all parameters which can be changed by the operator, the accelerating potential is the most important. The value of this parameter affects several characteristics of the analysis:

1. The intensity of the generated characteristic radiation increases with the overvoltage [Eqs. (9.2.24) through (9.2.29)]. The continuous radiation also increases in intensity as the potential E_0 increases. According to Kramers' law [Eq. (6.3.7)]

$$I(E)dE = kZ(E_0/E - 1)dE.$$

2. The depth of penetration also increases with the accelerating potential. This increase in penetration has three important consequences:

   a. The absorption of primary characteristic X-rays increases with increasing potential; hence, the uncertainty in the absorption also increases.

b. Due to the increased energy of the backscattered electrons, the backscatter correction factor, R, decreases.

c. Both the depth of penetration and the lateral diffusion increase. Therefore, the spatial resolution of the analysis deteriorates.

In the earlier days of microprobe analysis, when the efficiency of the X-ray spectrographs was low, it was usual to perform analyses at energies as high as 30 keV. However, as efficient spectrometers—particularly for soft radiation—became available, the advantages of softer lines and lower acceleration voltages became apparent. For instance, the line-to-background ratios for K-lines of elements of atomic number above 30 are considerably lower than those of the corresponding L-lines. Moreover, the analysis by means of hard K-radiation is complicated by the strong emission of fluorescent radiation excited by the continuum. These effects can be greatly reduced by using the L-lines, rather than the K-lines, of elements of atomic number above 30. However, the absorption uncertainty in the measurement of soft X-rays is high unless low excitation potentials are employed. As was pointed out, it is advantageous to lower the excitation potential until the absorption correction factor is higher than 0.6.

At which potential should one analyze a specimen which contains iron and magnesium? The wavelength tables indicate that only line (Mg Kα) can be used for magnesium. Since this is a line of long wavelength, the excitation potential should not be excessively high. However, the line commonly used for measuring the emission of iron (Fe Kα), is produced by ionization of the Fe K-shell, and the corresponding critical excitation potential is 7.11 keV. The best choice would be to determine magnesium at a potential below 10 keV, and iron at perhaps twice this potential. This, however, makes a simultaneous determination impossible. With efficient spectrometers, and high X-ray emergency angles, voltages between 12 and 15 keV will produce a sufficiently intense iron emission, without a large absorption uncertainty for the Mg Kα radiation. Alternatively, the iron could be determined by using the Fe Lα-line. This line, however, is very soft (13.6 Å) and may be affected by chemical shifts. Besides, mass attenuations at wavelengths longer than 10 Å are poorly known, and absorption uncertainties may be high. Therefore, the use of the Fe Lα-line is not recommended for general purposes.

At low overvoltages ($U_0 < 1.5$) the accuracy of the generation equation diminishes, and uncertainties in the values of the mean ionization potential become important. The risks of analysis at low voltages are, however, frequently overestimated. Our experience indicates that analyses of alloys can be performed at potentials below 10 keV, without risk of excessive errors in the generation equation, if lines of sufficient intensity are emitted by the elements of interest.

12.4 THE DETERMINATION OF ELEMENTS OF LOW ATOMIC NUMBER

The development of soft-X-ray spectrometers for the measurement of X-ray emissions from the elements of atomic number between three and nine has prompted efforts to apply to the quantitative determination of these elements the procedures described in Chapters 8 through 12. To date, the success in this effort has been moderate. We will discuss here some of the difficulties that arise in the use of long-wave X-ray emission for quantitation.

The accuracy of quantitation is, for the above-mentioned elements, limited by the experimental aspects of the measurement of the X-ray emission, by uncertainties in the models for correction, particularly in the absorption correction, and by limitations in our knowledge of the parameters which are required for the data reduction calculations.

It is frequently stated that a major difficulty resides in the fact that the emission of X-rays of elements having such low atomic numbers is exceptionally weak, due to their low fluorescence yield. A calculation of the expected generated primary intensities (Fig. 11.4), or the observation of energy-dispersive spectra in the soft X-ray region with a detector of reasonable efficiency clearly show the fallacy of this argument. Under appropriate conditions, the X-ray emission in this wavelength region is, in fact, quite high because of high overvoltages and low mean excitation potentials. Low observed X-ray intensities, particularly in older instruments, are mainly due to the low efficiency of the spectrometers used in the measurement.

A limitation arises, however, from the fact that the characteristic spectra of elements four to nine consist essentially of a single band, the width of which exceeds the resolution of conventional crystal spectrometers (see Fig. 5.20). Therefore, at any given wavelength position, the crystal spectrometer detects only a fraction of the total emission. Interferences of lines in higher orders of reflection are frequent (see Fig. 7.1) but can be minimized by pulse height selection. Interference by L-lines and bands in the first order, however, produces serious problems since no alternative to the K-bands of the low atomic number elements is available.

The situation is further complicated by the chemical shifts which affect the position, shape, and integral intensities of both the K-bands and the interfering bands of higher shells. There are even differences in the X-ray emission from allotropic forms of elements such as carbon, so that there is an uncertainty related to the choice of an elementary standard; for instance, the emission from diamond differs from the emission from graphite [12.18], [12.19]. Clearly, in a case in which the position and strength of both the analytical band and its interfering band are subject to such changes, we are faced with an almost insoluble problem.

The most serious challenges to quantitation of these elements arise, however, from the strong absorption of the long-wave X-rays within the excited target. As pointed out by Duncumb and Melford [12.20], in the case of very strong X-ray absorption the distribution in depth of X-rays which reach the specimen surface is limited by the absorption to a very shallow layer close to the surface, rather than being determined by the energy loss of the primary electrons. If we assume that in such a case the variations of the depth distribution and the energy loss of the electrons become insignificant as far as the X-ray emission is concerned, we obtain (Duncumb's "thin-film model"):

$$I = C\Phi(0)\,Q_q \int_0^\infty e^{-\chi z}\,dz = \frac{C\Phi(0)Q_q}{\chi}. \tag{12.4.1}$$

The relative X-ray intensity emitted from the specimen would thus be

$$k = C\frac{\Phi(0)^* \chi^s}{\Phi(0)^s \chi^*} \tag{12.4.2}$$

and the error produced in C by an error in χ^* is equal to

$$\frac{\Delta C}{C} = -\frac{\Delta\chi^*}{\chi^*} = \frac{-\Delta\mu^*}{\mu^*} \tag{12.4.3}$$

Similarly, the respective relative errors in C are also proportional to those in $\Phi(0)^*$, $\Phi(0)^s$, and χ^s. We recall that $\Phi(0)$ is the ratio of the X-ray generation within a thin layer at the target surface to that within a layer of equal thickness and composition, but not supported by target material under this layer. It is thus *not* subjected to the scale normalization expressed in Eq. (10.1.11), and it expresses the contribution to X-ray generation at the surface by backscattered electrons. In the absence of backscattered electrons $\Phi(0)$ would be equal to one.

According to Bishop [12.21], the thin-film model is strictly valid only when f_p is smaller than 0.01, but it is a useful approximation in cases where χ/σ is larger than three, as suggested by Love et al. [12.22]. It is therefore possible to draw from Eqs. (12.4.1) through (12.4.3) the following important conclusions as to quantitation with ultrasoft ($10\text{ Å} \leqslant \lambda \leqslant 100\text{ Å}$) X-rays:

1. The relative errors of predicted relative intensities are nearly proportional to the relative errors in the mass absorption coefficients. Therefore, high accuracy in the knowledge of the mass absorption coefficients is crucial to quantitation. (Note that Eq. (12.4.3) is the limiting case of Eq. (12.3.17), with $f \to 0$.)
2. For extremely high X-ray absorption, when the "thin-film model" is strictly valid, the shape of the depth distribution of generated X-rays, $\phi(z)$, is irrele-

vant, and the atomic number correction, or generation calculation, is reduced to knowing the values of $\Phi(0)$. Since the electrons are assumed not to have lost appreciable energy within the layer which emits X-rays, the stopping power and ionization cross-section terms in the atomic number correction cancel.

Some information on the values of $\Phi(0)$ was obtained from the tracer experiments discussed in Chapter 10. Equation (10.3.24) proposed by Reuter [10.15] is also used to predict the values of this parameter:

$$\Phi(0) \simeq R_0 \simeq 1 + 2.8(1 - 0.9/U_0)\eta. \tag{12.4.4}$$

It is difficult to estimate the accuracy of this prediction, as would be desirable since according to Eq. (12.4.2), relative errors in $\Phi(0)$ produce relative errors of the same magnitude in the emitted X-ray intensity.

A new formula for $\Phi(0)$ was obtained with the aid of Monte Carlo calculations by Love et al. [12.22].

Because the X-ray emission is concentrated in a very shallow target layer, the surface conditions of the specimen such as flatness and cleanliness are very important for analysis with soft X-rays. If insulating specimens must be coated, the coating must be very thin, and the thicknesses of the coatings on specimen and standard must match exactly. In the determination of carbon, the presence of this element in contamination spots formed due to the action of the beam is particularly undesirable, and the formation of the contamination layers must be prevented by the measures mentioned in Chapter 3. Similarly, superficial alteration of specimens exposed to air and moisture may affect the investigation of oxygen. Conversely, the fact that X-ray information produced in the long-wavelength range, particularly at low acceleration potentials comes from shallow depths, can be used deliberately when composition close to the specimen surface is of interest [12.24].

In view of the importance of the availability of accurate mass absorption coefficients in the region from 10 to 100 Å, it is useful to compare published values such as those reproduced in Table 12.4; which represent a selection of mass absorption coefficients found for 0 Kα radiation. It should be noted that the values proposed by Henke et al. [12.25], [12.26], [4.34] and Lukirskii et al. [12.27] were obtained in investigations not connected with electron probe microanalysis. On the contrary, those values proposed by Shiraiwa and Fujino [12.28], Kohlhaas and Scheiding [12.29], Love et al. [12.22], and Ruste and Gantois [12.30] were obtained with the aide of X-ray measurements of emissions from known targets in the electron probe microanalyzer. Since in these cases it was necessary to assume models for the absorption function, there is an obvious danger of a circular argument in which the parameters found with one model are an empirical accommodation, useful with this model only.

Table 12.4. Mass Absorption Coefficients for 0 Kα.

ABSORBER	H 1	He 2	Li 3	Be 4	B 5	C 6	N 7	O 8	F 9	Ne 10	Na 11	Mg 12
Henke et al. (1957) [12.25]	170	660	1450	2965	6130	10730	16270	983	1301	2379	3100	4592
Lukirskii et al. (1964) [12.27]				6670								
Henke et al. (1967) [12.26]						12200	17200	1440	1700	2600		
Kohlhaas and Scheiding (1970) [12.29]				4100								
Shiraiwa and Fujino (1970) [12.28]				4110				1250				5400
Henke and Ebisu (1974) [4.34]	58.2	455	1602	3922	7416	12380	17310	1200	1688	2582	3645	5174
Love et al. (1974) [12.22]								1300	1850	2750	3630	4950
Ruste and Gantois (1975) [12.30]	62	461	1496	3450	6595	11197	17590	1250	1866	2670	3692	4963

ABSORBER	Al 13	Si 14	P 15	S 16	Cl 17	Ar 18	K 19	Ca 20	Si 21	Ti 22	V 23	Cr 24	Mn 25	Fe 26
Henke et al. (1957)	5310	8040	9470	11190	12450	13540	15820	18030	18480	19820				
Lukirskii et al. (1964)	7960									8540				
Henke et al. (1967)				13000		15900								
Kohlhaas and Scheiding (1970)	7000	9100								38400	47800	2700	4200	4000
Shiraiwa and Fujino (1970)	7000	9100				16700		19000		19500		3700	4200	4700
Henke and Ebisu (1974)	6715	8790	10540	13010	14790	16160	19370	22030	23540	22140	24250	3143	3468	4001
Love et al. (1974)	6230	8140	9820	12400	14300	16100	20500	24600	26800	23200	26000	2470	2960	3630
Ruste and Gantois (1975)	6515	8383	10600	13201	14880	17000	19000	20400	22000	22000	23000	3343	3801	4300

In any case, the values of Table 12.4 clearly indicate a large uncertainty in μ, which will propagate into the analytical result according to Eq. (12.4.3).

A similar situation with regard to the mass absorption coefficients for the carbon K-emission is illustrated in Fig. 12.15, on the basis of values reported by Kohlhaas and Scheiding [12.31], Fornwalt and Manzione [12.32], Duncumb and Melford [12.20], and Henke and Ebisu [4.34].

It must also be recognized that in most practical situations the severity of the X-ray absorption will be such that the conventional absorption models may be inaccurate, while the "thin-film model" of Duncumb is not yet fully valid. It is therefore important to search for adequate models in this region.

The simplest solution would be that of an empirical calibration. As Duerr and Ogilvie [12.33] have shown, it is possible to use a simple graphic calibration curve for carbon, up to 1.6 wt % in ferrous alloys containing up to 20 wt % nickel. Such procedures are very accurate and simple for routine analysis within a well-defined simple matrix. Unfortunately, the required large number of stand-

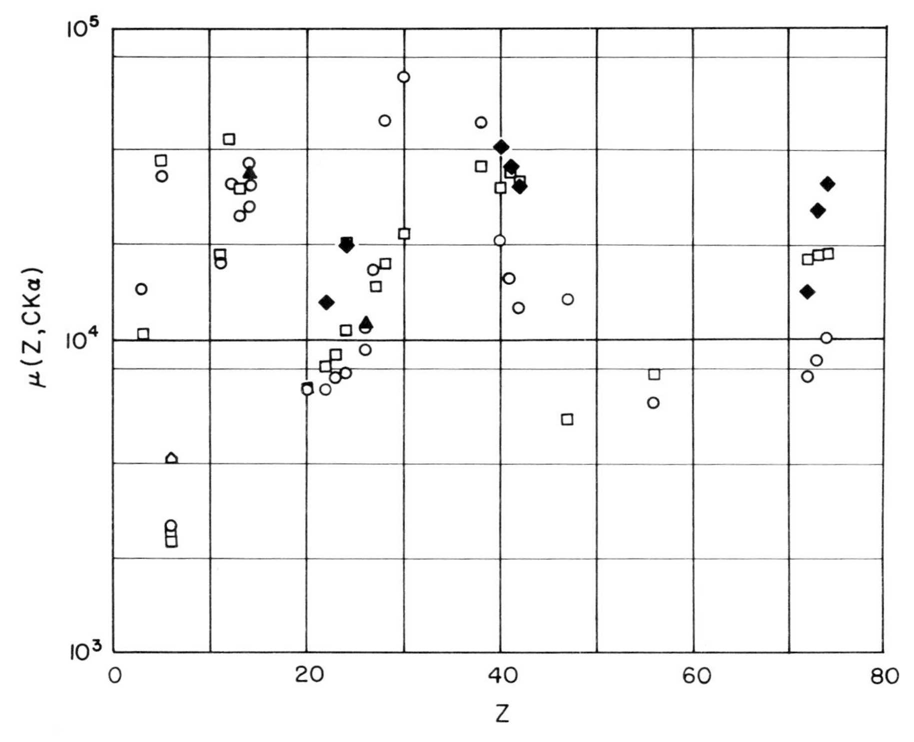

Fig. 12.15. Mass absorption coefficients for carbon K-emission. Triangles: Duncumb and Melford [12.20]; filled diamonds: Fornwalt and Manzione [12.32]; open circles: Kohlhaas and Scheiding [12.31]; open squares: Henke and Ebisu [4.34].

ards of known composition required to establish the calibration curve is seldom available.

Kohlhaas and Scheiding [12.29] as well as Shiraiwa and Fujino [12.28] modified or selected the parameters of the simple Philibert model [Eq. (10.3.23)], and adjusted the mass absorption coefficients so as to optimize the results of analyses of known materials. A comparison of this model with the "thin-film model" indicates that it is not adequate for strong X-ray absorption since at zero depth the value of $\phi(z)$ is zero. Hence the model fails in the region from which all of the soft X-rays are emitted. The same objection is valid against the quadratic model [Eq. (10.3.39)]. The importance of the model for strong absorption resides mainly in its giving the proper value at zero depth [i.e., $\Phi(0)$]. This point was made very effectively by Bishop [12.21] who showed that a simple rectangular distribution in depth of X-ray generation produced more accurate values of the absorption function than the abbreviated Philibert model (Fig. 12.16). The effectiveness of this model was verified by Love et al. [12.34]. At moderate absorption, however, this model may be in error by as much as 20%.

It follows from these considerations, that for high absorption a model such as Philibert's complete equation [Eq. (10.3.19)], the Wittry-Andersen-Kyser model [Eq. (10.3.57)], or a truncated quadratic model [Eqs. (10.3.49) or (10.3.52)] would be more appropriate. Love et al. [12.35] from a comparative study of the Andersen-Wittry model and the two Philibert models, found, in agreement with the above that the simple Philibert model fared worst. They also concluded that the full Philibert model was superior to the Andersen-Wittry model. Bishop [12.21] also indicated that the values of $\Phi(0)$ obtained by the Andersen-Wittry model are incorrect. Ruste and Gantois [12.30] proposed the use of the complete Philibert model, and of the values for μ we had mentioned previously.

Virtually all tests of the practicability of quantitative determination of the elements of atomic number below ten involved empirical adjustments of parameters which shed some doubt upon the general validity of the procedure. Both more accurate values of mass absorption coefficients and further painstaking experimental work on depth distribution of primary radiation would be desirable. In the meantime those who use the electron probe microanalyzer in practical applications will be well advised to determine such elements, wherever possible, by stoichiometry or by difference.

12.5 COMPUTER PROGRAMS FOR ELECTRON PROBE QUANTITATION

In view of the complexity and length of the theoretical data reduction calculation, the use of manual procedures or of graphic calibration curves is impractical except for the simplest binary systems. In the past, large computers were used for data reduction in the batch mode, or shared-time systems could be em-

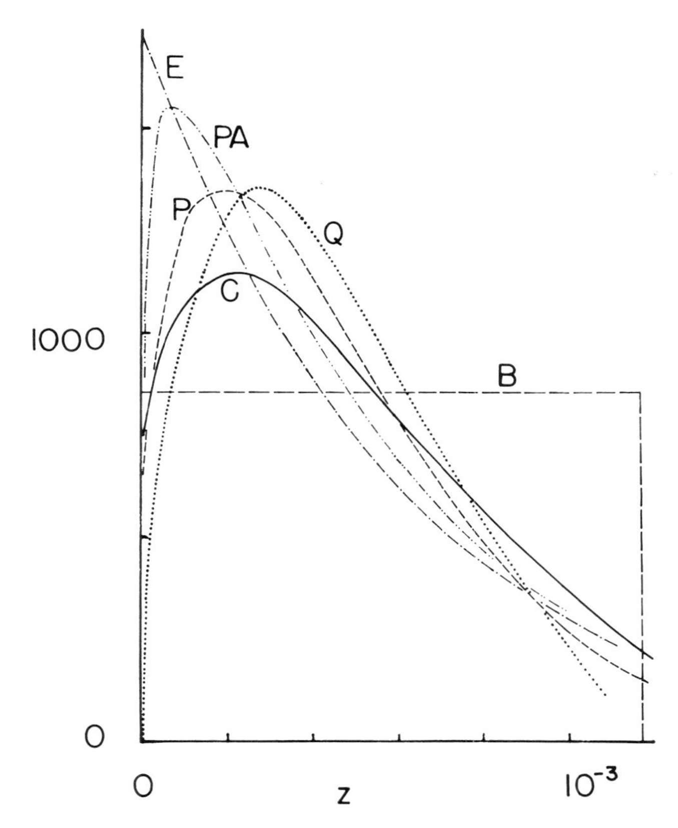

Fig. 12.16. Models for the depth distribution of the generation of primary radiation. P: Philibert [Eq. (12.3.17)]; PA: abbreviated Philibert model [Eq. (12.3.21)]; E: exponential model [$\phi(z) = \phi(0) \cdot \exp(-\sigma z)$]; Q: quadratic model [Eq. (10.3.46)]; C: experimental. All for copper target and Zn Kα tracer. V_0 = 29 kV.

ployed. At present, small computers of sufficient storage size and speed are available for the on-line performance of correction calculations with programs of satisfactory accuracy written in common languages such as Basic or Fortran. The cost of such computer facilities, including interface and typewriter output, is modest compared with that of the electron probe microanalyzer itself, and most modern instruments are equipped with such devices. The interfacing of computers with electron probes has, in turn, stimulated the development of programs for automated analysis in which various instrumental parameters are adjusted sequentially by the computer.

Forty computer programs written from 1963 to 1970 were reviewed by Beamann and Isasi [12.36]. Although programming for the electron probe has progressed rapidly since 1970, some of these programs are still used. Various criteria can be applied to comparing programs; some, such as the number of elements that can be handled, or the form of output of the results, are relatively trivial provided that the operator, or an associated programmer, are able to change such details of the program to suit their needs. Such programming skill, however, is highly desirable, because it is important that the operator be cognizant of the theoretical models embedded in the computation, and that he may, when required, change expressions, parameters, constants, or modes of output. Therefore, programs which are intended to be of use to persons other than the programmer must be extensively and carefully documented, with listings of the meaning of variables and other symbols and suggestions as to possible modifications. Without such documentation, it may be impossible to adapt a given program to an available computer, since computers differ, in spite of the standard languages used, in the way information is entered, and in other operative details.

In the process of adapting or modifying a program, the programmer may introduce mistakes. For this reason, it is important that test inputs be made available which are designed to generate a known output, so that the correctness of the program can be tested. The test inputs should be constructed so as to cover all options and variations of a program, and to provide clues as to the location of mistakes if such are present.

The most desirable characteristics of a program for data reduction are thus accuracy, versatility, economy in program length and storage requirements, proper documentation, the use of common users' languages such as Fortran or Basic, and the possibility of modification and the use of special options. Since the improvements in most aspects tend to augment the size of the program, there is no unique choice, and compromises are possible in many aspects. In the development or selection of a new system the user should consider the increasing speed of computers and the availability of storage devices which make oversimplifications of the data reduction process less and less desirable.

The Structure of Data Reduction Programs. In spite of differences in detail, the basic features of all theoretical programs for electron microprobe data reduction are the same. The core, or *main program*, centers around the iteration scheme we have described previously. Other sections of the program serve to enter the information required to initiate the iteration (*input*), to calculate functions, parameters, and constants which can be generated internally (*subroutines*), and to make the results of the calculation available to the analyst (*output*). The generation of functions and parameters in separate subprograms simplifies the modification of the correction models when required, and avoids duplications in programming when parameters—such as the mass absorption coefficients—are required in various parts of the main program. The execution of the program also requires certain decisions—such as performing corrections for characteristic fluorescence, determining elements by difference, etc.—which can be made either externally or internally. Parameters such as mass absorption coefficients can either be entered by the analyst, stored in the memory of the program or analytically developed within subroutines.

The limitations in accuracy inherent in the theoretical data reduction programs have already been discussed. The most a computer program can do to keep these limitations within reasonable bounds is to provide warnings when the conditions of analysis are conducive to excessive uncertainties. The program should produce warnings if the iteration convergence fails, if required parameters, e.g., the mass absorption coefficients, are not available, if the overvoltage is too small ($E_0/E_q < 1.5$), if the primary absorption within the target is excessive ($f_p < 0.7$), or if a line interference occurs.

In practice, *on-line programs* which receive measurement data directly from the microprobe and produce the results almost immediately after the measurement are preferable to those which require batch input and waiting time for the output. With on-line programs, in case of mistakes or large errors the measurement can be repeated immediately, and the results obtained at one point at the specimen can serve as guidance to the next steps to follow. However, on-line service often precludes the use of large programs such as COR. Programs for smaller computers contain more drastic simplifications, and the limited memory, speed, and sometimes accuracy (roundup errors can become significant in long calculations) must be carefully considered. The analyst must determine if the advantages of on-line procedures outweigh these drawbacks, which, as small computers become more efficient, are less and less important.

For a more detailed discussion of the relevant characteristics of correction programs I will use as reference four data evaluation procedures which were developed in the electron probe microanalysis laboratory of the U.S. National Bureau of Standards. These programs, written in Fortran 4, are extensively documented and are obtainable on request. Three of them are at present widely distributed and used. Their accuracy has been tested by the analysis of a wide range of specimens as well as by comparing among themselves the X-ray intensities predicted by each of the programs. The frequent reference to them does not necessarily imply that other programs, with which I am less familiar, may not equally serve the analyst.

The first program, MULTI8 [12.37] was first written in Basic for the use in a shared-time computer of limited computation and storage areas. It is essentially limited to the corrections provided within the *ZAF* approach, and to the mechanism of the iteration loop. Input must include, not only the experimental conditions and X-ray intensities, but also decisions concerning fluorescence corrections, mass absorption coefficients, and fluorescent yields. The structure of MULTI8 is shown in Fig. 12.17.

Years of experience with this program indicate that in general the errors introduced in the result by the program do not exceed ±3% of the weight fraction of major components. However, this program requires the input of relative

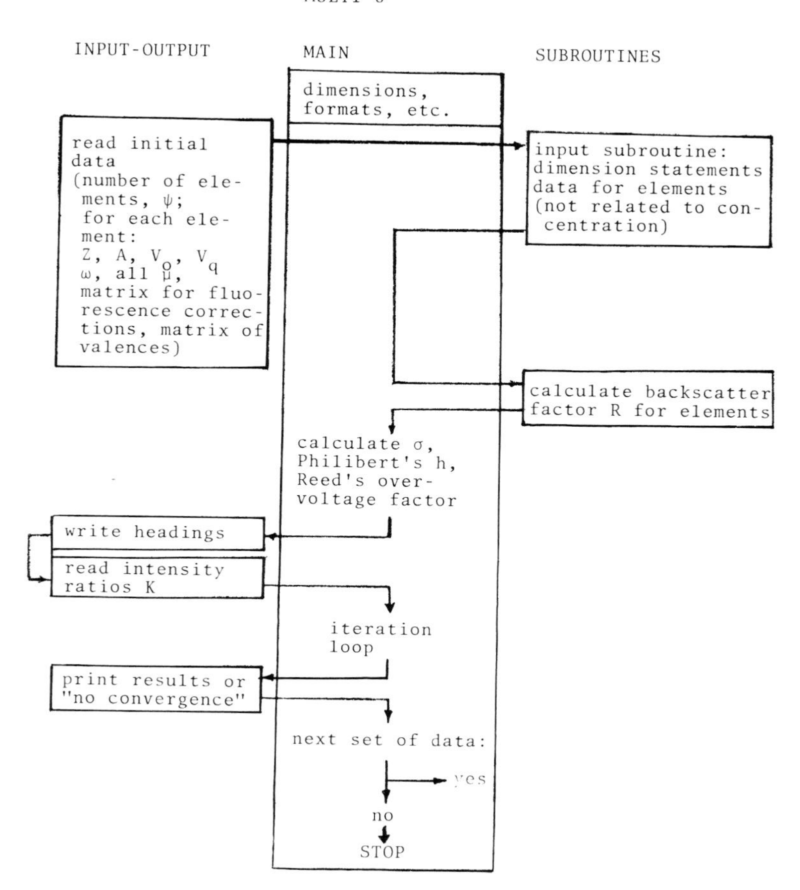

Fig. 12.17. Schematic of MULTI8.

X-ray intensities. Therefore, the background and dead-time corrections must be performed manually, at least with the original version. Moreover, the operator must input parameters such as the mass absorption coefficients from tables, and, after comparing positions of X-ray lines and absorption edges, he must decide if fluorescence corrections are to be performed. This input process is not only tedious and time consuming, but it is also a source of frequent mistakes and omissions.

The *ZAF* procedure in MULTI8 includes an atomic number correction in which the integration over the energy range of the penetrating electron is replaced by the calculation at an equivalent voltage, according to a fit by H. Yakowitz. The absorption model is that of Philibert [abbreviated, Eq.

(10.3.23)], with the expression for σ given in Eq. (10.3.29). The fluorescence correction is performed according to Reed [Eq. (11.1.43)]. No provisions exist for calculating the fluorescence from the continuum. The iteration procedure is that of Criss, described earlier in this chapter. Failure to converge has never been observed in the definitive version of the program.

Special options of MULTI8 include multiple fluorescent excitation of one or more lines by several elements, determination of one element by difference, and of one or more elements by stoichiometry. The use of compound standards can be added with a slight modification of the program.

The second program, COR [9.14], was written by Hénoc et al. Hénoc had previously [11.9] developed a formal procedure for integration of the fluorescence from the continuum [Eq. (11.2.1)], which was replaced in COR by a numerical integration. Sorting routines for determining the sequence of intervals of integration over wavelength were introduced, as mentioned in Section 11.2. Since these routines required the use of a large computer, the program was written in Fortran 4 for batch operation, and unnecessary simplifications of the corrections were avoided. COR is, to my knowledge, the most complete application of the theoretical quantitation procedure. It has been used by us as a tool to explore the effects of diverse models and parameters on the results. Moreover, it has also been applied without any difficulty to routine analysis, and can be used for this purpose with advantage as long as a batch operation is acceptable. All options previously mentioned in the discussion of MULTI8 are also available in COR.

A major advantage of COR over MULTI8 is in the input requirements. Parameters such as mass absorption coefficients are calculated in subroutines; others such as wavelengths are stored. A preliminary program was also developed which performs corrections of measured X-ray intensities for background and deadtime, and calculates the relative X-ray intensities. Most importantly, all decisions concerning the application of corrections for fluorescence by characteristic lines are made internally. Therefore, the program can be applied directly to experimental data, and the possibilities for input mistakes are greatly reduced. The structure of COR is shown in Fig. 12.18.

The third program, FRAME, is most useful for routine applications [4.42]. Written for on-line use, it conserves many features of COR, e.g., calculation of constants and parameters, and fluorescence decisions, but new simplifications developed with the aid of the experience gained with COR are introduced, and the correction for continuum fluorescence is omitted. This omission, and the replacement of tables by fitting algorithms, greatly reduces the required computer size, so that FRAME can be used on-line, with a small computer. It requires, in the Fortran version, about 4K works of computer core. A Basic version is also available. The economy in required memory is achieved by a carefully designed interaction between the common memory block and the subroutines,

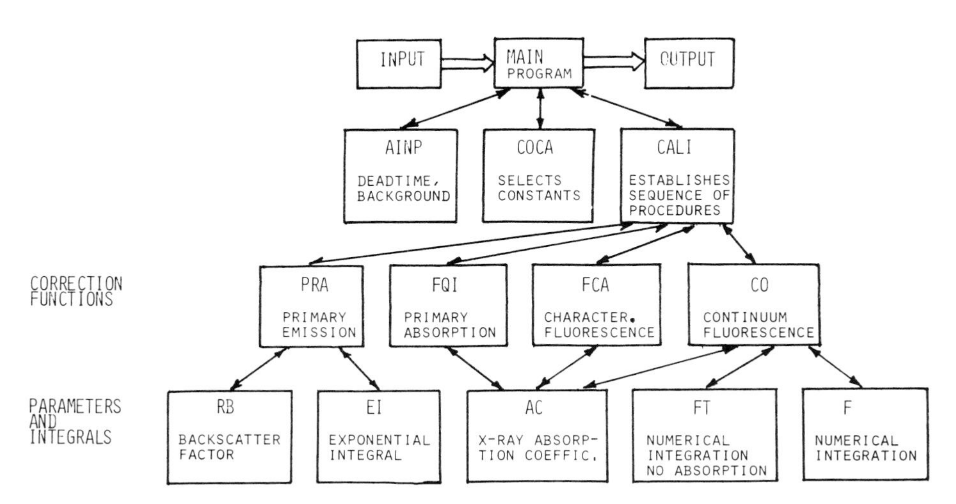

Fig. 12.18. Schematic of the data reduction program COR.

so that these are entered and deleted from the operating field of the computer as necessary (see Fig. 12.19).

A comparison of a large number of calculations of specimen composition (the set of data collated by Poole [12.11]) shows that, in spite of the radical change in program size, the differences in the results of FRAME and COR are small (Fig. 12.20), particularly when compared to the error distribution of the same analyses obtained with COR (Fig. 12.21). Further improvements may be achievable by changes in models and parameters as more information on these becomes available.

Not all simplifications proposed in the literature are acceptable. The reader should beware of totally empirical procedures tested on a few specimens, of atomic number corrections which do not vary with operating voltage, or of absorption corrections which ignore the inherent and unavoidable uncertainties which appear with large absorption losses. It requires more insight and experience to judiciously simplify the correction procedures than to develop a complex theory of data reduction.

The importance of the diverse aspects of data reduction, and the errors committed in them, depend to a large degree on the nature of the specimen. For instance, programs which do not contain provisions for the continuum fluorescence will produce seriously inaccurate results in special cases, such as the determination of elements of atomic number from 25 to 30, with the aid of K-lines, in a matrix of low average atomic number (see Table 11.4). In the analysis

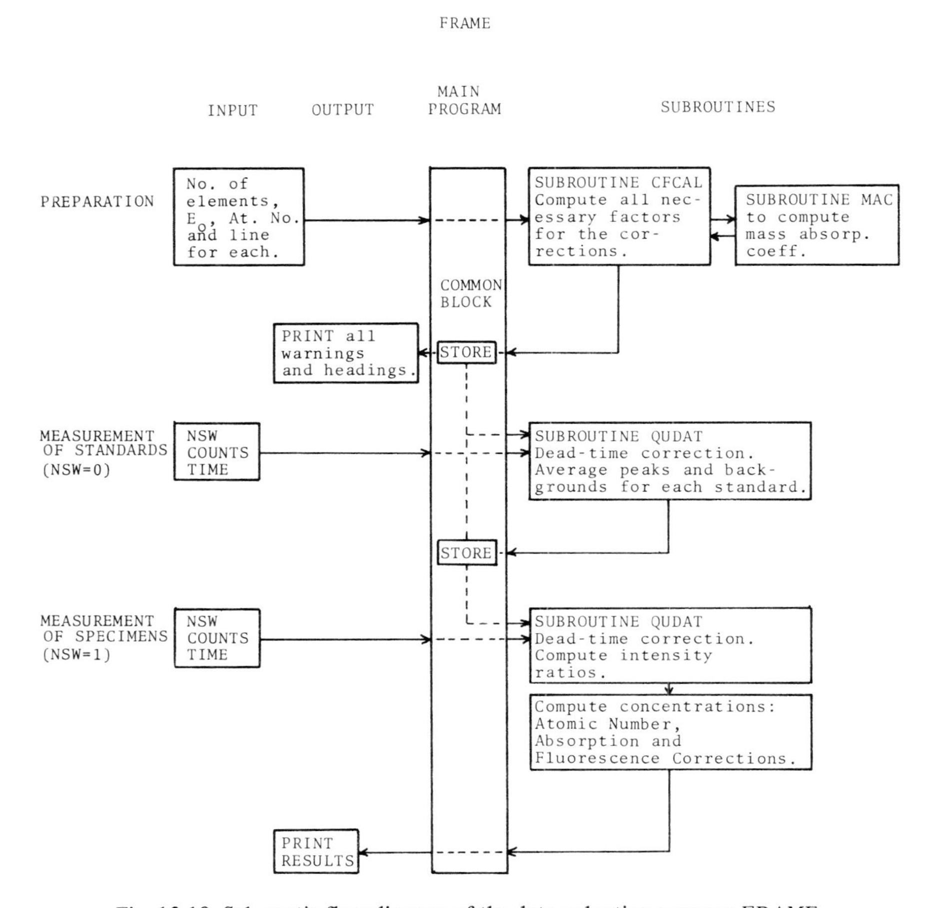

Fig. 12.19. Schematic flow diagram of the data reduction program FRAME.

of major components of brasses and ferrous alloys, the fluorescent emission is very important, while the atomic number and absorption effects can be handled easily. Conversely, in the analysis of silicate minerals, the absorption correction is very large, while fluorescence is usually a minor problem. Finally, in the analysis of sulfides, oxides, and carbides of heavy elements, the atomic number effects are, obviously, significant to a high degree.

The advantage of a complete and rigorous program such as COR is its versatility; all effects known to us to be significant are taken into account as effectively as possible. For this reason, many laboratories will use both types of programs: the complete program for initial testing of new systems, and the on-line program for routine use, after adaptation, where needed, to the specific systems to be handled.

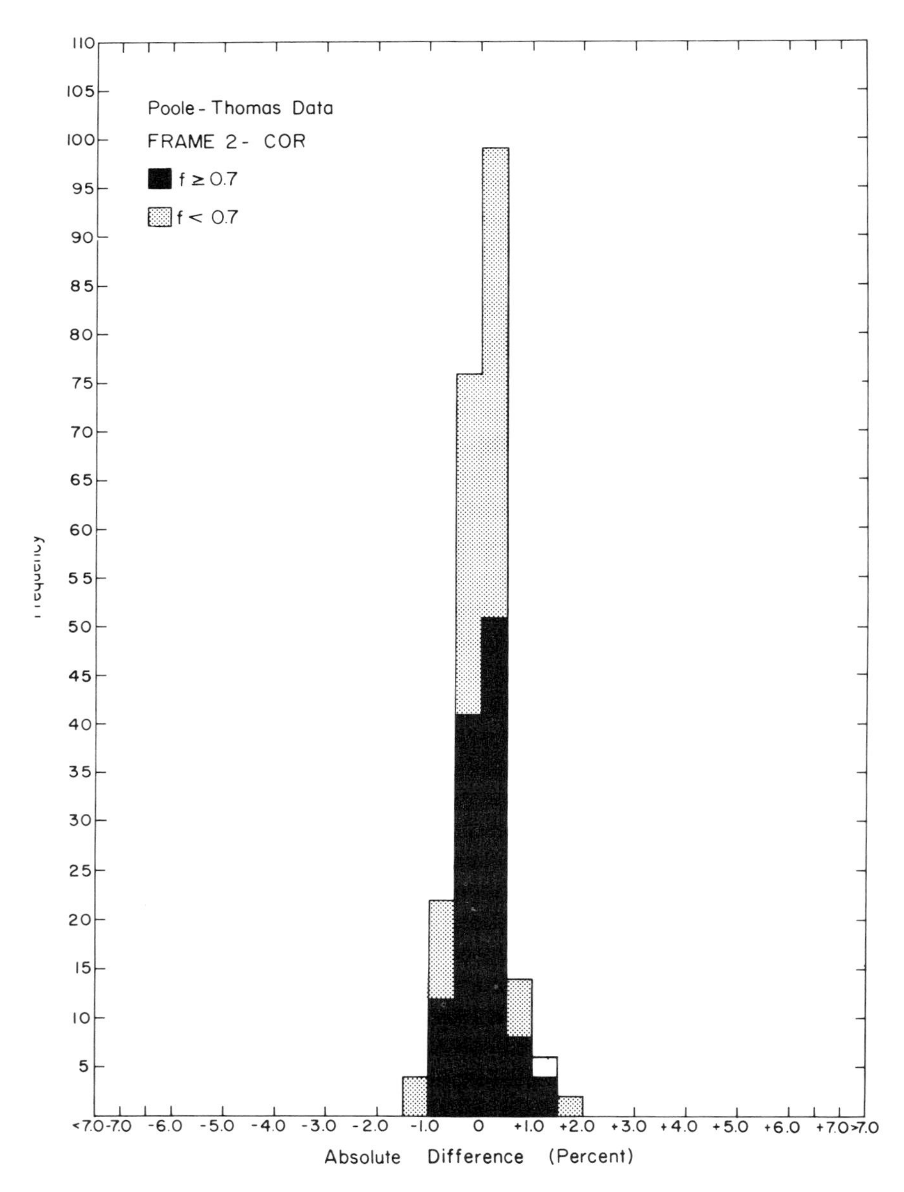

Fig. 12.20. Distribution of differences in results from Poole's data, between FRAME2 and COR.

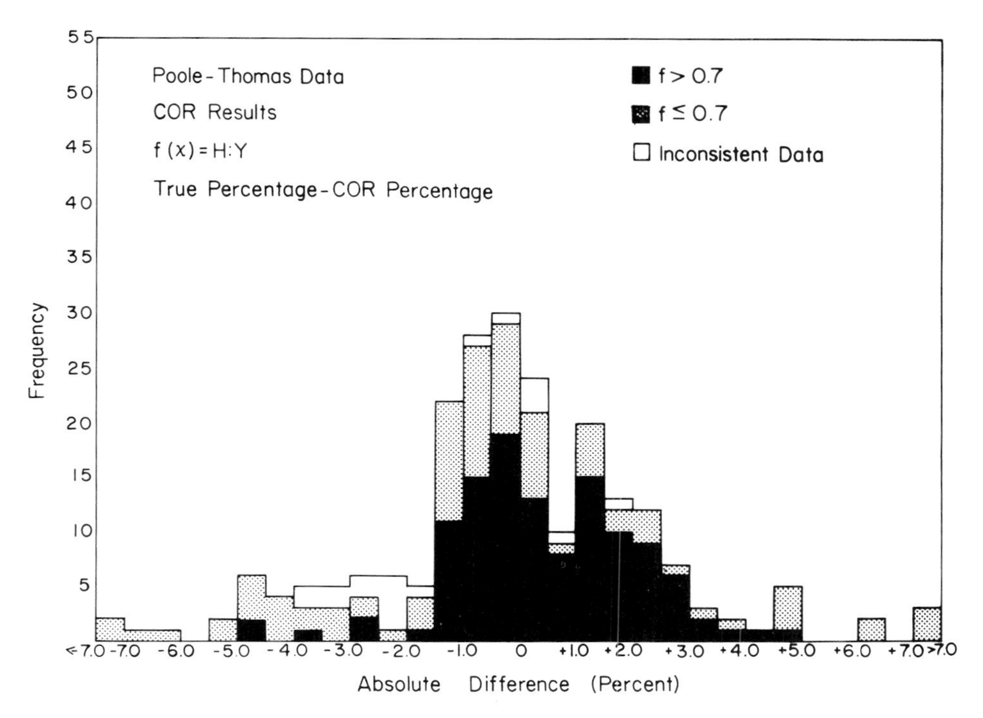

Fig. 12.21. Error distribution of the set of analysis shown in Fig. 12.20. Differences are between composition of specimens according to Poole [12.11], and that obtained by the program COR. For a detailed study see [8.6].

Output Variations and Modifications. The versatility to which we have referred is particularly obvious in the presentation of the analytical results. Programs, both large and small, can be adapted so as to present, on request, stoichiometric relations, graphic plots or statistical evaluations of a large number of measurements, expressions of results in diverse units (mass or atomic fractions), etc. However, while as far as computer requirements are concerned, such options are trivial in large programs, they increase the size of small programs, such as FRAME, which may have to be run on computers of limited capability. It should also be taken into account that the peripheral equipment used for complex output procedures is usually more expensive than the computer itself. Hence, if its use is justified, there is no good reason to diminish the efficiency of the system by choosing too small a computer.

The fourth NBS program, FRAME C, has been prepared for use with energy-dispersive spectrometers, and will be discussed later in this chapter.

Programs for the Hyperbolic Approximation. In systems where the coefficients of the hyperbolic approximation can be determined either by experience or theory, and where deviations from the approximation [Eq. (8.2.2)] are not significant, its use in an on-line program for small computers is particularly efficient and attractive [12.38]. These situations arise particularly in the analysis of minerals. The method, in conjunction with appropriate standards, has been tested on large numbers of analyses, and was found to be of good accuracy.

It is sometimes claimed that the hyperbolic approximation is more accurate than the theoretical procedure. In fact, with closely matching standards of high quality, it will usually produce more accurate results than the theoretical approach with the use of elemental standards. There is, however, no reason why matching standards whenever available should not be used in the theoretical procedure. It should also be noted that in programs such as COR and FRAME the hyperbolic approach is included in the iteration. If the theoretical calculation confirms the hyperbolic nature of the correction curve, the results would be identical to using only the hyperbolic approach. If, however, the theory predicts a deviation from the hyperbolic analytical curve, it is probably to be trusted more than the hyperbolic model. Certainly, the most accurate results will be obtained by combining the best standards with the most accurate data reduction model. The reason for using elementary standards is usually the unavailability of matching standards, or the excessive cost or effort which would be spent in their preparation. In this and similar aspects of the analysis, the economy or the procedure cannot be ignored in practice. The attractive feature of the hyperbolic approximation is its simplicity, and hence the possibility of application with small computing devices. As more efficient small computers become available, the strength of this economic argument tends to diminish.

12.6 QUANTITATION WITH THE SILICON DETECTOR

In 1959, Dolby [5.21], [5.22] described an energy-dispersive flow-proportional detector system which he used, in conjunction with an electronic network, to obtain signals for the K-bands of boron, carbon, nitrogen, and oxygen. Castaing and Pichoir also used energy-dispersion Ross filters to identify the X-ray emission of elements of low atomic number. The use of energy-dispersive techniques has been greatly extended by the development of the lithium-drifted silicon detector (see Subsection 5.3.2) which is now widely used, either in combination with crystal spectrometers or in scanning electron microscopes, as the sole means of X-ray measurement. The quantitative evaluation of energy spectra [12.39]-[12.42] requires the same conversion procedures as used for Bragg spectrometer measurement—either ZAF procedures or those based on the hyperbolic approximation. There are, however, added requirements which are due to the low resolution of the energy-dispersive spectrometer, and the consequent high background intensities and frequent line interferences.

Background Correction. Besides the low line-to-background ratios obtainable with the Si(Li) detector, difficulties may also arise from the closeness of the convoluted lines which frequently renders impossible the measurement of the background close to the lines of interest. Fiori et al. [6.13] have proposed a technique in which the intensities measured at two or more energy levels free of line interferences are used to reconstruct the continuous spectrum. In this procedure, the thicknesses of the gold and beryllium layers as well as that of the inert silicon layer on the detector surface are determined in a preliminary measurement on a carbon target. Then Eqs. (5.3.7) and (6.3.10) are combined to give the observed intensity of the continuum:

$$I' = \frac{\Omega}{4\pi}\, \epsilon\, e^{\Sigma - \mu_W z_W}\, (1 - e^{-\mu_d z_d}) \cdot \frac{I}{E}\, [k_1(E_0 - E) + k_2(E_0 - E)^2] f. \tag{12.6.1}$$

The absorption factor f in this equation is assumed to be equal to that given by Eq. (10.3.40). E is the mean energy of the region of interest of width ΔE in which the intensity $I'\Delta E$ is observed. The significance of all symbols is the same as in Eqs. (5.3.7) and (6.3.10). The two unknown variables in this equation are $(\Omega/4\pi)k_1$ and $(\Omega/4\pi)k_2$. If I' is measured for two values of E, we have two equations with two variables, and hence I' can be calculated for any other value of E as well. The continuous spectrum can thus be reconstructed, as illustrated in Fig. 6.6. More importantly, the background contribution to the regions of interest of the lines of elements to be determined can now be calculated and subtracted from the line intensity.

Special attention must be given to spectral artifacts. The line broadening and shifts at high count rates have been greatly reduced by instrumental improvements, but incomplete charge effects and escape peaks must also be taken into account. Therefore, in choosing the regions of interest for the background correction, energies just below that of a strong line should be avoided.

Line Interference. The overlapping of convoluted lines, rarely significant with Bragg spectrometers, is a limiting factor in energy-dispersive X-ray spectrometers. Several procedures have been proposed to resolve line interferences:

1. Intensities can be measured within energy regions of interest, and the interference coefficients determined. One can then calculate by means of a set of equations the interference-free intensities. This method was first used by Dolby [5.22] for the measurement of lines of elements of atomic number below ten, with a flow-proportional detector, and, for the Si(Li) detector, by Reed and Ware [12.44], in 1972.
2. It can be assumed that the distribution of intensities within a peak produced by the Si(Li) detector is Gaussian with respect to energy (see Fig. 5.35). A synthetic spectrum can be formed on a CRT screen and compared with that obtained from the detector, or the experimental spectrum can be "stripped"

by subtracting from it the postulated Gaussian peaks. The height of the Gaussian which is required to eliminate the corresponding peak from the original spectrum is proportional to the number of photons collected in this peak.
3. A mathematical procedure of deconvolution may be applied to the peaks of the experimental spectrum [12.42].

Regardless of the procedure followed, the treatment of line overlap is always limited by significant statistical uncertainty. Even in the absence of line interference, the high background levels observed with the Si(Li) detector limit the applications to concentrations above the trace level—typically above 0.1%. The shortcomings due to limited resolution and high background of the Si(Li) detector are, however, offset by the following advantages: its efficiency is greater by a factor of 10 to 100 than that of the crystal spectrometer, and the detector is nonfocusing so that it can observe signals from a larger field on the specimen than the crystal spectrometer. The simultaneous measurement of several lines with one detector is possible. In summary, this detector is particularly useful:

1. As an ancillary attachment to a scanning electron microscope, particularly for the analysis of small particles or thin films, and for conditions in which the beam current must be reduced below 10^{-9} Å and hence signals from crystal spectrometers become feeble.
2. As an auxiliary spectrometer for the simultaneous determination of several elements, in addition to crystal spectrometers. This combination extends the number of elements which can be observed simultaneously. It is important, however, that the count rates for both types of detectors be within the respective useful ranges. The common way to achieve this compatibility of spectrometers is to mount the silicon detector in such a fashion that, for feeble X-ray emission, it can be brought as close to the target as the configuration of the instrument permits, and be retracted to a greater distance when a strong X-ray source is observed (Fig. 12.22).
3. For absolute measurements.

The absence of complicated mechanical elements, the stability of the geometrical configuration, and the possibility of calculating with reasonable accuracy the solid angle covered by the detector as well as its efficiency make the Si(Li) a preferred tool for measuring the X-ray emission of a target on an absolute scale (i.e., photons per incident electron). Such measurements are, in the context of electron-opaque ("thick") specimens, mainly of theoretical interest. In the analysis of thin films and particles, however, the relation of X-ray emission to that from a thick specimen is less direct, and absolute measurements are useful in data reduction procedures as will be discussed in Chapter 13.

Fig. 12.22. ISI scanning electron microscope provided with a Si(Li) detector (left side). The detector assembly, including the liquid-nitrogen tank, is movable so that the distance between detector and specimen can be varied. The detector, D, at the tip of the rod protruding from the left into the specimen chamber, tilts downward. In the left lower corner is the Everhart-Thornley detector, which was removed to provide an unobstructed view at the X-ray detector. S: the specimen stage.

12.7 A PROGRAM FOR ENERGY-DISPERSIVE QUANTITATION

The program FRAME C [5.24], [12.46] is a modification of FRAME for use with a lithium-drifted silicon energy-dispersive X-ray detector in a small computer. The equipment for which this program was primarily adapted was a NOVA computer with a Basic interpreter, and a memory of 24K words, which also served to direct the operations of a 4096 channel pulse height analyzer. The program was therefore written in Basic; a Fortran 4 version was later produced as well. The memory of the analyzer was divisible into four quadrants of 1024 channels each. One of these quadrants is large enough to contain a complete pulse height spectrum. The analyzer was calibrated so that each channel corresponded to an energy range of 10 eV in the X-ray spectrum. Spectra could be transferred from the PHA memory to a magnetic tape, but no provisions existed for the external storage of information for the computer itself.

All procedures used in FRAME, including the calculation of line and edge energies, mass absorption coefficients, and other parameters, as well as the sorting routines for decisions on characteristic line fluorescence, were conserved in FRAME C. The *ZAF* and iteration procedures are also identical. In addition, the characteristics of the detector system presented new problems, mainly in the background correction, provisions for line overlap, and corrections for detector artifacts such as escape peaks.

The most important consideration in handling the energy spectra must be given to the low speed of operation of small computers which precludes even simple operations with all of the 1024 channels of each spectrum. For this reason it was decided to use regions of interest in FRAME C. One single region of interest was established for a peak of each element, and the rest of the spectrum was not utilized. The pulses within each region of interest were summed and this sum was the only signal element for the region used from this point on.

The method previously published by Fiori et al. [6.13] was used for the background prediction. This semiempirical approach provides, upon measurement of the emitted continuum intensity in at least two regions of interest free of characteristic lines, a continuous function which describes the generation of bremsstrahlung. However, the *emitted* continuum, after the correction for absorption with the target, contains discontinuities at the positions of absorption edges of the target materials. These jumps in the background level depend on the target composition and cannot be predicted from the fit. Therefore, the emitted background shape must be redetermined within each iteration loop of the *ZAF* procedure, and the background-corrected line emissions within each region of interest must be revised accordingly (see Figs. 12.23 and 12.24).

An important aspect of analysis with solid-state detectors is related to the fact that due to the overlapping of several lines the presence of an element may be overlooked, or a misidentification of a peak may occur. For this reason it was important to provide a visual comparison between the experimental spectrum and one constructed on the basis of the analytical result. It was therefore decided to reserve the first quadrant of the multichannel analyzer for the storage and display of the spectrum obtained from the specimen, the second quadrant for that of the standard spectra, and the third quadrant for the synthesis and display of a synthetic line spectrum after termination of the analysis. To pro-

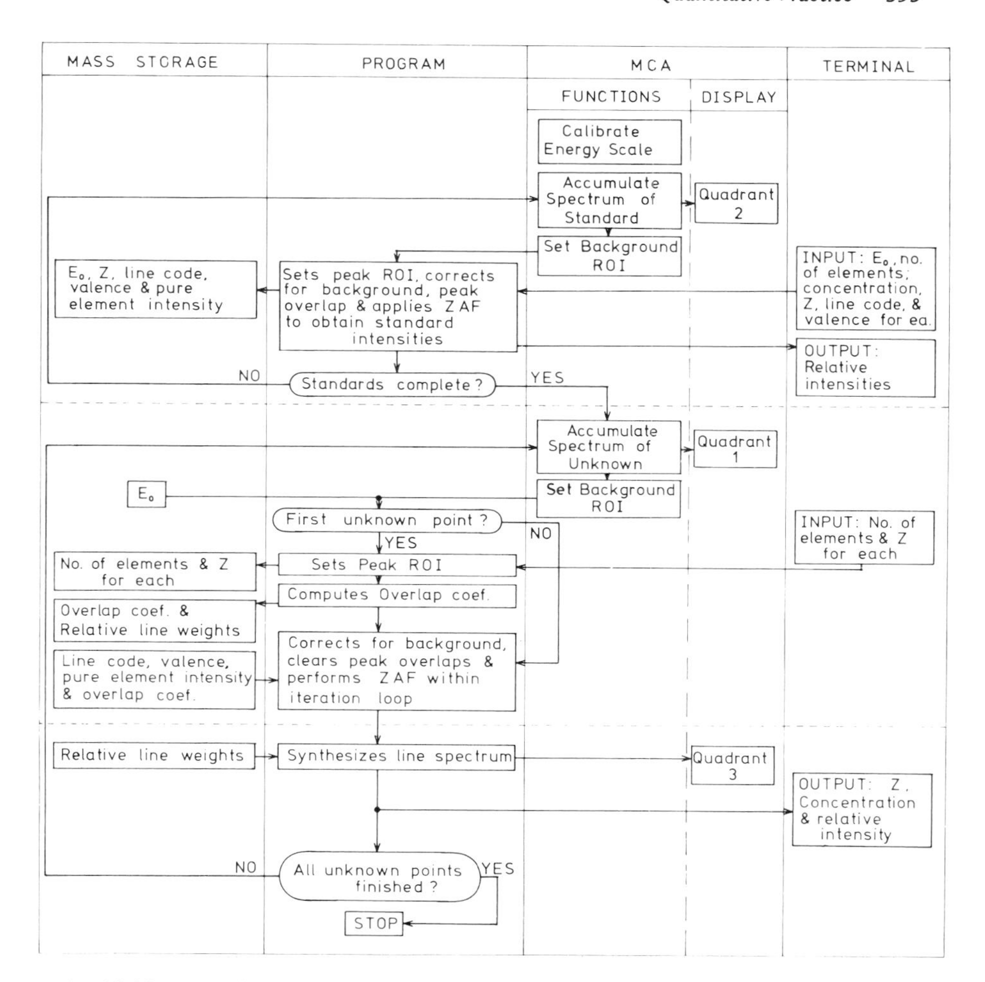

Fig. 12.23. Flow diagram for the program FRAME C, which applies *ZAF* correction to spectra obtained with the Si(Li) detector, displays accumulated and synthesized spectra on the oscilloscope, and prints the final results on a terminal. ROI:region of interest. "Points" refers to locations on the specimen to be analyzed. Background and peak overlap corrections are performed as described in the text.

duce such a spectrum in a reasonably short time, it was assumed that all peaks are strictly Gaussian, and to omit the reconstruction of the background. Figure 12.25 shows the simultaneous display of the analytical spectrum of the NBS glass K252 and the corresponding synthetic spectrum. The subtraction of the synthetic spectrum from the experimental spectrum provides a painfully critical test of success. The arrangement leaves the fourth quadrant conveniently free for the storage of information concerning standards.

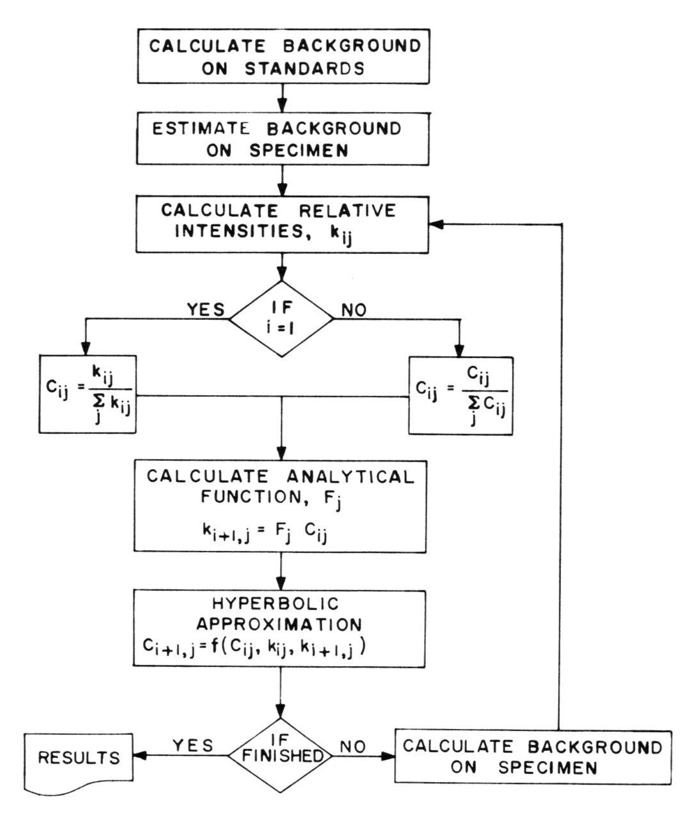

Fig. 12.24. Iteration loop of FRAME C, with updating of background. i is the number of the iteration.

At the start of the process, information on the detector efficiency and resolution (half-maximum width of the Mn Kα peak) as well as on incomplete charge characteristics of the detector is entered. The energy scale of the multichannel analyzer must also be calibrated. After these preliminaries, spectra from the standards are collected. Each standard may serve for several elements. Standard spectra may also be retrieved from the magnetic tape rather than produced at the time of the analysis. If such stored standards are used, the operator must make certain that between the taking of the standard spectra and those of the specimens the calibration of the analyzer has not changed. The specimen spectrum is obtained next, and the background correction is applied as discussed above. The required characteristics of the detector efficiency can be experimentally determined by means of a supplementary program.

Line Overlap. We will first discuss the case in which a line which has its own region of interest overlaps into the region of interest of a line of another element

(Fig. 12.26) The total intensity of this line, which we will call the aKα-line, is

$$Y_{aK\alpha} = I_{aK\alpha} \cdot \frac{\Omega}{4\pi} \cdot f(aK\alpha) \cdot i \cdot P(aK\alpha). \qquad (12.7.1)$$

The pulse height distribution of this signal is Gaussian. If $\sigma(aK\alpha)$ is the standard deviation of this distribution, then the part of $Y_{aK\alpha}$ which is within the limits of the region of interest for element a, $Y(a)$, is

$$Y(a) = \frac{Y_{aK\alpha}}{\sqrt{2\pi}\,\sigma(aK\alpha)} \int_{E_1}^{E_2} \exp\left[-\frac{1}{2}\left(\frac{E_{aK\alpha} - E}{\sigma(aK\alpha)}\right)^2\right] dE. \qquad (12.7.2)$$

In this equation, E_1 and E_2 are the limits of the region of interest for element a. We will call E_3 and E_4 the limits of the region of interest of the element b,

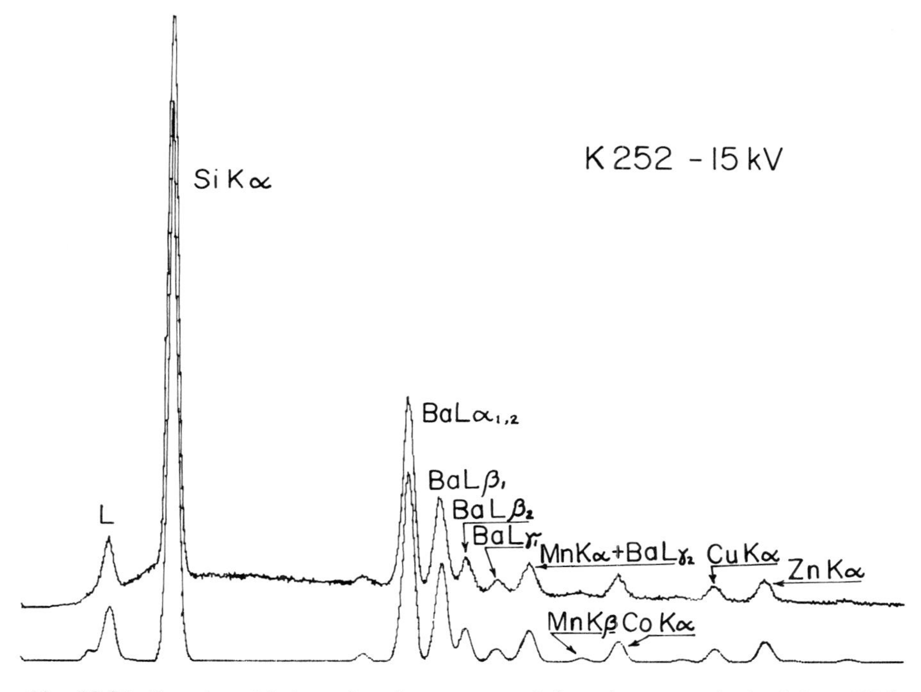

Fig. 12.25. Experimental (upper) and reconstructed (lower) spectra obtained from NBS glass K252. L:unresolved L-spectra of several elements. The composition of the glass (in weight fractions) is as follows: SiO_2: 0.40; BaO: 0.35; ZnO: 0.10; CoO: 0.05; CuO: 0.05; MnO_2: 0.05.

the line of which is overlapped by the line of element a. The intensity of the signal of element a falling within this region is

$$Y(ab) = \frac{Y_{a\mathrm{K}\alpha}}{\sqrt{2\pi}\,\sigma(a\mathrm{K}\alpha)} \int_{E_3}^{E_4} \exp\left[-\frac{1}{2}\left(\frac{E_{a\mathrm{K}\alpha} - E}{\sigma(a\mathrm{K}\alpha)}\right)^2\right] dE. \quad (12.7.3)$$

Hence, the "influence factor" by which the measured intensity of $a\mathrm{K}\alpha$ must be multiplied in order to obtain the line overlap into the region of interest of $b\mathrm{K}\alpha$, is:

$$\frac{Y(ab)}{Y(a)} = \int_{E_3}^{E_4} \exp\left[-\frac{1}{2}\left(\frac{E_{a\mathrm{K}\alpha} - E}{\sigma(a\mathrm{K}\alpha)}\right)^2\right] dE \Bigg/ \int_{E_1}^{E_2} \exp\left[-\frac{1}{2}\left(\frac{E_{a\mathrm{K}\alpha} - E}{\sigma(a\mathrm{K}\alpha)}\right)^2\right] dE. \quad (12.7.4)$$

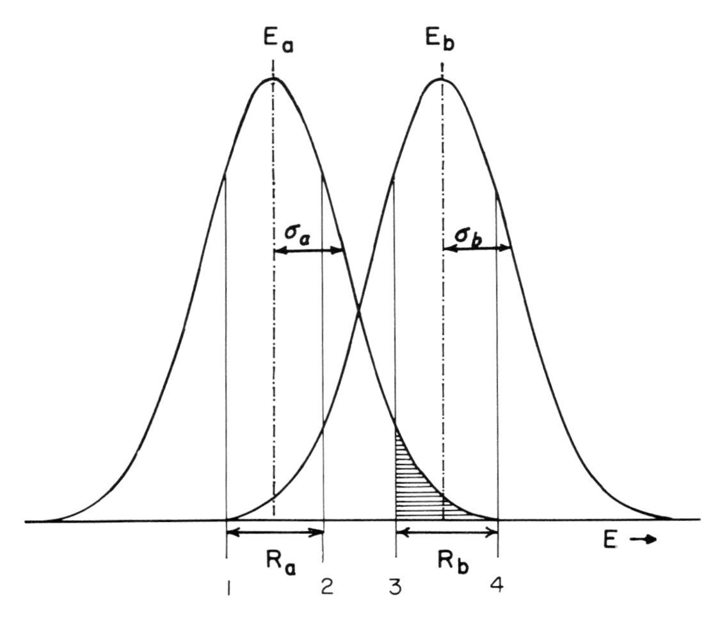

Fig. 12.26. The X-ray emission of the line of element a which has a region of interest R_a interferes with the region of interest R_b of element b. This interference (hatched region) can be calculated from the Gaussian centered around the energy E_a after the intensity of the peak at E_a and the standard deviation σ_a have been determined. Since the line at E_b interferes similarly with the measurement of the line at E_a, the correction procedure must be iterative.

Consider now the overlap of a line (viz., $a\mathrm{K}\beta$), which does not have a region of interest, with the region of interest of the line $b\mathrm{K}\alpha$ (Fig. 12.27). We can relate the generated intensities of $a\mathrm{K}\beta$ and $a\mathrm{K}\alpha$, which are produced by ionization of the same shell, by their relative transition probabilities, or "weights of lines" (see Subsection 4.2.5):

$$I_{a\mathrm{K}\beta} = I_{a\mathrm{K}\alpha}(p_{a\mathrm{K}\beta}/p_{a\mathrm{K}\alpha}).$$

The fraction of the signal $Y_{a\mathrm{K}\beta}$ which falls within the region of interest of the line $b\mathrm{K}\alpha$, $Y(a\mathrm{K}\beta, b)$, is:

$$Y(a\mathrm{K}\beta,b) = \frac{Y_{a\mathrm{K}\beta}}{\sqrt{2\pi}\,\sigma(a\mathrm{K}\beta)} \int_{E_3}^{E_4} \exp\left[-\frac{1}{2}\left(\frac{E_{a\mathrm{K}\beta} - E}{\sigma(a\mathrm{K}\beta)}\right)^2\right] dE \tag{12.7.5}$$

and

$$Y_{a\mathrm{K}\beta} = Y_{a\mathrm{K}\alpha} \cdot \frac{p_{a\mathrm{K}\beta}}{p_{a\mathrm{K}\alpha}} \cdot \frac{f(a\mathrm{K}\beta)}{f(a\mathrm{K}\alpha)} \cdot \frac{P(a\mathrm{K}\beta)}{P(a\mathrm{K}\alpha)} \tag{12.7.6}$$

so that

$$\frac{Y(a\mathrm{K}\beta)}{Y(a)} = \frac{p_{a\mathrm{K}\beta}}{p_{a\mathrm{K}\alpha}} \cdot \frac{f(a\mathrm{K}\beta)}{f(a\mathrm{K}\alpha)} \cdot \frac{P(a\mathrm{K}\beta)}{P(a\mathrm{K}\alpha)} \cdot \frac{\sigma(a\mathrm{K}\alpha)}{\sigma(a\mathrm{K}\beta)} \cdot \frac{\int_{E_3}^{E_4} \exp\left[-\frac{1}{2}\left(\frac{E_{a\mathrm{K}\beta} - E}{\sigma(a\mathrm{K}\beta)}\right)^2\right] dE}{\int_{E_1}^{E_2} \exp\left[-\frac{1}{2}\left(\frac{E_{a\mathrm{K}\alpha} - E}{\sigma(a\mathrm{K}\alpha)}\right)^2\right] dE}. \tag{12.7.7}$$

The distortion of the peak shape by the incomplete charge collection is handled by an empirical fit with two parameters which must be determined for each detector. The escape peaks are calculated by the formula of Reed and Ware [5.28]. The standard deviation of escape peaks is that which corresponds to their nominal energy rather than that of the parent peak. The calculation of the effects of overlap of escape peaks with peaks of other lines is analogous to that described above for the $\mathrm{K}\beta$-lines.

The accuracy of the correction for overlaps of lines which have no region of interest depends on the accuracy of the weights of lines. Therefore, some difficulties are observed in resolving overlaps of L-lines for which the weights are not accurately known. Further work in this area is expected.

Table 12.5 shows results obtained with the present version of FRAME C for a

Table 12.5. Energy-Dispersive Analysis of Silicate Minerals.

STANDARDS		Na K409[a]	Mg MgO	Al Al_2O_3	Si SiO_2	K OR-1[b]	Ca APATITE	Ti ELEMENT	Mn ELEMENT	Fe ELEMENT	Zn ELEMENT	Ba K371[a]
Specimen												
Anorthoclase	C	0.061	0.0001	0.108	0.309	0.029	0.0056	0.0001	–	0.0016	–	0.0013
(Nunivak)	F	0.064	0.0020	0.106	0.311	0.029	0.0073	NA	–	0.0013	–	0.0020
Rhodonite	C	–	0.0076	–	0.219	–	0.057	–	0.265	0.013	0.056	–
(Bald Knob)	F	0.0008	0.0093	0.0005	0.227	-0.0006	0.058	–	0.273	0.014	0.061	–
Tephroite	C	–	–	–	0.139	–	–	–	0.544	–	–	–
	F	–	–	–	0.144	–	–	–	0.541	0.0008	–	–
STANDARDS		K409[a]	K412[a]	K412[a]	K412[a]	OR-1[b]	K412[a]	TiO_2	TEPHROITE	K412[a]	ELEMENT	K371[a]
Anorthoclase (Nunivak)	F	0.064	0.0038	0.105	0.309	0.029	0.0073	–	–	0.0013	–	0.0021
Rhodonite (Bald Knob)	F	–	0.0096	0.0044	0.225	–	0.059	–	0.274	0.016	0.061	–
Basalt	C	0.019	0.041	0.074	0.238	0.0016	0.080	0.011	0.0017	0.092	–	–
(Juan de Fuca)	F	0.026	0.045	0.072	0.238	0.0019	0.079	0.011	0.0020	0.092	–	–

Note: All numbers are weight fractions of elements. V_0 = 15 kV, ψ = 52.5°. C:chemical analysis; F:FRAME 2. NA:not analyzed.

[a]NBS synthetic glasses.

[b]Geophysical Laboratory synthetic glass.

series of silicate minerals which were analyzed with the Si(Li) detector, at 15 kV, with an X-ray emergence angle of 52.5°. The efficiency of the overlap correction is shown more clearly in Table 12.6, in which pure elements and simple compounds were analyzed for elements which are not present in the specimens, and with whose lines the lines of the main component overlap. A perfect correction would produce an output of zero concentration in all cases. The only serious problem was observed in the overlap of Cu $K\alpha$ (8.05 keV) with Ta $L\alpha_1$ (8.15 keV) in which the distance between peaks is too small for effective resolution.

Analysis without Standards. The need for standards, and the use of relative X-ray intensities as a measure of concentrations, stem from the difficulties in the absolute calibration of crystal spectrometers. Another important factor is the variability in time of the beam intensity, as well as the spectrometer alignment and efficiency. These factors determine the time permissible between the measurement of X-ray intensities from the specimen and that from the standard.

The spectrometer stability is considerably improved if a Si(Li) X-ray detector is used. Therefore, we can use stored spectra for standardization, if the electron beam dose (i.e., the integration of the beam current over time) can be used as

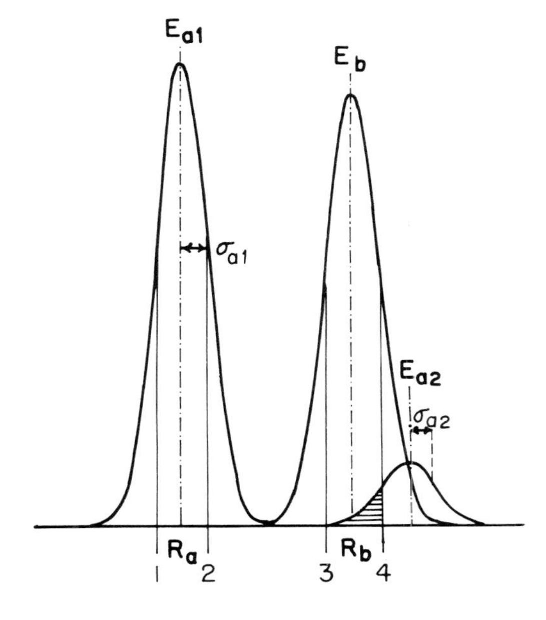

Fig. 12.27. The X-ray emission of the line of element a at the energy E_{a2} interferes with the region of interest R_b of element b. The line at E_{a2} does not have its own region of interest. Hence, to estimate its interference with the region R_b, another line of the same element, at the energy E_{a1}, is evaluated, the ratio of the intensities of lines E_{a1} and E_{a2} of the element a is calculated from theory, and the interference of E_{a2} with region R_b is then estimated.

a measure of specimen excitation. It is, in fact, sufficient to store or record the intensities of the lines of interest, corrected for background and line overlap. Such simplified procedures are of particular interest in scanning electron microscopy. In this technique, the prime objective is the characterization of the shape of the surface. Many specimens investigated with the scanning electron microscope are not flat electron-opaque bodies, and therefore the usual procedures of quantitation cannot be rigorously applied. Hence, most analyses of such specimens are semiquantitative. On the other hand, with the Si(Li) spectrometer, it is not difficult to predict, within an accuracy of a few percent, the intensity to be expected from elemental standards. It has been proposed [12.47], [12.48] that standard measurements can be omitted for such situations, and be replaced by storage in a table of prior readings of intensities of lines from elements, or even by calculation from theoretical first principles. Such a procedure would require a precise knowledge of the solid angle covered by the detector and its response function, as well as a precise measurement of the beam current. It appears preferable to measure the output from at least one X-ray emitter of known composition (e.g., and element). The ratios of X-ray emissions for differ-

Table 12.6. Overlap Corrections.

SPECIMEN	E_0 (kV)	ANALYZED FOR		OVERLAP WITH		MASS FRACTION FOUND BEFORE CORRECTION	MASS FRACTION FOUND AFTER CORRECTION
MgO	15	Al Kα	1.487	Mg Kβ	1.297	0.0026	0.0016
Al_2O_3	15	Mg Kα	1.254	Al Kα	1.487	0.0132	0.0058
Al_2O_3	15	Si Kα	1.740	Al Kβ	1.553	0.0039	0.0014
SiO_2	15	Al Kα	1.487	Si Kα	1.740	0.0072	−0.0010
Cr	15	Mn Kα	5.898	Cr Kβ	5.946	0.1311	−0.0024
Mn	15	Fe Kα	6.403	Mn Kβ	6.490	0.1045	0.0019
Fe	15	Co Kα	6.930	Fe Kβ	7.057	0.0653	0.0062
Cu	20	Zn Kα	8.638	Cu Kβ	8.904	0.0206	0.0039
Zr	15	Bi Mα	2.423	Zr Lβ_2	2.219	0.0214	−0.0113
				Lγ_1	2.302		
				Lγ_3	2.503		
Bi	15	Zr Lα_1	2.042	Bi Mα	2.423	0.0213	−0.0009
				Mβ	2.526		
				Mζ	1.901		
Zn	20	W Lα_1	8.396	Zn Kα	8.638	0.1329	0.0010
Ta	20	Cu Kα	8.047	Ta Lα_1[a]	8.145	0.1968	0.0700
				Ta La$_2$	8.087		

Note: The electron beam incidence angle is 90° and the X-ray emergence angle is 52.5°. Line energies are in KeV.

[a] The regions of interest for Cu Kα and Ta Lα overlap. These lines are too close to be separated well by this method.

ent elements can be measured accurately. Such ratios will be valid for a given detector and operating voltage, for the purpose of calculating the intensities of fictitious elemental standards for *ZAF*. They will remain constant as long as the absorption characteristics of the detector window do not change. If the region of interest procedure is used, the energy scale of the pulseheight analyzer must, of course, be rigorously calibrated. The window of the region of interest for the analysis must be exactly the same as that employed when the line intensity from the element of interest was compared with that of the element used as a beam intensity calibration standard at the time of analysis. The intensities, relative to those from the standard element, will of course vary with operating voltage, which therefore must be exactly the same for the calibration and the analysis.

Automated Analysis. Many applications of electron probe microanalysis require lengthy and repetitious measurements. It is often possible to perform these tedious tasks in an automated fashion. With full automation, and where a string of measurements can be preplanned, an unmanned operation can be sustained for many hours. But even if the operation is only partially programmable, automation can remove much of the tedium of the analysis, and the operator's attention can be directed to other more important and interesting aspects of the measurement process.

The aspects of electron probe microanalysis to which automation can be applied are listed in Table 12.7. A very important procedure is the stabilization of the beam current by means of a feedback loop. The monitor current which was discussed in Chapter 6 can serve to adjust the field strength of the condenser lens. Such an arrangement can establish a regime of beam emission which is essentially ($<1\%$) drift-free for several hours. Such stability is indeed required if a lengthy automated operation is to produce quantitative results.

Table 12.7. Microprobe Automation:

1. *Beam control*	a) Automatic focusing control b) Programmed step, line, and raster scan
2. *Specimen control*	a) Automatic step and area scan b) Control of specimen elevation c) Automatic reading of standards, automatic beam-current measurement (Faraday cage) and feedback d) Programmed feature search (e.g., to find particles, inclusions, fibers)
3. *Spectrometer, scaler, and output control*	a) Automatic scaler read-out sequence b) Automatic sequence of X-ray line readings c) Automatic background correction d) Programmed data reduction e) Formation of concentration maps

The two mechanical assemblies to which automation is most applicable are the crystal spectrometer and the specimen stage. A necessary prerequisite is good repeatability and lack of significant mechanical backlash of the mechanical elements of interest. If the spectrometers are already driven by pulsed stepping motors, it is quite easy to program alignment routines by means of which several X-ray lines can be measured in sequence by the same spectrometer. In fact, the reproducibility of spectrometers is often significantly improved when the manual operation is replaced by an automated procedure with peak searching when necessary. The changing of crystals on a spectrometer can also be programmed, but the errors arising from this operation may be substantial and must be carefully investigated. The counting sequences and times of scalers, and the settings of single-channel pulse height analyzers can easily be programmed as well, so that a completely automatic analysis sequence is established. The string of signals arising from the analysis can be stored for later evaluation, or the analytical result can be immediately calculated and stored, printed, or plotted.

The automatic movement of the specimen along a line [12.49] or in the form of a raster [5.2], [12.51], [12.52] can be programmed for the determination of microscopic homogeneity, or in order to characterize compositional changes along a path on the specimen. The movement with respect to the beam can either be achieved mechanically, or, within the limits of crystal focusing, by beam deflection. It is somewhat more complicated to program an irregular sequence of points on the surface of the specimen. This sequence must first be stored in a memory and then automatically followed [12.53], [12.54]. Devices for such a preprogrammed sequence of analysis locations were developed for the analysis of minerals [12.55] and of microdroplets [12.56], [12.57].

If a solid-state detector is used, new possibilities of automation arise, such as preliminary analysis with the Si(Li) detector, followed by more accurate analysis with the crystal spectrometers, as well as automatic storage and evaluation of energy-dispersive spectra. In this area considerable advances can be expected in the near future.

The major pitfall of automated electron probe operations is the possibility of oversight of unexpected signals or of flaws in the instrument operation that might have been corrected in a manual operating mode. For this reason, feedback, warnings of unusual instrumental conditions, and diagnostic statements should be liberally used in automated operations. Furthermore, specimens of known composition should frequently be run together with the unknowns.

In many types of analysis, the operator continuously adjusts the analytical sequence on the basis of his observations so that full preplanning of the sequence is impossible. In such cases, a partially automated operation will still be found very useful, in which, under the guidance of the operator, the computer performs adjustments and instrumental changes.

For a review of applications of the computer to electron probe microanalysis see Tixier [12.58].

12.8 APPLICATIONS OF QUANTITATIVE ANALYSIS

Electron probe microanalysis can be applied to the elemental characterization in the microscopic domain of virtually all solid bodies. Hence, a full list of actual and possible applications would be nearly inexhaustible. Since many references exist in which applications of one kind or another are described and discussed, I will not attempt to fully review the entire field. Table 12.8 indicates several areas to which electron probe microanalysis is commonly applied, with references to reviews or general treatises. We will here discuss some of these typical applications, without any pretense of completeness. For other examples of applied electron probe microanalysis the reader is referred to [12.80]-[12.88], as well as to the Proceedings of the Annual Conferences of the Electron Probe Analysis Society of America (after 1974 called Microbeam Analysis Society). Further useful information can be found in the annual reviews in the journal *Analytical Chemistry*, those of electron probe microanalysis colloquia of A.N.R.T. (in French) which are periodically published in the *Journal de Microscopie*, and in publications appearing in the journal *X-Ray Spectroscopy*. Bibliographic references are also reviewed in *Micronews*, the bulletin of the Microbeam Analysis Society.

Minerals. In no area has the electron probe changed the course of scientific investigation more incisively than in mineralogy. Keil [12.60], in a review published in 1973, cites six pages of new minerals, terrestrial or cosmic, which were first identified by electron probe microanalysis. In recognition of the role of the electron probe in mineral identification, the name Castaingite was proposed for a newly described mineral [12.87]. The distribution of trace and minor constituents in rocks and minerals is of particular interest, since such elements are used in establishing the past history of the material. As shown in Goñi and Guillemin [12.90], these elements can often only be observed in small regions at grain boundaries.

Although electron probe analysis was applied to many types of minerals (e.g., to sulfides [12.91], [12.92]), most work was performed on silicates, in which the predominant potential sources of analytical error are in the absorption correction and due to chemical shifts. For this reason, poor accuracy was obtained when elementary standards and the conventional *ZAF* method were first applied. Smith [12.59] concluded in 1965 that the use of calibration curves would be preferable to the *ZAF* procedure. In 1968, Bence and Albee [8.4] introduced their adaptation of the Ziebold-Ogilvie approach (see Chapter 8), which is now the most widely used technique. The possibility of executing this calculation on-line with very simple calculating machines has greatly contributed to its success. As the importance of errors stemming from the use of unreliable standards became clearer, a compromise solution was proposed in which the α-factors used in the hyperbolic approximation are calculated by a *ZAF* procedure. Thus, the number of standards which are required is reduced, but the method now combines the inaccuracies of both the *ZAF* and hyperbolic approaches [12.93].

The use of the *ZAF* approach was reinvestigated in 1969 by Sweatman and Long [4.25], who discussed the propagation of errors in the emergence angle of X-rays and in the operating potential, the choice of standards, chemical shifts, X-ray absorption coefficients, and the diverse improvements which were incorporated in the *ZAF* procedure. The results of analyses of silicates were also presented. The authors concluded that the *ZAF* procedure as used by them was a viable alternative to calibration curves and to the Bence-Albee technique. In 1977 Bence and Holtzwarth [12.94] reported that biases appeared in the analysis of the system MgO-Al_2O_3-SiO_2 [e.g., in olivines (Mg_2SiO_4), spinels ($MgAl_2O_4$), and pyropic garnet ($Mg_3Al_2Si_3O_{12}$)]. These observations indicated

Table 12.8. Summary of Applications and General References.

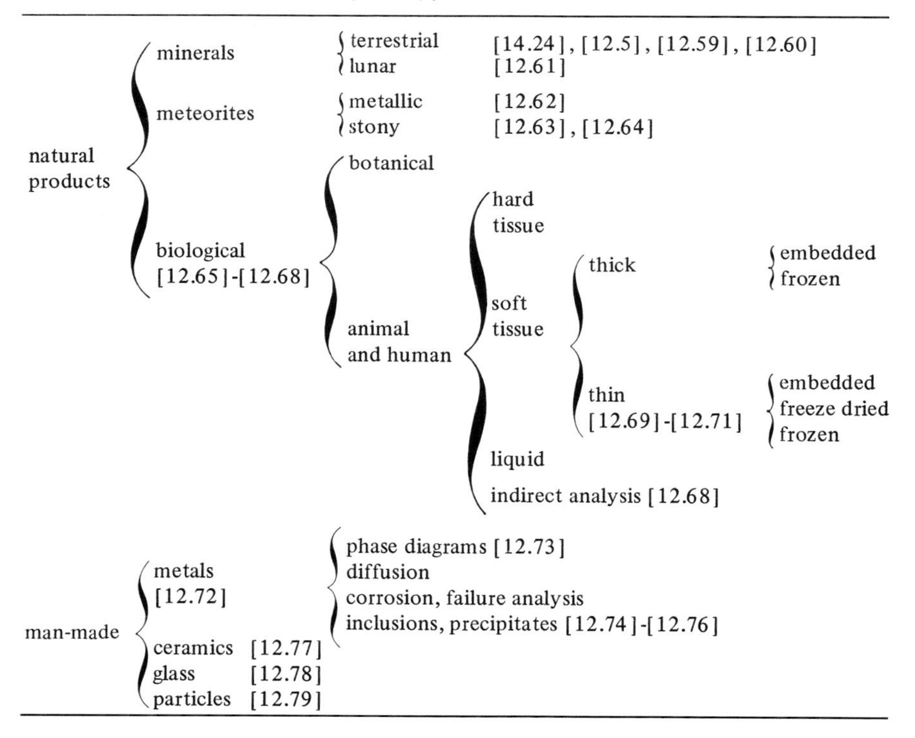

natural products	minerals	terrestrial	[14.24], [12.5], [12.59], [12.60]		
		lunar	[12.61]		
	meteorites	metallic	[12.62]		
		stony	[12.63], [12.64]		
	biological [12.65]-[12.68]	botanical			
		animal and human	hard tissue		
			soft tissue	thick	embedded
					frozen
				thin [12.69]-[12.71]	embedded
					freeze dried
					frozen
			liquid		
			indirect analysis [12.68]		
man-made	metals [12.72]	phase diagrams [12.73]			
		diffusion			
		corrosion, failure analysis			
		inclusions, precipitates [12.74]-[12.76]			
	ceramics [12.77]				
	glass [12.78]				
	particles [12.79]				

that the hyperbolic system failed to describe properly the variation of X-ray intensities emitted, as a function of composition. In view of these results, and of the availability of *ZAF* programs for small modern computers, the question as to the best procedure for mineral analysis is thus still unresolved.

Phase Diagrams of Metals. The usefulness of the electron probe microanalyzer for the study of alloys was established in Castaing's thesis [1.1]. The first area of application cited there is the diffusion of copper and zinc. Adda et al. [12.95] proposed the use of diffusion studies for establishing phase equilibrium diagrams, illustrating the procedure on several binary alloy systems. It is based on the fact that the interdiffusion of two end members of a binary system produces a specimen in which the composition varies continuously through solid solution ranges. In the two-phase regions of the diagram, abrupt composition changes are observed at the phase boundaries. The nature of the diffusion process was described in detail by Eifert et al. [12.96]. Figure 12.28, from [12.95], shows results obtained by Adda et al. In reference [12.97], a list of 37 publications on binary phase diagram studies by means of diffusion couples is given. The limitations of the method were discussed by Adda et al. Phases can fail to appear, or be insufficiently wide for analysis, because the kinetic process has not sufficiently approached equilibrium conditions. Goldstein and Ogilvie [12.73] compared this technique with the more conventional procedure, in which alloys within the two-phase composition range are annealed at the temperatures of interest, and the composition of the resulting phases is analyzed. These compositions indicate the extent of the solubility gap for the given temperature (Fig. 12.29). The procedure, though more tedious, is more accurate, and was used, in combination with other metallographic techniques where necessary, for the characterization of many binary [12.98], [12.99] (Fig. 12.29) and ternary [12.100], [12.101] systems. This subject, as well as other applications of electron probe analysis to metallurgy, are discussed in a review by Goldstein [12.72].

Meteorites and Lunar Rocks. Before the lunar landings, meteorites were the only extraterrestrial objects available for study. Their history is therefore a fascinating subject. Most meteorites are stones and contain small metallic inclusions. The composition and thermal history of these objects were analyzed by many investigators (see [12.65]). Several new minerals such as sinoite (Si_2N_2O) [12.103], [12.104] and ureyite ($NaCrSi_2O_6$) [12.105] were found in meteorites. The latter appeared as an accessory component in an iron metorite. Such inclusions in ferrous meteorites were the object of many studies, which were reviewed in 1970 by Bunch et al. [12.64].

The metallic meteorites consist of iron with varying amounts of nickel [12.64]. Most of them contain a characteristic structure formed mainly by the phases taenite and kamacite, of the iron-nickel system (Widmannstätten structure). Careful analysis of these phases and their boundaries reveals the thermal

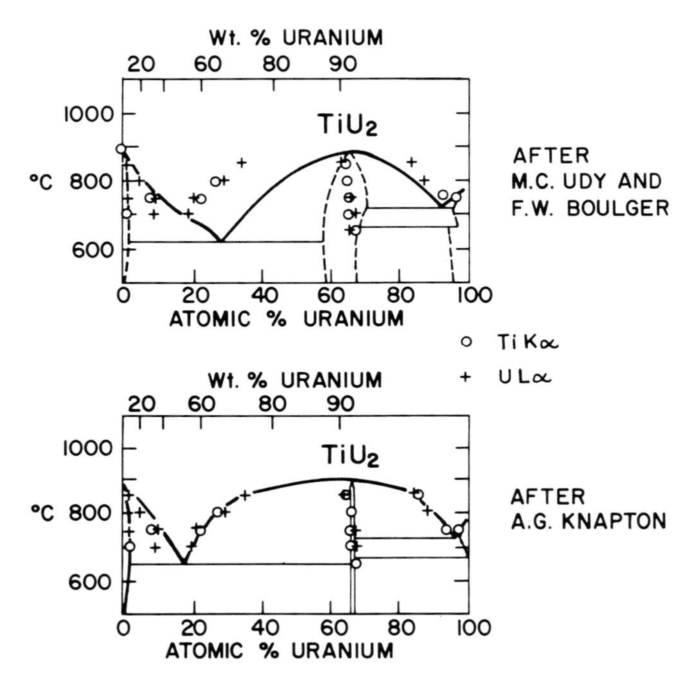

Fig. 12.28. Two proposed phase equilibrium diagrams for the system titanium-uranium, and the results of electron probe measurements on diffusion couples [12.95].

history of iron meteorites [12.62], [12.106], [12.107]. The cooling rate of the alloys is used, in turn, to estimate the size of the parent body from which the meteorites originated.

Many laboratories have participated in the study of the rocks returned from the moon [12.61], and the effort is still continuing. From the point of view of the analyst, this work has produced few novelties, except for the methods of specimen procurement and transportation.

Ceramics. The techniques used for minerals can also be applied to ceramic materials which usually are simpler in composition than minerals. The compatibility of ceramic seals with metals has been investigated. Reference [12.77] (in German) reviews many practical aspects of the analysis of ceramics.

Glass. Silicate glasses have several attractive properties as analytical specimens or standards. They are usually transparent, and inhomogeneity can frequently be recognized by the changes in color or in optical density (schlieren). A very large number of elements can be incorporated in them, and from lithium

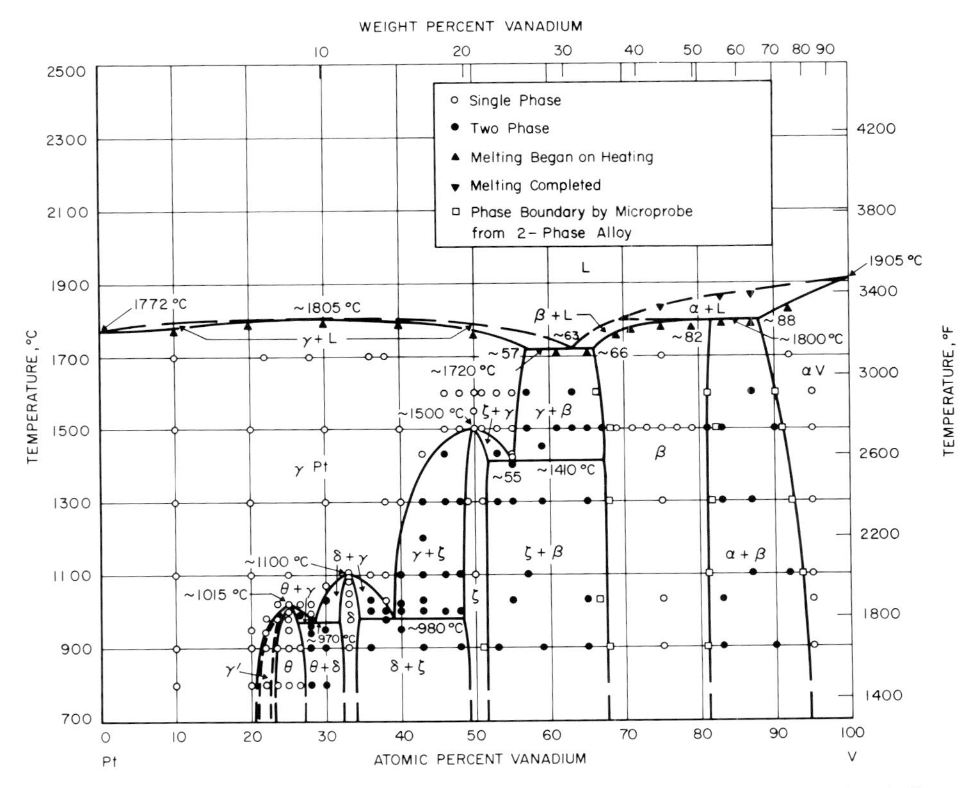

Fig. 12.29. Phase equilibrium diagram of the system vanadium-platinum. Square points indicate determinations by electron probe microanalysis [12.98].

borate to lead germanate, a large range of mean atomic numbers can be covered (see Fig. 12.10 and Tables 12.2 and 12.3). Problems in the excitation with electrons of glass are caused by the mobility of alkali ions [12.108], [12.109] and by the low thermal and electric conductivity of many glasses. These subjects will be treated in Chapter 16. The homogeneity of glasses was studied by Bondarev et al. [12.109] and by Adams et al. [12.110]. Homogeneity was also tested on the NBS glasses mentioned in Tables 12.2 and 12.3. The subject of glass analysis was reviewed by Kane [12.78] in 1973.

The analysis of thin films, small particles, and biological materials will be reviewed in Chapter 13, and applications involving scanning electron probe analysis are discussed Chapter 16.

12. REFERENCES

12.1 Springer, G., *Proc. 7th Int. Conf. on X-Ray Optics and Microanalysis*, Borovsky, I. and Komyak, N., Eds., Leningrad Mashinostroennie, 1976, p. 182.

12.2 Duwez, P., Willens, R. H., and Klement, Jr., W., *J. Appl Phys.* **31**, 1136 (1960).

12.3 Heinrich, K. F. J., Myklebust, R. L., Rasberry, S. D., and Michaelis, R. E., NBS Special Publ. 260-28, U.S. Government Printing Office, Washington, D.C., 1971.

12.4 Marinenko, R. B. and Heinrich, K. F. J., NBS Special Publ. 260-65, U.S. Government Printing Office, Washington, D.C., 1979.

12.5 Keil, K., *Fortschr. Miner.* **44**, 4 (1967).

12.6 Fairbairn, H. W. et al., U.S. Geological Survey Bull. 980, U.S. Government Printing Office, Washington, D.C., 1951.

12.7 Fairbairn, H. W. and Schairer, J. F., *Amer. Mineralogist* **37**, 744 (1952).

12.8 Mercy, E. L. P. and Saunders, M. J., *Earth Planet. Sci. Lett.* **1**, 169 (1966).

12.9 Estour, H., *Glastechn. Ber.* **45**, 499 (1972).

12.10 Poole, D. M. and Thomas, P. M., in *The Electron Microprobe*, McKinley, T. D., Heinrich, K. F. J., and Wittry, D. B., Eds., John Wiley & Sons, New York, 1966, p. 269.

12.11 Poole, D. M., NBS Special Publ. 298, Heinrich, K. F. J., Ed., U.S. Government Printing Office, Washington D.C., 1968, p. 93.

12.12 Ku, H. H., *J. Res. Nat. Bur. Stand.* **70C**, 263 (1966).

12.13 Heinrich, K. F. J., *Anal. Chem.* **44**, 350 (1972).

12.14 Heinrich, K. F. J. and Yakowitz, H., *Proc. 5th Int. Congr. on X-Ray Optics and Microanalysis*, Möllenstedt, G. and Gaukler, K. H., Eds., Springer, Berlin, 1969, p. 151.

12.15 Eisenhart C., *J. Res. Nat. Bur. Stand.* **67C**, 161 (1963).

12.16 Yakowitz, H. and Heinrich, K. F. J., *Mikrochim. Acta*, 182 (1968).

12.17 Heinrich, K. F. J. and Yakowitz, H., *Mikrochim. Acta*, 905 (1968).

12.18 Fischer, D. W. and Baun, W. L., *J. Chem. Phys.* **43**, 2075 (1965).

12.19 Weisweiler, W., *Mikrochim. Acta*, 145 (1972).

12.20 Duncumb, P. and Melford, D. A., *Proc. 4th Int. Conf. on X-Ray Optics and Microanalysis*, Castaing, R., Deschamps, P., and Philibert, J., Eds., Hermann, Paris, 1966, p. 240.

12.21 Bishop, H.E., *J. Phys. D, Appl. Phys.* **7**, 2009 (1974).

12.22 Love, G., Cox, M. G. C., and Scott, V. D., *J. Phys. D, Appl. Phys.* **7**, 2131 (1974).

12.23 Love, G., Cox, M. G. C., and Scott, V. D., *J. Phys.* **11**, 23 (1978).

12.24 Andersen C. A. and Hasler, M. F., *Proc. 4th Int. Conf. on X-Ray Optics and Microanalysis*, Castaing, R., Deschamps, P., and Philibert, J., Eds., Hermann, Paris, 1966, p. 310.

12.25 Henke, B. L., White, R., and Lundberg, B., *J. Appl. Phys.* **28**, 98 (1957); cited by Weisweiler, W., *Mikrochim. Acta*, 744 (1970).

12.26 Henke, B. L., Elgin, R. L., Lent, R. E., and Ledingham, R. B., *Norelco Reporter* **9**, 112 (1967).

12.27 Lukirskii, A. P. et al., cited by Weisweiler, W., *Mikrochim. Acta*, 744 (1970).

12.28 Shiraiwa, T. and Fujino, N., *Japan. J. Appl. Phys.* **9**, 976 (1970).

12.29 Kohlhaas, E. and Scheiding, F., *Arch. Hüttenwesen* **41**, 97 (1970).

12.30 Ruste, J. and Gantois, M., *J. Phys. D, Appl. Phys.* **8**, 872 (1975).

12.31 Kohlhaas, E. and Scheiding, F., *Proc. 5th Int. Conf. on X-Ray Optics and Microanalysis*, Möllenstedt, G. and Gaukler, K. H., Eds., Springer, Berlin, 1969, p. 193.

12.32 Fornwalt, D. E. and Manzione, A. V., *Norelco Reporter* **13**, 39 (1966).

12.33 Duerr, J. S. and Ogilvie, R. E., *Anal. Chem.* **44**, 2361 (1972).

12.34 Love, G., Cox, M. G. C., and Scott, V. D., *J. Phys. D, Appl. Phys.* **9**, 7 (1976).

12.35 Love, G., Cox, M. G. C., and Scott, V. D., *J. Phys. D, Appl. Phys.* 7, 2142 (1974).
12.36 Beamann, D. R. and Isasi, J. A., *Anal. Chem.* **42**, 1540 (1970).
12.37 Heinrich, K. F. J., Myklebust, R. L., Yakowitz, H., and Rasberry, S., D., NBS Special Tech. Note 719, U.S. Dept. of Commerce, Washington, D.C., 1972.
12.38 Chodos, A. A., 8th EPASA Conf., Paper No. 45 (1973).
12.39 Tenny, H., *Metallography* **1**, 221 (1968); also **3**, 338 (1970).
12.40 Heinrich, K. F. J., U.S. NBS Tech. Note 502, 30 (1969).
12.41 Myklebust, R. L. and Heinrich, K. F. J., in *Energy Dispersive X-Ray Analysis: X-Ray and Electron Probe Analysis*, Russ, J. C., Ed., ASTM Special Tech. Publ. 485, American Society for Testing and Materials, Philadelphia, Pa., 1971, p. 232.
12.42 Russ, J. C., in *Energy Dispersive X-Ray Analysis: X-Ray and Electron Probe Analysis*, Russ, J. C., Ed., ASTM Special Tech. Pub. 485, American Society for Testing and Materials, Philadelphia, Pa., 1971, p. 154.
12.43 Beaman, D. R. and Solosky, L. F., *Anal. Chem.* **44**, 1598 (1972).
12.44 Reed, S. B. and Ware, N. G., *J. Phys. E, Scient. Instrum.* **5**, 1113 (1972).
12.45 Statham, P. J., *X-Ray Spectrometry* **5**, 16 (1976).
12.46 Myklebust, R. L., Fiori, C. E., and Heinrich, K. F. J., *Proc. 8th Int. Conf. on X-Ray Optics and Microanalysis*, Ogilvie, R. and Wittry, D. B., Eds., Boston, Mass., 1977 (in print).
12.47 Russ, J. C., *Proc. 9th Ann. Conf. EPASA*, Ottawa, Canada, 1974, p. 22.
12.48 Russ, J. C., *Proc. 8th Int. Conf. on X-Ray Optics and Microanalysis*, Ogilvie, R. and Wittry, D. B., Eds., Boston, Mass., (in print).
12.49 Heinrich, K. F. J., *Adv. X-Ray Analysis* **6**, 291 (1962).
12.50 Jones, M. P. and Shaw, J. L., *10th Int. Mineral Processing Congr.*, Paper 31, Institution of Mining and Metallurgy, London, 1973.
12.51 Heinrich, K. F. J., *Proc. 5th Int. Congr. on X-Ray Optics and Microanalysis*, Möllenstedt, G. and Gaukler, K. H., Eds., Springer, Berlin, 1969, p. 415.
12.52 Chodos, A. A. and Albee, A. L., *Proc. 7th Int. Conf. on X-Ray Optics and Microanalysis*, Borovsky, I. and Komyak, N., Eds., Leningrad Mashinostroennie, 1976, p. 241.
12.53 Grinton, G. R., *J. Austr. Inst. Metals* **17**, 188 (1972).
12.54 Roinel, N., Guernet, J., Lepareur, M., Richard, J. P., Robin, G., and Morel, F., *Proc. 7th Int. Conf. on X-Ray Optics and Microanalysis*, Borovsky, I. and Komyak, N., Eds., Leningrad Mashinostroennie, 1976, p. 331.
12.55 Hadidiacos, C., Geophysical Laboratory, Washington, D.C., private communication, 1977.
12.56 Roinel, N., Richard, J., and Robin, G., *J. Microscopie* **18**, 285 (1973).
12.57 Moher, T. and Lechene, C. P., *Biosci. Commun.* **1**, 314 (1975).
12.58 Tixier, R., *J. Microscopie Biol. Cell.* **22**, 303 (1975).
12.59 Smith, J. V., *J. Geology* **73**, 830 (1965).
12.60 Keil, K., in *Microprobe Analysis*, Andersen, C. A., Ed., Wiley-Interscience, New York, 1973, p. 189.
12.61 *Proc. 2nd Lunar Science Conf.* (3 vols.), Levinson, A. A., Ed., M.I.T. Press, Cambridge, Mass., 1971.
12.62 Wood, J. A., *Icarus* **3**, 429 (1964).
12.63 Wood, J. A., *Icarus* **6**, 1 (1967).
12.64 Bunch, T. E., Keil K., and Olsen, E., *Contrib. to Min. Petrology* **25**, 295 (1970).
12.65 Robison, W. L., in *Microprobe Analysis*, Andersen, C. A., Ed., John Wiley & Sons, New York, 1973, p. 271.
12.66 *Microprobe Analysis as Applied to Cells and Tissues*, Hall, T. A., Echlin, P., and Kaufmann, R., Eds. Academic Press, London, New York, 1974.

12.67 *Biological Microanalysis*, Echlin, P., Galle, P., Eds., Société Francaise de Microscopie Electronique, Paris, 1975. (*J. Microscopie Biol. Cell.* **22**, No. 2-3, 121-512 (1975).)
12.68 Coleman, J. R., in *Practical Scanning Electron Microscopy*, Goldstein, J. I. and Yakowitz, H., Eds., Plenum Press, New York, 1975, Ch. 13, p. 491.
12.69 *Proc. 1st Symp. High Resolution Spectroscopy and X-Ray Microanalysis in Biology*, Nottingham, 1971; *Micron.* **3**, 81-147 (1972).
12.70 *Thin-Section Microanalysis*, Russ, J. C. and Panessa, B. J., Eds., EDAX Lab., Raleigh, N.C., 1972.
12.71 Hall, T. A. and Cosslett, V. E., *Proc. 6th Int. Conf. on X-Ray Optics and Microanalysis*, Shinoda, G., Kohra, K., and Ichinokawa, T., Eds., University of Tokyo Press, Tokyo, Japan, 1972, p. 809.
12.72 Goldstein, J. I., in *Electron Probe Microanalysis*, Tousimis, A. J. and Marton, L., Eds., Academic Press, New York, 1969, p. 245.
12.73 Goldstein, J. I. and Ogilvie, R. E., *Proc. 4th Int. Conf. on X-Ray Optics and Microanalysis*, Castaing, R., Deschamps, P., and Philibert, J., Eds., Hermann, Paris, 1966, p. 594.
12.74 Yakowitz, H. and Heinrich, K. F. J., *Metallography* **1**, 55 (1968).
12.75 Kiessling, R. and Lange, N., *J. Iron and Steel Inst.* **201**, 1016 (1963).
12.76 Kiessling, R., Bergh, S., and Lange, N., *J. Iron and Steel Inst.* **200**, 914 (1962).
12.77 Obst, K. H., Münchberg, W., and Malissa, H., *Elektronenstrahl-Mikroanalyse (ESMA) zur Untersuchung basischer feuerfester Stoffe*, Springer, Vienna, New York, 1972.
12.78 Kane, W. T., in *Microprobe Analysis*, Andersen, C. A., Ed., Wiley-Interscience, New York, 1973, p. 241.
12.79 Bayard, M., in *Microprobe Analysis*, Andersen, C. A., Ed., Wiley-Interscience, New York, 1973, p. 323.
12.80 *The Electron Microprobe, Proc. Symp. by the Electrochem. Soc.*, McKinley, T. D., Heinrich, K. F. J., and Wittry, T. D., Eds., John Wiley & Sons, New York, 1966.
12.81 *X-Ray Optics and X-Ray Microanalysis, Proc. 3rd Int. Symp. on X-Ray Optics and X-Ray Analysis*, Pattee, H. H., Cosslett, V. E., and Engström, A., Eds., Academic Press, New York, 1963.
12.82 *X-Ray Optics and Microanalysis, Proc. 4th Int. Symp. on X-Ray Optics and Microanalysis*, Castaing, R., Deschamps, P., and Philibert, J., Eds., Hermann, Paris, 1966.
12.83 *Proc. 5th Int. Congr. on X-Ray Optics and Microanalysis*, Möllenstedt, G. and Gaukler, K. H., Eds., Springer, Berlin, 1969.
12.84 *Proc. 6th Int. Conf. on X-Ray Optics and Microanalysis*, Shinoda, G., Kohra, K., and Ichinokawa, T., Eds., University of Tokyo Press, Tokyo, Japan, 1972.
12.85 *X-Ray Optics and Microanalysis, Trans. 7th Int. Conf. on X-Ray Optics and Microanalysis*, Borovsky, I. B. and Komyak, N., Eds., Leningrad Mashinostroennie, 1976.
12.86 *Proc. 8th Int. Conf. on X-Ray Optics and Microanalysis*, Boston, Mass., 1977 (in press.
12.87 *Microprobe Analysis*, Andersen, C. A., Ed., Wiley-Interscience, New York, 1973, p. 189.
12.88 *Electron Probe Microanalysis*, Tousimis, A. J. and Marton, L., Eds., Academic Press, New York, 1969.
12.89 Schüller, A. and Ottemann, J., *N. Jb. Miner. Abh.* **100**, 317 (1963).
12.90 Goñi, J. and Guillemin, C., *Bull. Soc. Franc. Miner. Crist.* 87, 149 (1964).
12.91 Klemm, D. D., *N. Jb. Miner. Abh.* **103**, 205 (1965).
12.92 Rucklidge, J. and Stumpfl, E. F., *N. Jahrb. Mineral., Monath.* 61 (1968).
12.93 Albee, A. L. and Ray, L. S., *Anal. Chem.* **42**, 1408 (1970).
12.94 Bence, A. E. and Holtzwarth, W., Ref. [12.86], Paper 38.

12.95 Adda, Y., Beyeler, M., Kirianenko, A., and Maurice, F., *Mem. Scient. Rev. Metallurg.* **58,** 716 (1961).
12.96 Eifert, J. R., Chatfield, D. A., Powell, G. W., and Spretnak, J. W., *Trans. AIME* **242,** No. 1, 66 (1968).
12.97 Heinrich, K. F. J., ASTM Special Tech. Publ. 349, American Society for Testing and Materials, Philadelphia, Pa., 1964, p. 163.
12.98 Waterstrat, R. M., *Metallurg. Trans.* **4,** 455 (1973).
12.99 Waterstrat, R. M., *Metallurg. Trans.* **4,** 1585 (1973).
12.100 Langenscheid, G. and Naumann, F. K., Ref. [12.83], p. 467.
12.101 Widge, S. and Goldstein, J. I., *Metallurg. Trans.* **8A,** 309 (1977).
12.102 Doan, A. S. and Goldstein, J. I., *Metallurg. Trans.* **1,** 1759 (1970).
12.103 Andersen, C. A., Keil, K., and Mason, B., *Science* **146,** 256 (1964).
12.104 Keil, K. and Andersen, C. A., *Natur* **207,** 745 (1965).
12.105 Frondel, C. and Klein, Jr., C., *Science* **149,** 742 (1965).
12.106 Short, J. M. and Andersen, C. A., *J. Geophys. Res.* **70,** 3745 (1965).
12.107 Goldstein, J. I., *J. Geochim. Cosmochim. Acta* **31,** 1001 (1967).
12.108 Vassamillet, L. F. and Caldwell, V. E., *J. Appl. Phys.* **40,** 1637 (1969).
12.109 Bondarev, K. T., Minakov, V. A., and Zaikina, A. A., *Izv. Akad. Nauk SSSR, Neorgan. Mat.* **1,** 963 (1965).
12.110 Adams, R. V., Rawson, H., Fischer, D. G., and Worthington, P., *Glass Technol.* **7,** 98, (1966).

PART V. SPATIAL ASPECTS OF ELECTRON-PROBE MICROANALYSIS

13.
Spatial Distribution of X-Ray Generation

13. Spatial Distribution of X-Ray Generation

An important characteristic of electron probe microanalysis is the spatial distribution of the excited volume, which makes possible the characterization of a microscopic region within a large specimen without previous physical separation. The distribution in depth of the generation of X-rays has already been discussed in Chapter 10, but only for the purpose of determining the absorption losses of X-ray signals on emergence from the interior of the specimen. Now we will study the parameters which determine the dimensions both in depth and laterally of the region excited by the primary electrons, and hence the spatial resolution of electron probe microanalysis. These dimensions are determined by the following:

1. the diameter and intensity distribution of the impinging electron beam;
2. the size, shape, and density of the specimen in the region of electron beam impact;
3. the penetration, deceleration, and scattering of the primary electrons within the specimen (electron diffusion);
4. the absorption of X-rays within the target;
5. the effect of secondary radiation excited by either continuous or characteristic X-rays.

Of these effects, the electron diffusion is usually the determining factor in resolution within a thick (i.e., larger in all directions than the penetration of electrons) specimen.

Spatial resolution both in depth and sideways is particularly important when small objects, such as particles of dust and inclusions, are analyzed. Depth resolution is also crucial in the analysis of thin films. If the excited volume cannot be contained within the objects or phases to be analyzed, then the accuracy of quantitation may be severely compromised. It is particularly difficult, or even impossible, to determine, or even detect, a minor or trace constituent in such objects if it is present at high concentrations in their surroundings.

13.1 DIFFUSION OF ELECTRONS WITHIN THE TARGET

The extensive literature dealing with the penetration of electrons in solid matter has been reviewed by Schumacher [2.2]. Among the quantities of interest are the depth of penetration along the impact direction, the length of the true path, the scattering probability distributions and scattering angles, and the loss of energy of the penetrating electrons, as a function of pathlength or of depth of penetration. A more recent communication by Everhart and Hoff [13.1] can be summarized, as far as it is related to our subject, as follows:

By appropriate scaling, a universal curve for energy-dissipation pathlength versus normalized energy of the electrons can be established. The total pathlength, r_e of the electrons can be derived from the stopping power equation of Bethe [Eq. (9.2.2)], and represented by an equation of the form

$$r_e = k \cdot E_0^n \tag{13.1.1}$$

in which the constant k and the exponent n vary slowly with atomic number. For aluminum-silicon systems, the authors found, between 5 and 25 keV, that $r_e = 4.0 \times 10^{-6} E_0^{1.75}$ g/cm^2 (with E_0 given in kiloelectron volts). They obtained a normalized energy-dissipation curve which has a form not unlike the function $\phi(z)$ described in Chapter 10, and to which it is closely related. The experiments they and other authors performed confirm the validity of Bethe's law in the energy domain of interest in electron probe microanalysis. Cosslett and Thomas [13.2] obtained similar relations between the electron pathlength and the excitation voltage. It should be taken into account, however, that the Bethe equation estimates the loss of energy along the trajectory, which is larger than the depth of penetration into the specimen because the paths are not straight. Therefore, the range of depth cannot be simply deduced from Bethe's law, without taking into account the effects of scattering of the electrons.

13.1.1 Depth Range of X-Ray Generation

Although the range in depth of X-ray generation, z_r, is related to the electron range, it is not identical. The generation of X-rays depends on the ionization cross sections, which vary with energy, and tend to zero as the electron energy approaches the critical excitation energy, E_q. Since on the average the electron energy diminishes with increasing depth of penetration, the depth of X-ray generation is shallower than that of electron penetration. The difference can be taken into account by replacing, in Eq. (13.1.1) the factor E_0^n by the term

$E_0^n - E_q^n$, which also appears in the calculation of absorption loss of primary radiation (Section 9.4). For instance, Castaing [13.3] proposed (in terms of operating and critical excitation potentials):

$$z_r = 0.033(V_0^{1.7} - V_q^{1.7})\frac{1}{\rho}\frac{A}{Z}\ \mu\text{m} \simeq 6.5 \times 10^{-6}(V_0^{1.7} - V_q^{1.7})\ \text{g/cm}^2.$$

Similar expressions were given by various authors (e.g., [12.23]); they can also be derived from the differential depth distribution function, $\phi(z)$. If we use a simple exponential approximation [Eq. (10.3.34b)] and calling p the fraction of photons which are produced below the depth z_r, we obtain:

$$1 - p = \frac{(a_1\gamma)^{-1}\int_0^{z_r} \exp\left(-\frac{z}{a_1\gamma}\right)dz}{(a_1\gamma)^{-1}\int_0^{\infty} \exp\left(-\frac{z}{a_1\gamma}\right)dz} \tag{13.1.3}$$

so that

$$-\left|\exp\left(-\frac{z}{a_1\gamma}\right)\right|_0^{z_r} = 1 - p; \quad z_r = -a_1\gamma \cdot \ln p. \tag{13.1.4}$$

Table 13.1 shows the values of $z_r/a_1\gamma$ and of z_r/γ for various values of p, and assuming that $a_1 = 2.6 \times 10^{-6}$ g/cm².

With the exponent $n = 1.65$ in the formulation of $\gamma = V_0^n - V_q^n$, the following definition of the range is in close agreement with those of other authors:

$$z_r = 7 \times 10^{-6}\ \gamma\ \text{g/cm}^2. \tag{13.1.5}$$

Although according to Eq. (13.1.3), 6.8% of the primary photons would be generated below this range, in reality this fraction is smaller, since at low energies the electrons lose energy faster than predicted by Bethe's law.

The values of z_r can be read from the nomogram on Fig. 13.1. It can be observed that shallow depths of generation cannot be obtained for lines of short wavelength (large E_q), by lowering the operating voltage since no useful intensities are obtained at low overvoltages.

Table 13.1.

p	0.50	0.37	0.10	0.05	0.01
$z_r/a_1\gamma$	0.69	1.0	2.3	3.0	4.6
$10^6 z_r/\gamma$	1.8	2.6	6.0	7.8	12.0

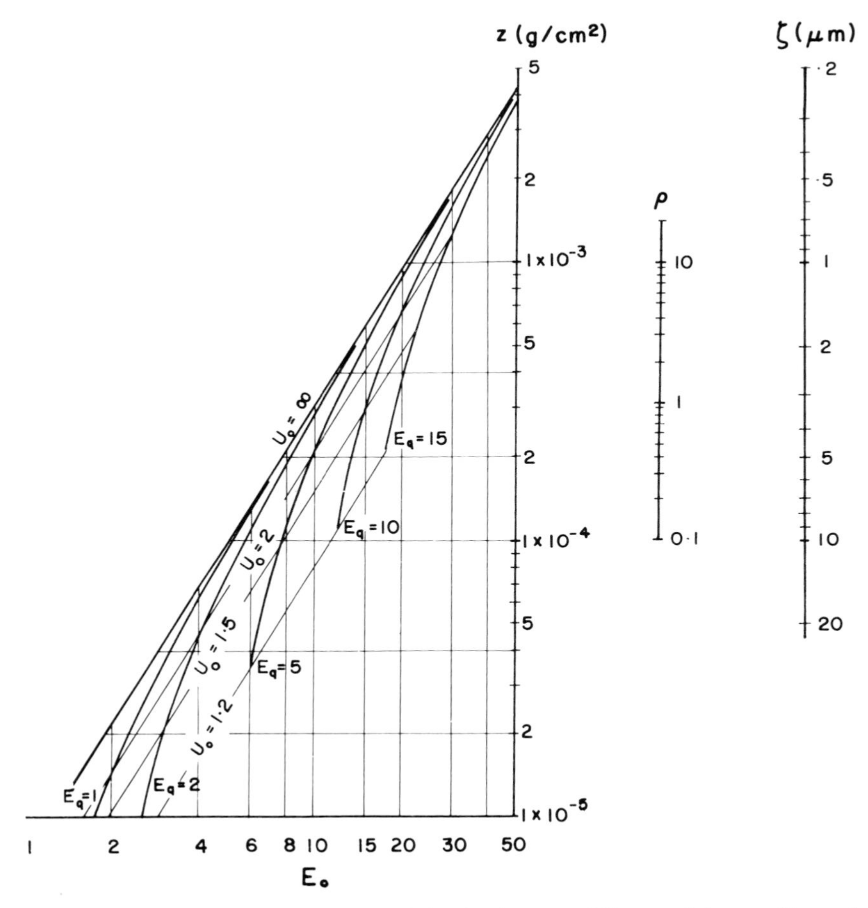

Fig. 13.1. Nomogram for the calculation of the depth range of X-ray emission, z_r. For each pair of either E_0 and E_q, or E_0 and U_0, a value of z_r is found. Tracing a line through the point at the z axis and the value of target density, the linear depth is found at the right-hand scale.

If the primary X-rays are strongly absorbed by the target, the effective range of emission is also a function of the absorption coefficient and of the X-ray emergence angle. If p' is the fraction of *emitted* photons which were generated below z'_r, we obtain:

$$1 - p' = \frac{\int_0^{z'_r} \exp - z\left(\frac{1}{a_1\gamma} + \chi\right)dz}{\int_0^{\infty} \exp - z\left(\frac{1}{a_1\gamma} + \chi\right)dz} = \frac{\int_0^{z'_r} \exp - z(\sigma + \chi)dz}{\int_0^{\infty} \exp - z(\sigma + \chi)dz}$$

$$z'_r = \frac{-a_1\gamma \ln p}{1 + a_1\gamma\chi} = \frac{-\ln p}{\sigma + \chi} = -f_p a_1 \gamma \ln p = -\frac{f_p \ln p}{\sigma} = f_p \cdot z_r. \tag{13.1.6}$$

This equation illustrates the convergence of the emergence pattern towards Duncumb's thin-film model as the absorption factor f_p decreases [Eq. (12.4.1)].

Some models of X-ray emission are based on the concept of a mean depth of X-ray generation. There are, however, several similar but not identical definitions to be distinguished [13.4]. The *mean depth of generation of primary X-rays* is defined as follows:

$$\bar{z} = \frac{\int_0^\infty \phi(z) \cdot z \cdot dz}{\int_0^\infty \phi(z) dz}. \tag{13.1.7}$$

To illustrate this and other depth definitions, we can at this point use a simple exponential model for distribution in depth [Eq. (10.3.34)]. We obtain:

$$\bar{z} = a_1\gamma = 2.6 \times 10^{-6}\gamma.$$

This parameter is not to be confused with the *median depth of generation* (i.e., the depth, above which half of the X-ray photons are generated), which is the case given by Table 13.1, with $(p = 0.5) : x_{\text{med}} = 1.8 \times 10^{-6}\gamma$. The *mean depth of generation of emergent X-rays* is given by:

$$\bar{z}' = \frac{\int_0^\infty \phi(z) \exp(-\chi z) z dz}{\int_0^\infty \phi(z) \exp(-\chi z) dz} = f_p a_1 \gamma = f_p \bar{z}. \tag{13.1.8}$$

Finally, an *effective depth of X-ray production* can be defined as the depth at which an X-ray emission from a single point would have to take place if such an emission were to be attenuated the same as the real emission. From Beer's law we obtain for the effective depth, $\tilde{z}$,

$$f_p = e^{-\chi\tilde{z}} \text{ so that } \tilde{z} = \frac{-\ln f_p}{\chi} = \frac{\ln(1 + a_1\gamma\chi)}{\chi}. \tag{13.1.9}$$

The effective depth is the "mean depth of X-ray production" used in Green's thesis [6.2, p. 120] and by Andersen and Wittry [10.23]. For moderate absorption we can use the approximation: $\ln(1 + y) = y - y^2/2 + y^3/3 - \ldots$ so that

$$\tilde{z} \simeq a_1\gamma - \frac{1}{2}a_1{}^2\gamma^2\chi \cdots = \bar{z} - \frac{1}{2}\bar{z}^2\chi \ldots . \qquad (13.1.10)$$

Unlike $\bar{z}$, the parameters $\bar{z}'$, and $\tilde{z}$ depend on χ and hence on the X-ray emergence angle.

The mean depth of generation is, according to Eq. (13.1.7), approximately one-third of the range of generated X-rays, regardless of operating condition, line, or target composition.

When instead of the exponential model a more sophisticated model is used for $\phi(z)$, the relations developed above are still equally valid, although the numerical value of the diverse mean depths will slightly increase.

It would appear that in view of the broad and assymetric distribution in depth of the primary radiation any model of X-ray absorption based on a "mean depth of X-ray generation" is didactically misleading. We prefer to achieve the same purpose by the introduction of the energy-dependent scaling term γ without assuming a simplification which is so clearly contrary to the physics of X-ray generation.

13.1.2 Lateral Distribution of X-Ray Generation

The lateral resolution of the X-ray signal (and of other signals) from a flat and electron-opaque target depends on both the lateral distribution of electrons within the beam of primary electrons, and of the diffusion of the electrons penetrating the target. We will first consider the effects of electron diffusion. As described in Chapter 2, the initial direction of the penetrating electrons is randomized due to scattering. Therefore, a high density of ionizations is observed in a funnel-shaped region beneath the point of impact of the electron beam (which we consider of negligible lateral dimensions), centered around the original direction of the beam. After spreading due to scattering events, the electrons cover an approximately spherical volume which intersects the surface (Figs. 2.2 and 2.3). This description is in accordance with both the observation of an elongated region of intense energy deposition beneath the point of electron beam impact [13.5], and that of an approximately spherical bound of the electron diffusion (see Fig. 2.2). It also makes plausible the simple model of Archard and Mulvey [9.39], [10.12] in which all electrons penetrate to a point the depth of which decreases with increasing atomic number, and spread from there isotropically in all directions.

These considerations are valid over the range of operating potentials commonly used; it follows that the lateral spread of primary electrons—and hence of sites of excitation of primary X-ray photons is, over this potential span, nearly proportional to the depth range. For the heaviest elements, for which scattering can be assumed to take place at the specimen surface, the diameter of the region emitting X-rays would be twice the depth range, and for the lightest elements it would be slightly larger than the depth range. As can be verified by Monte Carlo calculations, the lateral distribution of sites of primary X-ray generation from a beam of negligible width is a bell-shaped curve, peaking at the point of impact of the primary electron beam (Fig. 13.2). It resembles a Gaussian distribution; however, it differs from a Gaussian by the well-defined lateral range.

Unfortunately, the fairly sharp demarcation of the region of primary excitation can be perturbed by the width of the real electron beam, by the wide and ill-defined spatial range of secondary excitation, and, at least in certain instrumental configurations, by doubly backscattered and stray electrons which excite X-ray emission outside the desired lateral range of the primary beam. Artifacts produced by these effects render difficult the determination at the trace level of elements which are abundant in the vicinity of the analyzed region. A simple test for stray electrons consists of directing the focused electron beam into the center of a very small aperture located above a hole drilled into a metal block. If the beam is well focused, it is entirely absorbed within the Faraday cage thus formed, and the X-radiation observed should be negligible. In practice, intensities as large as 1% or more have been observed in some instruments, and sometimes even a strong X-ray signal from the material surrounding the aperture diaphragm was observed. The emission of such parasite signals renders impossible the analysis of small particles and inclusions, and must be stopped by the insertion of apertures at appropriate positions in the electron optics. The only limitation of this test is in the fact that little can be stated about the quality of the beam below the diameter of the aperture.

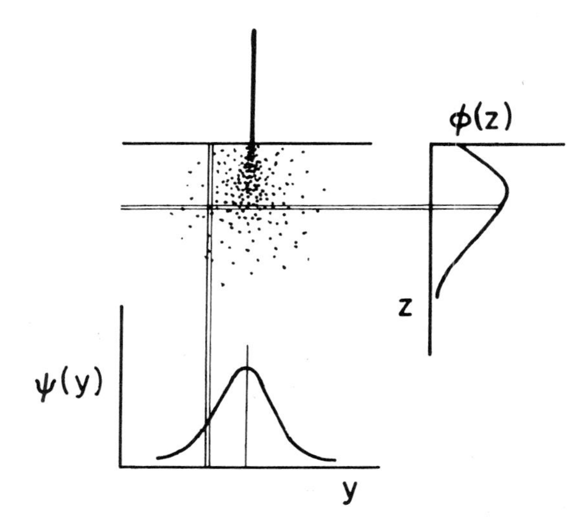

Fig. 13.2. The lateral diffusion of electrons produces a lateral distribution function for X-ray generation, $\phi(y)$, similar in shape to a Gaussian distribution.

Particular attention must be given to the production of fluorescence due to the continuum which may take unexpected dimensions in specimens of low average atomic number. These effects can sometimes be avoided by using soft lines (e.g., Zn $L\alpha$ instead of Zn $K\alpha$), and operating at low voltages, so that, at the same time, the volume of primary excitation is reduced. Double backscattering may occur when the pole piece of the objective lens is very close to the specimen. It can be reduced by covering this or other closely located instrument components with a material of low atomic number, such as beryllium or colloidal graphite. The effects of signal emission due to double backscattering, or of other stray electrons within the specimen area, are more severe in the use of solid-state detectors in the energy-dispersive mode, since in the crystal spectrometer the signals generated outside the focal circle are strongly attenuated. Wherever feasible, tests for spurious signals should therefore be performed with energy-dispersive spectrometers.

13.2 THE WIDTH OF THE ELECTRON BEAM

It is not difficult to design electron optics of such characteristics that in the analysis of semi-infinite specimens at conventional (10-20 kV) operating potentials a useful X-ray signal is produced by an electron beam of a diameter much smaller than that of the region of electron diffusion. In that case, the electron diffusion essentially defines the spatial resolution of the X-ray signal. However, some objects to be analyzed are smaller than the potential region of electron diffusion, in depth only ("thin film") or in more than one dimension ("slender fibers" and "small particles") (Table 13.2, Fig. 13.3). The lateral dimension of measurement of a small particle is given by its diameter. Nevertheless, it is useful to diminish the beam size so as to increase the obtainable signal and to reduce X-ray signals from surrounding objects. But a fine beam is even more desirable in the analysis of thin unsupported inhomogeneous films such as biological thin sections. If the section is sufficiently thin, and particularly if the material is of low atomic number, the diffusion of electrons is greatly reduced, and hence high resolution can be obtained if a fine electron beam is available.

As discussed in Chapter 3, the beam cross section at the specimen surface is strongly affected by the aberrations of the objective lens, especially the spherical aberration which is proportional to the cube of the aperture angle α [Eq. (3.2.8)]. Hence, the diameter of the focused beam is also proportional to α^3 [Eq. (3.2.9)], and this diameter can be diminished by using a smaller final aperture and hence a smaller angle α. This change, however, also reduces sharply the beam current which is proportional to the solid angle subtended at the level of the aperture, and hence to α^2. If the drop in beam current is not objectionable, the beam diameter can be reduced, in high-resolution instruments, to less than 10 nm (100 Å). In this operational mode, other defects such as the chromatic aberration (which is proportional to α), and diffraction effects (pro-

Table 13.2. Shapes of Targets.

						EXAMPLES
Target	homogeneous	larger than electron penetration		surface flat	("semi-infinite")	polished thick specimen
				surface irregular	("SEM specimen")	fracture specimen
		smaller	uniform film	unsupported		evaporated layers
				supported		coatings on alloys
			nonuniform thin section	"unsupported"		biological thin section
				supported		microdroplets
			particle	"free" (on thin film)		inclusions on extraction replicas, asbestos fibers
				embedded		inclusions in alloy
	inhomogenous: (phase boundaries)			deconvolution		grain boundaries in materials and alloys

portional to $1/\alpha$) become significant, and depending on their magnitudes, there is an optimum aperture angle at which the beam diameter is minimal ([13.6, p. 80]). To estimate this minimum, and the corresponding aperture size, the diameters calculated for each aberration are added in quadrature to the Gaussian (i.e., aberration-free) beam diameter, d_g, which depends solely on the demagnification of the virtual electron source [Eq. (3.2.8)]:

$$d_{\text{total}}^2 = d_g{}^2 + \Sigma\, d_{ab}{}^2. \tag{13.2.1}$$

Each of the component diameters is expressed as a function of α. The resulting equation is then differentiated to find d_{min}.

In the case of X-ray analysis, it should be noted that the generation and detection of X-rays are both very inefficient. The number of characteristic photons produced per incident electron from a solid specimen of a pure element is typically on the order of 10^{-3} (see Fig. 11.4). For an element present at a mass fraction of 0.01, it would be approximately 10^{-5}. For a small particle or thin film, this number could be much smaller still. Furthermore, the product of detector efficiency and solid angle covered by a Si(Li) detector is usually 10^{-2} to 10^{-4}, and for a crystal spectrometer, on the order of 10^{-4} to 10^{-6}. Therefore, a beam dose on the order of 10^8 electrons, equivalent to $10^8 \times 1.6 \times 10^{-19}$ C

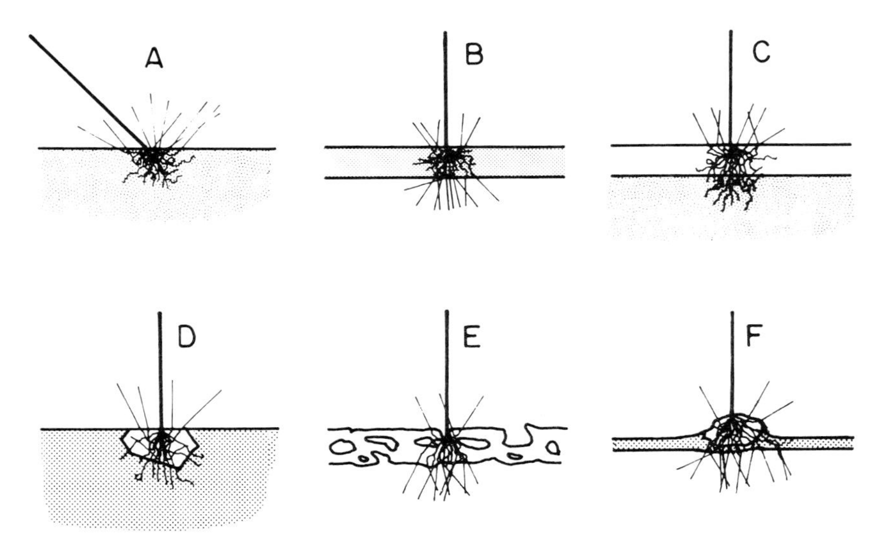

Fig. 13.3. Some cases not covered by the classical *ZAF* correction. (A) Semi-infinite specimen, inclined electron beam (SEM configuration). (B) Unsupported uniform thin film. (C) Supported film. (D) Embedded particle. (E) Film of irregular thickness (biological tissue section). (F) "Free" particle, supported by a very thin film of low mean atomic number.

or 1.6×10^{-11} A · sec must be expended, on the average, to produce a single X-ray signal pulse from a massive elemental target. (This number may be lower by a factor of ten for the Si(Li) detector, and perhaps higher by the same factor for a crystal spectrometer.) But, according to the Poissonian count distribution, the precision of an X-ray measurement, under stable conditions, depends on the number of photons that were collected [see Eqs. (6.2.1), (8.2.13), and (12.3.6)]. We learn, for instance, from Eq. (12.3.6) that even in the absence of background, we must obtain 10^4 pulses in order to have a Poissonian relative standard deviation of 1%. Let us assume that we are satisfied with a counting period of 100 sec. Hence, the count rate for collecting 10^4 pulses in this time period must be 100 counts/sec, and the calculated beam current is in the order of 10^{-9} A. Currents of this magnitude require an aperture of such size that the defects other than the spherical aberration can be ignored. Therefore, we can write for the beam diameter:

$$d_{\text{total}}^2 = d_g^{\,2} + d_s^{\,2} \tag{13.2.2}$$

where, d_g is the aberration-free Gaussian beam diameter at the specimen level, and d_s is the diameter of an image of a point source which is determined by the spherical aberration. We will now derive d_g, as a function of the aperture α.

According to Richardson's Eq. (3.2.13), the current density at the filament tip is $j_0 = K(1 - r)T^2 \exp(- e\phi/kT)$. For metals, $K = 120\ \text{A/cm}^2\,\text{deg}^2$. The reflection coefficient r is 0.05 and can be neglected here. The current density at the specimen is, according to Langmuir's Eq. (3.2.18):

$$j_g = j_0 \frac{eV_0}{kT} \alpha^2 \simeq j_0 \left(\frac{11.6 \times 10^6}{T} V_0 \right) \alpha^2 \ \text{A/cm}^2. \tag{13.2.3}$$

Here, α is measured in radians, and the filament temperature T is measured in degrees Kelvin. The operating potential, V_0, is measured in kilovolts.

The beam current i_b is equal to

$$i_b = \frac{\pi}{4} d_g^{\,2} j_g = \frac{\pi}{4} d_g^{\,2} j_0 V_0 \frac{e}{kT} \alpha^2 \tag{13.2.4}$$

so that

$$d_g^{\,2} = \frac{4}{\pi} \cdot i_b \cdot \frac{kT}{eV} \cdot \frac{1}{j_0 \alpha^2}.$$

Calling $\overline{N}$ the number of collected photons, t the time of collection, P the detector efficiency, and $\Omega/4\pi$ the detector aperture, we get, according to Eq. (8.2.8):

$$\overline{N} = i_b \cdot t \cdot I \cdot f \cdot (\Omega/4\pi) \cdot P.$$

As usual, I is the efficiency of X-ray photon generation (photons/electron), and f is the absorption factor. Hence

$$d_g^{\,2} = \frac{4\pi}{\Omega P} \cdot \frac{\overline{N}}{I \cdot f \cdot t} \cdot \frac{4}{\pi} \cdot \frac{kT}{eV_0} \cdot \frac{1}{j_0} \cdot \frac{1}{\alpha^2}. \qquad (13.2.5)$$

For d_s we have $d_s = (1/2)C_s\alpha^3$ [see Eq. (3.2.9)], or $d_s^{\,2} = b\alpha^6$ with $b = (1/4)C_s^2$ so that

$$d^2_{\text{total}} = d_g^{\,2} + d_s^{\,2} = a\alpha^{-2} + b\alpha^6. \qquad (13.2.6)$$

The equation has a minimum for d_{total}: $d_{\text{min}} = 1.24\, a^{3/8} C_s^{1/4}$ and the aperture required to achieve this minimum diameter is

$$\alpha_{\text{opt}} = (a/3b)^{1/8} = 1.04 a^{1/8} C_s^{-1/4}.$$

This last equation summarizes all effects which contribute to determine the Gaussian beam diameter required for X-ray analysis. Note that

$$a = \frac{4}{\pi} \cdot i_b \cdot \frac{kT}{eV_0} \cdot \frac{1}{j_0} = \frac{16}{\Omega} \cdot \frac{\overline{N}}{t \cdot I \cdot f \cdot P} \cdot \frac{kT}{eV_0} \cdot \frac{1}{j_0}$$

and that therefore the minimum beam diameter increases with the 3/8th power of i_b, of $1/V_0$, and of $1/j_0$, as well as with the 1/4th power of the spherical aberration, C_s. Wells [13.6] states these conditions as follows:

$$d_{\text{min}} = 0.7 C_s^{1/4} \beta^{-3/8} i_b^{\,3/8} \text{ cm}$$

and

$$i_b = 2.6 C_s^{-2/3} \beta\, d_{\text{min}}^{\;8/3} \text{ A}. \qquad (13.2.7)$$

Here,

$$\beta = j_g/\pi\alpha^2 = \frac{j_0}{\pi}\left(1 + 11.6\frac{V_0}{T}\right) \simeq \frac{j_0}{\pi}\,\frac{11.6 V_0}{T}$$

is the brightness of the system (i.e., current/(area × solid angle) in the units $A/(cm^2 \cdot sr)$). Wells gives the following values for β at an operating potential of 25 kV:

tungsten hairpin: 3×10^4 - $1.5. \times 10^5$ (depending on T);
lanthanum hexaboride: 10^5 - 5×10^6 (depending on radius of tip);
tungsten field emission: 2×10^8.

To permit a focusing condition close to optimal, instruments designed to work over a wide range of beam currents must therefore have several interchangeable apertures of a variety of diameters. It can also be appreciated from these calculations that the analysis of thin films and small particles is often severely limited by the low efficiency of X-ray production and detection. For this reason, closely coupled solid-state detectors are desirable in such situations, particularly in the analysis of thin biological sections of soft tissue.

If we apply the beam current of 10^{-9} A, found necessary to detect a 1% level of an element in a semi-infinite specimen under average conditions, to Eq. (13.2.6), with a spherical aberration of $C_s = 2$ cm, and a value of $\beta = 5 \times 10^4$, we obtain a minimum diameter of 0.08 μm. For a beam current of 3×10^{-8} A, which is typical for many electron probe analyses, this minimum diameter increases to 0.21 μm. These values agree reasonably with practical experience. The analyst will, of course, determine the necessary minimum currents by experiment rather than from theoretical calculations of detector efficiency and X-ray productions. Once he has defined this current, he can determine the corresponding minimum diameter and the appropriate aperture size from Eq. (13.2.6).

It is commonly assumed that the cross section of the beam of an electron probe always has a Gaussian intensity distribution. This assumption is a good approximation when a point source is imaged by an electronic lens of significant spherical aberration, such as the objective lens of the electron probe. In some instances the wings of the distribution can be clipped by an aperture, at the position of a focus, before the next step of demagnification; however, if, as in the typical electron probe, the last lens is of short focal distance, the effects of its spherical aberration cannot be reduced by this procedure.

If, at the other extreme, a limiting aperture is used to reduce the intensity of the electron beam by a large factor (Fig. 3.24) and the objective lens beneath the limiting aperture is not set to sharply focus the beam, the resulting image in the specimen plane is a well-delineated circle of practically uniform density. In analytical practice, intermediate situations between these extremes may be present. Therefore, it cannot always be assumed without proof that the cross section of the beam is Gaussian. Further complications of the distribution can occur due to the anisotropic emission of electrons from the emitter tip [13.7].

We will not enter the argument on how to compare the diameter of a Gaussian distribution with that of a disk of uniform density, which is discussed in [13.6] and only introduces slight changes in the numerical values of the calculations we have described in this section.

13.3 DETERMINATION OF THICKNESS AND COMPOSITION OF THIN LAYERS

We will now consider the characterization of specimens such as thin films, small inclusions, biological sections, and small particles. These targets have in common that their dimensions are smaller than that of the region of excitation in a semi-infinite target. The simplest type of such a "small" target is a thin homogeneous layer of uniform thickness below the electron range. Such a film can either be supported by some material of different composition, or unsupported. The purpose of the electron probe measurements can be the determination of the layer thickness, or of its composition, or of both. The phenomena of interest in the analysis of a supported film are shown schematically in Fig. 13.4.

Sweeney et al. [13.8] first used the electron probe for measuring the thickness of layers of several metals in the thickness range from 0 to 3000 Å, evaporated upon glass substrates. The authors attempted to use Castaing's depth distribution curves for calibration, but found that the results were in significant error, and hence, they suggested that empirical calibration curves be used.

Cockett and Davis [13.9] also used Castaing's depth distributions, and were therefore restricted to an operating potential of 29 kV. They used radiation both from the film and from the substrate to determine film thicknesses, and they estimated that for copper, films as thin as 4 to 8 Å could be detected, and that the upper limit of the method was about 2 μm.

More recently, Bishop and Poole [13.10] published a simple graphic method for estimating the thickness of thin supported films which is also based on the $\Phi(z)$ function. The authors employed a model used previously by Bishop [13.11], [10.30] in Monte Carlo calculations. They assume that the depth distribution is the same for the film and the substrate. Hence, the fraction of ionizations above the boundary at the depth z_t is equal to

$$p(t) = \frac{\int_0^{z_t} \phi(z)dz}{\int_0^{\infty} \phi(z)dz}. \tag{13.3.1}$$

If absorption can be neglected and substrate and film have similar atomic numbers

$$k_a = p_a(t) \qquad k_b = (1 - p_b(t)) \tag{13.3.2}$$

where a is the element of the film and b is the element of the substrate. The integral curves representing $p(t)$ depend on $\bar{Z}$, E_0, and E_q. For a generalization, they are scaled with respect to the Bethe range. To justify this scaling, the authors invoke the observations of Spencer [13.12] concerning the scaling of

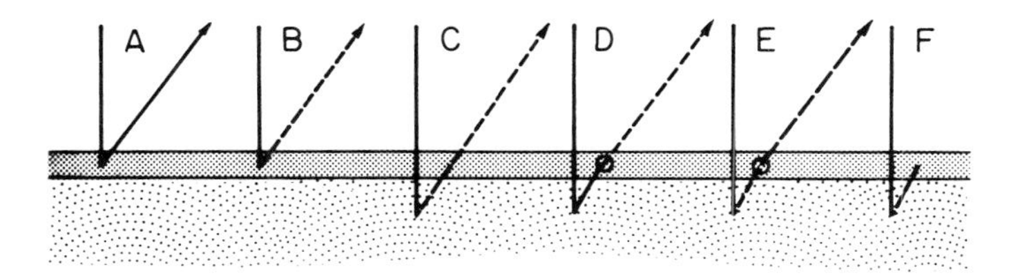

Fig. 13.4. Events of interest for the analysis of a thin film supported by a substrate. (A) Electrons are backscattered within the film. (B) Characteristic primary X-rays are produced within the film. (C) Characteristic and continuous X-ray photons are produced in the substrate and are emitted through the film. (D) Electrons backscattered by the substrate produce X-ray emission within the film. (E) X-rays produced within the substrate excite secondary X-ray emission in the film. (F) X-rays produced within the substrate are attenuated by absorption within the film.

dimensions of electron scattering with regard to the Bethe range [13.13], [13.14]. When an absorption correction is needed, they resort to Green's model of a point source, so that $f = \exp(-\chi z_m)$ in which case z_m is the median depth of ionization ($p = 0.5$). The depth $\bar{z}$ would correspond to $p/2$ for the surface layer, but it is sufficient to assume that all X-rays are generated at half the thickness of the layer:

$$p = k \frac{\exp(-\chi\bar{z})}{\exp(-\chi \cdot \tfrac{1}{2}z_t)}. \qquad (13.3.3)$$

They assumed that the point source model is acceptable for $f \geqslant 0.85$.

The authors, and other investigators as well, pointed out that in thin films the effects of indirect radiation due to primary radiation generated in the film are negligible, because the fraction of primary radiation absorbed in the film is much smaller than in a semi-infinite target (Fig. 13.5). Changes in the mean atomic number have, however, a significant effect because a significant fraction of the ionization within the surface layers is due to backscattered electrons [Eq. (10.3.3)], and electron backscattering varies considerably with atomic number. As a consequence, the mean pathlength at the surface, R_0, varies with atomic number [Eq. (10.3.24)]. Bishop and Poole proposed that when the atomic number of the substrate differs from that of the film, an interpolation between the respective curves on their graphs should be performed. The effect of changes in backscattering due to differences in atomic number between layer and support had also been noted previously by Cockett and Davis [13.9].

This substrate effect was more clearly defined by Hutchins [13.15] in the development of a method for determining film thickness in the order of 10^{-5} g/cm^2. This author measured the ratio of X-ray emissions from the supported film of thickness z_t, $I'(t)$, and from a massive (semi-infinite) target of the same

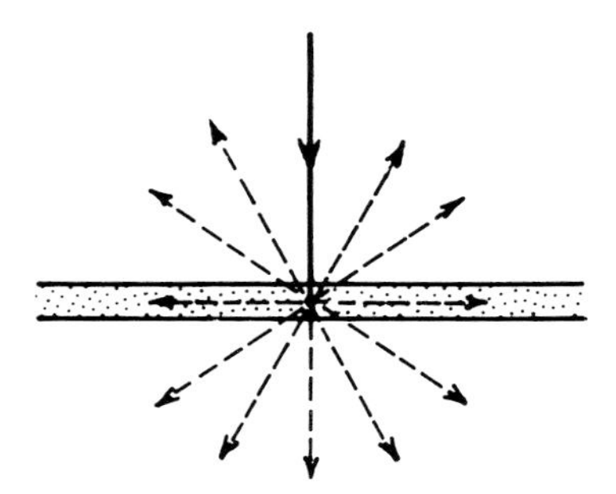

Fig. 13.5. Most X-ray photons generated within a thin film (broken lines) leave the film without being absorbed. Therefore, the fluorescent radiation excited by them is negligible.

material, $I'(\infty)$, which is used as a standard. The ratio is expressed as a combination of three parameters:

$$I'(t)/I'(\infty) = \Phi(0) \cdot I_u'(t)/I'(\infty). \qquad (13.3.4)$$

$\Phi(0) = I(0)dz/I_u(0)dz$ is the ratio of the emission from an infinitesimally thin target layer at the specimen surface to that from an identical unsupported layer. $I_u'(t)$ is the intensity which would be emitted from the film of thickness t if the film were unsupported by any substrate. If we call $\Phi_u(z)$ the depth distribution of emission for a free film scaled so that $\Phi_u(0) = 1$, we get

$$I'_u(t) = K \int_0^{z_t} \Phi_u(z)\exp(-\chi z)dz. \qquad (13.3.5)$$

For the semi-infinite target used as standard we can use the same distribution function and proportionality factor:

$$I'(\infty) = K \int_0^{\infty} \Phi_u(z)\exp(-\chi z)dz. \qquad (13.3.6)$$

For the parameter $\Phi(0)$ which expresses the effect of backscatter from the substrate, Hutchins proposed the following:

$$\Phi(0) = 1 + 2.0\eta_s. \qquad (13.3.7)$$

Here, η_s is the backscatter coefficient of the substrate. If the atomic number of the substrate is below 20, the correction is changed to $1 + 1.7\,\eta_s$. A comparison of experimental values of $\Phi(0)$ obtained by Hutchins with those of Castaing and Descamps [9.1] shows considerable disagreement.

Since the author concerns herself only with very thin films, a simple approach assuming a linear increase of $\Phi_u(z)$ with z suffices:

$$\Phi_u(z) = 1 + B \cdot z, \qquad B \ll 1. \tag{13.3.8}$$

The value of B can be obtained from the initial slope of depth distribution curves after substrate correction, from theoretical calculations on scatter, or from curve fitting on observations of emissions from thin films. Hutchins offers the following typical values:

V_0 (kV):	10	20	30
B (cm²/g):	9000-12,000	4000-5000	2000-3000

Great accuracy in the value of B is not needed, since the effect of this parameter is of second order. After integrating Eq. (13.3.5), and expanding into a series, we get

$$I_u'(t) = K[z_t + z_t^2(B/2 - \chi/2) - z_t^3(B\chi/3 - \chi^2/6) + \cdots] \tag{13.3.9}$$

which we can reduce to its first two terms. Combining the above expressions, we write:

$$I'(t)/I'(\infty) = (1 + 2\eta_s) \cdot [z_t + 1/2 z_t^2(B - \chi)] \Big/ \int_0^\infty \Phi(z)dz \cdot f_p. \tag{13.3.10}$$

From this equation, the thickness of the film z_t can be obtained iteratively:

$$z_t = (I'(t)/I'(\infty)) \Big/ \left\{ \left[1 + z_t\left(\frac{B}{2} - \frac{\chi}{2}\right)\right] \int_0^\infty \Phi_u(z)dz \cdot f_p \right\}. \tag{13.3.11}$$

If the integral in this equation could be evaluated, the calculation could be performed without further measurements. Instead Hutchins proposes that the "normalization factor" $(\int_0^\infty \Phi_u(z)dz \cdot f_p)^{-1}$ be obtained by applying the equation to the measurement of X-ray emission from a film of known thickness. It can also be obtained from the initial slope of $I'(t)/I'(\infty)$ versus z, after substrate correction.

Hutchins' procedure eliminated the need for direct reference to the experimental $\Phi(z)$ curves which at the time the procedure was published were available for an operating potential of 29 kV only.

Djurić and Cerović [13.16] extend Hutchins' method to the analysis of thin alloy films, with or without substrate. Calling K_i the normalization factor for element i, $(\int_0^\infty \Phi_u(z)dz \cdot f_p)^{-1}$, we can write for the intensities from thin films of elements a and b, compared with those of massive elements,

$$I'(ta)/I'(a) = K_a(1 + 2.0\eta_s)[z_t + 1/2z_t^2(B - \chi_a)]$$
$$I'(tb)/I'(b) = K_b(1 + 2.0\eta_s)[z_t + 1/2z_t^2(B - \chi_b)] \quad (13.3.12)$$

and the factors K_a and K_b can be determined from films of known thickness. For an alloy foil of thickness z_t the ratios of relative intensities are:

$$\frac{I'^*{}_a(t)/I'(a)}{I'^*{}_b(t)/I'(b)} = \frac{K_a}{K_b} \cdot \frac{C_a}{C_b} \cdot \frac{z_t + 1/2z_t^2\,[B - \chi(^*a)]}{z_t + 1/2z_t^2\,[B - \chi(^*b)]}. \quad (13.3.13)$$

For thin films, the term quadratic in the mass thickness z_t can be omitted. We obtain:

$$\frac{I'(ta)/I'(a)}{I'(tb)/I'(b)} = \frac{K_a z_{ta}}{K_b z_{tb}} = a \qquad \frac{I'^*{}_a(t)/I'(a)}{I'^*{}_b(t)/I'(b)} = \frac{K_a C_a}{K_b C_b} = b. \quad (13.3.14)$$

In these equation, z_t is the total mass thickness for the alloy foil, and the mass thicknesses of the element foils of a and b can differ from each other, although we have not subscripted them in Eq. (13.3.12). It can be observed that the substrate correction has canceled out of Eqs. (13.3.13) and (13.3.14). We therefore obtain the mass fractions of a and b as follows:

$$C_a/C_b = b/a \text{ and } C_a + C_b = 1. \quad (13.3.15)$$

In principle, this technique can also be applied to multielement films.

Colby [11.5] used his program for data reduction, MAGIC, to analyze films in the thickness range from 18 $\mu g/cm^2$ to 50 $\mu g/cm^2$. He applied the method to the compositional characterization of films of known thickness of silica, alumina, and silicon nitride on semiconductor materials. For films of thickness above 50 $\mu g/cm^2$ he used conventional *ZAF* techniques, lowering the operating voltage of 4-6 kV.

In his theoretical treatment, the depth of penetration of the primary electrons is tied to their energy by:

$$z_r\,(g/cm^2) = 3.3 \times 10^{-6}\,(A/Z)E^{1.5} \quad (13.3.16)$$

He claims that experimental data by Cosslett and Thomas support such a power law. The energy of the penetrating electrons is then scaled to the depth of penetration. In particular, if a thin film of material is deposited upon a substrate, the electrons at the interface are supposed to all have the same energy, regardless of their direction, or previous trajectory (Fig. 13.6). It is also assumed that the fraction of electrons returning from the substrate into the film can be expressed by the backscatter coefficient of the substrate, η_s. The number of ionizations per penetrating electron in the film is, according to Colby,

$$n_t = (N_{av}C_a/A_a)R_t\left[\int_{E_t}^{E_0} Q/S_t \cdot dE + \eta_s \int_{E'_t}^{E_t} Q/S_t \cdot dE\right] \tag{13.3.17}$$

where S_t is the stopping power of the film of thickness z_t, R_t is the corresponding backscatter factor (which is not considered altered by the limited thickness of the layer), E_t is the energy of the electron crossing, in either direction, the interface between film and substrate, and E'_t is the energy of an electron which, after returning through the film, is backscattered. The ratio of the X-ray intensity generated in the film to that generated in a semi-infinite target of the element a is thus:

$$k_a = C_a \frac{R_t\left[\int_{E_t}^{E_0} (Q/S_t) \cdot dE + \eta_s \int_{E'_t}^{E_t} (Q/S_t) \cdot dE\right]}{R_a \int_{E_q}^{E_0} (Q/S_a) \cdot dE}. \tag{13.3.18}$$

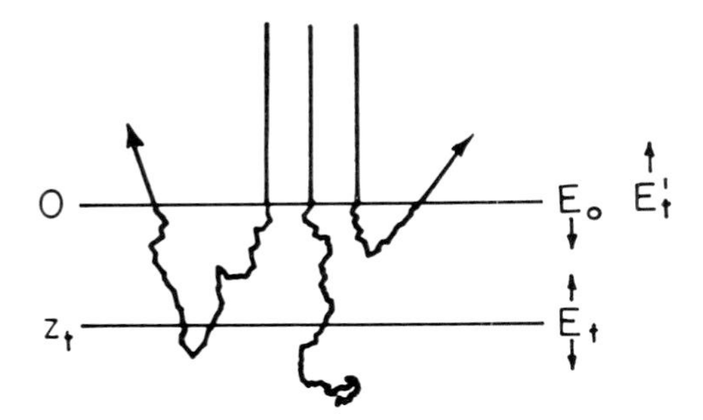

Fig. 13.6. In Colby's model for the analysis of thin films, the energy loss of the electrons within the film (from depth 0 to z_t) is a function of depth only. Energy losses in the substrate are neglected, so that the energy of the electron at the left of this figure is the same (E_t) when it crosses the depth z_t, in descending as well as in ascending direction.

The energies E_t and E'_t must now be calculated, by scaling the range obtained from Eq. (13.3.16):

$$E_t = \left(E_0^{\,3/2} - \frac{z_t(\mathrm{g/cm^2})}{3.3 \times 10^{-6}} \cdot \frac{Z}{A}\right)^{2/3}$$
$$E'_t = \left(E_0^{\,3/2} - \frac{z_t(\mathrm{g/cm^2})}{1.65 \times 10^{-6}} \cdot \frac{Z}{A}\right)^{2/3} \tag{13.3.19}$$

To calculate the absorption in the film, the Philibert formula is used [Eq. (10.3.23)], with a value for σ in which the energy E_t is introduced:

$$\sigma = \frac{4.5 \times 10^5}{E_0^{\,1.67} - E_t^{\,1.67}}. \tag{13.3.20}$$

A similar balance of backscatter and transmission coefficients is used in a method of thickness measurement for thin films on substrates by Yakowitz and Newbury [13.17]. An empirical model for $\phi(z)$, matched with experimental and depth distribution curves and those obtained by Monte Carlo calculations, combines a parabolic fit for the region of high $\phi(z)$ values near the surface, and an exponential fit for deeper regions. Absorption within the film is assumed to vary linearly from the surface to the full X-ray emission range. For compositional analysis, the authors assume that no energy loss or scattering of electrons occurs within the film, so that each component can be treated as an independent layer, and the relation between X-ray production and film thickness is linear. This assumption is valid for very thin films only; however, those made in the thickness measurement should be applicable over the full X-ray emission range.

Reuter [10.15] used his revision of Philibert's procedure (see Chapter 10) to establish a technique for the analysis of thin films, including their thickness. Some aspects of his procedure follow that of Rydnik and Borovsky [13.18], who also established an analytical expression for distribution in depth which includes the effects of backscattering from the substrate. From Eq. (13.3.1), introducing the absorption of X-rays in the target, and considering that the depth distribution of the film, ϕ_t, may differ from that of the substrate, we obtain the general expression:

$$k_t = \frac{I'(t)}{I'(\infty)} = \frac{\int_0^{z_t} \phi_t(z)\exp(-\chi z)dz}{\int_0^{z_r} \phi(z)\exp(-\chi z)dz}. \tag{13.3.21}$$

The changes to Philibert's depth distribution model were already mentioned:

1. The mean pathlength within the layer dz is made equal to $R_0 = 1 + 2.8(1 - 0.9/U)\eta$; at complete diffusion, $R_\infty = 3$, rather than 4, as postulated by Philibert.
2. The factor k in Eqs. (10.3.6) through (10.3.9), and Eq. (10.3.13), which was obtained by Philibert from Bothe's scattering law, is obtained from a fit to experimental data by Cosslett and Thomas [10.16], on the angular distribution of electrons transmitted thorugh a thin film. Reuter obtains:

$$k = 6.27 \times 10^4 Z^{3/2}/(E_0 A).$$

3. The electron transmission initially follows a linear attrition law: $n_z = n_0(1 - 4 \times 10^4 Z^{1/2} z E_0^{-1.7})$ until the depth of complete diffusion is reached. This change is proposed on the basis of experimental evidence from various authors.
4. The ionization cross section varies with energy according to: $Q = U^{-0.7} \ln U$. Energy losses are according to Bethe.

The backscatter coefficient η used in the calculation of R_0 in the thin film is calculated, according to Rydnik and Borovsky, as follows:

$$\eta_{\text{eff}} = \eta_s \frac{n_z}{n_0} + \eta_t\left(1 - \frac{n_z}{n_0}\right) \tag{13.3.22}$$

in which n_0 and n_z are, respectively, the number of electrons at the surface, and at depth z.

On the basis of this model, Reuter produced two data reduction programs. The first program gives $\Phi(0) = R_0$ and the Bethe range, and calculates for a series of steps in electron energy E the ionization cross section, the depth z, the ratio n_z/n_0, and the exponential of $(-kz)$ for Eq. (10.3.6). In the output, $\Phi(z)$ is listed for all steps of depth z, as is the integrated generated intensity. By evaluating the integral with and without absorption one obtains the absorption factor f_p.

In the second program, the mass thickness z_t of the layer of element a is calculated by iterating with various values of z_t until the calculated relative intensity is equal to that which was measured. Thin composite films can also be analyzed with this program. Experimental and calculated values of film thickness for various film compositions and substrates are presented.

Unsupported Films. Specimens such as thin etched metal foils and biological thin sections can be analyzed without being affixed to a substrate. They can also be mounted on substrates for easier handling, since many such free films are very fragile. In such a case, however, the X-ray spectrum of the supporting material adds to that of the objects to be analyzed, thus increasing the limits of detection.

If a support is required, it is advantageous to use very thin organic films which emit no characteristic lines and little continuum. If a supporting stub must be used, it is advisable to choose a material of low atomic number, such as beryllium. If the films can be analyzed without substrate–e.g., mounted on metal grids–the theory of analysis is greatly simplified, the limits of detection remain low, and the spatial resolution improves since the electrons do not diffuse significantly in very thin targets. The lateral electron diffusion is of a dimension similar to the thickness of the film. A further advantage accrues in the microscopic observation, since the topography of a very thin film cannot be observed by scanning electron microscopy because of lack of contrast. Therefore, transmission electron microscopy is required for this purpose. The analysis of such electron-transparent specimens is best performed in instruments which are outfitted with a transmission microscope section.

The conventional electron probe is also inadequate in another aspect. The lateral spread of the electrons within a thin target diminishes with increasing electron energy. But due to the limited thickness of the specimen, the depth of X-ray emission is unchanged. Since, in general, the emitted X-ray intensity also increases with operating voltage, one obtains simultaneously better spatial resolution and sensitivity. Accelerating voltages of 60-100 kV are therefore commonly used in the analysis of thin films in transmission electron microscopes, but these are not available in conventional electron probe microanalyzers.

To overcome these limitations, Duncumb [13.19] and Cooke and Duncumb [13.20] designed a family of instruments called electron microscope-X-ray microanalyzer (EMMA) which partake of the characteristics of both the electron transmission microscope and the electron probe microanalyzers (Fig. 13.7). In these instruments the specimen can be illuminated by a relatively wide electron beam and a transmission image on the screen can be used to center upon a region of the specimen of interest for analysis. The beam can then be focused into the dimensions usual in electron probe analysis (0.1 μm). The maximum operating potential in newer EMMA instruments is 100 kV, and the resolution in the transmission electron microscopy mode is ~20 Å. Energy-dispersive and crystal spectrometers can be used to measure the X-ray emission. Another option is that of a transmission microscope attachment to an electron probe (Fig. 13.8). The limitation in this case is in the maximum operating voltage of the instrument. Either instrument can also be used for electron diffraction patterns, which are particularly useful in the characterization of small particles. For these, again, an electron acceleration potential of at least 80-100 kV is desirable. For this purpose, energy-dispersive X-ray spectrometers can be attached to conventional transmission electron microscopes. Such instruments are called *analytical transmission electron microscopes* [13.22], [13.23], [13.121]. The microscope must be modified so that the electron beam which normally floods the entire specimen area to be depicted can be focused to 1 μm or less in diam-

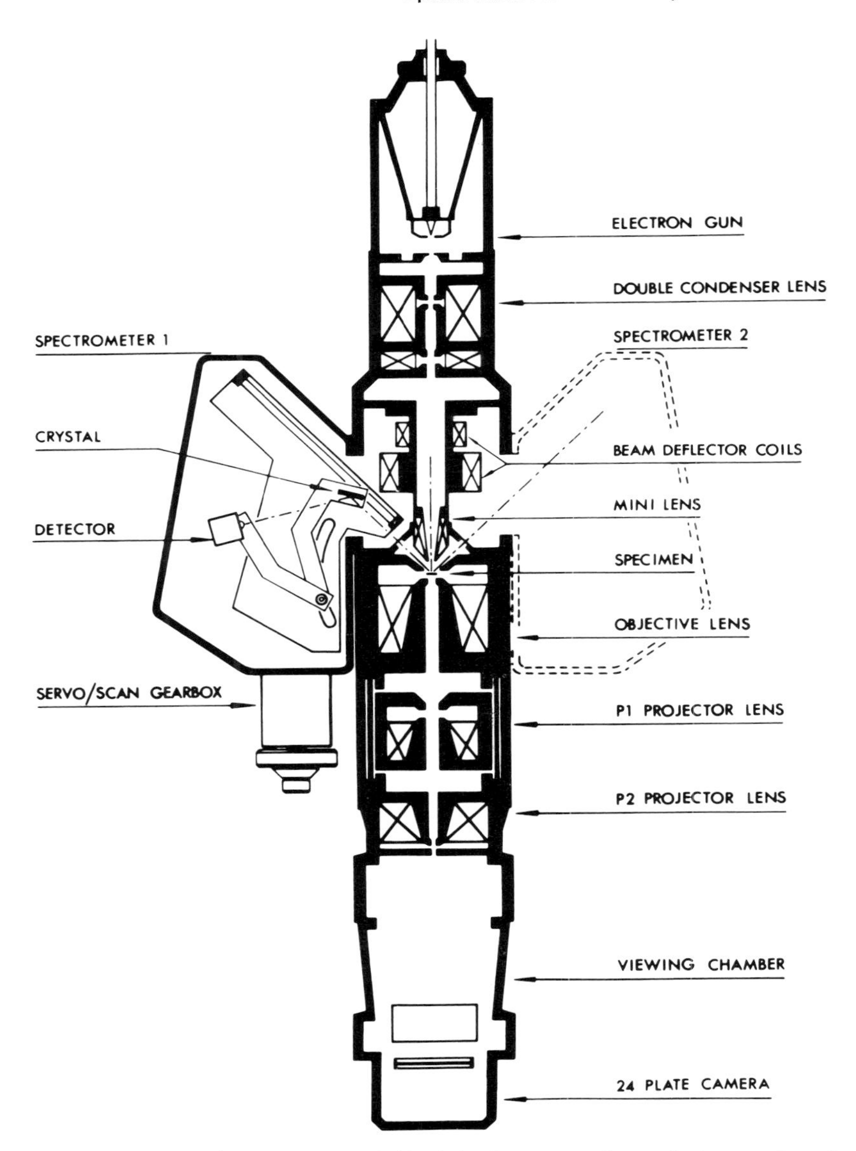

Fig. 13.7. EMMA III/IV. The upper half of the instrument is an electron probe microanalyzer, with double condenser lens, beam deflector coils, a minilens used as an objective lens, and two crystal spectrometers. This upper section also serves as illumination for the electron microscope section, which has an objective lens, two projector lenses, a viewing chamber and a camera, all located below the specimen level. (Courtesy of P. Duncumb.)

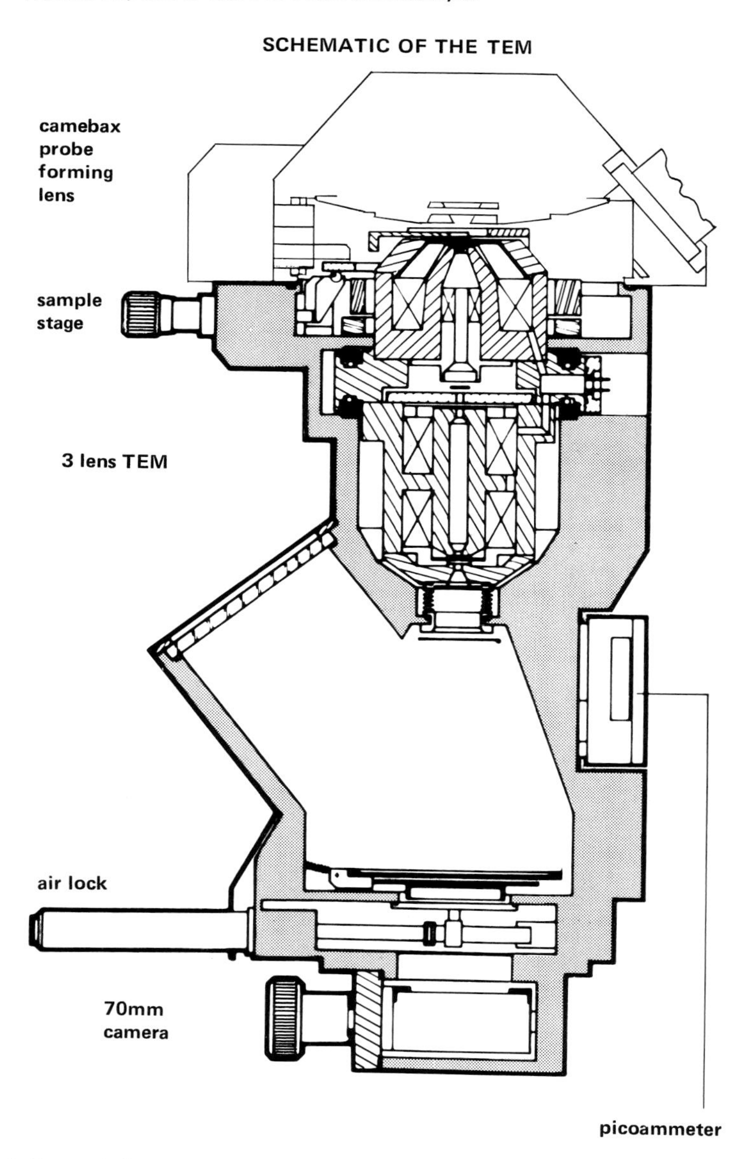

Fig. 13.8. Transmission electron microscope attachment for a scanning electron microscope-electron probe microanalyzer. (Courtesy of CAMECA.)

eter. Spurious X-ray signals from pole pieces, specimen holders and other instrument parts irradiated by stray electrons, as well as interference by fluorescent X-radiation produced in the specimen region are frequently observed. Such parasite signals, which raise the X-ray signal background and produce spurious peaks, increase in importance as the operating voltage is raised, due to the transmission of X-rays and electrons through aperture diaphragms. These spectral artifacts may ultimately limit the level to which the operating voltage can be raised usefully [13.24].

If the film to be analyzed is so thin that electron decelerating and scatter can be neglected as far as X-ray production is concerned, the theory of quantitation is considerably simplified [13.25]. Applying Eq. (9.2.9) to a multielement target of thickness z_t (g/cm^2) traversed normally by the electron beam, we obtain the number of ionizations per electron:

$$n_a = C_a(N_{av}/A_a) \cdot Q_q \cdot z_t \tag{13.3.23}$$

which multiplied by the fluorescent yield and the weight of the corresponding line gives the production of X-ray photons of that line. The specimen thickness to which this simple scheme is applicable increases with operating voltage, as indicated by the decrease with increasing voltage of the diffusion factor B in Hutchins' Eq. (13.3.8), and this is another good reason for analyzing thin films at high operating potentials.

The theoretical and experimental conditions for quantitative analysis of thin specimens were reviewed by Tixier [13.26] in 1972 and a simple theory for such analyses with the use of semi-infinite standards was described by Tixier and Philibert [13.27]. Combining Eqs. (9.2.11), (9.2.26), and (13.3.23), we obtain for the ratio of generated X-ray intensities for the element a:

$$\frac{I(t)}{I(\infty)} = \frac{2C_a N_{av}\ \pi e^4 (Z/A) E_q^{-2} \cdot (\ln U_0/U_0) \cdot z_t}{R_a\{U_0 - 1 - (\ln M/M)\ [\mathrm{li}(U_0 M) - \mathrm{li}(M)]\}} \tag{13.3.24}$$

where $M = 1.166E_{qa}/J_a$, li is the symbol for the logarithmic integral defined in Eq. (9.2.21), and all element-dependent parameters (such as Z, A, R, etc.) are those for the element a. The authors rewrite this equation as follows:

$$\frac{I(t)}{I(\infty)} = C_a \frac{2\pi e^4 N_{av}}{E_{qa}} \left[\frac{\ln U_0}{E_{qa}} \cdot \frac{S(a)}{R(a)}\right] \cdot z_t \tag{13.3.25}$$

where

$$S(a) = (Z_a/A_a)\left\{U_{0qa} - 1 - \left[\frac{\ln M}{M}(\mathrm{li}\,(U_{0qa}M) - \mathrm{li}\,(M))\right]\right\}^{-1}$$

The term $(\ln U_{0a}/E_{qa})(S(a)/R(a))$ is independent of specimen composition. There is no scattering correction for the thin foil, and secondary excitation can also be neglected. For the emitted intensity ratio, absorption corrections are, however, necessary. The generated intensity of the semi-infinite elemental target must be multiplied with the corresponding absorption factor, $f_p(a)$. If the film is very thin, we can calculate the absorption loss for it by assuming that $\phi(t)$ is constant with respect to depth, so that we obtain:

$$f(t) \simeq \exp(-1/2\chi z_t) \simeq 1 - 1/2\chi z_t + \cdots, \qquad \chi z_t < 0.1. \qquad (13.3.26)$$

For foils in which the absorption correction is larger, the authors postulate that $\phi(t)$ diminishes exponentially with depth:

$$\phi(t) \propto \exp(-\sigma z) \qquad (13.3.27)$$

and we obtain for the correction:

$$f(t) = \frac{1 - 1/2(\sigma + \chi) \cdot z_t + \cdots}{1 - 1/2\sigma \cdot z_t + \cdots}. \qquad (13.3.28)$$

In each case the integral was expanded in series and the unnecessary terms were dropped. Even these approximate absorption corrections are often unnecessary.

The limit of validity of the concepts we have developed is given by the thickness at which the diffusion of electrons becomes significant. This diffusion process starts at the surface of the specimen, and the electron backscatter coefficient from a free film is a good indicator of its progress (Fig. 13.9). After an initial rise in slope, η increases almost linearly until it converges to the level for a massive specimen of the same mean atomic number. The extrapolation of the virtually linear slope of this level provides a reference point (D in Fig. 13.9) for the film thickness at which complete diffusion is reached. At this point, the transmission of electrons through the film has dropped to about one-half of the beam current. The depth of D does not vary greatly with the atomic number of the target, but increases with accelerating voltage:

E_0 (kV):	10	20	40
t_D (g/cm^2):	110	350	1000

The value of the backscatter factor R varies with depth in a very similar manner. The most probable loss of energy of the electrons in an aluminum film 100 μg/cm^2 thick, at an operating potential of 20 kV, is less than 2%, and such a loss should not affect the accuracy of thin-film analysis, since the ionization cross section changes slowly at overvoltages above three [13.26].

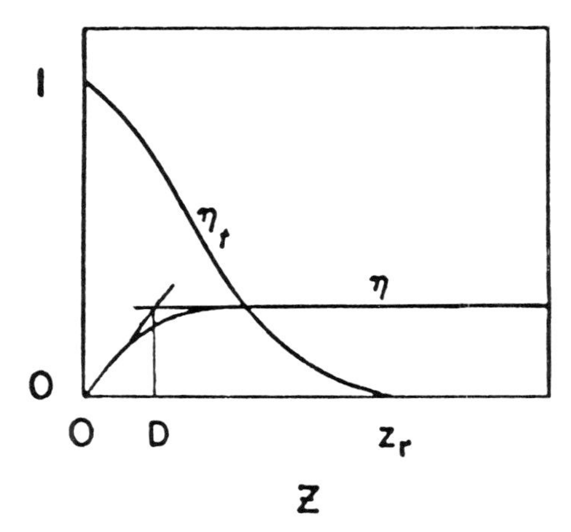

Fig. 13.9. Backscattered electrons (η) and transmitted electrons (η_t), as a function of depth, scaled to the electron range, z_r. These relations are virtually independent of the atomic number of the absorber and the energy of the electrons, E_0, provided that η_t is scaled to the backscatter coefficient of the respective massive target. D is the depth of complete electron diffusion.

In the practice of thin-film analysis, particularly in the transmission electron microscope, the use of elemental standards, either massive or thin, is cumbersome. Cliff and Lorimer [13.28], Nasir [13.29], and Goldstein and Williams [13.22] have advocated basing the analysis on the ratios of observed X-ray intensities from the elements in the specimen. For thin foils, and with Si(Li) detectors, this technique can be quite accurate, since according to Eq. (13.3.23), it only depends on the knowledge of the ionization cross section, Q, the fluorescent yield, ω_q, the weight of the observed line, p_{ql}, and the efficiency of the detector, P. The aperture of the detector, $\Omega/4\pi$, cancels in the comparison of the line intensities. Thus, one obtains:

$$k_t = \frac{I'(t)_a}{I'(t)_b} = \frac{C_a}{C_b} \cdot \frac{A_b}{A_a} \cdot \frac{Q_a}{Q_b} \cdot \frac{\omega_{qa}}{\omega_{qb}} \cdot \frac{p_{qla}}{p_{qlb}} \cdot \frac{z_{ta}}{z_{tb}} \cdot \frac{P_a}{P_b} \cdot \qquad (13.3.29)$$

The ratios for elements of equal mass thickness are often tabulated with reference to silicon, and if they are known for a given detector and operating voltage, the analysis can be performed without standards. If both intensities are measured from the same film, the thickness z_t also cancels (standardless analysis). The method works satisfactorily except for radiation of large wavelength such as Na Kα, for which X-ray absorption in the film is excessive.

13.4 THE ANALYSIS OF BIOLOGICAL TISSUE

Electron probe microanalysis of biological material, particularly of soft tissue, presents special challenges to the analyst and it was therefore necessary to develop special theoretical and experimental procedures which we will discuss here.

Among the difficulties in the analysis of soft biological tissue are the following:

1. Most elements which are of high enough atomic number to produce useful X-ray lines are present in soft organic tissue at trace levels. The composition of the organic matrix itself cannot be determined since the major constituents are elements of low atomic number (C, H, O, N).
2. For the study of many important processes, a resolution of 0.5 μm or better is required. But, because of the combination of the low density of biological tissue and the dimensions of cells and subcellular organelles, satisfactory spatial resolution of electron probe analysis (especially of dried tissue) is seldom achieved unless the specimen is cut into thin sections.
3. In the organized biological tissue, the local mass thickness of sectioned specimens varies greatly from one point to another.
4. Biological soft tissues are poor conductors of heat and electricity, they are prone to mechanical deformation in preparation, and to damage under the impact of the electron beam.
5. The features of unstained biological tissue are usually difficult to observe by conventional light or scanning electron microscopy procedures, so that orientation on the specimen may be difficult.
6. Artifacts due to migration of elements in the interval between the end of cellular life and the analysis of the prepared tissue occur frequently, particularly with elements present in the live tissue as electrolytes (e.g., sodium and potassium), which play an essential role in fundamental biological processes.

There are, of course, many instances in which some of these difficulties are absent, and we will discuss special approaches at a later point, after having treated the most typical situation which involves the quantitative analysis of a thin section of soft tissue. The subject of the quantitation procedure itself has been treated extensively by Hall and co-investigators [13.30]-[13.34]. Hall proposes using the continuous radiation as a monitor and measure of total local mass. The atomic number dependence of the continuum is not a serious obstacle since the average atomic number of dried soft tissue does not vary over a large range.

Assume that the specimen is observed by two detectors, one of which (subscript 1) is tuned to the line element a to be used in the determination, while the other (subscript 2) receives mainly, or exclusively, a portion of the continuous radiation. Since continuous radiation also is emitted from the element to be determined, and background is also emitted within the acceptance range of the spectrometer tuned to the line, both detectors receive contributions due to all elements present in the specimen. For the derivation of the algorithms to be used, we will distinguish the element to be measured, a, from the matrix, m, which, by definition, is the sum of all elements except a. Not considered part of the matrix is the supporting film (usually a nylon film weighing about 40 $\mu g/cm^2$). The signals emitted by this support, B_1 and B_2, can be determined separately, and are subtracted from the respective signals, Y_1 and Y_2. Calling k_{ij} the efficiency factors of the detector system for element (or matrix) I and detector j, we obtain:

$$\begin{aligned} \underline{Y_1} - B_1 &= \underline{k_{a1} M_a} + k_{m1} M_m \\ \underline{Y_2} - B_2 &= k_{a2} M_a + \underline{k_{m2} M_m} \end{aligned} \qquad (13.4.1)$$

where M_a and M_m denote the masses per unit area (g/cm^2) of the element a and of the matrix m. The prominent contributions are underlined in Eqs. (13.4.1).

Besides B_1 and B_2, the following parameters can be measured:

$$r_{spec} = \frac{Y_1 - B_1}{Y_2 - B_2},$$

which is the support-corrected intensity ratio from the two detectors for the specimen, $r_{el} = k_{a1}/k_{a2}$ which is the ratio obtained from the two detectors when a thin film of the pure element a is irradiated, and $r_{mat} = k_{m1}/k_{m2}$ which is the corresponding ratio obtained from a thin film of the matrix material, in the absence of the element a. This ratio can be obtained from a thin film of Aquadag; as long as no characteristic lines are observed, the composition of the matrix is not crucial.

Equations (13.4.1) and the definitions of the ratios shown above can be used to develop the following equation, which forms the basis of the *relative method* developed by Hall et al.:

$$\frac{M_a}{M_m} = \frac{r_{spec} - r_{mat}}{r_{el} - r_{spec}} \quad \frac{k_{m2}}{k_{a2}} \qquad (13.4.2)$$

It should be noted that the concentration of element a is equal to $M_a/(M_m + M_a)$, which only for low concentration of a is virtually equal to M_a/M_m; furthermore, that the method does not lead to absolute concentration levels, unless the factor k_{m2}/k_{a2} can be determined, or measurements on a standard of known composition can be performed. On the other hand, for low concentrations, $r_{el} \gg r_{spec}$, so that $r_{el} - r_{spec}$ is virtually constant and its determination is not required in the relative method. Hence, measurements on the elemental foil are unnecessary in such a case. Practical limitations of the method were discussed by Hall and Werba [13.31].

An absolute method of film analysis, developed by Marshall for his doctoral thesis, is described in [13.30]. According to Kramers,

$$I(E_c)dE_c \cdot dE = k \cdot Z \cdot (dE_c/E_c) \cdot dE \tag{13.4.3}$$

where E is the energy of the incident electron and E_c is the energy of the X-ray photon from the continuum excited by the electron. If, for a massive target, this equation is integrated over the range of E from E_0 to E_c, Eq. (6.3.7) is obtained. The above equation is valid for a thin film, in which the incident electrons do not lose a significant fraction or their energy. To restate the equation in terms of film thickness, we calculate the trajectory of the electron corresponding to the energy loss dE by means of Bethe's energy loss Eq. (9.2.4):

$$I(E_c)dE_c \cdot dx = \frac{2k\pi e^4}{E_c \cdot E} \sum N_i Z_i^2 \cdot \ln \frac{1.17E}{J_i} dE_c \cdot dx \tag{13.4.4}$$

with summation over all i elements in the specimen. $N_i = N_{av}/A_i$ is the number of atoms of element i in 1 g of the specimen. Hence, $N_i \cdot dx$ is the number of atoms per cm^2 of element i. Because we have assumed that the film is so thin that scatter and energy loss are negligible, E is equal to the incident electron energy, E_0, and dx is equal to the film thickness, z_t. If we assume that the absorption of X-rays within the specimen and the fluorescent excitation are also negligible, we get for the observed continuous X-ray intensity (photons/electron), between E_c and $E_c + dE_c$,

$$I'^{*}_c = (\Omega_c/4\pi) \cdot P_c \cdot I(E_c)dE_c \cdot z_t \,. \tag{13.4.5}$$

It is important that within this energy interval of continuum radiation no characteristic radiation may be collected. The parameters Ω_c and P_c refer to the spectrometer on which the continuum is being measured. If at the same time another spectrometer measures the characteristic emission of a line of element a, we obtain, from the ionization cross section according to Worthington and Tomlin [Eq. (9.2.12)]:

$$I'_a = (\Omega_a/4\pi) \cdot P_a \cdot \omega_q P_{ql} \cdot k' E_q^{-2} U^{-1} \ln U (C_a/A_a) \cdot z_t \tag{13.4.6}$$

where Ω_a and P_a are the aperture and efficiency of the spectrometer which measures the characteristic line of element a, ω_q and p_{ql} are the fluorescent yield and weight of line corresponding to this line, and $C_a \cdot z_t$(g/cm^2) is the effective thickness of the element a. The authors propose that in view of the small variations of composition of soft tissue, A_i, Z_i, and J_i can be replaced by average values, $\bar{A}$, $\bar{Z}$, and $\bar{J}$. After lumping into one constant, K, all parameters and constants which do not depend on the specimen composition, we obtain for the ratio of the two intensities:

$$\frac{I'_a{}^*}{I'_c{}^*} = K \cdot \frac{\bar{A}}{A_a} \frac{C_a}{\bar{Z}^2 \ln(1.17E_0/\bar{J})}. \quad (13.4.7)$$

The constant K can be determined by means of measurements on a thin film of element a. If the corresponding intensities are $I'(a)$ and $I'_c(a)$, we get the following formula for the absolute method:

$$C_a = \frac{I'_a{}^*/I'_c{}^*}{I'(a)/I'_c(a)} \cdot \frac{A_a}{\bar{A}} \cdot \frac{\bar{Z}^2 \ln(1.17E_0/\bar{J})}{Z_a{}^2 \ln(1.17E_0/J_a)}. \quad (13.4.8)$$

The authors proved the validity of their proposed method through measurements on alloy films of known composition. They obtained an estimated accuracy of 8% relative to concentration.

A method for quantitative analysis of biological material (called BICEP) in which a greater specimen thickness is permissible was proposed by Warner and Coleman [13.35]. These authors have adapted Colby's thin-film program MAGIC to this purpose, introducing options for entering as known the concentrations of the matrix elements which cannot be determined by X-rays (C, H, O, N). Philibert and Tixier's [9.3] version of the *ZAF* procedure was incorporated, and provisions had to be made for measuring the local specimen thickness, which is assumed to be known in Colby's program. For this purpose, Warner and Coleman use a substrate such as aluminum or silicon which emits characteristic radiation. The attenuation of this emission serves to estimate local mass thickness. Eshel [13.36] studied the mutual effects of variation in concentration of elements such as Ca, K, Cl, S, and P in the BICEP procedure, and came to the conclusion that their variations within the usual concentration ranges did not affect the intensity of emission of other elements. On the contrary, the determination of mass thickness was found very important to the success of the analysis. The absorption losses for magnesium and sodium were found to be very strong ($\geqslant$20%).

A method for the preparation and analysis of thick specimens of soft biological tissue was proposed by Ingram et al. [13.37]. Following the suggestion of Andersen and Hasler [12.24], these authors work at an accelerating potential of 10 kV in order to keep the volume of excitation within reasonable bounds, and they reduce specimen damage by working at beam currents as low as 50 nA. Their specimen preparation procedure consists of rapidly freezing the tissue, drying in vacuum for several weeks, fixing with osmium tetroxide vapor, embedding in Epon resin, microtoming into sections 3 μm thick (these are electron-opaque at 10 kV), mounting on a quartz slide, and coating with a 200-Å layer of evaporated carbon. In this process, the distribution of diffusible elements can be seriously disturbed.

Ingram et al. used solutions of serum albumin for the calibration of their signals. Von Rosenstiel et al. [13.38] employ solutions of agar-agar for this purpose. Andersen and Hasler [12.24] proposed the use of organometallic compounds, while Warner and Coleman [13.35] showed that chlorine in parylene polymer can be determined with good accuracy with the aid of a sodium chloride standard. Obviously, the controversy of simple standards against matching standards is well and alive in biological microanalysis. An extensive review of the standardization problem was published in 1975 by Spurr [13.39].

A fairly complete list of references of the analysis of massive biological specimens, including hard tissue (bones and teeth) is contained in the publication by von Rosenstiel et al. [13.38] on the measurement of electrolyte in kidney sections.

When the continuous emission from the specimen is taken as an indicator of total specimen mass, precautions must be taken to avoid or minimize alterations of the mass thickness by deposition of contaminant material, or of beam-induced loss of organic mass. That such a loss can be significant has been established by Stenn and Bahr [13.40] and by Hall and Gupta [13.41]. In view of the use of the continuum for thickness calibration, this subject should be further investigated. Shuman et al. [13.42] have used Hall's method with a Si(Li) detector. In their publication, they discuss the procedure as well as several artifacts, including mass loss due to radiation damage.

13.4.1 The Analysis of Biological Fluids

In 1967, Ingram and Hogben [13.43] demonstrated the use of the electron probe microanalyzer in characterizing the concentrations of electrolytes in microscopic droplets deposited on quartz and dried and coated with carbon. This method was further developed and applied to the analysis of biological fluids by Morel and Roinel [13.44]-[13.47] and by Lechene et al. [13.48]-[13.51]. Roinel [13.46] describes the technique as follows: volumes of 50-200 pl are drawn from a sample kept under paraffin oil by a capillary pipette and deposited on a beryllium planchette. The droplets are washed by chloroform, dried at room temperature, and stored until shortly before the analysis, when they are rehydrated, and lyophilized (freeze-dried) at $-40°$ C under vacuum. The round residual spots contain the solid in fine dispersion in a circle of about 80-μm diameter which is irradiated by a defocused electron beam, with an operating potential of 15 kV and a beam current of 200 nA. The evaluation of the analysis is performed with calibration curves obtained by evaporating microdroplets from synthetic solutions. These curves are linear at the concentration of the elements (Fe, Na, K, Cl, P, and Mg) found in fluid extracted from renal tubules by micropuncture. The significance to research in renal physiology is discussed by Morel [13.47]. The technique followed by Lechene is similar [13.49]. The droplets are, however, cooled in isopentane to $-160°$ C, after pipetting and washing with *p*-xylene, and then freeze-dried at $-70°$ C. The analysis is performed with a potential of 11 kV, and a beam current of 300 nA. Analyses from drops as small as 10 to 80 pl are reported [13.50], [13.51]. Lechene and collaborators applied this technique to renal fluid, and also to other fluids of the mammalian reproductive system. Rick et al. [13.52] have used an energy-dispersive Si(Li) detector for the analysis of microdroplets. They deposit the liquid on thin organic films. Their line-to-background ratios are lower than those obtainable with Bragg spectrometers.

Related to this type of analysis is the investigation of single cells such as sperm cells [13.53], red blood cells [13.54], [13.55], cellular components [13.57]-[13.59], and inclusions [13.60]-[13.63]. In such work, the determination of diffusible elements [13.54]-[13.56] is again considerably more difficult than that involving fixed elements [13.53], [13.57], [13.63], and no definitive method seems to have evolved as yet.

13.4.2 Indirect Electron Probe Microanalysis

In 1961, Birks and Seebold [13.64] demonstrated the precipitation in alloys of elements present at great dilution (sulfur in iron, in the case illustrated) by means of a diffusion process with another metal (niobium). The authors suggested that such reactions could be used to investigate the presence of trace elements. Although this line of procedure was not followed in metallurgical specimens by other investigators, several procedures involving precipitation reactions in biological tissues with identification of the precipitate in situ were proposed [13.65]-[13.68]. While in such intratissular reactions the question of migration of components before precipitation must be carefully considered, such problems do not exist in precipitation reactions carried out in microdroplets [13.69]-[13.71]. This technique enables the analyst to extend electron probe microanalysis of organic compounds such as urea [13.71]. Indirect analysis both in tissue and microdroplets was reviewed in [13.50], [13.51].

13.4.3 Applications of Electron-Probe Microanalysis to Biological Problems

A review of the very extensive literature of biological applications of the electron probe would go beyond the scope of this book. The reader may use for this purpose the following sources. General reviews: [12.66]-[12.71], [13.50], [13.51]; for early references: [13.72] (before 1963); calcium in biological systems: [13.73]; for preparatory methods: Coleman and Terepka [13.74]; for mineralized tissue: [13.75], [13.76].

13.5 THE ANALYSIS OF SMALL PARTICLES [13.122]

The quantitative determination of small particles presents problems similar to those which apply to the analysis of thin sections of biological specimens. The following aspects must be carefully considered:

1. The particles must be affixed to a substrate, and usually coated to provide conductivity. Apart from the manipulative difficulties in mounting particles, the analysis is complicated by the need to separate signals produced by the substrate and coating from those emitted by the specimen.
2. The total mass of the particle, and its density, are usually unknown. The particle shapes are frequently irregular, and the depth of a particle (i.e., its dimension in the direction of the beam) may be difficult or impossible to estimate.
3. Unlike most biological specimens, particles of practical interest may have a wide range of matrix composition, average atomic number, and density. They are frequently composite or inhomogeneous, and usually contain one or more elements of low atomic number which cannot be accurately determined even in a flat specimen.

A particle is, in this context, considered small when its dimensions in any direction are below the range of the beam electrons under the conditions of the measurement. The analysis of a large, flat-topped particle can be performed as if it were a flat polished electron-opaque specimen, provided that the volume of excitation is fully contained within it. Even if the top surface is not parallel with the supporting surface, the common analytical procedures can be used, provided that the angle of the specimen surface with respect to the electron beam can be estimated with reasonable accuracy. Such a situation is common in the analysis of objects in scanning electron microscopes, with solid-state detectors. But when the potential volume of excitation exceeds that of the particle, difficulties arise from the backscattering of electrons from the substrate into the particle, and from the X-rays emitted by the substrate. In order to minimize these effects, substrates of low atomic number (beryllium or pyrolytic graphite), or thin-film substrates, such as employed for biological specimens, should be used [13.77].

An extreme case is provided by a particle the depth of which is so small, compared with the electron range, z_r, that the deceleration and scattering of the primary electrons can be neglected. Such a particle can thus be handled in the same way as a very thin film (e.g., by the technique of Tixier and Philibert). The absorption correction can be omitted. If it is applied, the estimated particle diameter should replace z_t in Eq. (13.3.26) and the parameter χ should be replaced by the mass absorption coefficient μ.

If the absorption of the X-rays can be neglected, then the observed signal intensity is proportional to the mass present of the respective element. To assure uniform excitation of the particle, it is best to focus the beam as sharply as possible and to scan it in the form of a frame slightly larger than the particle; in this manner the effect of density distribution within the beam will be minimized.

In most cases, the desired information is the weight fraction of the element being determined, C_a. Similarly to Tixier's development for thin films, we can write for the particles, that the number of ionizations per electron which traverses a particle is

$$n = C_a \frac{N_{av}}{A_a} Q_q \bar{z}_p = C_a \frac{N_{av}}{A_a} Q_q \cdot \frac{\rho_p v_p}{a_p} \tag{13.5.1}$$

where $\bar{z}_p$ (g/cm^2) is the average pathlength of the primary electron, ρ_p (g/cm^3) is the density of the particle, v_p is its volume, and a_p is the cross section of the particle. In the general case, neither the density nor the volume of the particle are known; the main problems in the analysis of a particle is to estimate its mass which is the product of these two unknowns.

The simplifying assumptions implicit in the above model hold only for particles much smaller than the electron range. When backscatter, deceleration of the electrons, and X-ray absorption are significant, it becomes important how the particle mass is geometrically distributed. One can thus express the mass uncertainty by two parameters: the density and a shape parameter.

Hoffmann et al. [13.78] have given a treatment of this problem in which the particle to be analyzed is covered by an area raster scan so that it is uniformly irradiated. An imaginary "equivalent particle," which contains only the element to be determined is used to derive an estimate of the analyte concentration in the specimen, C_a. This "equivalent particle" has uniform depth, $z_p(e)$ such that Castaing's first approximation is valid when this imaginary particle is compared with the specimen:

$$C_a = I_t^*/I_t(e). \tag{13.5.2}$$

To calculate the unknown concentration, we must first establish what relation the parameters of the equivalent parameter must have with those of the specimen to be analyzed. Then we will derive the ratio of the intensities $I_t(e)$ (from the equivalent particle) and $I(a)$ (generated within a massive specimen of the element a).

The relations between the equivalent and the real particles postulated by the authors are:

1. The average particle depth $\bar{z}_t$ and the depth of the equivalent particle $z_t(e)$ are scaled to the electron ranges in the respective materials:

$$\bar{z}_t/z_t(e) = z_r/z_r(e). \tag{13.5.3}$$

2. The average area projection of the real particle, $\bar{a}_t$, and the area projection of the equivalent particle, $a_t(e)$ are scaled to the squares of the electron ranges in the respective materials:

$$\bar{a}_t/a_t(e) = [z_r/z_r(e)]^2. \tag{13.5.4}$$

If the volume of the real particle, v_t, is known, $\bar{a}_t$ can be calculated from: $\bar{a}_t = v_t/\bar{z}_t$.

3. The circumference of the projection of the real particle and that of the equivalent particle are scaled to the respective electron ranges:

$$u_t/u_t(e) = z_r/z_r(e). \tag{13.5.5}$$

Let us call a_s the area swept by the electron beam scan. We also assume that the particle depth is larger than the range of electrons, so that no electron transmission occurs, and that there is no X-ray intensity loss through scattering of the electrons out through the sides of the particle. For the relative intensity obtained from such a particle, referred to a massive element standard, we would then obtain:

$$k_t = I_t/I(a) = C_a I_t(e)/I(a) = C_a \cdot a_t(e)/a_s. \tag{13.5.6}$$

Since, however, for a small particle, the thickness of the equivalent particle is smaller than the corresponding range ($z_t(e) < z_r(e)$), some electrons are transmitted, and the corresponding loss of X-ray production must be taken into account by means of a corrective factor, f_z. Furthermore, some electrons leave the equivalent particle through the sides (sidescatter); the corresponding X-ray intensity loss requires another corrective factor, f_y. Therefore, we obtain

$$C_a = k_t/[(a_t(e)/a_s) \cdot f_z \cdot f_y]. \tag{13.5.7}$$

The factor, f_z, was determined experimentally, as a function of the scaled film thickness, z_t/z_r, by measurements on supported films from several metals on several substrates. The authors indicate that a universal curve of f_z versus z_t/z_r can be used, and represent this curve in a graph.

The sidescattering factor, f_y, was also determined experimentally:

$$f_y = 0.31 - 0.008 \cdot z_r(e) \cdot u_t(e)/a_t(e). \tag{13.5.8}$$

This equation is assumed to be valid for the range $0.2 < z_t(e)/z_r(e) < 2$. For a large particle this factor should become equal to one, which is, however, impossible. It is not clear where the limits of validity of this method can be drawn. The authors state that the ratios of concentrations of elements can be determined much more accurately than the concentrations themselves.

The authors propose that the dimensions of the particle first be measured by scanning electron microscopy and then converted into the parameters of the equivalent particle. To do this, however, one must know that densities of all materials involved, since the linear (μm) dimensions are what is being measured; but the range contains, implicitly or explicitly, the density of the material. Hence, a density uncertainty is hidden in this procedure; this is a difficulty inherent in all procedures in which particle mass is estimated by optical procedures. The authors propose, to overcome this problem, that one should estimate the specimen density by $\rho^* = \Sigma_i C_i \rho_i$. How this should be applied when some constituents are nonmetallic (e.g., in oxides), is not clear.

As in the case of thin films, the parameters affecting emitted X-ray intensities for several elements from the same particle are similar, particularly for emissions at close wavelengths. Therefore, the shape and density factors largely cancel in the determination of ratios of line intensities from the constituent elements of the particle. But, a procedure based on ratios of intensities of lines from different elements requires that all elements emit measurable X-rays or that stoichiometric relations can be assumed for those elements which cannot be measured. Semiquantitative procedures based on ratios of relative intensities, without further corrections, have been used frequently in the characterization of particles. The technique is discussed by Armstrong and Buseck [13.79] who, however, add considerable refinement to the evaluation of such results.

These authors largely follow a classical *ZAF* approach for the treatment of the generation calculation. Concerning the backscatter correction factor, they observe that at any point of a particle, it can be approximated by that of an electron-opaque specimen of surface tangent to the particle at this point. They further indicate that the backscatter correction factor for an inclined specimen is related to that for a specimen normal to the beam by a factor K:

$$R(\epsilon) = K \cdot R(90^\circ)$$

which tends to cancel in the ratio of two relative intensities. Hence, the overall backscatter factor, integrated over the entire particle will largely cancel in the estimate of C_a/C_b.

For the absorption of primary X-rays Armstrong and Buseck use Reuter's procedure for thin films [10.15], with a modification which takes into account the shape factor, i.e., the change in particle cross section $a(z)$ as a function of depth, and hence of the intensity distribution function $\Phi(z)$:

$$\Phi'(z) = \frac{a(z)}{a(o)} \Phi(z).$$

The pathlength over which the absorption occurs must also be modified according to the shape of the particle. The authors have calculated the effects of particle shape for several configurations which they deem to be typical for practical specimens, and they show that, for semiquantitative determinations, the indeterminacy in the intensity ratios due to shape variation is tolerable; analyses of anorthite particles indicate errors of ±8% relative.

The fluorescence due to the continuum, and usually also that caused by characteristic lines, is ignored.

In the procedure of Armstrong and Buseck, the density uncertainty is eliminated by measuring the ratios of line intensities from several elements. In view of the problems with assumptions of stoichiometry, and the possibility of errors or spurious counts it would be desirable to complement the measurements by further signals. The measurement of soft X-rays (such as the O K-band) can be affected by large absorption if the particle is larger than about 0.1 μm. A procedure using the emission of continuous radiation, such as that used by Hall et al., is preferable. This, however, requires the mounting of the particles on thin electron-transparent films, particularly when they are small and of low mean atomic number. With such a mounting technique, the continuous radiation emitted from the specimen assembly can be greatly diminished. Moreover, the elimination of scattering of electrons from the substrate greatly reduces the volume from which radiation is emitted, and hence produces a considerable increase in resolution which is conserved when the operating potential is raised. A method for determining the mass of very small particles was proposed by Lamvik and Langmore [13.80]. It is based on the measurement of currents formed by electrons in transmission.

The analysis of small particles and fibers is gaining increased attention in view of the interest in the identification of asbestos and other mineral fibers as potential health hazards [13.61], [13.81]-[13.87]. For such materials, a complete identification must include topographic and crystallographic characterization as well as elemental analysis. Because the width of an individual chrysotile asbestos fiber is 200-300 Å, high-resolution microscopy is needed. The electron diffraction, in turn, cannot be efficiently performed at operating potentials below 50 kV [13.87]. Transmission electron microscopes are at present best suited for this task, although scanning electron microscopy and electron probe microanalysis have also been used. For X-ray emission, an operating potential of 80-100 kV is advantageous, and at such potentials selected area electron diffraction becomes a powerful ancillary technique for the determination of lattice parameters, while the use of transmission electron microscopy often helps significantly in the orientation of the electron beam on the specimen for diffraction and spectroscopic analysis.

Inclusions and Precipitates. A quite different situation arises in the analysis of inclusions embedded in a massive matrix. If the diameter of the inclusions is larger than 2-3 μm, the common *ZAF* procedure for massive specimens can be followed, after careful polishing of the compound target [12.74], so that a flat surface is obtained. Because it is impossible to judge the depth of the remaining part of the inclusion embedded in the matrix, the possibility that part of the matrix will also be excited cannot be excluded. The situation is particularly confusing in the analysis of a thin film or shallow particle atop or embedded within a substrate which contains elements in common with the object to be analyzed. This problem is often found in the characterization of precipitates and inclusions in alloys. The spectrometric information indicates these elements, but the analyst cannot be sure that these elements are components of the film or particle. This problem was considered by Satta et al. [13.88] who suggest that the continuous background be measured across the absorption edge or edges of the substrate element in question. They use in their calculations a simplified model for $\phi(z)$ which assumes that the depth distribution is constant up to a maximum depth as does Bishop [12.21], and they show on the graphs obtained from such calculations that a heterogeneous target can be distinguished from a homogeneous specimen by this technique. Similar results can be obtained from measurements of a pair of lines straddling an absorption edge of the element of interest.

In such situations it is usually more advantageous and simpler to extract and remove the inclusions and precipitates from the matrix [13.89], [13.90]. The particles are then analyzed while mounted on a thin film. This mode of analysis is much more sensitive than excitation of the particles within the matrix, even in cases where particle and matrix have no elements in common, because the background is greatly reduced. The analyst must, of course, be certain that the composition of the particles is not altered during extraction.

13.6 STATISTICAL MODELS OF ELECTRON-TARGET INTERACTION (MONTE CARLO METHOD)

In the *ZAF* procedure, the probability of primary X-ray generation depends on the balance between the deceleration and the ionization cross section at the instantaneous energy of the electron along its trajectory. The pathlength is eliminated in the calculation; therefore, the positions of the electron are not calculated. The absorption correction, which depends on the depth of penetration, and the loss of electrons through backscattering, must be estimated in independent calculations which are based on empirical observations obtained from flat homogeneous and electron-opaque specimens. These models cannot be easily or accurately modified to take into account different specimen configurations. The difficulties we have observed in the theoretical treatment of the X-ray emission from thin films and small particles indicate the need for a complementary model of electron-target interaction which provides a more detailed description of the actual events. Such a description is provided by the Monte Carlo calculations [13.91], which are based on the simulation of individual electron trajectories. These were first used, in connection with electron probe microanalysis, by Green [6.2], [13.92].

The simulation requires physical models for predicting the changes in both the energy and location of the electron moving within the target. To the degree that such changes can be quantitatively described, we can derive from Monte Carlo calculations a complete data reduction procedure, as proposed by Pascal [13.93], or calculate partial aspects of the *ZAF* procedure, such as the $\phi(z)$ and f_p absorption functions. Such calculations are not applied routinely because of their length and cost, and they are not absolutely foolproof, because of the uncertainties in the models and parameters involved in the calculations. Some of these are identical to the uncertainties found in the *ZAF* procedure. The Monte Carlo procedure is very useful, however, particularly in those situations which are not adequately covered by the conventional *ZAF* procedure, such as oblique electron beam incidence, the analysis of thin free standing or supported films [13.94]-[13.97], the simulation of steep concentration gradients and determination of the excited volume [13.98], [13.99], and the characterization of small particles [13.100], [13.101] (Fig. 13.10). Some situations of this type are shown in Fig. 13.3. The Monte Carlo technique is also widely used to gain insight into the behavior of electrons in the target relevant to scanning electron microscopy [13.102] such as the emission of low-loss backscattered electrons [13.103]. A monograph covering a wide range of subjects related to Monte Carlo calculations was issued in 1977 by the National Bureau of Standards [13.104]. For further review of the subject, the reader is referred to [13.105], [13.106].

The Monte Carlo procedure is based on the simulation of a large number of trajectories of primary electrons within the target and the averaging of the resultant effects. The name of the technique is derived from the use of random numbers in the simulation of individual interactions of the primary electron with target atoms. For such events, the statistical probability distributions must be known for each parameter which defines the event (e.g., a scattering angle). The range over which this parameter can vary may then be divided into an arbitrary number, n, of regions of equal probability (see Fig. 13.11, in which $n = 10$). Let $f(x)dx$ be the probability density function, which is the probability that a particular value of x falls between x_i and $x_i + dx$. Then, the probability that the value of x is below a given value is $F(x) = \int_{x_0}^{x} f(x)dx$; and $\int_{x_0}^{x_n} f(x)dx = 1$. In these equations, x_0 and x_n are the lowest and highest values possible of the variable x. From the n regions of equal probability, one region is chosen for each event, by means of a random number between 0 and 1. The central value of the variable x in this region is chosen for the event of reference. The summation of a large number of events reproduces the probability distribution. To minimize

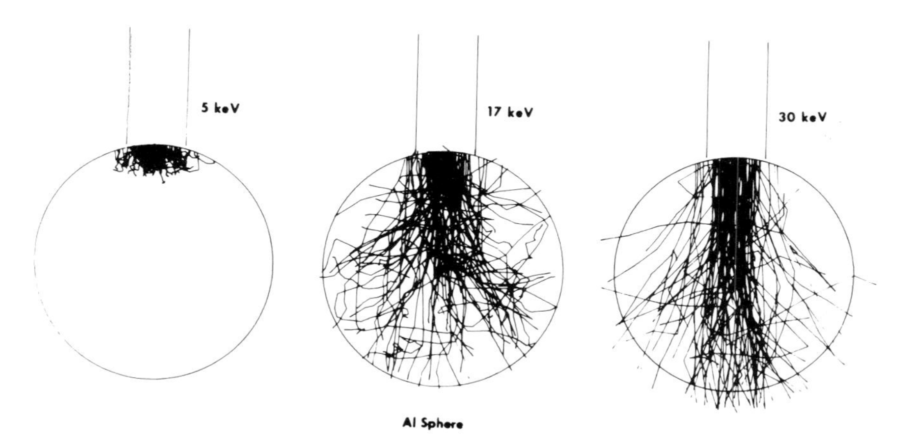

Fig. 13.10. Aluminum spheres, 2 μm in diameter, irradiated by an electron beam of 0.5 μm in diameter. At high voltages, transmission and sidescatter of the electrons becomes significant. (From Newbury and Yakowitz [14.7].)

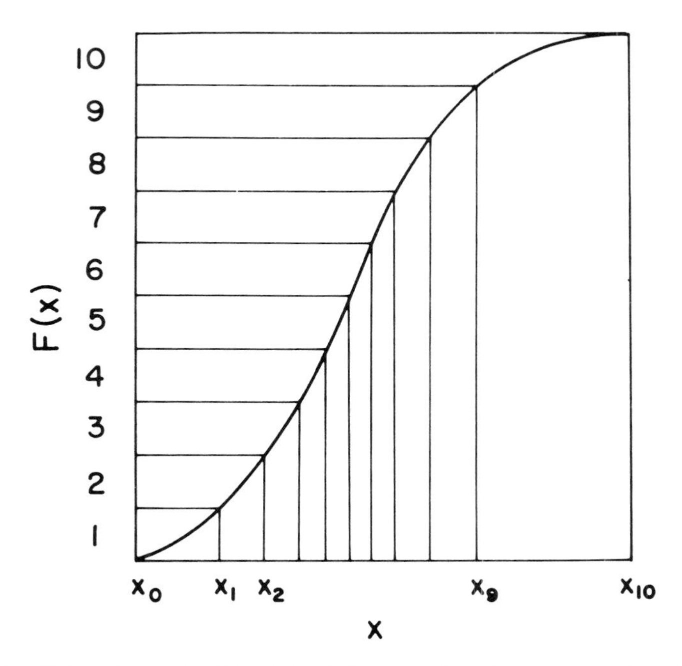

Fig. 13.11. $F(x_i)$ is the probability that x is between the values of x_0 and x_i. In the case shown in the figure, there are ten regions of equal probability. For each simulated event, one of these regions is chosen with the aid of a random number between 0 and 10. (After Hénoc and Maurice [13.94].)

the uncertainties in the result, it is important to select the right size of regions of equal probability, depending on the required resolution in the values of the parameter in question, and the number of events which can be calculated in a reasonable time.

The procedure becomes particularly useful when several random events combine, since then simple probability plots can no longer be constructed. This is the case in the simulation of electron trajectories in which several scattering acts can occur, and each scattering act is defined by more than one variable parameter.

The primary electron trajectories are conspicuously characterized by large angular trajectory changes due to collisions with target atoms (Fig. 2.3). The effect of such changes in the electron trajectory on the depth of X-ray production was first pointed out by Archard [13.107]. Most of these interactions are elastic, i.e., they do not change the kinetic energy of the beam electron. Therefore, in most Monte Carlo calculations it is assumed that all scattering events which produce directional changes are elastic, and that between such collisions the electron travels in a straight line, losing energy continuously, according to Bethe's law [Eq. (9.2.2)]. It is also assumed that the specimen is amorphous so that effects of lattice orientation can be neglected. The trajectory of each electron is determined in a number of steps which are calculated sequentially. The conceptually simplest procedure is to make each step equal in length to the mean free path of the electron between elastic collisions. Each step can then be assumed to contain one elastic scattering event. (single-scattering model). In reality, the distances between scattering acts are distributed according to a Poissonian distribution. The mean free path can be obtained from the total cross section for elastic collisions, σ_{el}, by

$$\Lambda = \frac{1}{n\,\sigma_{el}} \qquad (13.6.1)$$

where n is the number of atoms per unit mass of target material. For a multi-element target

$$\Lambda^* = (N_{av} \sum_i \frac{C_i}{A_i} \sigma_{el,i})^{-1}. \qquad (13.6.2)$$

Reimer [13.111], instead proceeds in small (10-1000 $\mu g \cdot cm^{-2}$) steps and changes the path direction after each step if the calculated free path

$$\Lambda = -\ln R/(\sigma_{el} + \sigma_{in}) \qquad (13.6.3)$$

has been exceeded. As this equation indicates, the pathlength is modified in each step by means of the random number R ($0 < R < 1$), and the inelastic scattering is explicitly included (σ_{in}) in the calculation of the free path. The cumulative effect of inelastic interactions is treated as a continuous process, according to Bethe's law [Eq. (9.2.2)].

The change of direction of the scattered electron is defined in terms of the angle of deflection, β, and the azimuthal angle, γ (Fig. 13.12). All azimuthal angles are equally probable; hence, the range of 2π is divided into an arbitrary (e.g., 20) number of sectors of equal size, and a random number is used to decide which sector applies to a given event.

For the elastic scattering cross section, a form of the screened Rutherford model [13.104] is used by many authors, e.g., [2.5], [13.11], [13.94], [13.102], [13.109], [13.110]. The angular cross section $d\sigma/d\Omega$ is obtained as the product of a total cross section, σ_E, and an angular distribution function, $f(\beta)$. These are ([13.104, p. 83]):

$$\begin{aligned} \sigma_E &= \left(\frac{Ze^2}{E}\right)^2 \frac{\pi}{4\alpha(1+\alpha)} \frac{Z+1}{Z} \\ f(\beta) &= \frac{\alpha(1+\alpha)}{\pi} \frac{1}{(1+2\alpha-\cos\beta)^2}\,. \end{aligned} \qquad (13.6.4)$$

The screening constant α is equal to $3.4 \times 10^{-3} Z^{2/3} E^{-1}$. The angular cross section is thus

$$\frac{d\sigma}{d\Omega} = \sigma_\beta = \frac{Z(Z+1)e^4}{4E^2} \cdot \frac{1}{(1+2\alpha-\cos\beta)^2}\,. \qquad (13.6.5)$$

The factor $(Z + 1)/Z$ in the formula for σ_E was introduced in the cross section for elastic scatter in order to dispense with the necessity for a separate calculation of inelastic scatter as performed by Reimer. More complex expressions for the screening constant were proposed by other authors (e.g., [13.110]) in order to improve over the imperfect fit with experimental evidence found, for instance, by Bishop [13.111]. Other authors (e.g., Green [13.92] and Curgenven and Duncumb [2.5]) chose to empirically adjust the Rutherford cross section to improve the fit. Reimer, however, indicated ([13.104, p. 45]) that Mott cross sections calculated by various authors are preferable to the Rutherford cross section, and that they produce significantly different angular cross sections.

For the calculation of the scattering cross section, as well as for that of the ionization cross section, it is necessary to determine the mean energy of the

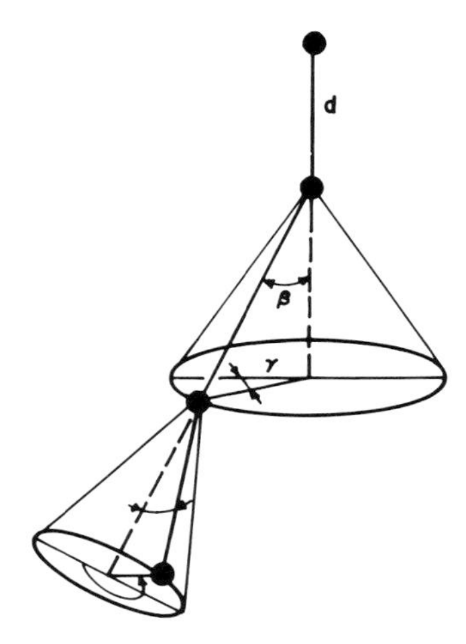

Fig. 13.12. Scattering is defined by the angle of deflection, β, the azimuthal angle, γ, and the distance between scattering events, d.

electron at the particular step. This is usually done by applying to Bethe's equation a numerical method by Runge-Kutta [13.112].

For most purposes, and with most computers, the single-scattering procedure we have described is very time consuming and expensive. Therefore, simpler procedures have been devised in which the cumulative effect of various scattering acts within a large step is expressed by a single formula [13.11] (*multiple-scattering model*) (Fig. 13.13). In this simulation, the pathlength is divided into an arbitrary number of segments. In each of these segments, one single hypothetical scattering act is assumed to occur. The parameters of this hypothetical event are adjusted so that, on the average, it would alter the path of the electron in the same way as the collisions which would occur in a real target along the same length of electron trajectory. The steps into which the trajectory is divided can be a fixed fraction of the Bethe electron range, or they can be adjusted according to the mean electron energy in each so that the number of real collisions in each segment is approximately the same.

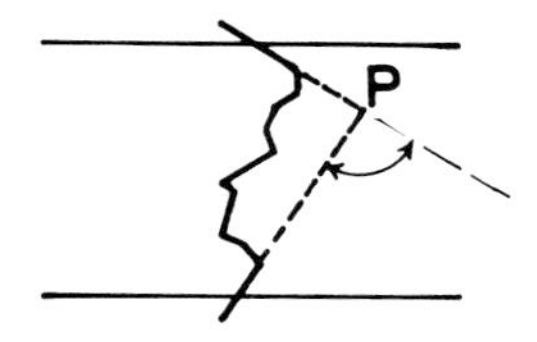

Fig. 13.13. Within each segment of the multiple-scattering model, the directional change due to all interactions is assumed to be caused by a single hypothetical event, at point P.

A very simple multiple-scattering model based upon an unscreened Rutherford scattering model, with empirical adjustment, was designed by Curgenven and Duncumb for use on small computers [2.5]. The adjustment forbids scattering into an angle smaller than a threshold set by the adjustment. This model was used by several investigators; Love et al. [13.113] employed it for the determination of $\Phi(z)$ curves.

The properly adjusted multiple-scattering model reproduces well most experimental results, with the exception of those in which the interactions at very small depth below the specimen surface are significant. When this is the case, it is advantageous to use a single-scattering model, or a mixed model [13.114] in which a single-scattering model is used when the electron is close to the surface, and a multiple-scattering model is used at greater depth.

In the application of the Monte Carlo method, the shape of the target, or, in a composite target, of each phase in the target, must be defined by a mathematical expression of its spatial boundaries. Whenever an electron, after having entered the target, crosses these boundaries, it is considered backscattered or transmitted, depending on its direction. If interactions within the target are only of interest, such electrons can be eliminated from further consideration (Fig. 13.14). For the case in which outscattered electrons are of interest, the spatial, angular, and energy distributions can be determined, in addition to the backscatter or transmission coefficient. One of the attractive features of the Monte Carlo method is the ease with which a given model can be adapted to a particular target shape, by changing the expression for the target boundaries. The applications of the method have included a wide range of situations and signals, including primary [11.2], [13.104], [13.114] and secondary [11.2] characteristic X-rays, electron backscatter [13.11], [13.88], [13.104], [13.113], [13.114], transmission [13.11], [13.116] and energy distributions [13.11], [13.92], [13.105], [13.116]. When X-ray production is calculated, the electron energy

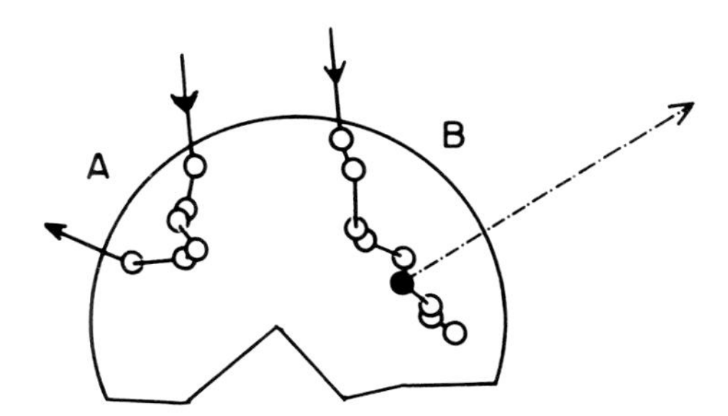

Fig. 13.14. A mathematical expression must be given to the surface boundary of the target. Whenever this boundary is crossed for a second time, the electron is considered backscattered (A). The boundary also determines the length of the X-ray path (broken line) from the point of generation to the surface, and hence, the absorption correction (B).

in each segment is used to determine the ionization cross section. We recall that the production of characteristic X-rays is an infrequent event. It usually takes hundreds of electrons to produce a single X-ray photon. To imitate nature in this respect would be intolerably wasteful in the Monte Carlo calculation. One may arbitrarily increase by an appropriate factor the ionization cross section, or simply record the fractional probability of X-ray production for each segment of the electron path. However, as can be observed by the clustering of sites of X-ray production on graphs, the sites of X-ray production along the same electron path are not statistically independent. Once the energy of the penetrating electron is below the critical excitation potential for the respective X-ray line, the ionization cross section drops to zero, and the further fate of this electron is of no interest to us. On the other hand, as pointed out by Reimer [13.111], high-energy secondary electrons are produced within the target, though with relatively low probability. These electrons can generate a significant fraction of X-ray photons of low energy, such as aluminum K, which are not normally taken into account in the *ZAF* procedures. The distribution in depth of X-ray generation, $\Phi(z)$, has been calculated by several authors [11.2], [13.92], [13.94], [13.99], [13.113], [13.115], as has the absorption function f_p (e.g., [13.104, p. 119], [13.115]), and the distributions obtained with the best programs available presently are probably more reliable than some of the experimental results. Undoubtedly, the quality of data reduction procedures can be improved on the basis of carefully tested Monte Carlo procedures. However, the investigator must remain aware of the uncertainties in the parameters entering the procedure. One should be particularly careful of adjusted procedures, in which a property is inadvertently built into the model, to be later "confirmed" by its results. As pointed out, simple procedures with empirical adjustment can be used for many applied purposes. However, successful models have been established from theory or with independently determined parameters, without empirical fitting. When models so constructed provide a satisfactory match to experimental results obtained with electron probe instruments, the microanalyst has good reason for confidence in the soundness of the procedure. As the refinement of the Monte Carlo models increases, the gap between theoretical prediction and experimental verification is rapidly closing, and therefore there is ground for optimism.

13. REFERENCES

13.1 Everhart, T. E and Hoff, P. H., *J. Appl. Phys.* **42**, 5837 (1971).

13.2 Cosslett, V. E. and Thomas, R. N., *Brit. J. Appl. Phys.* **15**, 1283 (1964).

13.3 Castaing, R., Ref. [3.22], p. 353.

13.4 Hanson, H. P. and Salem, S. L., *Phys. Rev.* **124**, 16 (1961).

13.5 Everhart, T. E, Herzog, R. F., Chung, M. S., and Devore, W. J., Ref. [12.84], p. 81.

13.6 Wells, O. C., Boyde, A., Lifshin, E., and Rezanowich, A., *Scanning Electron Microscopy*, McGraw-Hill, New York, 1974.

13.7 Sewell, P. B. and Ramachandran, K. N., in *Scanning Electron Microscopy/1977*, Johari, O., Ed., IITRI, Chicago, 1977, p. 17.

13.8 Sweeney, W. E., Seebold, R. E., and Birks, L. S., *J. Appl. Phys.* **31**, 1061 (1960).

13.9 Cockett, G. H. and Davis, C. D., *Brit. J. Appl. Phys.* **14**, 813 (1963).

13.10 Bishop, H. E. and Poole, D. M., *J. Phys. D, Appl. Phys.* **6**, 1142 (1973).

13.11 Bishop, H. E., *Brit. J. Appl. Phys.* **18**, 703 (1968).

13.12 Spencer, L. V., *Phys. Rev.* **98**, 1597 (1955).

13.13 Bishop, H. E., Rep. AERE-R 7158, Mater. Devel. Div., Atomic Energy Research Establishment, Harwell, England, 1972.

13.14 Poole, D. M., Bishop, H. E., and Preece, E. R., Rep. AERE-R 7159, Mater. Devel. Div., Atomic Energy Research Establishment, Harwell, England, 1972.

13.15 Hutchins, G. A., Ref. [12.80], p. 390.

13.16 Djurić, B. and Cerović, D., Ref. [12.83], p. 99.

13.17 Yakowitz, H. and Newbury, D. E., *Scanning Electron Microscopy/1977, Proc. of the 9th Ann. Scanning Electron Microscopy Symp.*, Yohari, O., Ed., IITRI, Chicago, 1976, p. 151.

13.18 Rydnik, V. I. and Borovsky, I. B., *Zavodskaya Lab.* **34**, 960 (1968).

13.19 Duncumb, P., Ref. [12.80], p. 490.

13.20 Cooke, C. J. and Duncumb, P., Ref. [12.83], p. 245.

13.21 Chandler, L. A., *X-Ray Microanalysis in the Electron Microscope*, North-Holland Publishing Co., Amsterdam, 1977.

13.22 Goldstein, J. I. and Williams, D. B., *Scanning Electron Microscopy/1977*, Vol. I, IIT Res. Inst., Johari, O., Ed., Chicago, 1977, p. 651.

13.23 Joy, D. C. and Maher, D. M., *Scanning Electron Microscopy/1977*, Vol. I, IIT Res. Inst., Johari, O., Ed., Chicago, 1977, p. 325.

13.24 Russ, J. C., *Scanning Electron Microscopy/1977*, Vol. I, IIT Res. Inst., Johari, O., Ed., Chicago, 1977, p. 335.

13.25 Theisen, R., Ref. [4.39], p. 20.

13.26 Tixier, R., Inst. de Rech. De la Sidérurgie Française (Paris) Report IRSID-MET.PHY. 8Q5, 1972.

13.27 Tixier, R. and Philibert, J., Ref. [12.83], p. 180.

13.28 Cliff, G. and Lorimer, G. W., *Proc. 5th Eur. Congr. on Electron Microscopy*, Inst. Phys., Bristol, 1972, p. 140; also *J. Spectroscopy* **103**, 203 (1975).

13.29 Nasir, M. J., *J. Microscopy* **108**, 79 (1976).

13.30 Marshall, D. J. and Hall, T. A., *Brit. J. Appl. Phys. (J. Phys. D) Ser. 2,* **1**, 1651 (1968).

13.31 Hall, T. A. and Werba, P., Ref. [12.83], p. 93.

13.32 Hall, T. A., Anderson, H. C., and Appleton, T. C., *J. Microscopy* **99**, 177 (1973); also Hall, T. A. and Peters, P. D., Ref. [12.66], p. 229.

13.33 Hall, T. A., in *Quantitative Electron Probe Microanalysis*, Heinrich, K. F. J., Ed., NBS Special Publ. 298, National Bureau of Standards, Washington, D.C., 1968, p. 269.

13.34 Hall, T. A., *J. Microscopie Biol. Cell.* **22**, 271 (1975).

13.35 Warner R. R. and Coleman, J. R., *Micron.* **4**, 61 (1973); also Warner, R. R. and Coleman, J. R., *Micron.* **6**, 79 (1975).

13.36 Eshel, A., *Micron.* **5**, 41 (1974).
13.37 Ingram, F. D., Ingram, M. J., and Hogben, C. A. M., *J. Histochem. Cytochem.* **20**, 716 (1972).
13.38 von Rosenstiel, A. P., Kriz, W., Höhling, H. J., and Schnermann, J., *Mikrochim. Acta*, 697 (1971).
13.39 Spurr, A. K., *J. Microscopie Biol. Cell.* **22**, 287 (1975).
13.40 Stenn, K. and Bahr, G. F., *J. Ultrastruc. Res.* **31**, 526 (1970).
13.41 Hall, T. A. and Gupta, B. L., Ref. [12.66], p. 147; also *J. Microscopy* **100**, 177 (1974).
13.42 Shumann, H., Somlyo, A. V., and Somlyo, A. P., *Ultramicroscopy* **1**, 317 (1976).
13.43 Ingram, M. J. and Hogben, C. A. M., *Anal. Biochem.* **18**, 54 (1967).
13.44 Morel, F. and Roinel, N., *J. Chim. Phys.* **66**, 1084 (1969).
13.45 Morel, F., Roinel, N., and le Grimmelec, C., *Nephron* **6**, 350 (1969).
13.46 Roinel, N., *J. Microscopie Biol. Cell.* **22**, 261 (1975).
13.47 Morel, F., *J. Microscopie Biol. Cell.* **22**, 479 (1975).
13.48 Lechene, C. P., *Proc. 5th Nat. Conf. EPASA*, New York, 1970, Paper 32A.
13.49 Lechene, C. P., Ref. [12.66], p. 351.
13.50 Lechene, C. P. and Warner, R. R., *Ann. Rev. Biophys. Bioeng.* **6**, 57 (1977).
13.51 Lechene, C. P., *Amer. J. Physiol.* **232**, F 391 (1977).
13.52 Rick, R., Horster, M., Dörge, A., and Thurau, K., *Pflügers Arch.* **369**, 95, 1977.
13.53 Hall, T. A., Ref. [12.82], P. 679.
13.54 Roinel, N. and Passow, H., *FEBS Lett.* **41**, 81 (1974); also *J. Microscopie* **22**, 475 (1975).
13.55 Kimzey, S. L. and Burns, L. C., *Ann. N.Y. Acad. Sci.* **204**, 486 (1973).
13.56 Lechene, C. P., Bronner, C., and Kirk, R. G., *J. Cell Pathol.* **90**, 117 (1977).
13.57 Sutfin, L. V., Holtrop, M. E., and Ogilvie, R. E., *Science* **164**, 947 (1971).
13.58 Coleman, J. R., Nilsson, J. R., Warner, R. R., and Batt, P., *Exper. Cell. Res.* **76**, 31 (1973).
13.59 Nilsson, J. R. and Coleman, J. R., *J. Cell Sci.* **24**, 311 (1977).
13.60 Galle, P., Ref. [12.82], p. 670.
13.61 Zeedijk, H. B., *Mikrochim. Acta*, 977 (1973).
13.62 Ferin, J., Coleman, J. R., Davis, S., and Morehouse, B., *Arch. Environ. Health* **31**, 113 (1976).
13.63 Capitant, M., Goñi, J., Rose, Y., and Roujeau, J., *Bull. Soc. Franc. Miner. Crist.* **87**, 300 (1964).
13.64 Birks, L. S. and Seebold, R. E., ASTM Special Tech. Publ. 308, American Society for Testing and Materials, Philadelphia, Pa., 1961.
13.65 Hale, A. J., *J. Cell Biology* **15**, 427 (1962).
13.66 Sims, R. T. and Marshall, D. J., *Nature* **212**, 1359 (1966).
13.67 Podolsky, R. J., Hall, T. A., and Hatchett, S. L., *J. Cell Biol.* **44**, 699 (1970).
13.68 Tandler, C. J., Libanati, C. M., and Sanchis, C. A., *J. Cell Biol.* **45**, 355 (1970).
13.69 Bonventre, J. V. and Lechene, C. P., *Proc. 9th Ann. Microbeam Analysis Society (MAS) Conf.*, Ottawa, Canada, 1974, Paper 8.
13.70 Beeuwkes, III, R. and Rosen, S., *J. Histochem. Cytochem.* **23**, 828 (1975).
13.71 Beeuwkes, III, R., *35th Ann. Proc. El. Micr. Soc. Am.*, Boston, Mass., Bailey, G. W., Ed., 1977, p. 358.
13.72 Tousimis, A. J., Ref. [12.81], p. 539.
13.73 Terepka, A. R., Coleman, J. R., Armbrecht, H. J., and Gunter, T. E., in *Symposia of the Soc. for Exper. Biology* **30**, Calcium in biological systems. Cambridge University Press, Cambridge, England, 1976, p. 117.

13.74 Coleman, R. J. and Terepka, A. R., in *Principles and Techniques of Electron Microscopy, Biological Applications,* Vol. 4, Hayat, M. A., Ed., Van Nostrand Reinhold Co., New York. 1974, Ch. 8, p. 159.
13.75 Mellors, R. C., Solberg, T., and Huang, C. Y., *Lab Invest.* **13**, 183 (1964).
13.76 Frazier, P. D., *Archs. Oral Biol.* **12**, 25 (1967).
13.77 Bhalla, R. J. R. S. B., White, E. W., and Roy, R., *J. Luminescence* **6**, 116 (1973).
13.78 Hoffmann, H. J., Weihrauch, J. H., and Fechtig, H., Ref. [12.83], p. 166.
13.79 Armstrong, J. T. and Buseck, P. R., *Anal. Chem.* **47**, 2178 (1975).
13.80 Lamvik, M. K. and Langmore, J. P., *SEM/1977*, Vol. 1, IITRI, Chicago, Yohari, O., Ed., p. 401.
13.81 Langer, A. M., Rubin, I. B., and Selikoff, I. J., *J. Histochem. Cytochem.* **20**, 723, (1972).
13.82 Langer, A. M., Rubin, I. B., Selikoff, I. J., and Pooky, F. D., *J. Histochem. Cytochem.* **20**, 735 (1972).
13.83 Ruud, C. O., Barrett, C. S., Russel, P. A., and Clark, R. L., *Micron.* **1**, 115 (1976).
13.84 Champness, P. E., Cliff, G., and Lorimer, G. W., *J. Microscopy* **108**, 231 (1976).
13.85 Beaman, D. R. and File, D. M., *Anal. Chem.* **48**, 101 (1976).
13.86 *First FDA Symp. on Electron Microscopy of Microfibers*, Pennsylvania State University, 1976, Asher, I. M. and McGrath, P., Eds., U.S. Government Printing Office, Washington, D.C., 1977.
13.87 *Proc. Workshop on Asbestos, Definitions and Measurement Methods*, Gravatt, C. C., LaFleur, P. D., Heinrich, K. F. J., Eds., National Bureau of Standards, Washington, D.C., 1978.
13.88 Satta, K., Takeoka, T., and Oda, Y., Ref. [12.84], p. 357.
13.89 Henry G., Philibert, J., Plateau, J., and Weinryb, E., *J. Microscopie* **2**, 505 (1963).
13.90 Fleetwood, M. J., Higginson, G. M., and Miller, G. P., *Brit. J. Appl. Phys.* **16**, 645 (1965).
13.91 Berger, M. S., in *Methods in Computational Physics*, Vol. 1, Alder, B., Fernback, S., Rotenberg, M., Eds., Academic Press, New York, 1963.
13.92 Green, M., *Proc. Phys. Soc.* **82**, 204 (1963).
13.93 Pascal, B., *J. Microscopie* **8**, 276 (1969).
13.94 Hénoc, J. and Maurice, F., Ref. [12.84], p. 113.
13.95 Hénoc, J., *J. Microscopie* **15**, 289 (1972).
13.96 Bolon, R. B. and Lifshin, E., in *Scanning Electron Microscopy II*, Johari, O., Ed., IITRI, Chicago, 1973, p. 285.
13.97 Kyser, D. F. and Murata, K., *IBM. J. Res. Develop.* **18**, 352 (1974); also Ref. [13.99], p. 129.
13.98 Shimizu, R. and Shinoda, G., *Osaka U. Fac. of Eng. Tech. Rep.* **14**, 897 (1964).
13.99 Shinoda, G., Murata, K., and Shimizu, R., NBS Special Publ. 298, Heinrich, K. F. J., Ed., 1968, p. 155.
13.100 Fiori, C. E., Heinrich, K. F. J., Myklebust, R. L., and Darr, M., NBS Special Publ. 422, Vol. 1, 1283 (1976).
13.101 Yakowitz, H., Newbury, D. E., and Myklebust, R. L., *Scanning Electron Microscopy/1975*, IITRI, Chicago, Johari, O., Ed., Part I, p. 93.
13.102 Shimizu, R. and Everhart, T. E., *Optik* **36**, 59 (1972).
13.103 Murata, K., *J. Appl. Phys.* **45**, 4110 (1974).
13.104 NBS Special Publ. 460, Heinrich, K. F. J., Newbury, D. E., and Yakowitz, H., Eds., U.S. Dept. of Commerce, 1976.
13.105 Shimizu, R., Ikuta, T., and Murata, K., *J. Appl. Phys.* **43**, 4233 (1972).
13.106 Shimizu, R., *Technol. Repts., Osaka Univ.* **27**, 69 (1977).

13.107 Archard, G. D., *2nd Int. Symp. on X-Ray Optics and Microanalysis*, Engström, A., Cosslett, V. E., and Pattee, H. H., Eds., Elsevier, Amsterdam, p. 331.
13.108 Wentzel, G., *Z. Phys.* **40**, 590 (1927).
13.109 Bishop, H. E., Ref. [12.82], p. 112.
13.110 Shimizu, R., Nishigori, N., Murata, K., Ref. [12.84], p. 1284.
13.111 Reimer, L., Ref. [13.100], p. 45.
13.112 Wilson, M. V., *Introduction to Numerical Analysis*, Cambridge University Press, 1966, p. 55.
13.113 Love, G., Cox, M. G. C., and Scott, V. D., *J. Phys. D, Appl. Phys.* **10**, 7 (1977).
13.114 Newbury, D. E., Myklebust, R. L., and Heinrich, K. F. J., *Proc. 8th Int. Conf. on X-Ray Optics and Microanalysis*, Ogilvie, R. and Wittry, D. B., Eds., to appear.
13.115 Murata, K., Okamachi, M., Gennai, N., Shimizu, R., and Kato, S., Dept. Appl. Phys., Osaka Univ. Tech. Rep. 19, 337 (1969).
13.116 Murata, K., Matsukawa, T., and Shimizu, R., Ref. [12.84], P. 105.
13.117 Drescher, H., Reimer, L., and Seidel, H., *Z. Angew. Phys.* **6**, 331 (1970).
13.118 Murata, K., Shimizu, R., and Shinoda, G., Osaka Univ. Tech. Rep. 16, 121 (1966).
13.119 Matsukawa, T., Murata, K., and Shimizu, R., *Phys. Stat. Sol. (b)* **55**, 371 (1973).
13.120 Shimizu, R., Tech. Reps., Osaka Univ. 27, 69 (1977).
13.121 Hren, J. J., Goldstein, J. I., and Joy, D. C. eds., *Introduction to Analytical Electron Microscopy*, Plenum Press, New York, 1979.
13.122 *Characterization of Particles,* Heinrich, K. F. J., Ed., NBS Special Publication 533 (1980).

14. Scanning Electron Microscopy

In the preceding chapters the X-ray emission from a homogeneous region of the specimen excited by a stationary focused electron beam has been discussed. The spatial definition of single-point analysis has also been given. But the full power of the microscopic aspects of analysis with an electron beam becomes apparent only when the *variation of composition* of a specimen, as a function of its spatial coordinates, is investigated. Procedures will now be described in which the position of the electron beam relative to the specimen is changed during the experiment (scanning). Such procedures range from point-by-point analysis along a line on the specimen surface to the area scanning techniques used in scanning electron microscopy and scanning electron-probe X-ray analysis in which images of the specimen surface are produced. A listing of variables in the scanning technique is given in Table 14.1. Since line scanning is usually performed for the collection of X-ray signals, it will be discussed in the next chapter.

14.1 THE SCANNING ELECTRON MICROSCOPE (Fig. 14.1)

In the scanning electron microscope, a finely focused electron beam is moved in a raster over a region of a solid specimen. A signal excited by the electron beam, such as secondary electrons, is used to produce an image of the specimen surface by modulating the brightness of an oscilloscope which scans in synchronism with the scanning electron microscope. The main purpose of scanning electron microscopy (SEM) is the characterization of the shape of the specimen surface, although other properties of the specimen can be revealed as well by ancillary techniques.

Several books on scanning electron microscopy are available in which high-resolution scanning electron microscopy and related special techniques are described in great detail [13.6], [14.1]-[14.3]. A large amount of useful information can also be obtained from the Proceedings of the Annual SEM meetings organized by O. Johari [14.4] and those of the Electron Microscopy Society of America [14.5]. Only those simple scanning operations which are

Table 14.1. Variables in the Scanning Technique.

Scans (topography)	line	continuous		
		discontinuous		
	area	line raster	isomorphic	signal deflection
		point raster (linear sequence or pseudorandom)	not isomorphic	no
Signal (contrast)	pulse	unmodified		
		selective		
	continuous	unmodified		
		modified	smoothed	isotropic
			enhanced contrast	isotropic
			gamma	isotropic
			concentration mapping	isotropic
			differentiated	dynamic
			filtered	dynamic
			higher order transforms	dynamic

frequently performed in connection with electron-probe microanalysis will be discussed here. The handling of X-ray signals for the formation of images is the subject of Chapter 15.

The scanning electron microscope is very similar in principle and construction to the electron-probe microanalyzer. For sharper focusing of the electron beam and hence better spatial resolution, it is provided with an additional condenser lens. To deflect the electron beam, a double set of scanning coils is built into the column above the center of the objective lens (see Figs. 3.5, 3.30, and 14.1). The first set of coils deflects the beam from the optical axis, and the second set of coils deviates it in the opposite direction, so that the beam passes again through the optical axis at the level of the pole-piece gap and hence does not suffer field aberration. If an aperture is provided at this level, the second set of coils ensures passage of the beam through this aperture. To provide scanning in two orthogonal directions, a quadruple winding can be arranged around a ferromagnetic core (Fig. 14.2). If the coils must be arranged close to the field of the final lens, the use of magnetic cores should be avoided.

Typical scanning electron microscopes are not provided with Bragg X-ray spectrometers, but Si(Li) detectors are frequently attached to them. Because the functions of these detectors are not as sensitive to the position of the specimen along the optical axis as those of the Bragg spectrometers, it is not

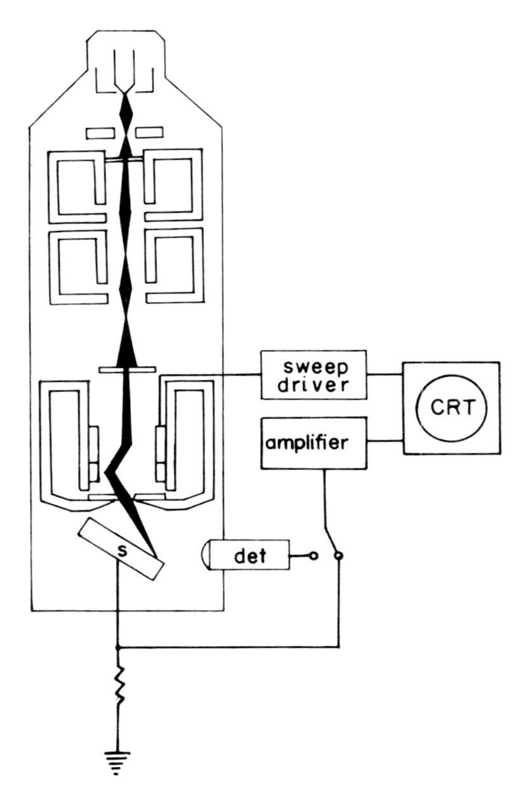

Fig. 14.1. Schematic of scanning electron microscope. Note the double condenser lens and the final aperture in the gap of the objective lens. det:secondary electron detector; s:specimen; CRT:cathode-ray tube.

necessary to install an optical microscope. The most common signal used in scanning electron microscopes is the emission of secondary electrons, and all scanning electron microscopes are therefore provided with detectors for low-energy electrons.

Such electrons are easily influenced by electrostatic and electromagnetic fields. Their emission can be inhibited by charging the specimen positively [9.21], or by interposing in their path a negatively charged wire screen. A positive charge can be used to attract them toward the detector, and by means of an electrostatic field, signal collection over a very wide angle of emission can be achieved.

The most commonly used detector, described by Everhart and Thornley [14.6], is based upon the collection by a scintillator of the secondary electrons attracted toward the detector by a positively charged grid. The electrons which

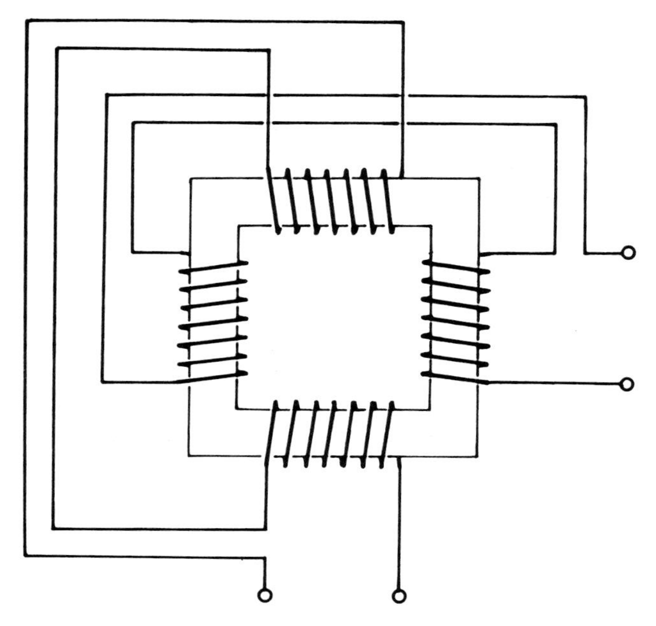

Fig. 14.2. Scanning coils on a ring of ferrite.

have passed the grid are further accelerated, by charging the scintillator positively (Fig. 14.3). The accelerated electrons produce within the scintillator a flash of light which is transmitted through a lightpipe to the photocathode of a photomultiplier tube. The electrons attracted by the wire mesh acting as a positive grid travel toward the detector along curved paths, covering a large solid angle at the level of emission. Therefore, this detector has a high collection efficiency and is thus attractive for high-resolution image formation, since a small beam can only be obtained at low beam currents. (For a resolution on the order of 10 nA, the permissible beam current obtained with a tungsten filament gun is on the order of 10^{-11} A.)

The intensity of emission of secondary electrons increases with the angle of tilt, $\pi - \epsilon$, if the electron beam is inclined with respect to the specimen surface (Fig. 14.4) [14.7]. Furthermore, the imaging of odd-shaped objects requires freedom in the angular orientation of the specimen. For these reasons, the typical stage for secondary electron microscopy allows for freedom in the angular orientation, as well as for precision in the angular setting of the specimen. It is desirable that the inclination of the specimen can be changed without displacing it from the optical axis of the microscope (eucentric stage). The specimen chamber of a scanning electron microscope should be as spacious as

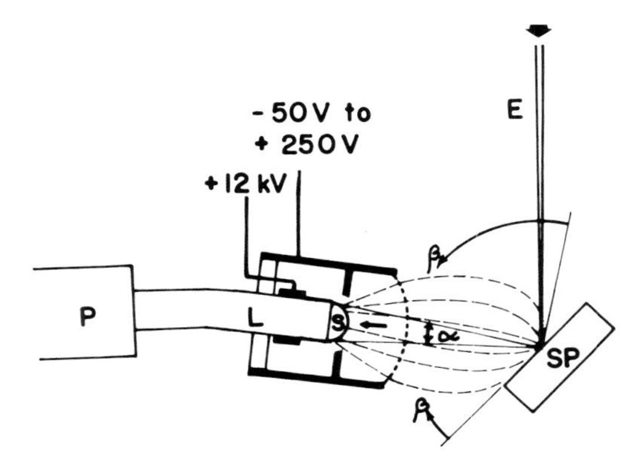

Fig. 14.3. Everhart-Thornley detector for secondary and backscattered electrons. E:electron beam; L:light pipe; P:photomultiplier tube; S:scintillator. When the housing of the detector is charged negatively, only the backscattered electrons are collected through the solid angle α. When the housing is charged positively, secondary electrons are also collected through the larger solid angle β. All electrons are further accelerated by the positive charge on the scintillator which is coated with a conductive layer.

possible, to accommodate a variety of stages and specimen shapes and sizes, and it should be provided with several ports for the attachment of detectors and feedthroughs for electrical connections. The specimen chambers of scanning electron microscopes are seen in Figs. 12.22 and 14.5.

Scanning coils are now used in all electron-probe microanalyzers, and X-ray detection has been proven to be a powerful auxiliary function in microscopes. Instruments are now available which combine the crystal X-ray spectrometers and the optical microscope of the electron-probe microanalyzer with the resolution, scanning capabilities, and secondary electron detection of the scanning electron microscope (Fig. 14.6). Such instruments frequently have interchangeable holders for specimens of various sizes, and for the performance of both scanning electron microscopy and electron-probe microanalysis. They may also be provided with light transmission optics, light spectrometers for the measurement of cathodoluminescent emission, and transmission electron microscope attachments.

14.2 SCANNING IMAGES

The area covered by the electron beam scan on the specimen surface can be considered as a two-dimensional array of points. It is possible, and advantageous for

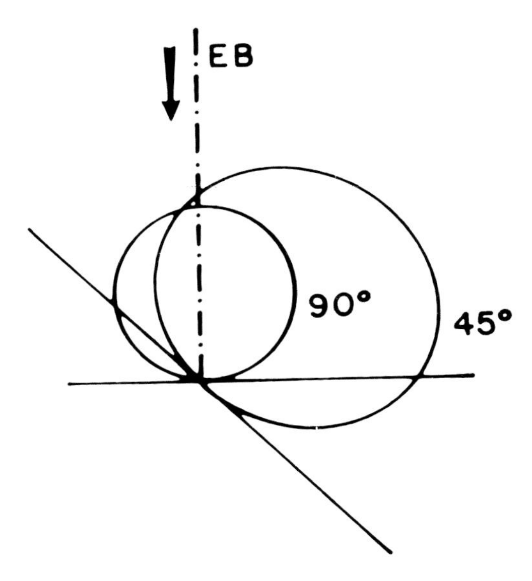

Fig. 14.4. The intensity of emission of secondary electrons increases with specimen tilt, but their distribution in space obeys a cosine law even with a tilted target surface.

a numerical evaluation by computers of measurements over the area, to actually produce such a point raster. However, many instruments use analog procedures to generate the sweep raster which then consists of a series of continuous parallel lines, produced by two ramp (sawtooth) voltages of unequal frequency. Call x and y the two orthogonal directions on the specimen surface of the scan. If the generator for direction x produces n_x sawteeth per second, and the y generator produces n_y sawteeth per second ($n_x > n_y$), the resulting frame has n_x/n_y lines per frame, and the scanning speed is n_y frames per second. Although it is possible to gather the information required for the image in more than one sweep, it is simplest to produce an image in a single sweep. In such a case, the statistics of the signal to be gathered must be taken into account when the speed in the y direction is chosen.

If one signal such as secondary electrons is collected in an area scan, then each point of the array is characterized by three associated parameters: the two orthogonal positional coordinates x and y, and the measured signal intensity $I(x,y)$. In the past the information gathered in a scan was usually too large to be stored in electronic memory devices, although now such storage is feasible. The information is of little use without a simplifying mode of presentation, either in the form of a statistical table, or in a graphic mode. A *scanning image* is a two-

Fig. 14.5. Specimen chamber of the CAMEBAX electron-probe-scanning electron microscope. H:movements in the plane of the specimen surface; E:elevation; R:rotation; S:specimen holder for electron-probe microanalysis; A:aperture manipulator.

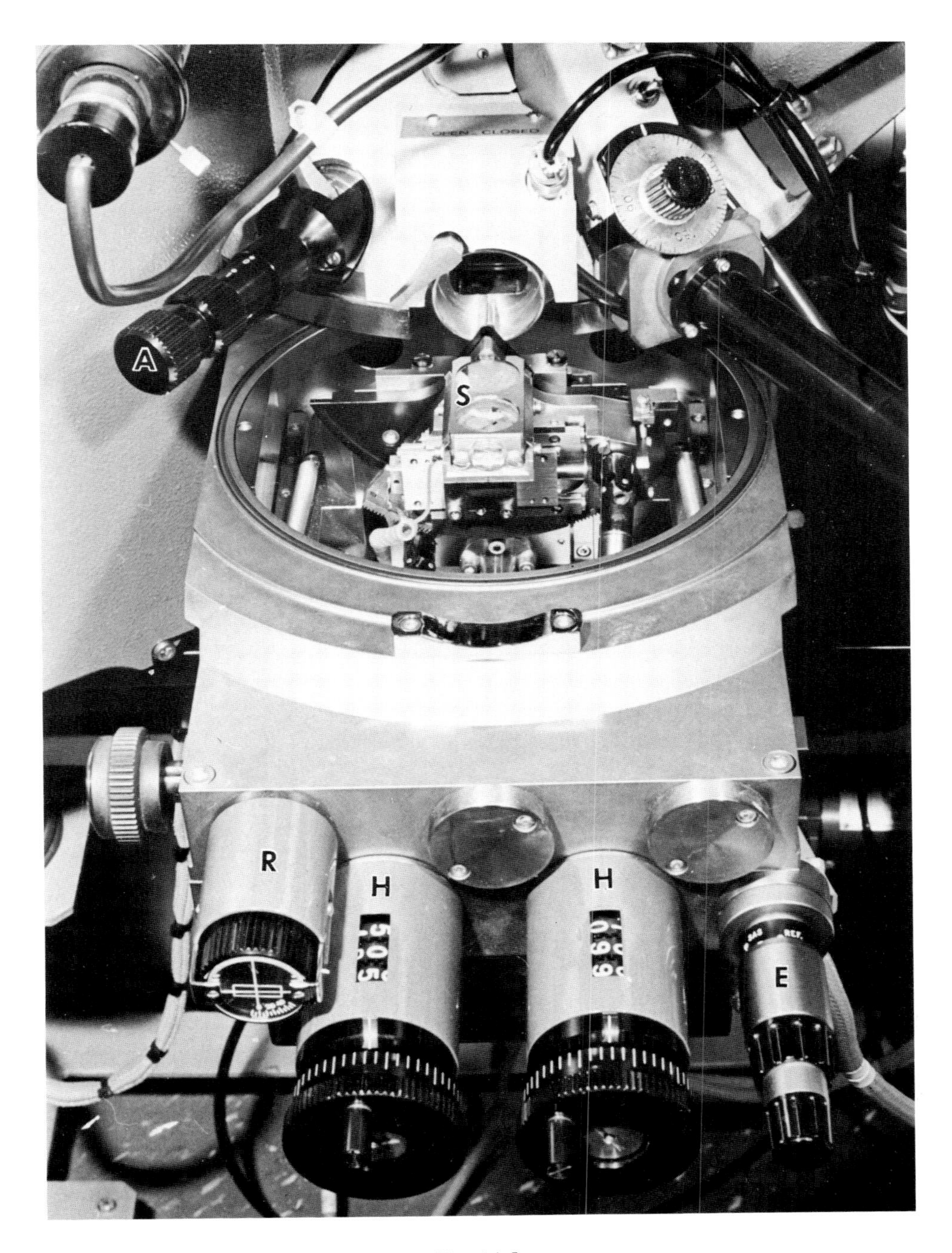

Fig. 14.5.

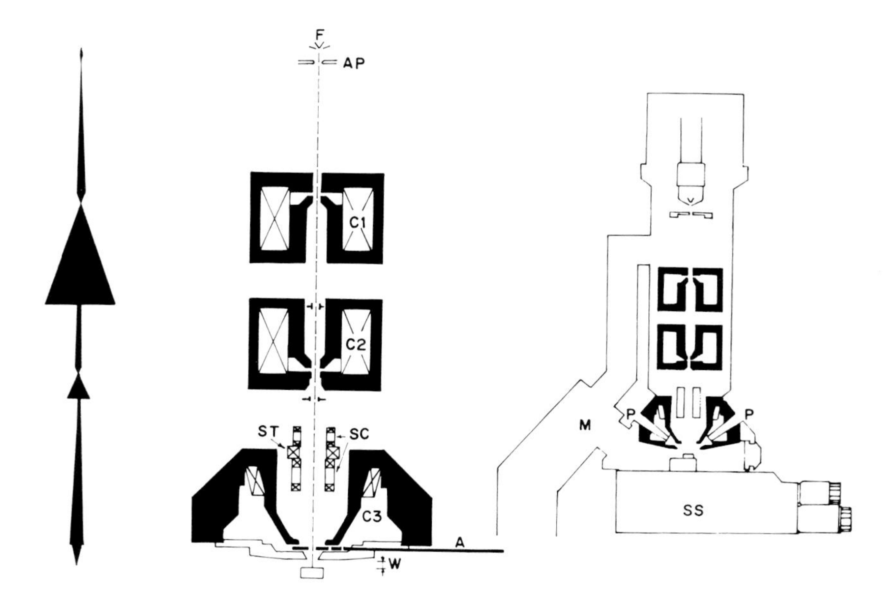

Fig. 14.6. Electron-optical paths (left), electron optics (center), and assembled column of the CAMEBAX combined electron probe-scanning electron microscope. Note the double condenser lens (C1 + C2), and ports for crystal X-ray detectors (P). F:filament; AP:anode plate; ST:stigmator coils; SC:scanning coils; C3:objective lens; W:working distance; M:vacuum manifold; SS:specimen stage; A:aperture holder.

dimensional representation of the scan in which the coordinates of each point of the raster determine those of the corresponding point on the representation. The signal is usually employed to determine the gray level on this point, which can range from black to white. The image can be obtained by means of a cathode-ray tube (CRT), the beam of which scans in synchronism with the electron probe. A permanent record of the image can be obtained by means of a photographic camera focused upon the screen of the cathode-ray tube.

If the image position of any point on a flat specimen is determined only by the two coordinates x and y on the specimen, we will call such an image isomorphic. If the specimen is not flat, the image obtained by the same procedure will be isomorphic with the parallel projection in the direction of the beam of the scanned specimen surface. If, in addition, the dimensions on the specimen surface are all reproduced at the same proportion in the image, and hence all angles on the specimen match the corresponding ones on the image, the latter is geometrically similar to the specimen.

Images of surfaces are sometimes formed by the y-deflection technique, in which the signal is applied to the vertical position of the respective picture ele-

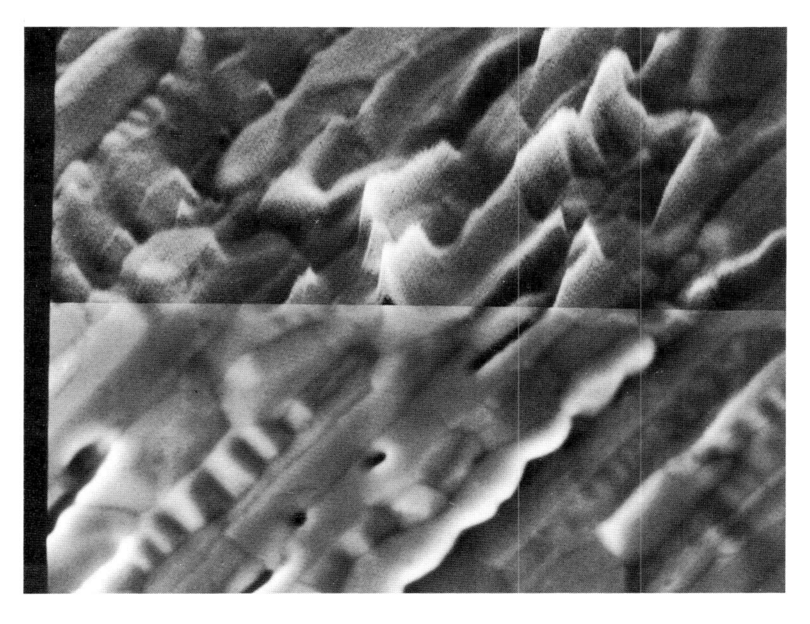

Fig. 14.7. Surface of the filament of a fluorescent lamp after use. Secondary electron image. Width of the field: 12.5 μm. The addition of y deflection (upper half) to the brightness modulation causes distortion of the topographic representation of the filament surface. See, for comparison, Fig. 14.37. E_0 = 20 keV.

ment, rather than to its gray value, or in addition to the gray-value modulation [14.8] (Figs. 14.7 and 14.8). To provide an appearance of three-dimensionality, the frame dimension y (vertical dimension) may, on the image, be projected at an angle and sometimes foreshortened. If the number of lines per frame in the y-deflection mode is small so that the individual scan traces can be distinguished in the image, the effect is that of a superposition of parallel line scans (Fig. 14.8). At a greater line density, the effect can be confusing since the signal interacts with the location parameters. Since points of the image cannot be unequivocally related to the corresponding points on the specimen, such a procedure should be avoided. It is particularly undesirable for the imaging of objects which are not flat, since the effects of true dimensions, projection, and signal, cannot be distinguished (Fig. 14.9). For the enhancement of small-signal differences, the first derivative of the signal (to be discussed in Section 14.5) is more useful since it conserves the isomorphism of the image [14.9].

The ratio of the metric distance between two image points to that between the corresponding points on the specimen is called the *magnification* of the image. If the scanned area and its image are geometrically similar, then the magnification is the same in all directions (isometric image). Such an image can only

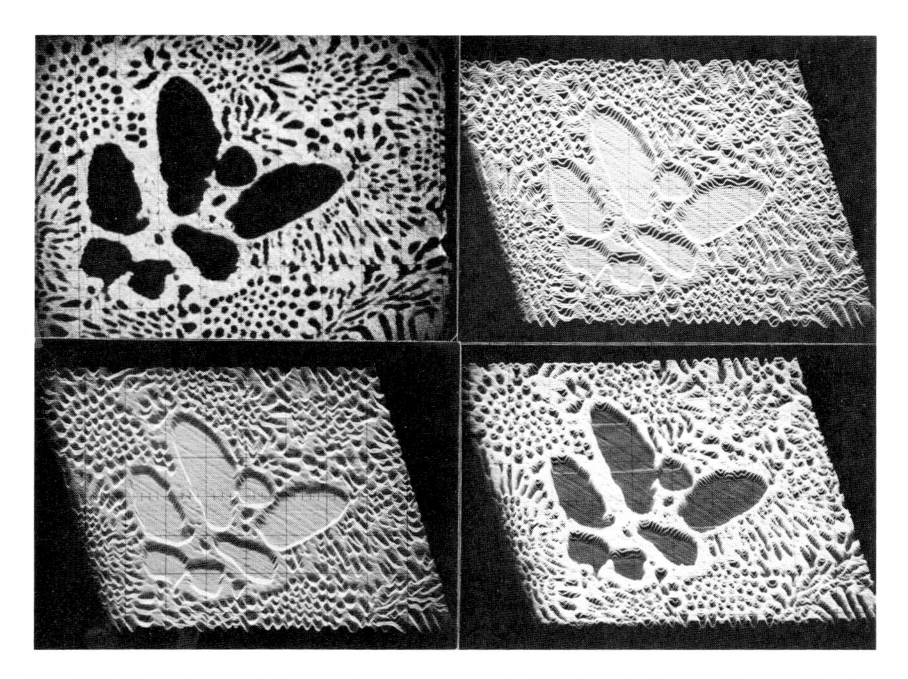

Fig. 14.8. Target current images from a region of diffusion of silver and brass. The large islands in the center are brass. Width of the scanned area: 800 μm. Upper left: inverted target current image. Upper right: *y* deflection, 150 lines per frame. Lower left: *y* deflection, 300 lines per frame. Lower right: same, with additional brightness modulation.

be obtained from flat surfaces. In the imaging of three-dimensional objects, the primary electron beam produces a virtually parallel projection of the specimen (the angular changes of the beam during the scan are very small). Hence, the nominal magnification is valid only in the plane normal to the beam. When specimens are observed at an inclination to the beam, to increase the production of secondary electrons and thus obtain a better image, the resulting foreshortening of the image can be canceled by adjusting the scanning width of the primary beam so that an isometric image of a flat surface is obtained. In such a procedure, however, the image of three-dimensional objects is distorted (Fig. 14.10). In addition to the foreshortening correction, the focusing of the beam should also be dynamically adjusted during the scan so that a sharp image of the entire scanned area is obtained (Fig. 14.11). But even without dynamic adjustment, the depth of focus of the scanning electron micrograph is much larger than that of an optical microscope, because the aperture of the electron beam, which determines the depth range, is much smaller than that of an optical microscope objective lens (Fig. 14.12).

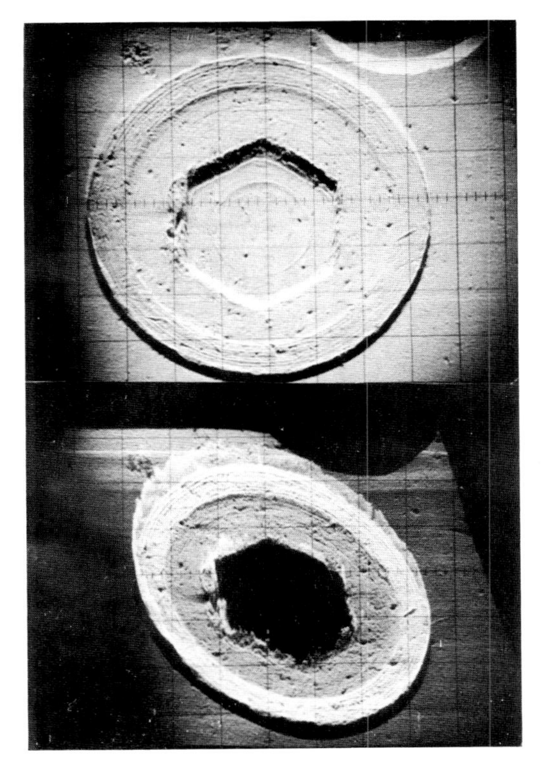

Fig. 14.9. Secondary electron images of the head of an Allen screw, 1 cm in diameter. Above: y deflection only. Below: y deflection, tilt of y axis, and brightness modulation.

14.3 SIGNALS FOR SCANNING ELECTRON MICROSCOPY [14.7], [14.10]

(see Table 14.2.)

14.3.1 Secondary Electrons [14.11]

As indicated in Fig. 2.5, the specimen bombarded by primary electrons emits electrons of an energy below 50 eV, called secondary electrons, through various mechanisms of energy transfer. The ratio of secondary to primary electrons for a given target material, configuration, and operating voltage, can be defined analogously to the backscatter coefficient [Eq. (9.3.1)], as a measure of secondary electron emission:

$$\delta = i_s/i_b. \tag{14.3.1}$$

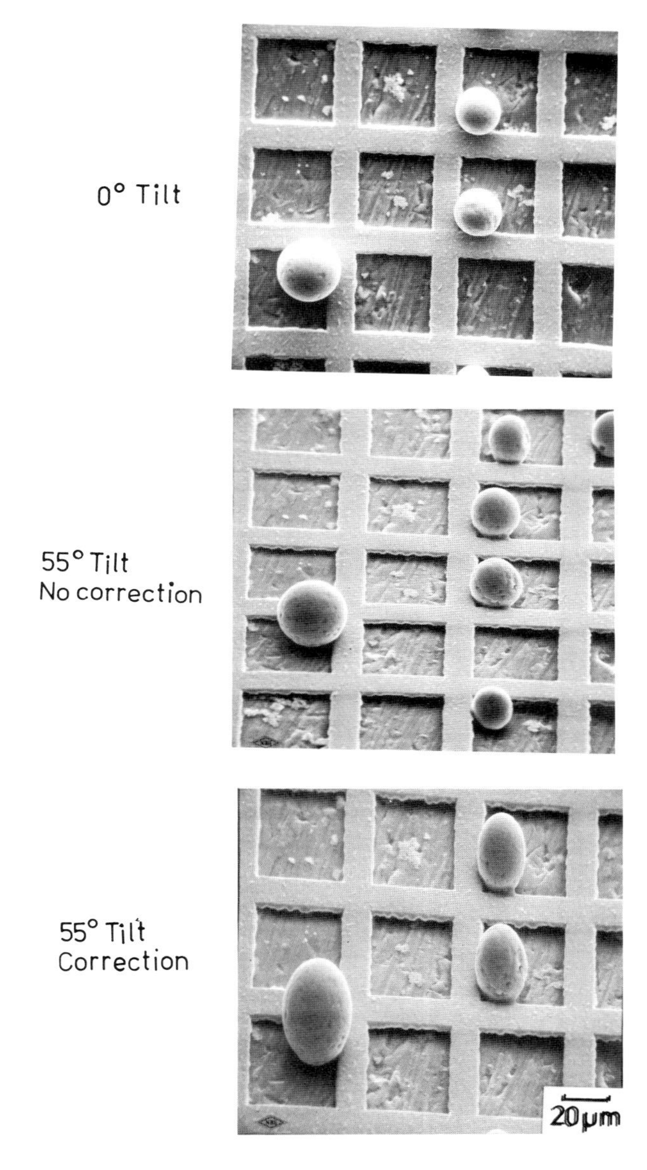

Fig. 14.10. Scanning electron micrograph of glass spherules on a copper grid. Operating voltage: 20 kV. Width of scanned field: 400 μm. Top: normal beam incidence upon the plane of the grid. Center: oblique beam incidence (ϵ = 55°). Bottom: same, with foreshortening correction. The grid appears in correct proportions, but the three-dimensional spherules are distorted.

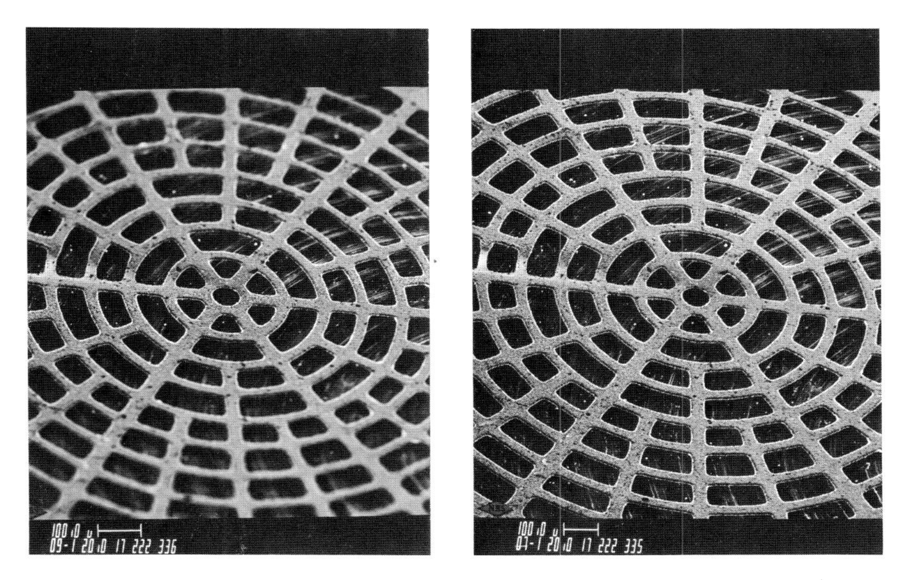

Fig. 14.11. Secondary electron image of a grid at a 45° inclination. Left: without dynamic focus adjustment. Right: with adjustment. V_0 = 20 kV. The bar below the image is 100 μm long. (Courtesy of D. Ballard, NBS.)

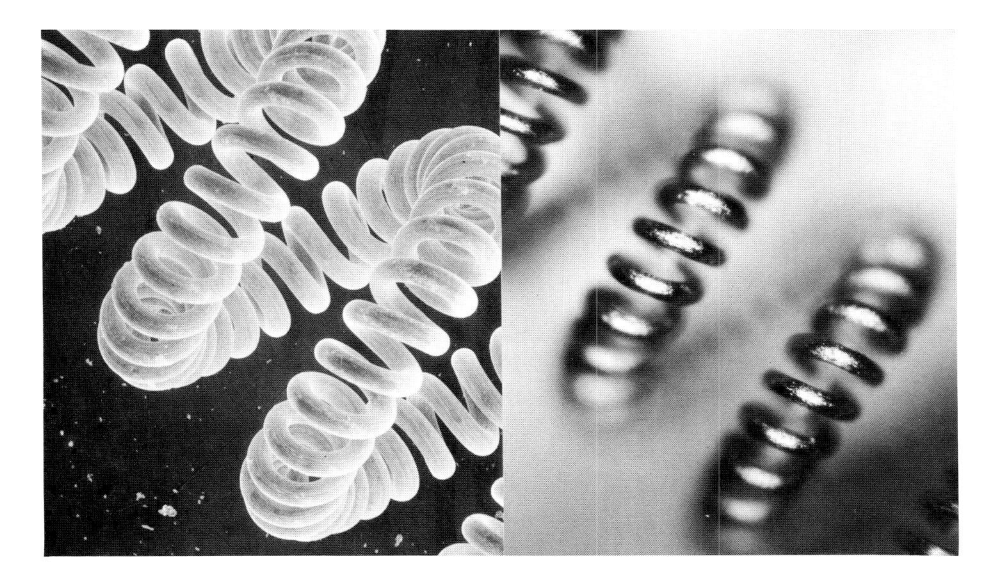

Fig. 14.12. Depth of field of scanning electron microscope (left) and light microscope (right). Distance between large loops of the filament: 0.6 mm.

TABLE 14.2. Signals for Electron Beam Instruments.

Signal	Detector	Current Range (A)	Application
Secondary electrons	Everhart-Thornley	10^{-6}-10^{-12}	surface configuration, electronic and magnetic fields [14.12]
Backscattered electrons	Everhart-Thornley	10^{-6}-10^{-10}	surface configuration, atomic number, magnetic contrast [14.12]
	silicon detectors	10^{-6}-10^{-11}	same (less directional)
Low-energy loss backscattered electrons	Wells [13.6]	10^{-6}-10^{-11}	high-resolution surface topography
Target current	amplifier	10^{-6}-10^{-11}	surface configuration, atomic number, magnetic contrast [14.12]
Induced target current	amplifier	10^{-6}-10^{-10}	semiconductor functions
Cathodoluminescence	camera, eye, monochromator	10^{-6}-10^{-12}	trace elements, defects in semiconductors, photoradiative recombination
Electron diffraction patterns (Kikuchi lines)	film Everhart-Thornley, specimen current, silicon detectors	10^{-6}-10^{-10}	lattice dimensions orientation crystal perfection [14.13]
X-ray diffraction (Kossel lines)	camera	10^{-6}-10^{-8}	lattice (requires single crystal of 20-40 μm diameters) [14.14]

The energy distribution of secondary electrons peaks close to 3 eV, and is negligible for most purposes above 20 eV (Fig. 14.13). Such electrons are produced throughout the volume of specimen excitation, but, due to their low energy, only those generated close (20-300 Å) to the specimen surface can leave the specimen and be detected. For this reason, the secondary electron coefficient, δ, after reaching a maximum around an operation potential of approximately 1 kV, diminishes with further increases of the potential (Fig. 14.14), to 0.2 or less for metals at 20 kV. This behavior is in agreement with the stopping power formula of Bethe, which indicates that the energy deposition into a thin layer diminishes with increasing electron energy.

Secondary electrons emitted by the specimen have thus been produced close

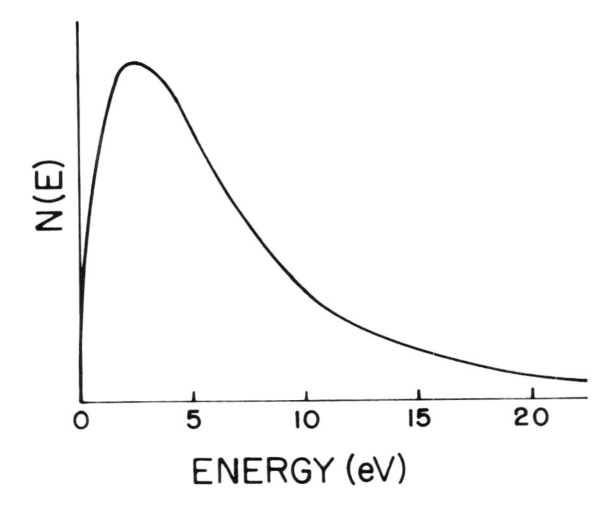

Fig. 14.13. Energy distribution of secondary electrons.

to the intercept of the primary electron path with the specimen surface. This is the cause for the high resolution of secondary electron scanning images. However, secondary electrons are also formed at the sites of exit of backscattered electrons (Fig. 14.15). Given the angular spread of the exiting electrons, their path through the layer which can emit secondary electrons is larger than that of the primary electrons entering the specimen at normal beam incidence. There-

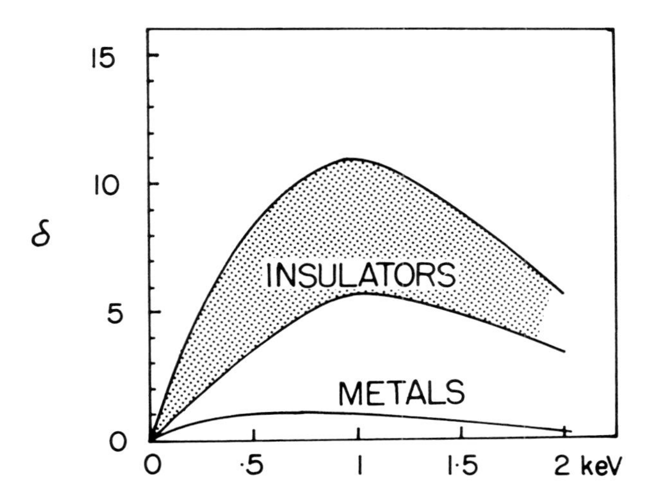

Fig. 14.14. Emission coefficient for secondary electrons of insulators and metals, at normal electron beam incidence angle upon massive targets, as a function of operating potential.

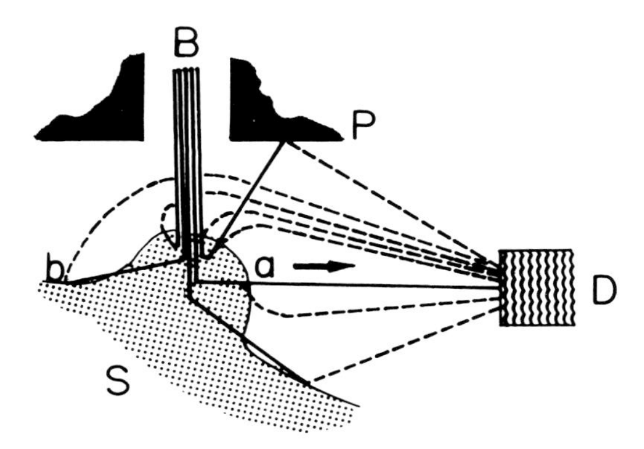

Fig. 14.15. Secondary electrons reach the detector D from a region of the specimen S close to the impact of beam B, but also from farther regions, due to the exit (a) or reentry (b) of backscattered electrons. Secondary electrons can also be produced by the impact on instrument components (P) of backscattered electrons. Schematic not drawn to scale.

fore, the production of secondary electrons by the exiting backscattered electrons exceeds that at the entrance of the primary electrons in the specimen, sometimes by factors as large as four or five.

The secondary electron coefficient does not depend on the specimen composition in so clean-cut a manner as the backscatter coefficient. Secondary electron emission from insulators is larger than that from conductors (Fig. 14.14). At operating potentials around 1 kV, the coefficient δ may substantially exceed unity. If the specimen is conductive, the specimen current may therefore reverse direction. In uncoated insulators, the emission of electrons can produce a local positive charge, which forms an electrostatic field of sufficient strength to reduce the emission of the low-energy secondary electrons. An equilibrium is thus obtained at low-operating voltages, so that insulating specimens can be irradiated for indefinite periods without application of a conductive layer, or the discharge of the specimen through a specimen current. Since at conventional operating voltages (20-30 kV) δ is typically close to 0.1 for a normally incident electron beam, the use of conductive coatings is necessary.

If it is assumed, as an approximation, that the secondary electrons escaping from beneath a surface normal to the primary beam are emitted isotropically from the mean depth of emergence, one can predict a cosine distribution of the emerging electrons, as a function of emergence angle:

$$I(\psi) = \frac{I}{\pi} \cos \psi \tag{14.3.2}$$

where I is the total emission, and $I(\psi)d\Omega$ is the number of electrons emitted into an element of solid angle $d\Omega$, at the mean angle ψ with the normal to the specimen surface. This angular distribution is virtually independent of the electron beam incidence angle ϵ.

Analogously, one obtains for the variation of δ as a function of the electron beam incidence angle ϵ (see Fig. 10.28) the function:

$$\delta_\epsilon = \delta_0 \sec \epsilon \qquad (14.3.3)$$

which agrees reasonably with experiments. Furthermore, increased emission can be observed from tips and specimen edges where the probability of emergence of low-energy electrons is large (Fig. 14.16). The increase of secondary electron emission with specimen inclination is one of the reasons for preferring, in the scanning electron microscope, an inclined specimen position. Such a configuration also provides the space necessary for the detector assembly (Fig. 14.1).

The secondary electron emission can be strongly affected by local electric charges or magnetic fields. Therefore, regions or objects of low electric conductivity may produce artifacts in scanning images (Figs. 14.17 and 14.18).

14.3.2 Backscattered Electrons [14.15]

Due to their high energy, backscattered electrons (see Section 9.3) are neither significantly deflected nor stopped by weak electrostatic or electromagnetic fields. If the Everhart-Thornley detector is placed in line of sight of the specimen, it will also detect the backscattered electrons emitted within the small solid angle α shown in Fig. 14.3. If the grid of the detector is charged negatively, so as to reject the secondary electrons, the image obtained will be entirely produced by backscattered electrons (Fig. 14.19). Other types of detectors, such as silicon detectors, are also used. These detectors are, in principle, similar to

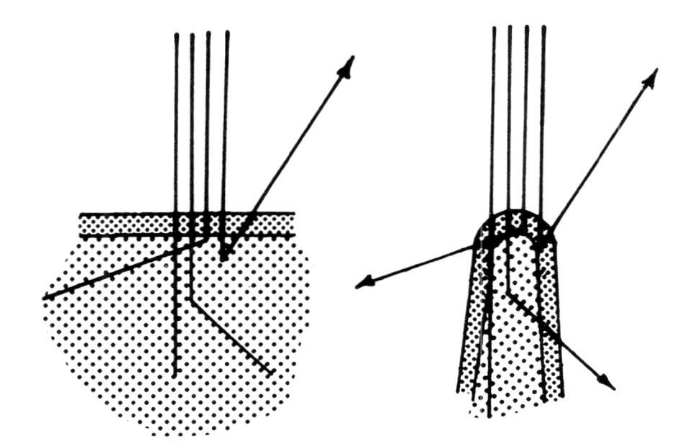

Fig. 14.16. Primary electrons entering salient parts have a greater probability of being backscattered. They also travel for a longer distance close to the surface, where they can excite secondary electrons.

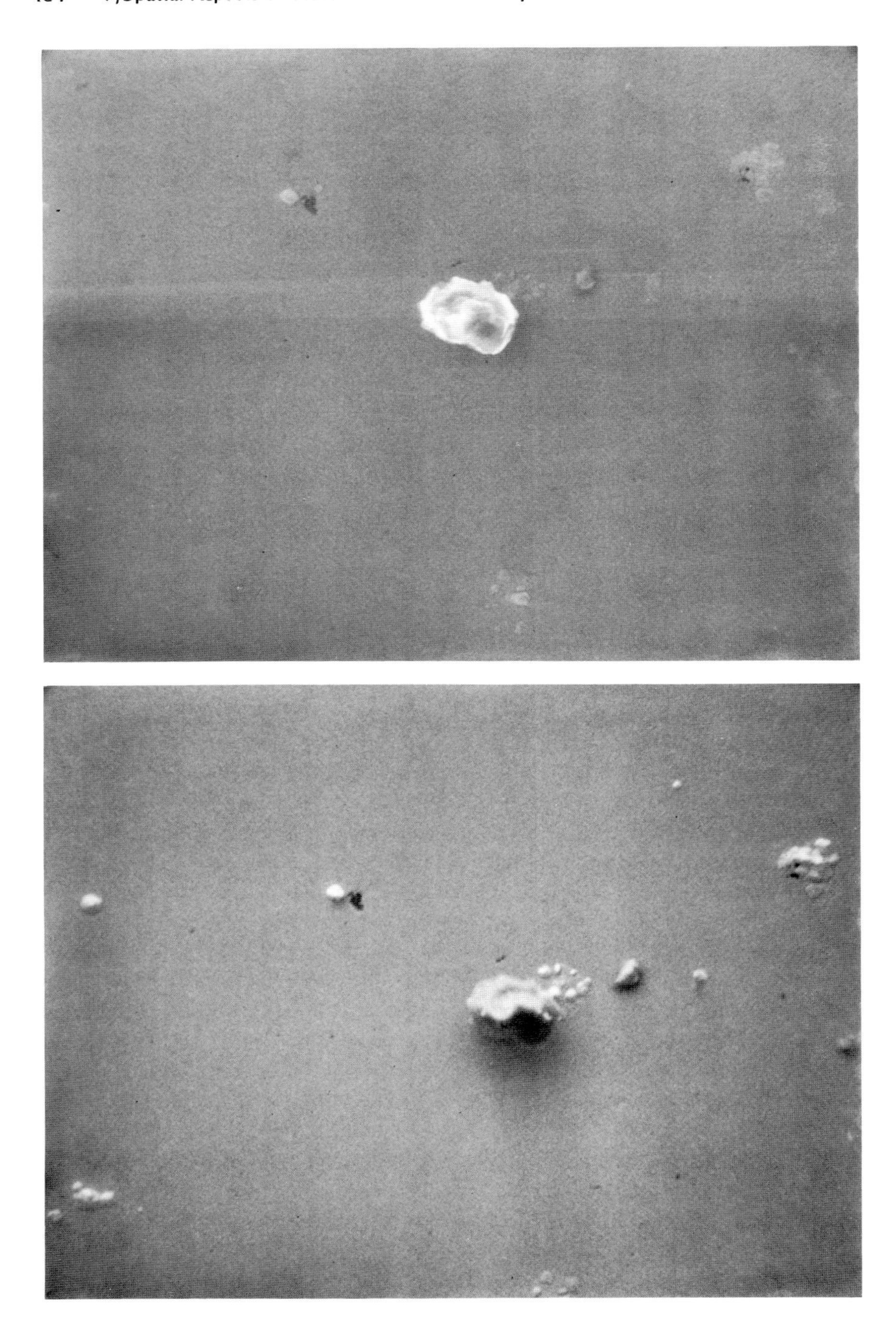

Fig. 14.17a. Dirt particles on Daguerrotype plates. Above: E_0 = 15 keV. Scanned area: 66 × 53 μm. Secondary electrons. Below: same, backscattered electrons.

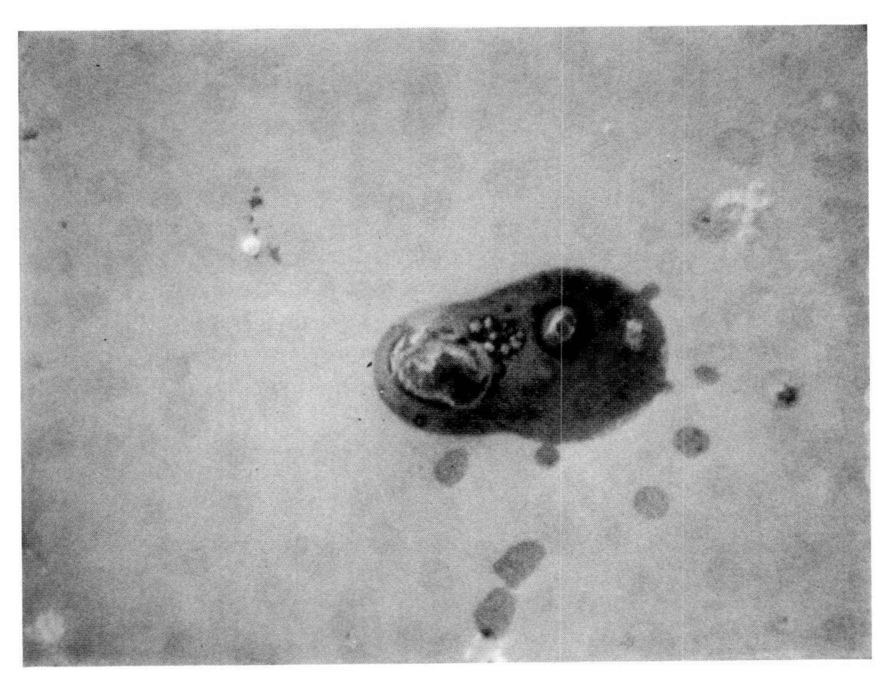

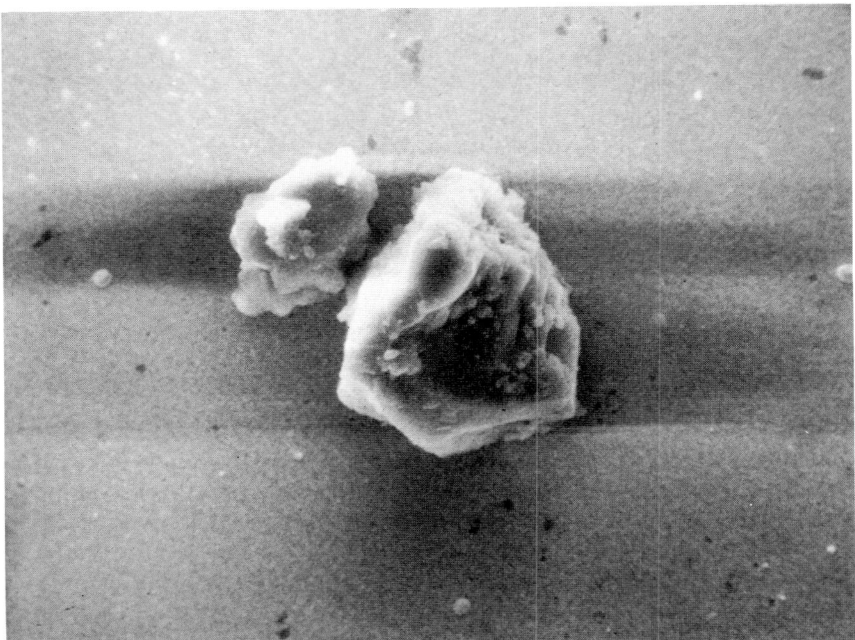

Fig. 14.17b. Above: Same particles as Fig. 14.17a, E_0 = 2 kV, secondary electrons. Below: large silica particle which charges electrostatically, distorting the gray values of the background due to suppression of low-energy (secondary) electrons. E_0 = 17.4 keV. Area scanned: 80 × 60 μm.

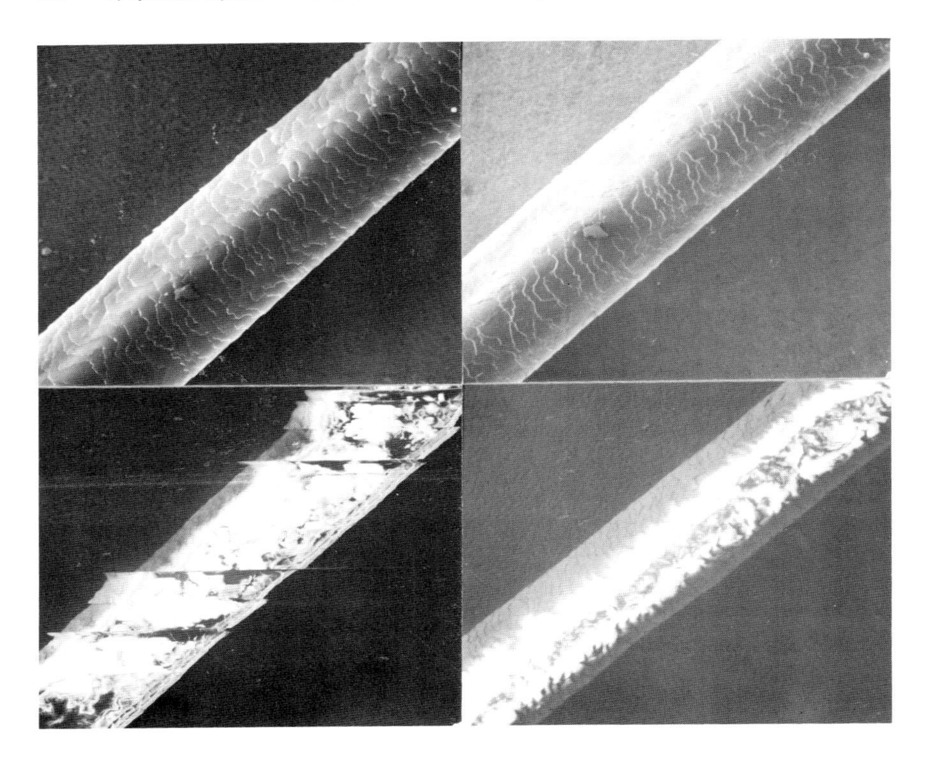

Fig. 14.18. Secondary electron images of a human hair. Width of field: 125 μm. Upper left: coated hair, E_0 = 20 kV. Upper right: coated hair, E_0 = 2.5 kV. Lower left: uncoated hair, E_0 = 20 kV. Lower right: same, at 2.5 kV. At 20 kV, periodic discharges from the uncoated hair distort the image contours. This does not occur at 2.5 kV, but local charges still distort the brightness values of the image.

those used for X-rays, except that no attempt is made to optimize the energy resolution. They can operate at room temperature, and, contrary to the Everhart-Thornley detector, there is no danger of interference by visible light. They are inexpensive, and cover a large solid angle; they also produce a larger signal pulse per electron than the Everhart-Thornley detector, and the pulse is, within statistics, proportional to the energy of the electron.

The backscatter coefficient η of most targets, especially of high atomic numbers, is larger than the corresponding secondary electron emission efficient. However, the backscatter signal collected in the Everhart-Thornley detector is composed of fewer electrons than the signal from emission of secondary electrons. This difference is due to the fact that the high-energy backscattered electrons are not significantly deviated from a straight path by the fields close to the detector, and hence are collected over a much smaller solid angle than the secondary electrons (Fig. 14.3). For this reason, the image from backscattered

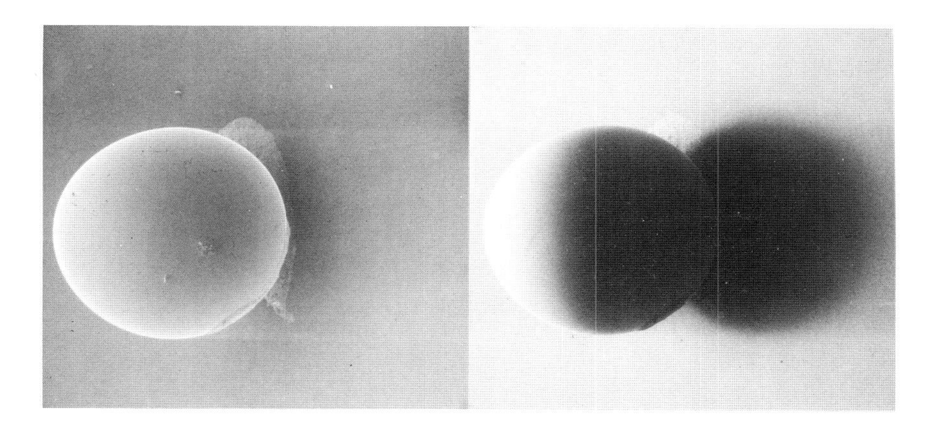

Fig. 14.19. Steel sphere on an oblique plane. Left: secondary electrons observed with Everhart-Thornley detector, with grid charged positively. Right: same, backscattered electrons, with grid charged negatively. Sphere diameter: 0.1 mm. Operating voltage: 20 kV. Slope of background: 45°. (Courtesy of H. Yakowitz, NBS.)

electrons tends to be grainy unless a fairly large beam current (10^{-9} Å) is used. Moreover, due to the straight path of these electrons, no signal is received by the detector from regions of the specimen for which the line of sight is blocked, and such regions appear deeply shaded on the image (Figs. 14.19 and 14.20).

Because of their greater energy, backscattered electrons can emerge from

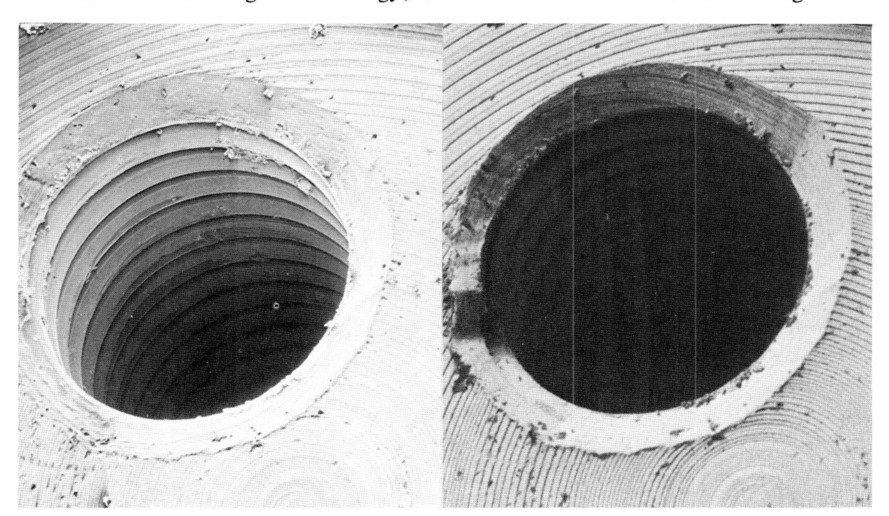

Fig. 14.20. Scanning electron micrographs of hole with thread. Left: secondary electrons. Right: backscattered electrons. Operating voltage: 20 kV. The diameter of the hole is 1 mm.

greater depths within the specimen than the secondary electrons, but at the magnifications used in electron-probe microanalysis, the loss of resolution caused by electron penetration is imperceptible. As shown by Shinoda et al. [13.99] in 1968 by means of Monte Carlo calculations, the mean depth of penetration of the backscattered electrons is for all atomic numbers less than half that of the absorbed electrons, and the range in depth of the backscattered electrons is smaller than the total range in the same proportion (Figs. 14.21 and 13.9). Oatley [14.2, p. 153] pointed out that a metal film of one-fifth of the electron range has a backscatter coefficient half that of a thick specimen. The lack of resolution often observable on backscatter images is usually due to the collection of an insufficient number of electrons.

Wells showed that greatly improved spatial resolution can be obtained by collecting the backscattered electrons from a target surface placed at a low angle with the electron beam, and if only those electrons are used for image formation which have lost not more than a few hundred electron volts [14.15], [14.16]. These *low-loss electrons* are almost always the result of single wide-angle scattering events which occur close to the specimen surface. The low-loss backscatter images cannot be produced, however, at the high electron beam impact angles commonly used in X-ray analysis.

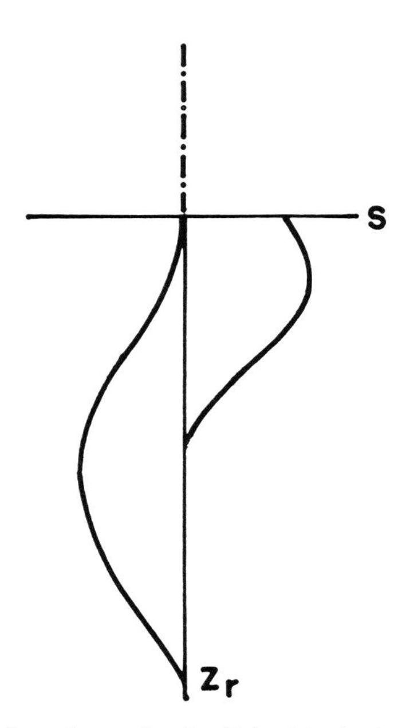

Fig. 14.21. Distribution of maximum depth attained by backscattered (right) and absorbed (left) electrons. If these distributions are scaled with the electron range, z_r, they change little with atomic number and operating potential [13.99].

The strong dependence of the backscatter coefficient on the atomic number of the target was discussed in Section 9.3. Because of this effect, images of phases of different composition usually have different brightness values (Fig. 14.22). However, since the emission of backscattered electrons also depends on the incidence angle of the electron beam on the specimen, and the shape of the specimen close to the point of entry of the beam (Fig. 14.23), targets which are not perfectly flat show effects of both composition and topography in the backscatter image. Kimoto and Hashimoto [14.17] showed that these effects can be separated to a large degree. They used two silicon detectors placed at opposite sides of the electron beam. These detectors have, for reasonably flat specimens, virtually the same response for the effects of atomic number, but differ in their response to moderate topography. The sum of the output signals of the two detectors emphasizes the compositional effects, while the difference provides an image virtually free of them, and which is essentially determined by the surface configuration (Fig. 14.24).

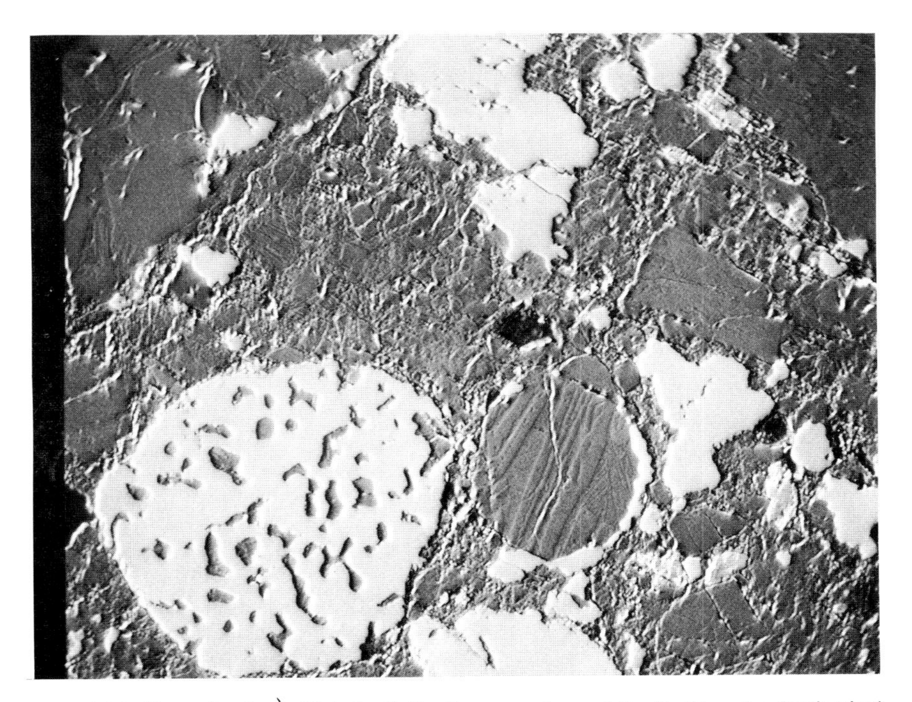

Fig. 14.22. Meteorite from Djahala, India (loan courtesy of the Smithsonian Institution). Backscatter signal obtained from a scan of 800 × 1000 μm. E_0 = 20 keV, specimen current: 0.5 × 10^{-7} Å. The metallic phase of this meteorite is brighter on the image than the oxide phases, due to the atomic number dependence of the emission of backscattered electrons.

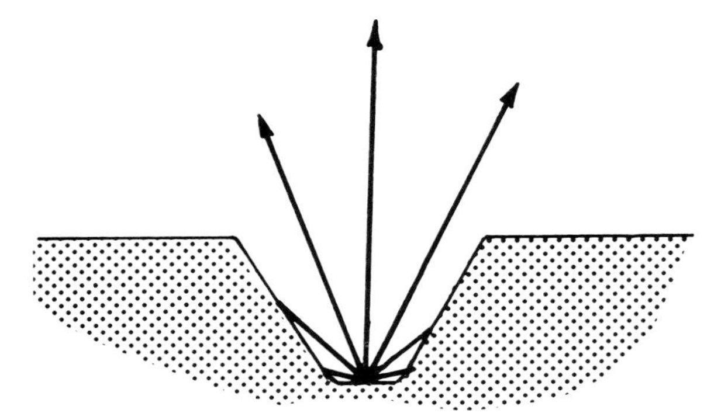

Fig. 14.23. Some backscattered electrons generated in fissures and holes are reabsorbed in the specimen.

The dependence of the backscatter coefficient on atomic number decreases as the inclination of the electron beam with regard to the normal to the specimen surface increases.

In comparing backscatter signals with those obtained from secondary electrons, one observes a tendency toward greater contrast, which is related to their rectilinear propagation. The secondary electron image is not, in fact, a signal

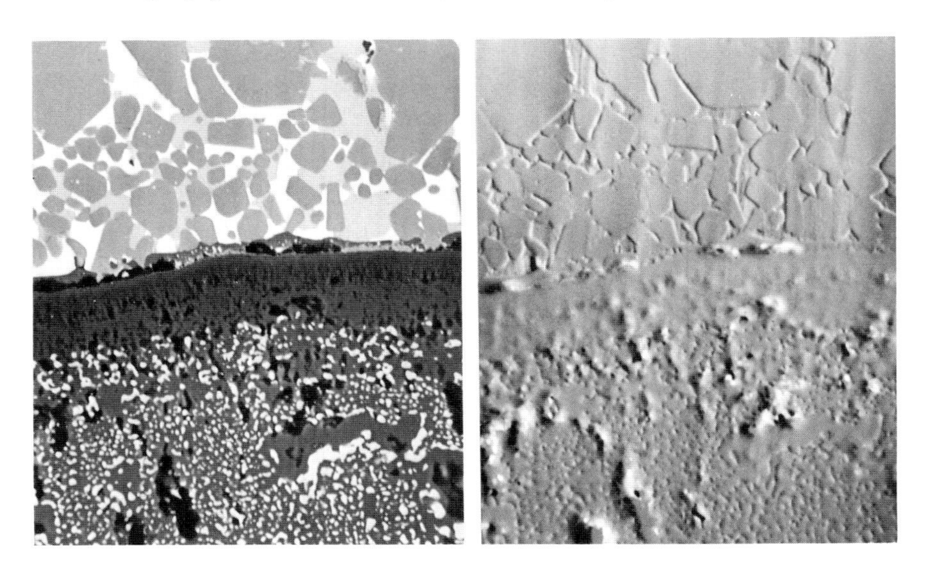

Fig. 14.24. Sum (left) and difference (right) of signals from two backscatter detectors [14.17]. The difference image shows topographic effects only, and is remarkably similar to a first-derivative image (compare with Fig. 14.41). The specimen is a solder joint of nickel ferrite prepared with silver solder. Width of field: 0.1 mm. (Courtesy of JEOI)

particularly rich in information. Its wide use is mainly due to the fact that satisfactory images can be obtained from this signal at low beam currents.

14.3.3 Specimen Current [14.18], [14.19]

The current which is collected by the specimen bombarded by the electron beam (specimen current or target current) can be used to produce scanning images (Fig. 14.25). The signal is a current of the same order of magnitude as the beam current, and it must therefore be amplified by a low-noise amplifier before it can be applied to the formation of the image. The first stage of amplification should be placed close to the specimen, and parasitic noise, particularly from alternating current fields, must be reduced as far as possible by shielding and elimination of ground loops. With modern solid-state electronics, useful images can be obtained from currents as small as 10^{-13} A. Hence, specimen current images can be obtained at all conditions of practical interest in electron-probe microanalysis [14.20].

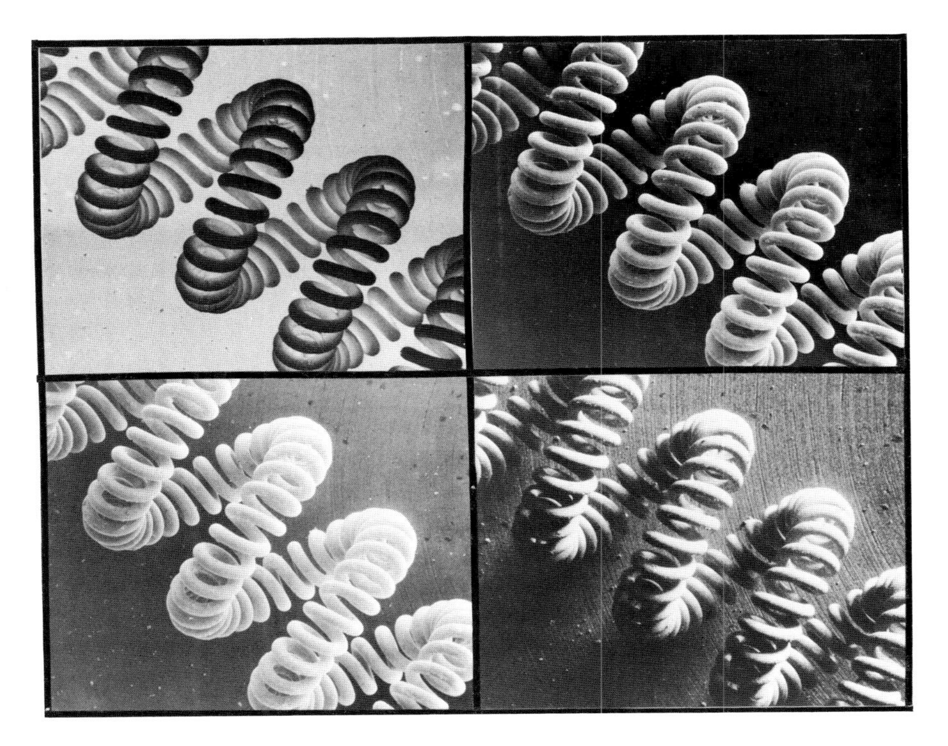

Fig. 14.25. Same filament of a fluorescent lamp as in Fig. 14.12. Upper left: target current image. Upper right: inverted target current image. Lower left: secondary electron image. Lower right: backscattered electron image.

The specimen current is the excess of all currents entering the specimen (the primary beam, and doubly backscattered or secondary electrons from instrument components) over those leaving it through emission of backscattered or secondary electrons [14.21] (Fig. 14.26). The emission and collection of secondary electrons, and hence the specimen current, can be affected by nearby fields, or by charging the specimen positively or negatively [9.30] (Fig. 14.27). At conventional operating potentials, the strongest component of the signal is electron backscattering, and for qualitative considerations one can assume that the specimen current is the difference between beam current and electron backscatter current. Since in this signal the backscattering into all directions above the specimen is included, the specimen current image lacks the directionality of the backscatter image obtained with a detector of small aperture, and hence the strong effects of apparent directional illumination which are so characteristic of backscatter images of three-dimensional objects.

To use the specimen current for image formation, a large fraction of the signal which does not vary from point to point of the image must be subtracted, and then the signal should be inverted, so that the areas of highest specimen current become black, and those of lowest specimen current become white.

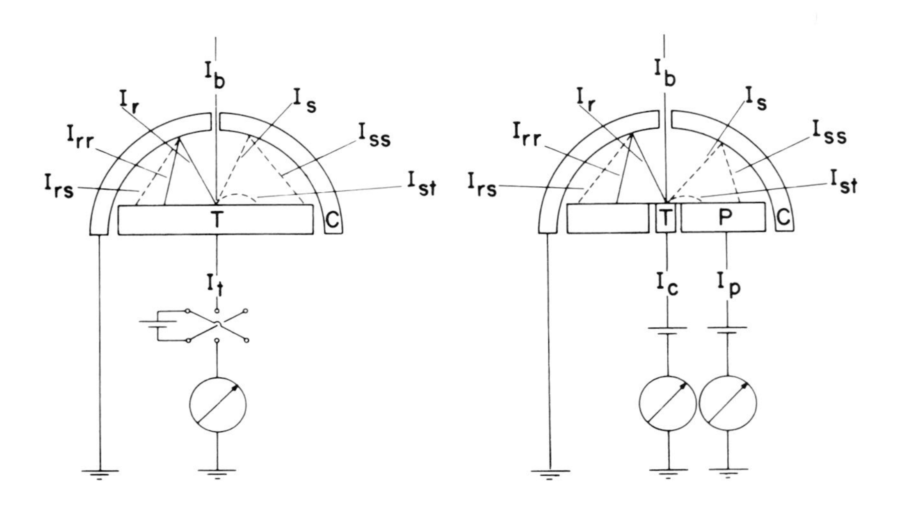

Fig. 14.26. Components of the target current, I_t. I_b: beam current; I_r: backscattered electron current; I_{rr}: doubly backscattered electron current; I_{rs}: secondary electrons produced by backscattered electrons on instrument components; I_s: secondary electrons; I_{ss}: secondary electrons caused by low-energy electrons from the target; I_{st}: low-energy electrons returning to the target. Some of these components can be observed separately if the target is divided into a small central zone, and a large peripheral zone, and the respective currents are measured separately (right). All measured currents can be observed with different biases of the target or target zones [9.30], [14.21].

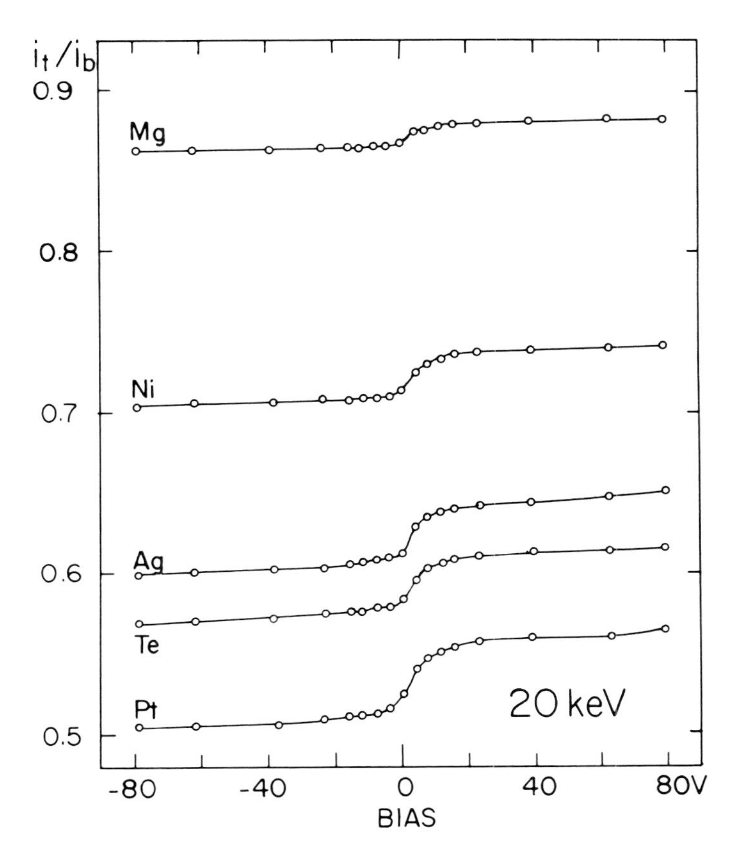

Fig. 14.27. Variation of the ratio between target current and beam current, as a function of target element and bias. (From [14.21].)

The reason for this inversion is in the fact that the specimen current varies inversely with the backscatter; therefore, the direct specimen current image often gives the impression of a negative, or of an image in transmission (Fig. 14.25).

The variation from point to point on the specimen of the backscatter into the entire hemisphere above the specimen is the main cause of specimen current variation. The specimen current is in effect a very efficient collector of the backscatter signal, and for this reason the images are, at the currents used in microanalysis, free of statistical graininess. At low operating potential, the effects of secondary emission are more significant. But even at high operating potentials, specimen current images of three-dimensional objects resemble secondary electron images rather than conventional backscatter images, because of their lack of directionality of collection (Fig. 14.28). On the other hand, specimen current images of flat objects containing regions differing in mean atomic number are

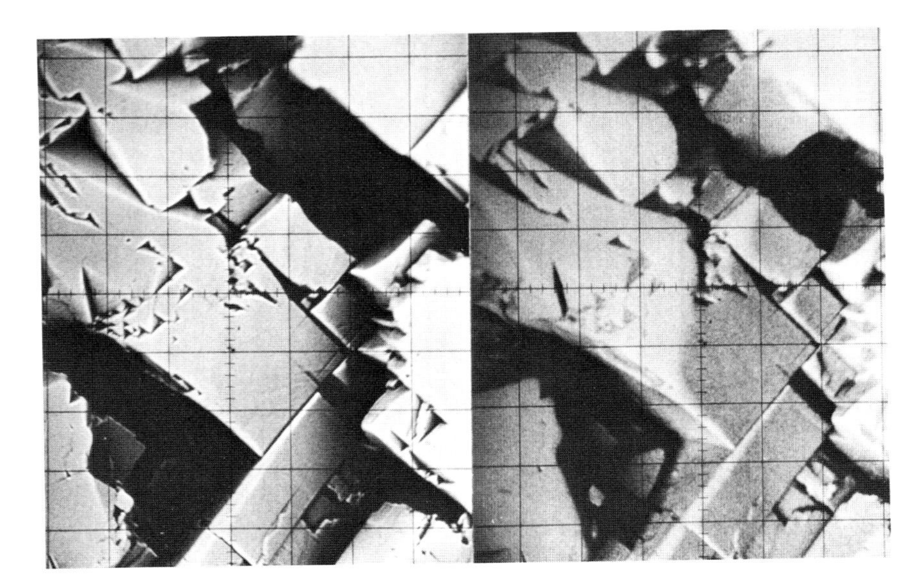

Fig. 14.28. The specimen current image (left) appears illuminated from above, while that obtained with a backscatter detector (right) shows a lateral directional effect. Specimen is a fractured surface of galena. E_0 = 20 kV. Specimen current: 5×10^{-8} Å. Scanned area: 400 μm × 480 μm.

similar to backscatter images (Figs. 14.8, 14.29, 14.42). In either case, the specimen current image, particularly with its transforms which will be discussed next, carries a maximum of information, and its usefulness as an adjunct of electron-probe microanalysis has usually been underestimated. As shown by Newbury [14.19], the spatial resolution of specimen current images is very close to that of secondary electron images (Fig. 14.30).

14.3.4 Cathodoluminescence

The emission of light from nonmetallic targets excited by the primary electron beam is called cathodoluminescence. The cathodoluminescence of ionic crystals is due to the excitation of luminescence centers, and that of semiconductors to radiative recombination of excess carriers. The mechanism can be started by impact ionization of atoms of the lattice or of impurity atoms. The subject is discussed in detail in [14.22] and [14.23]. The intensity and spectral composition of the light vary greatly with the composition of the specimen. When the impurities of the matrix are involved in the process, cathodoluminescence can be a very sensitive indicator of their presence. In rare earth oxides, 50 parts per billion (mass) of some elements can be detected [14.24]. In general, the energy levels involved are disturbed by the presence of ions; therefore, the typical spec-

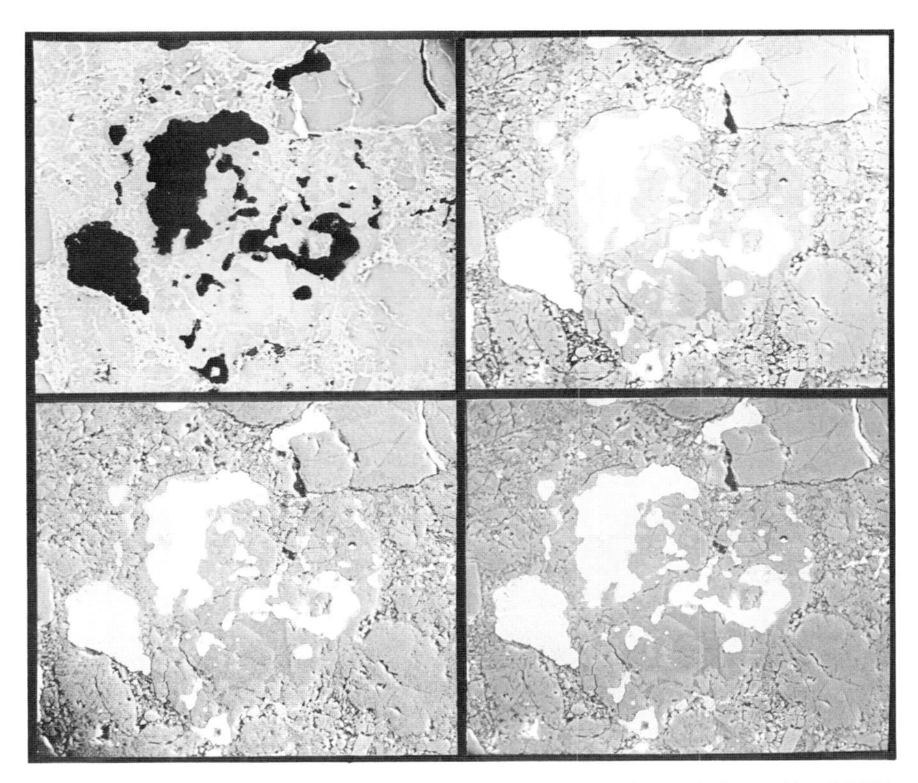

Fig. 14.29. Same specimen as in Fig. 14.22. Area scanned: 400 × 350 μm. E_0 = 30 kV. Upper left: specimen current image. Upper right: same, inverted. Lower left: backscattered electrons. Lower right: secondary (+ backscattered) electrons. The lower images taken with an Everhart-Thornley detector. In all images, brightness contrast is due to the atomic number difference between the silicate phases (gray) and a metallic (Fe, Ni) phase. Minor brightness differences between silicates of different iron concentrations can also be seen. (Scans by D. E. Newbury, NBS.)

trum consists of broad bands. Exceptions occur, however, such as with rare earth emitters which yield sharp lines.

Although the emission of light is a sensitive indicator of the conditions which cause it, it is very difficult or impossible to apply this phenomenon to quantitative purposes as is done with X-rays. Most uses of cathodoluminescence are qualitative (Figs. 14.31 and 14.32). For instance, in minerals, cathodoluminescence may be related to trace impurities which provide information on the formation and alteration of the specimen. The quenching of the emission by small amounts of iron, for instance, accounts for differences in brightness which permit distinguishing primary and secondary carbonates. Changes in color also indicate the structure of some fossil materials (Fig. 1, color plate I) [14.25].

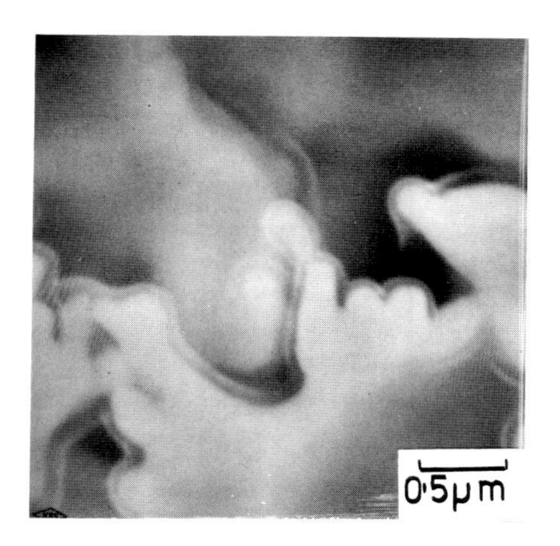

Fig. 14.30. Resolution of scanning electron microscopy with secondary electrons (left), and inverted specimen current (right). The double contours are due to the buildup of carbonaceous contamination. The resolution of the two images is virtually identical. (Courtesy of D. E. Newbury [14.19].)

Wittry and Kyser [14.26], [14.27] have used cathodoluminescence in the near infrared as a means of localizing defects in semiconductor materials such as gallium arsenide (see also [14.28], [14.29]). The emission can also be used to measure changes in the concentration of free electrons, to map dislocations and surface conditions, and to observe variations in the concentration of dopants.

Cathodoluminescence can be observed visually and photographed through the optical microscope of the electron-probe microanalyzer [14.25], [14.30]. In that case it is unnecessary to use a finely focused electron beam; rather, the field of observation of the microscope is flooded with a broad beam. The resolution of the spatial distribution thus depends essentially on the quality of the microscope. For such observations the analyst can use an inexpensive attachment to an optical microscope which floods its field with an unfocused electron emission of relatively low excitation potential [14.30]. But the luminescence can also be observed with an appropriate detector, and, if necessary, with a monochromator, while the focused electron beam performs a raster scan. The signal is then used for the formation of scanning images in the same fashion as other signals [14.24], [14.31]-[14.35]. Such a procedure becomes necessary:

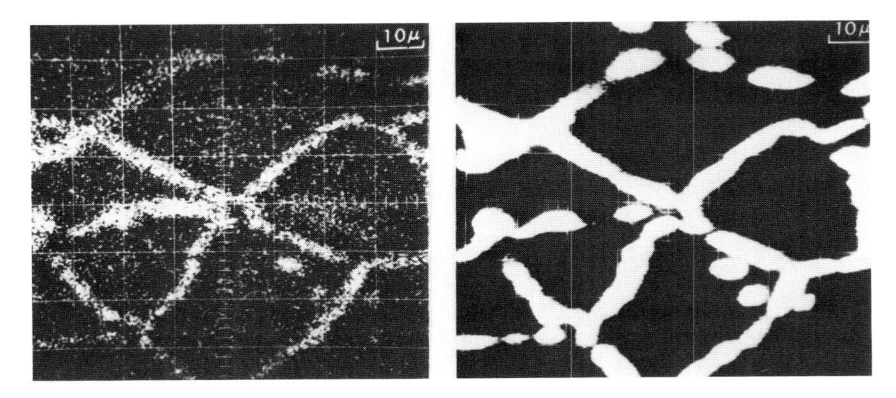

Fig. 14.31. Observation of the precipitation of zirconium oxide at the grain boundaries of an oxidized specimen of a niobium alloy, by scanning imaging of the Zr $L\alpha_1$ radiation (left) and of the cathodoluminescent emission (right). (From [12.97].)

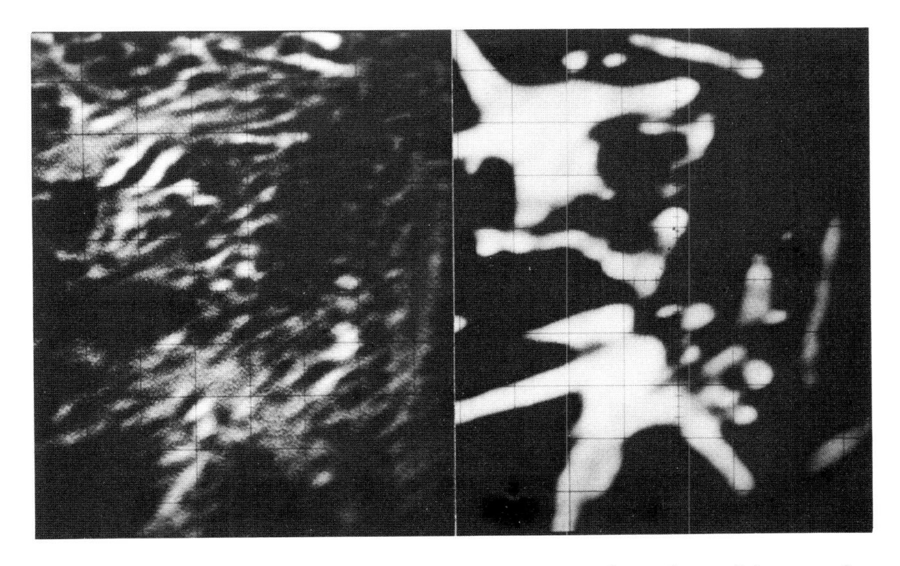

Fig. 14.32. Sapphire crystals embedded in a nickel-base alloy, observed in a specimen current image (left) and in cathodoluminescence scanning image (right). Scanned area: 55 μm × 70 μm. E_0 = 20 keV. (From [14.18].)

a. if the instrument has no light microscope, as is the case with most scanning electron microscopes;
b. when the radiation is outside the spectral range of visibility [14.26], [14.27], [14.29];
c. if the spectral composition of the cathodoluminescent emission must be analyzed spectroscopically [14.24], [14.34];
d. when the spatial resolution of the optical microscope is insufficient [14.32].

14.4 ARTIFACTS AND SIGNAL TRANSFORMS

With the exception of X-ray signals, all signals used to form scanning images are obtained as, or transformed into, continuous electrical currents. The statistical noise of such signals due to the random distribution of the individual events and to thermal electronic noise, can be a significant limitation, particularly at low beam currents. As mentioned in Section 6.6, the smoothing and reduction of statistical fluctuations at low pulse rates requires large time constants, and hence, in order to conserve spatial resolution, low scanning speeds. Therefore, recording a good image from weak signals may require several minutes per frame. However, for preliminary orientation and visual observation a high-speed scan (TV scan) which produces a coherent visual image on a screen with a fast-decay phosphor is very useful. Such scans require fast electronics, and a relatively large signal flux. When the fields of the scanning coils overlap with those of the objective lens, hysteresis effects may be observable at high scanning speeds. Image distortions at such speeds may also be due to deficient scan-drive amplifiers, or to slow or saturated signal amplifiers (Figs. 14.33].

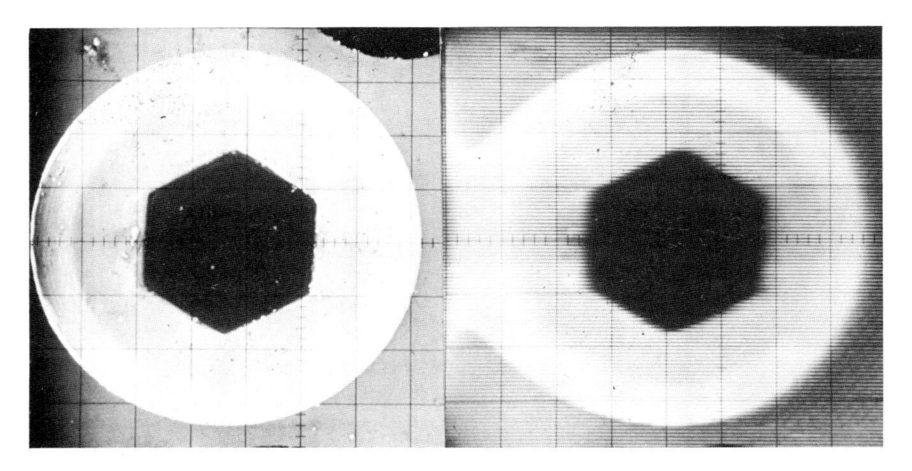

Fig. 14.33. Specimen current image of the head of an Allen screw. Width of field: 6 mm. V_0 = 20 kV. Left: inverted image obtained at slow scanning speed. Right: same, at TV scan speed. The distortions and blurring are due to hysteresis in the scanning system.

Another type of noise which is frequently found within the specimen current signal is induced cyclic noise which may produce a moiré pattern superposed upon the scanning image (Fig. 14.34). Such noise can be caused by a variety of sources, including glow discharge lights, microscope light transformers, and also cyclic variations in power supplies for beam and lenses. Mechanical vibrations can also be a problem, especially at high magnifications (Fig. 14.35).

We will now discuss the function which relates the signal level to the gray level it produces on the image. Let us assume that the signal is free of significant dynamic distortions such as overshoots, tailing, or saturation of a component, so that the speed and direction of the scan do not visibly affect the image. A strictly linear relation between the signal and the corresponding gray level is theoretically simplest but has no special merit, since the observer is incapable of quantitatively judging the brightness of an image element. Rather, he observes *relations* of brightness of distinct picture elements. He cannot distinguish more than about ten levels between black and white on a photographic reproduction. These levels form a logarithmic scale of intensities (i.e., the ratios of intensities between discernible brightness levels are roughly independent of the absolute intensities). Hence, the available information must be compressed into these gray levels, and is not evaluated quantitatively. Furthermore, neither the response of brightness of the oscilloscope screen, nor its photographic reproduction conserve linear relations with the signal.

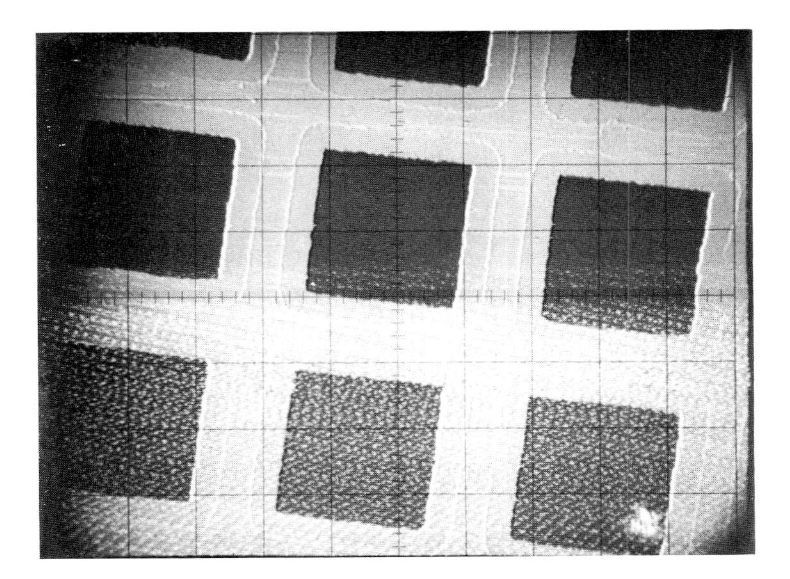

Fig. 14.34. Specimen current scan of a copper grid; periodic noise superposed on lower half of the image. E_0 = 20 keV. Width of scanned area: 360 μm.

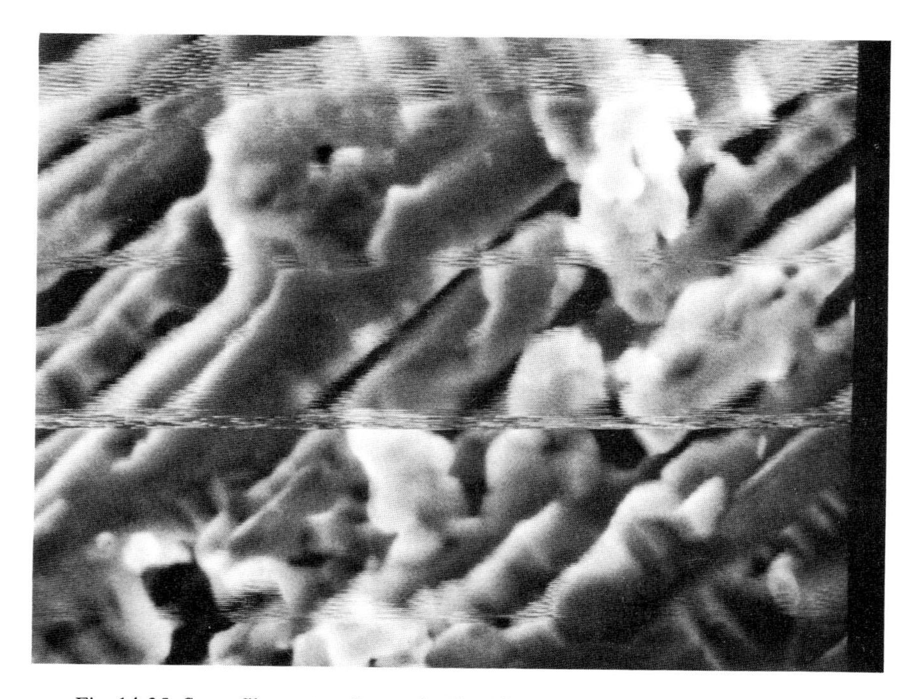

Fig. 14.35. Same filament surface as in Fig. 14.7. Effect of mechanical vibrations.

The purpose of images produced in scanning electron microscopy is to produce an interpretable image of the specimen surface. Any signal transformation which conserves the relevant information and aids in the interpretation of the image is legitimate. In discussing deliberate electronic signal transformations for image formation, we will call the original signal an input (I), and its transform an output (O). The two are related by a transform function, which can be continuous or discontinuous, isotropic or anisotropic with respect to the image dimensions, monotonic or not, and time dependent or not. The simplest signal transform is the *inversion* of the signal ($O = -I$), after which the brightness of the image diminishes as the signal increases. This inversion is used with target current signals (Fig. 14.25). Another important transform is the *subtraction of a constant background level*, used to increase the contrast of the significant part of the signal [expanded contrast, black level adjustment; $O \propto (I - \text{constant})$]. If we wish to emphasize small contrasts in the presence of much higher signal levels, we can use the *gamma transform* in which the relation between input and output is as follows [14.2]:

$$O \propto I^{1/\gamma}. \tag{14.4.1}$$

If $\gamma > 1$, then low-signal levels are expanded, and high-signal levels are compressed; the contrary occurs when $\gamma < 1$. The cathode-ray tube itself introduces to the signal displayed by it a transformation of this type, with γ being approximately equal to 2.5. The practical use of these and other transforms is aided by an oscilloscope display of the output, on a vertical scale, for a linear scan across the scanning field.

With discontinuous transform functions, discrete ranges of signal intensity can be displayed in distinct gray tones. The functions can be monotonic or not. In the latter case, one range of signal can, for instance, be displayed white while levels above and below are shown in black, or three levels (white, gray, and black) can be used (Fig. 14.36). The discontinuous mapping of signal ranges is mostly used for X-ray signals: color display of the output can also be employed to advantage [12.49].

Among the time-dependent or dynamic transforms, the most important one is the *first derivative* of the input with respect to time, by itself or, more frequently, mixed with the input [14.9]:

$$O = a \cdot I + b\frac{dI}{dt} \qquad (14.4.2)$$

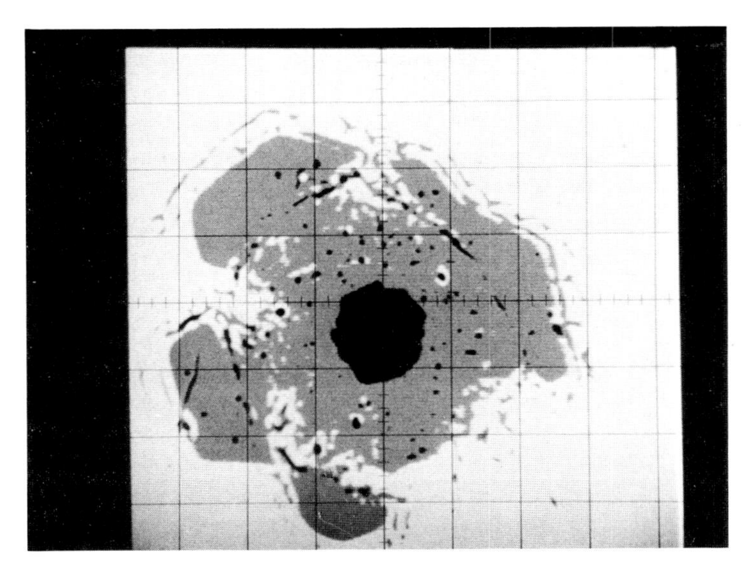

Fig. 14.36. Three-level target current concentration map of the diffusion of tantalum into tin (see Fig. 15.15). Width of field: 160 μm.

with t denoting time. This transform increases contrast at boundaries between phases, edges and holes, and, in proper mixture with the input, produces an apparent increase in spatial resolution (Fig. 14.37). The partly differentiated signal emphasizes sudden signal changes, and hence minor topographic features. The combination of background subtraction and inversion, often also with addition of the first derivative, is almost always used for specimen current signals. Background subtraction and addition of the first derivative are also useful for other signals. More complex transforms were discussed and experimentally produced by Fiori et al. [14.36]. A simple circuit for mixing the direct signal and its first derivative is shown in Fig. 14.38.

In an image showing only topographic features, the first derivative, without admixture of the original signal, produces a strong enhancement of surface details but the lack of the original signal levels causes the images of three-dimensional objects to appear flattened (Fig. 14.39). By a proper mixture of the detector signal with its first derivative, small topographic features in such objects can be enhanced in the image, while the appearance of volume is conserved (Fig. 14.40).

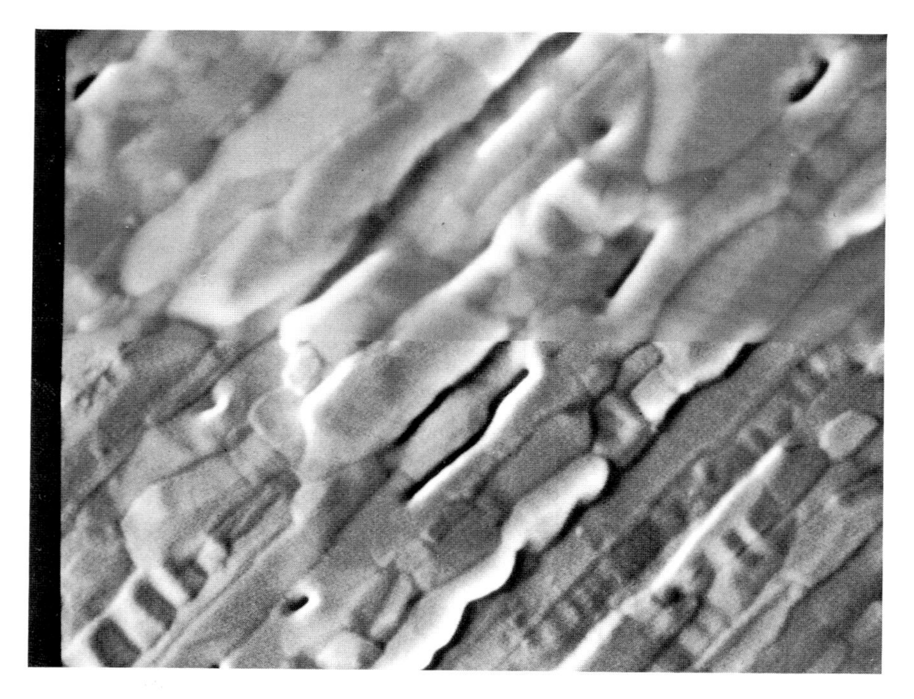

Fig. 14.37. Increase in contrast and boundary definition through addition of first derivative (below) to a secondary electron image (above) of the surface of the filament of a used fluorescent lamp. Diameter of field: 12.5 μm. E_0 = 20 keV.

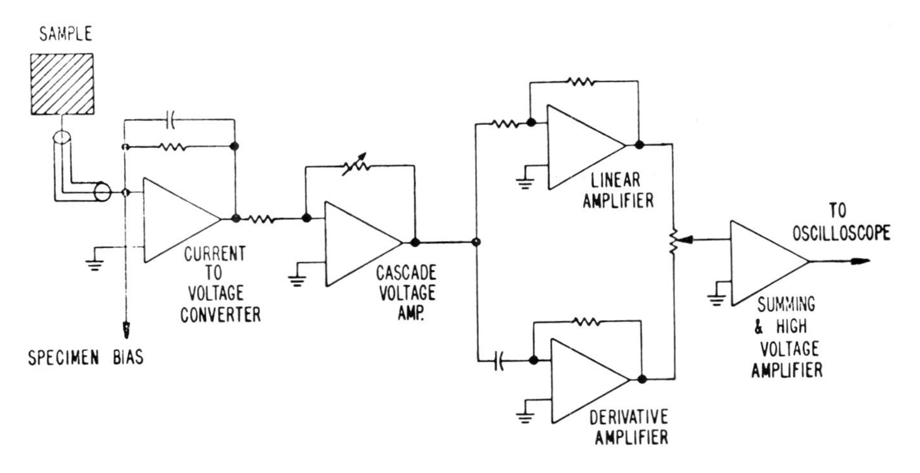

Fig. 14.38. Specimen current amplifier combining linear and derivative signal [14.20].

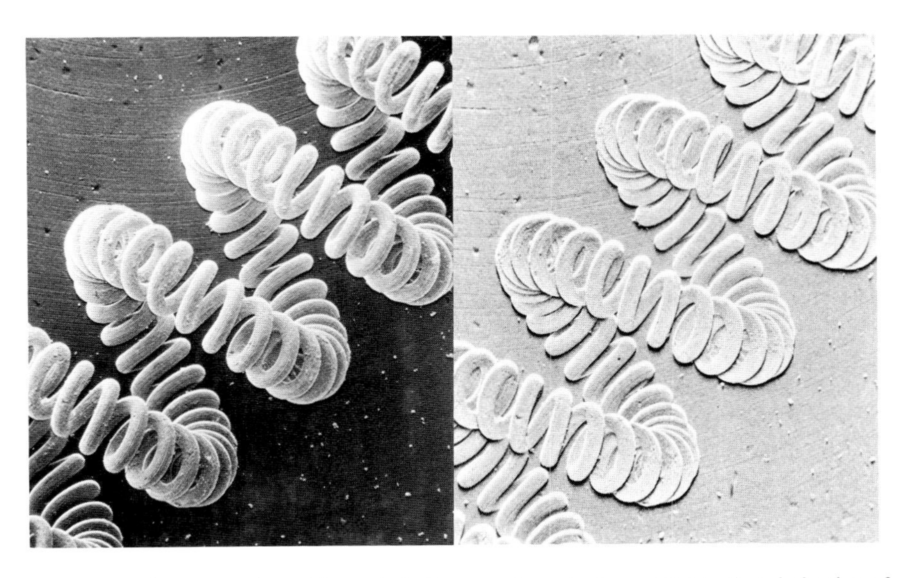

Fig. 14.39. Fluorescent lamp filament. Left: secondary electrons. Right: first derivative of the same signal. Three-dimensional objects imaged with the first derivative take an appearance of flatness.

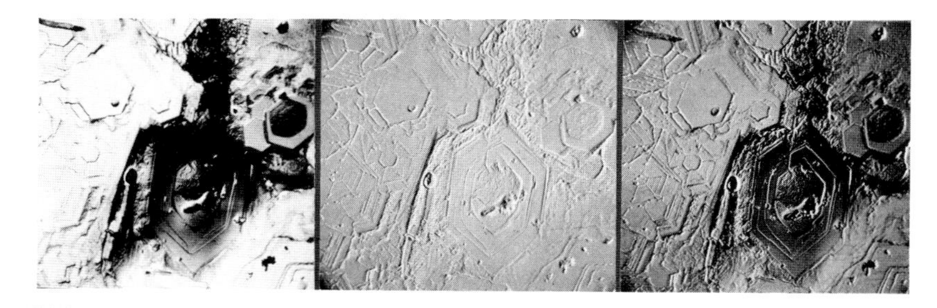

Fig. 14.40. Specimen current images of a chromium boride deposit with iron impurities in the interstitial spaces. Left: inverted target current scan. Center: derivative specimen current scan. Right: signal mixed from inverted specimen current and its first derivative. The contrast in the original signal is partly topographic and partly atomic number contrast.

In specimen current or backscatter images of flat specimens which have large differences in atomic number (Fig. 14.41), the first derivative of the signal can be used to advantage. In this case, the atomic number effects in emission are eliminated, but their changes at phase boundaries clearly mark these boundaries, as if the specimen had been etched. Hence, an appearance of topographic features is created. If topographic features were present in the original image, these are depicted as well. In a combination of the specimen current signal with its first derivative, the atomic number contrast can be attenuated so that other features of the images can also be observed. If one of the phases is much brighter than the rest, the simultaneous use of the gamma transform will be useful (Fig. 14.42).

14.5 INTERPRETATION OF IMAGES OF THREE-DIMENSIONAL OBJECTS

In view of the remarkable similarity between scanning images and the macroscopic view of real objects, such images are usually easy to interpret. The analyst should, however, be aware of the fact that the mechanisms of formation of the scanning image are quite different from those of vision. In the visual process, the distribution of image elements depends on the position of the observer with respect to the observed objects, while the gray values (and colors if present) depend on the nature and position of the light source and the properties of the object surface. In the scanning image, instead, the distribution of elements of the image depends on the position of the virtual electron source with respect to the object; therefore, the electron source takes the place of the observer. The gray tones in turn depend on the position and nature of the detector—which in the analogy assumes the role of the light source—and the properties of the object surface. To establish an analogy between observation by eye and formation of a scanning image, the roles of the observer and the illuminating source must

Fig. 14.41. Upper left: inverted specimen current image of a silicate rock (basalt from Disco Island), showing contrast between phases of different mean atomic numbers. The white elongated area is ilmenite ($FeO \cdot TiO_2$); other light-colored areas are iron silicates. Dark areas are silicates of light elements. Upper right: first derivative of specimen current. Lower left: Fe Kα X-ray scan. Lower right: Al Kα X-ray scan. V_0 = 20 kV. Field width: 180 μm.

therefore be exchanged (Fig. 14.43). Shadows in the scanning image are regions out of the line of sight of the detector; the curved paths of secondary electron produce the impression of diffused light because the directionality of detection is reduced. This impression is also present when the target current signal is used because in this case the virtual "illumination" is distributed over the entire hemisphere above the specimen surface. To complete the analogy with vision, we may consider that in ordinary scanning electron microscopy with rare exceptions (Fig. 14.44), the signals come virtually from the object surface, so that the objects appear opaque in the image, just as most objects in everyday life are opaque.

The interpretation of an image of an obviously flat surface is very simple, since any pattern must be due to local changes of the target properties, such as in the main atomic number for backscatter signals, in emission of cathodoluminescence, or of magnetic fields, in the corresponding techniques. In a similar

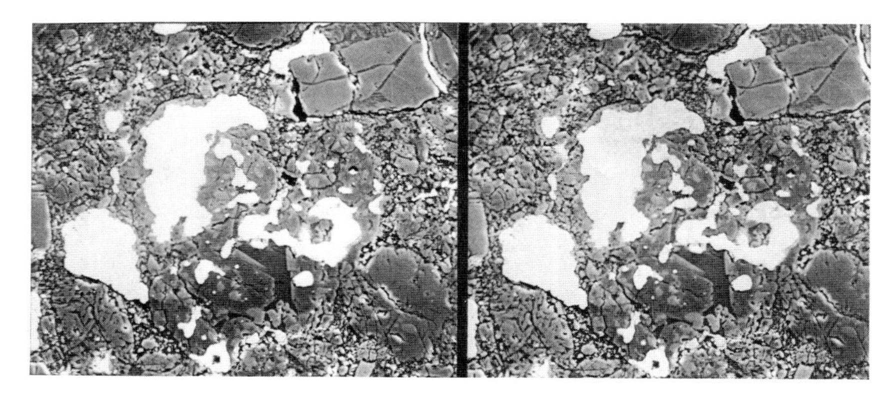

Fig. 14.42. Left: same target and conditions of scan as Fig. 14.29. Right: same, with gamma transform which compresses the brightness of the metallic phase so that variations in the gray value within this phase become visible. (Scans by D. E. Newbury, NBS.)

fashion, decidedly three-dimensional objects are easy to recognize, since the features of illumination, foreshortening, and even sometimes the lack of sharpness of features above or below the plan of focusing, can be properly interpreted (Fig. 14.45).

When the shape of the depicted object is not obvious, the observer tends to interpret it according to his visual experience. A field of basically uniform gray tone is interpreted as a horizontal plane viewed from above, or slightly raised toward the far side. Dark regions are interpreted to be entrant, and light areas as salient. Illumination from the front, or obliquely from the front, is considered more natural than illumination from behind the observer. If several of such indications reinforce each other, the interpretation is confirmed. If, to the contrary,

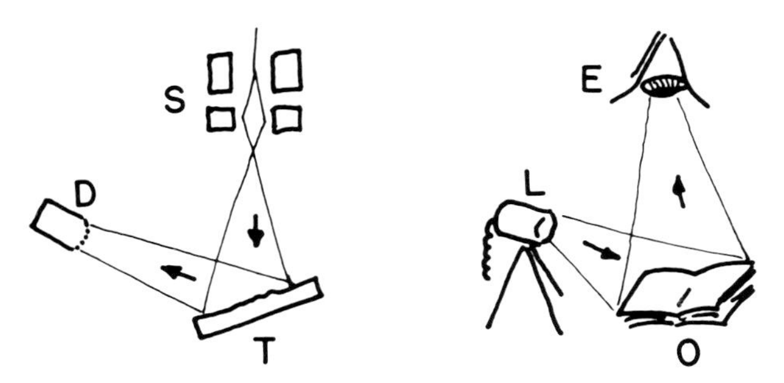

Fig. 14.43. Analogy between scanning electron microscopy and vision. Concerning the spatial distribution and illumination, the scanning system S is analogous to the eye E, the target T is analogous to the observed object O, and the detector D is analogous to the light source L. The direction of the electrons is, however, reversed with respect to that of the light in vision.

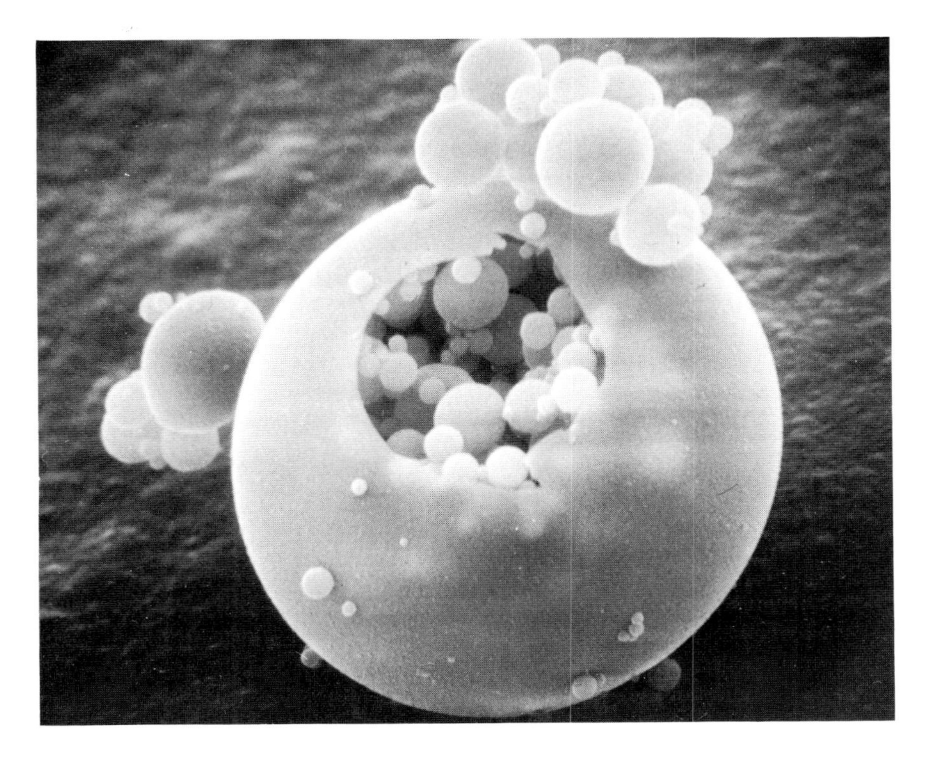

Fig. 14.44. The shell of this fly-ash particle is so thin that it is translucent to the electron beam, and signals from the small spheres inside it can reach the detector. (Courtesy of ETEC Corp.).

they contradict each other, then the image can become ambiguous; a crack can be interpreted as a ridge, or a hole as a tip. The interpretation frequently changes as the image is turned upside down, or various observers may give different interpretations (Figs. 14.46-14.48). Such ambiguities should be avoided whenever possible.

One may expect that the combination of atomic number contrast and topographic contrast in three-dimensional specimens could also cause much confusion. This is not, however, the case, because of the analogy with visual images of objects differing in brightness or color. For instance, the backscatter image on Fig. 14.49 (upper left), can be interpreted without any difficulty, while the X-ray scanning image of the same object (lower right) is much more ambiguous, because of insufficient information on the shape of the specimen.

If the shape of three-dimensional objects and ensembles of objects is of particular interest, the best solution is to obtain stereo image pairs. These can be produced by changing the specimen tilt 3 to 8° between exposures. In some

Fig. 14.45. Decidedly three-dimensional objects are easy to recognize. Head of ant. (Courtesy of E. Lifshin, General Electric Co.)

instruments stereo pairs can also be obtained by changing electronically the angle of impact of the electron beam between taking the two images. The large depth of focus of scanning images enables us to obtain impressive three-dimensional vistas of the microscopic world, and inexpensive stereo viewers are available for this purpose.

As we have seen in the discussion of the double-detector technique for backscatter images by Kimoto and Hashimoto, signal contrast depends both on variations in the signal emissivity (emissivity contrast), and on variation of emission intensity as a function of direction (directional contrast). The emissivity contrast of backscattered electrons is strongly dependent on the composition of the target material, while the directional contrast depends exclusively on the shape of the specimen surface. In secondary electron signals, the directional contrast is attenuated by the collection of the signal over a wide angle, and the effects of emissivity contrast are usually insignificant, particularly for coated specimens, although it can be used in the study of semiconductor devices in

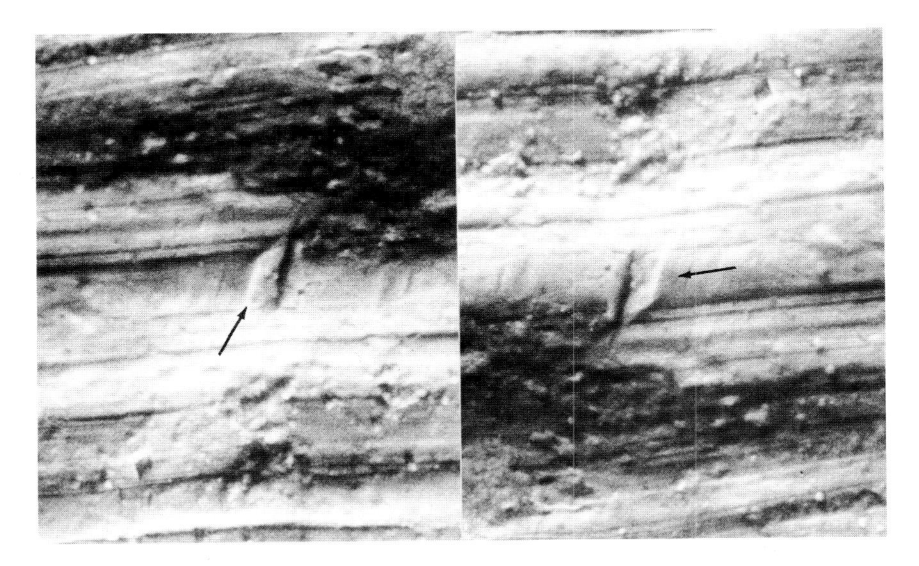

Fig. 14.46. The feature at the center of these images is usually perceived as protruding in the left image, and as entrant in the right image. The two images are identical, but differ in orientation.

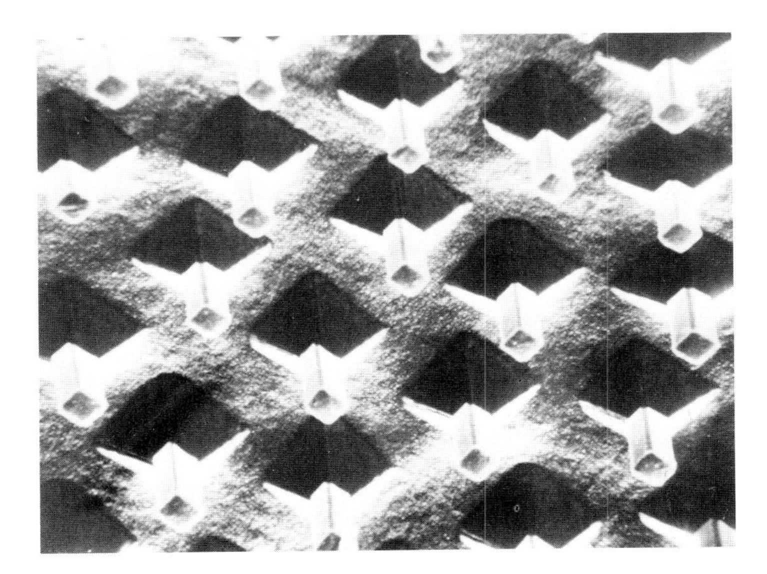

Fig. 14.47. Square rods emerging from the openings in a metal grid? (see Fig. 14.48).

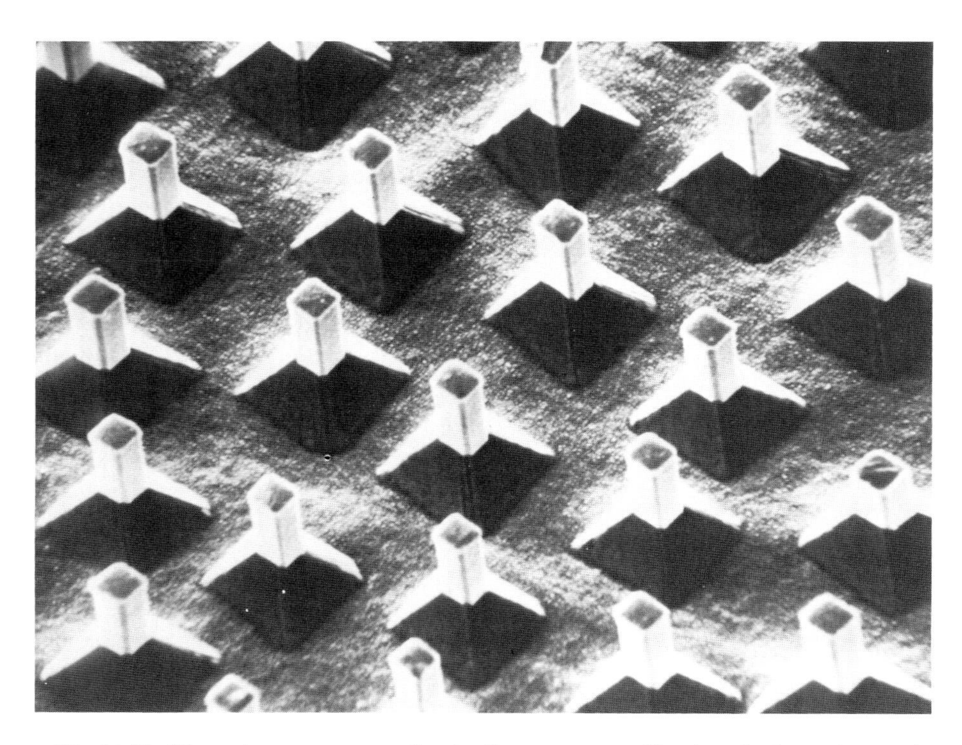

Fig. 14.48. The rods are seen correctly standing upon pyramids when the image is turned 180°. The specimen is a composite of TaC fibers in a nickel-base superalloy. Width of the scanned field: 20 μm. (Courtesy of E. Lifshin, General Electric Co.)

which the secondary electron emission can be made to depend upon local charges (Fig. 14.50). Similarly, in a variation of the specimen current technique, the emissivity of parts of semiconductor devices can be stimulated by the electron beam which produces hole-electron pairs in semiconductor material. These charges can be collected by p-n junctions and produce an induced current in an external circuit [14.37], [14.38] (Fig. 14.51).

The same as any other report of results of electron probe microanalysis, the scanning image is useless unless appropriate information concerning the specimen and the operating conditions is added. The following is a minimum of information which should be permanently affixed to each image: a clear identification of the specimen, and, where significant, of the region scanned, the magnification of the image, the operating voltage and beam (or specimen) current, and the nature of the signals displayed, the date, reference to the page of the notebook, and initials of the operator if several operators have access to the instrument. Some of this information can be reproduced on the oscilloscope screen by means of letter and number generator attachments (Fig. 14.52). A

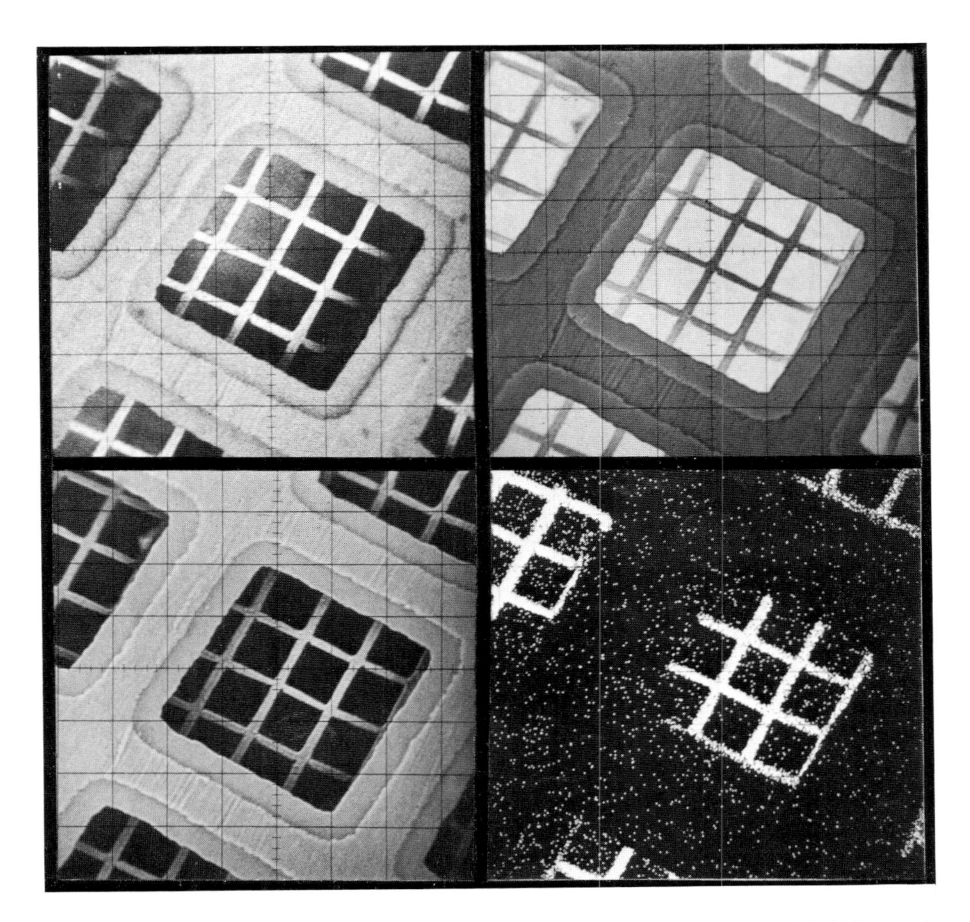

Fig. 14.49. Backscatter image (upper left), specimen current image (upper right), inverted specimen current image (lower left), and Ag Lα_1 X-ray image of a fine silver grid beneath a coarse copper grid on an iron substrate. Operating potential: 30 kV. Scanning width: 150 μm. Contrast on the first three images is mostly topographic, but increased backscatter of electrons on the silver grid is observable.

rubber stamp applied on the back of the exposure, and filled out by the operator is very useful. Much information generated in long hours turns out later to be useless because of incomplete documentation. Whoever directs an analytical laboratory must be absolutely inflexible in demanding that the operators properly document their work. Specimens must, of course, be identified by a number of entry. Indicating the magnification by a multiplier can be misleading if the image is copied at a size other than the original (for instance, in publications and slides). It is preferable to use a bar in the image, the length of which is an appro-

Fig. 14.50. Variation of emission of secondary electrons in the scanning image of a transistor device, as a function of local potentials.

priate metric subunit, or to indicate the size of the area scanned. The beam current, measured by means of a Faraday cage, is the best indicator of dose of electrons. The target current varies, for the same beam current, as a function of the specimen composition. In instruments which provide a monitor current (see Section 6.6 and Fig. 3.24), the intensity of this current must be converted into

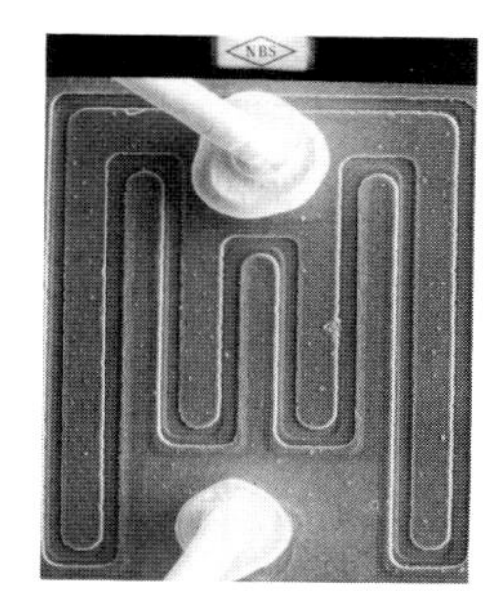

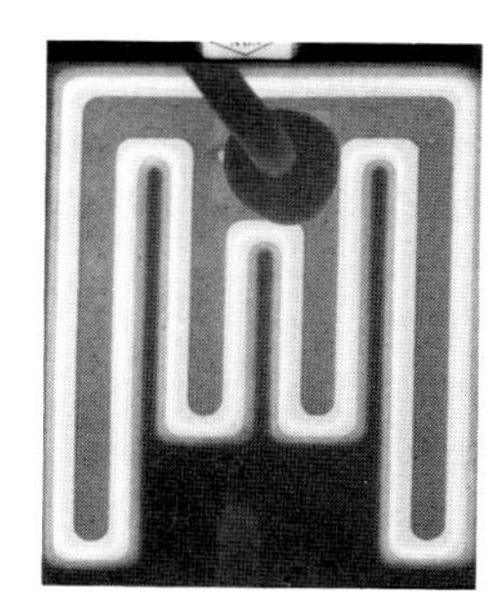

Fig. 14.51. Induced current emission in a transistor device (EBIC). Center: no induced current. (Courtesy of W. Keery, NBS.)

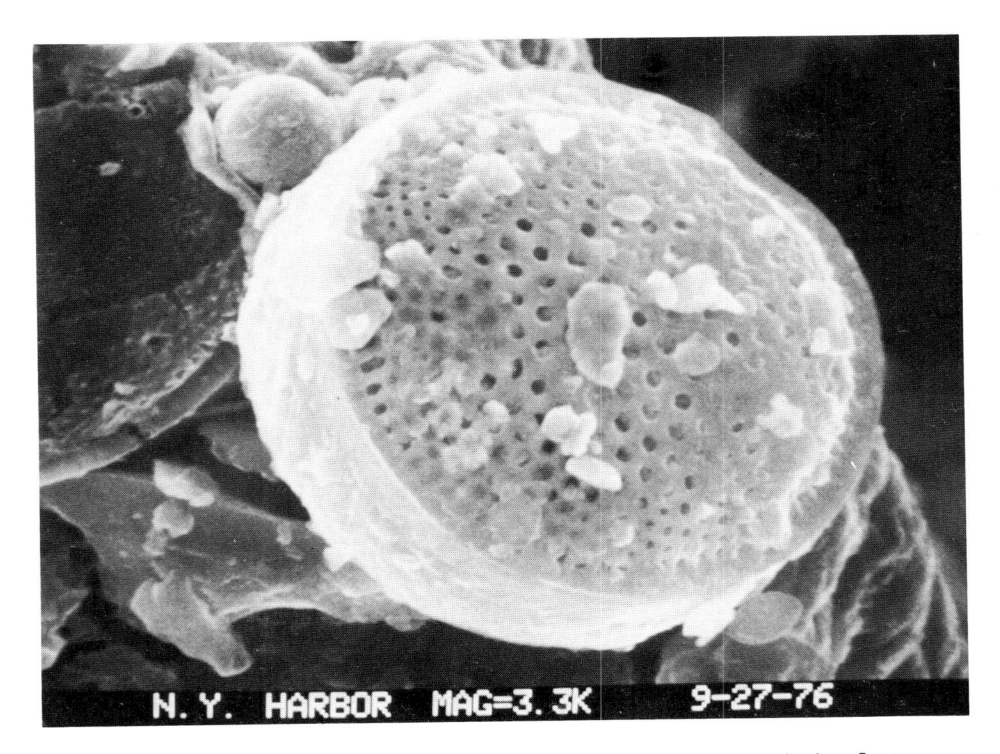

Fig. 14.52. Secondary electron micrograph of diatomea found in New York harbor. Letters and numbers were generated on the screen of the cathode-ray tube. Operating potential: 17.4 kV.

beam current for the records, since the ratio of beam current to monitor current differs from instrument to instrument, and may vary in the same instrument after changing the relevant apertures.

At this point we should also advise the operator to carefully fix fast-development exposures which require such treatment. If a part of an exposure has been inadvertently left without fixative, it will bleach in short time. Many useful images are destroyed in time by lack of proper fixation.

14. REFERENCES

14.1 Thornton, P. R., *Scanning Electron Microscopy*, Chapman & Hall, London, 1968.

14.2 Oatley, C. W., *The Scanning Electron Microscope*, Cambridge University Press, Cambridge, England, 1972.

14.3 Goldstein, J. I. and Yakowitz, H., *Practical Scanning Electron Microscopy*, Plenum, New York, 1975.

14.4 *Scanning Electron Microscopy 1968* to *Scanning Electron Microscopy/1977*, *Proc. Symp. on Scanning Electron Microscopy*, Johari, O., Ed., IIT Res. Institute, Chicago, Ill., 1968-1977.
14.5 *Proc. Confs. of the Electron Microscopy Society of America.*
14.6 Everhart, T. E. and Thornley, R. F. M., *J. Sci. Instr.* **37**, 246, (1960).
14.7 Newbury, D. E. and Yakowitz, H., Ref. [13.104], p. 15.
14.8 Kelly, T. K., Lindqvist, W. F., and Muir, M. D., *Science* **165**, 283 (1969).
14.9 Heinrich, K. F. J., Fiori, C. E., and Yakowitz, H., *Science* **167**, 1129 (1970).
14.10 Newbury, D. E., *Scanning Electron Microscopy/1977*, Vol. 1; also Ref. [14.4], p. 553.
14.11 Bruining, H., *Physics and Application of Secondary Electron Emission*, Pergamon Press, London, 1954.
14.12 Fathers, D. J., Jakubovics, J. P., Joy, D. C., Newbury, D. E., and Yakowitz, H., *Phys. Stat. Sol.* **20**, 535 (1973); also **22**, 609 (1974).
14.13 Coates, D. G., *Phil. Mag.* **16**, 1179 (1967).
14.14 Yakowitz, H., Ref. [12.87], p. 383.
14.15 Wells, O. C., *Scanning Electron Microscopy/1977*, Vol. 1, p. 747.
14.16 Wells, O. C., *Appl. Phys. Lett.* **16**, 151 (1970).
14.17 Kimoto, S. and Hashimoto, H., Ref. [12.80], p. 480.
14.18 Heinrich, K. F. J., *Scanning Electron Probe Microanalysis*, NBS Tech. Note 278, 1967.
14.19 Newbury, D. E., *Scanning Electron Microscopy/1976*, Vol. 1, p. 173.
14.20 Yakowitz H., Fiori, C. E., and Newbury, D. E., *Scanning Electron Microscopy/1973*. Vol. 1, p. 173.
14.21 Heinrich, K. F. J., *Adv. X-Ray Analysis*, Univ. of Denver, Colo. **7**, 325 (1963).
14.22 Kniseley, R. N. and Laabs, F. C., Ref. [12.87], p. 371 (44 references).
14.23 Remond, G., *J. Luminescence* **15**, 121 (1977).
14.24 Kneiseley, R. N., Laabs, F. C., and Fassel, V. A., *Anal. Chem.* **41**, 50 (1969).
14.25 Smith, J. V. and Stenström, R. C., *J. Geology* **73**, 627 (1965).
14.26 Wittry, D. B. and Kyser, D. F., *J. Appl. Phys.* **35**, 2439 (1965); also **36**, 1387 (1965), and **38**, 375 (1967).
14.27 Wittry, D. B., Ref. [12.87], p. 123.
14.28 Borot, M., *L'Onde Eléctrique* **45**, 1204 (1965).
14.29 Hosoki, S. and Okano, H., Ref. [12.84], p. 589;
Ishikawa, A., Uchikawa, Y., and Maruse, S., Ref. [12.84], p. 597;
Hijikigawa, M., Koba, M., and Inoguchi, T., Ref. [12.84], p. 603.
14.30 Weiblen, P., *Advances in Applied Spectroscopy*, Vol. 4, Plenum, New York, 1965, p. 245.
14.31 Davoine, F. and Pinard, P., *Electron Microscopy 1964*, *Proc. 3rd Eur. Reg. Conf. on Electron Microscopy*, Prague, Czechoslovak Acad. Sci., 1964, p. 113.
14.32 Shaw, D. A., Wayle, R. C., and Thornton, P. R., *Appl. Phys. Lett.* 8, 289 (1966).
14.33 Bernard, R., Davoine, F., and Pinard, P., *C.R. Acad. Sci.* **248**, 2564 (1959).
14.34 Davey, J. P., Ref. [12.82], p. 566.
14.35 Remond, G., Kimoto, S., and Okozumi, H., Ref. [12.84], p. 611.
14.36 Fiori, C. E., Newbury, D. E., Yakowitz, H., and Heinrich, K. F. J., *Scanning Electron Microscopy/1974*, p. 167.
14.37 Gonzales, A. J. *Scanning Electron Microscopy/1974*, p. 941.
14.38 Galloway, K. F., Leedy, K. O., and Keery, W. J., *IEEE Trans. Parts, Hybrids and Packaging* **PHP-12**, 231 (1976).

15. Scanning Electron-Probe Microanalysis [14.18], [15.1]

The usefulness of the electron probe for practical applications was greatly increased when Cosslett and Duncumb applied the technique of scanning electron microscopy to forming images with X-ray signals [15.1]. The speed at which scanning electron-probe microanalysis provides information on the distribution of elements over a microscopic area is of great practical importance. X-ray area scans may be performed prior to quantitative analysis of selected spots. In many cases, however, they provide all the information needed for solving a problem. Scans may be used, for instance, to identify a precipitate, an inclusion, or an oxide phase formed at the surface of an alloy. Whenever the knowledge of the elemental concentrations is not required with accuracy, the electron-probe microanalyzer can be used as a *scanning X-ray microscope* (i.e., as a scanning electron microscope with X-ray signals).

Most of the concepts developed in the previous chapter are also applicable to scanning electron-probe microanalysis by means of X-rays. However, to understand the limitations of the method, consideration must be given to the intended use of scanning electron-probe microanalysis and to the characteristics of the X-ray signal (Table 15.1).

The X-ray signals used in area scans are from lines characteristic of the elements which emit them. Neglecting the background contribution from the continuum and possible line interferences we presume that all observed signal pulses are due to the presence of the emitting element at the point of impact of the electron beam. On the contrary, in scanning electron microscopy, varying levels of the signal are usually emitted from all regions of the specimen, and the feature used in the image formation is the variation of a signal such as target current or secondary electrons. The signals are usually not specific, and their mere presence cannot be interpreted in any way. In the case of the target current signal, for instance, we observe signal changes above a large background, the nature and intensity of which is ignored. The background is simply eliminated

Table 15.1. Comparison of Scanning Electron Microscopy (SEM) and Scanning Electron-Probe Microanalysis (SEPMA).

	SEM	SEPMA
Signals	nonspecific	specific for elements
Efficiency of signal production	usually high	typically less than 5×10^4 photons/sec
Spatial resolution (width and depth)	typically 100-500 Å	typically 1-3 μm
Use of signal	qualitative	intensities interpreted as proportional to concentrations
Sensitivity to specimen position	moderate	critical

by applying a bias (black level), or by differentiating the signal, or both. When we deal with a characteristic X-ray signal, we would also like to eliminate the background signal, as was discussed in Section 6.4. But in view of the low yield of X-ray signals, and of the variability of the intensity of continuous radiation with the atomic number of the target, it would present a formidable task to make an accurate background correction for each point of the raster. We will deal later with some simple attempts to handle this *background problem.*

Statistical limitations are highly significant in the accurate evaluation of line intensities, particularly if quantitative interpretations are sought. As will be seen later, the low signal production efficiency also affects the *modes of presentation* of the signal, and it is responsible for the long time periods required for the performance of a scan. Finally, in view of the possible presence of many elements in a specimen region, any one single scan, representing one element distribution only tells part of the story, and the possibility of combining several images of the same area produced with X-ray signals from different elements into one image for easier interpretation must be considered.

The spatial distribution of X-rays generated by a small electron beam was studied in Chapters 10 and 13. We recall that except for cases of strong X-ray absorption or high critical excitation potentials almost the entire volume excited by electrons emits observable X-rays. Therefore, under comparable conditions the spatial resolution of X-ray signals is significantly poorer than that of secondary electron signals, target current, or backscattered electrons, which do not emerge from great depth within the target. The situation is still less favorable when secondary X-ray photons are also emitted.

Another limitation can arise from the focusing properties of the Bragg spectrometers. Usually, area scans are achieved by movement of the electron beam rather than of the specimen. When the beam deflection carries the point of beam incidence on the specimen beyond the focal region of the spectrometer, the observed X-ray intensity diminishes (Chapter 5). The band within which the focusing condition is achieved can be seen on X-ray area scans over large regions of a homogeneous specimen (Fig. 15.1).

Before characterizing a specimen area by means of X-ray scans, we must know which elements are present and of interest to us. If the specimen composition is a priori unknown, the scans must be preceded by a qualitative spectrographic characterization. As we have seen, this qualitative analysis can be a lengthy operation even for a single point; to perform it for all points of the area to be scanned is practically impossible. We can, of course, obtain a qualitative analysis of the entire area, by quickly rastering it during a conventional crystal spectrometer wavelength scan or by defocusing the electron beam. However, elements present at a few small regions in the scanned area, particularly if at low concentrations, may not be found in this procedure. Quick qualitative tests with the Si(Li) detector may be of great value. Another important preliminary step which may disclose regions and phases of interest for a more detailed investigation is a systematic observation of the specimen surface by means of target cur-

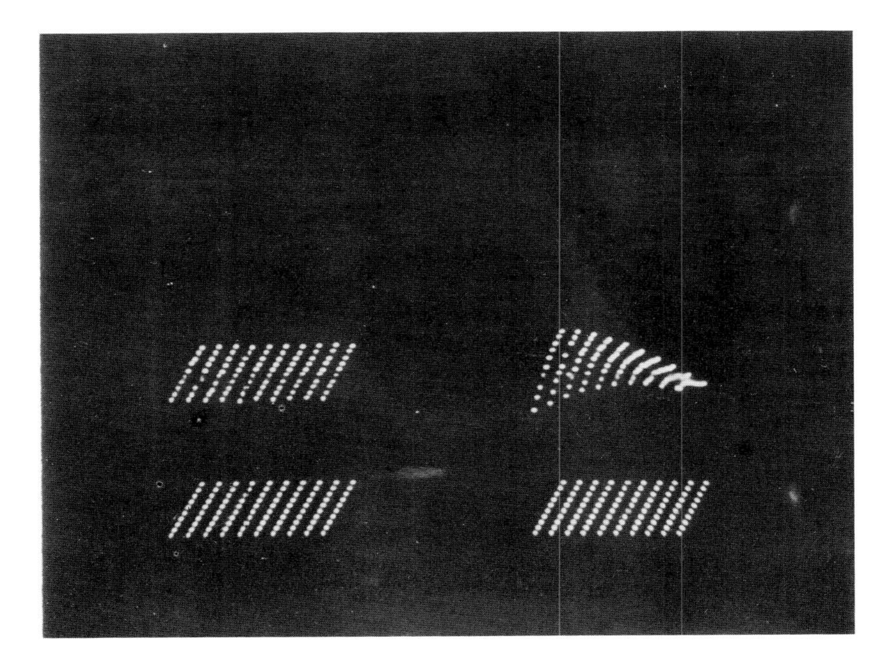

Fig. 15.1. Grid scan of Nb $L\alpha_1$ on a niobium specimen, with energy-dispersive (left) and wavelength-dispersive (right) spectrometers. Distances between corresponding points of upper and lower grids are proportional to signal intensity. The crystal spectrometer focuses along a diagonal band on the scanned grid, which is 440 μm wide.

rent, backscattered electrons and secondary electrons, and sometimes also by cathodoluminescence (i.e., observation of the specimen during the scan, with the illumination turned off, through the light microscope).

In most cases the specimen is not a complete unknown to the analyst, and the desired analytical information can be obtained with a small number of observations or scanning images. However, the unexpected presence of extraneous elements should never be discounted. It is very difficult to make in each analytical situation the right choice of procedure, ranging from quantitation with a static electron beam to semiquantitative line scans, area scans, or a combination thereof. Such decisions require the skill and judgment that characterize an experienced operator.

We will next discuss in more detail the limitations to the information content of the X-ray area scan related to spatial resolution, defocusing of spectrometers, counting statistics of X-ray photons and interference by lines and background, and we will describe some solutions proposed to deal with them.

15.1 RESOLUTION AND SCANNING DIMENSIONS

In the typical X-ray area scan the electron beam is moved in a line raster, covering one or several scanning frames (i.e., raster excursions). The X-ray detector pulses, after amplification and often after single-channel pulse height selection, are used to modulate the brightness of the oscilloscope screen, which, in the absence of a pulse, is kept dark. We will call this procedure a *standard pulse-recording scan.* Variations and alternatives to this technique will be discussed later.

Because the magnification of the scanning image is equal to the ratio of the scanning excursion of the beam of the cathode-ray tube (picture tube) to that of the electron beam on the specimen, the useful range of magnifications is determined by the resolution of the picture tube and that of the electron beam as well as by the largest permissible scanning excursion. Although display systems of high resolution can be used with the electron-probe microanalyzer, the picture tube used in some instruments cannot resolve more than 400-500 lines. The use of high-resolution cathode-ray tubes would not improve the quality of X-ray images because of the statistical restrictions we will discuss below. The width of the display screen is usually 8-10 cm; hence, the width of each line of scan is about 0.2 mm. A picture element could not be smaller than this width. If, on the other hand, the size of the smallest distinguishable picture element were to become larger than 1 mm, we would call the picture unsharp.

The range of primary X-ray generation is given in Eq. (13.1.5):

$$z_r = 7 \times 10^{-6}(V_0^{1.65} - V_q^{1.65})\, g/cm^2.$$

The width of the region emitting X-rays is about the same as the range: for typical materials and operating conditions, both are usually 1-3 μm, and even worse if a significant fraction of the signal is fluorescent. If the operating potential is lowered, the dimensions of the excited volume can be reduced (Fig. 15.2). Unfortunately, the intensities of the emitted X-ray lines also diminish. This limitation often precludes working at potentials below 20 kV. With a lateral signal dispersion of 2 μm, and a picture element equal to or smaller than 1 mm, a reasonably sharp image could not be obtained unless the magnification is less than 500X (from 2 μm to 1 mm). To use to advantage the high-resolution cathode-ray tubes the magnification ought to be even lower. In practice most X-ray scanning images have a picture resolution worse than 1 mm, and such unsharp images may still convey adequately the required information. We are here merely comparing the image-formation quality of the X-ray signal with that

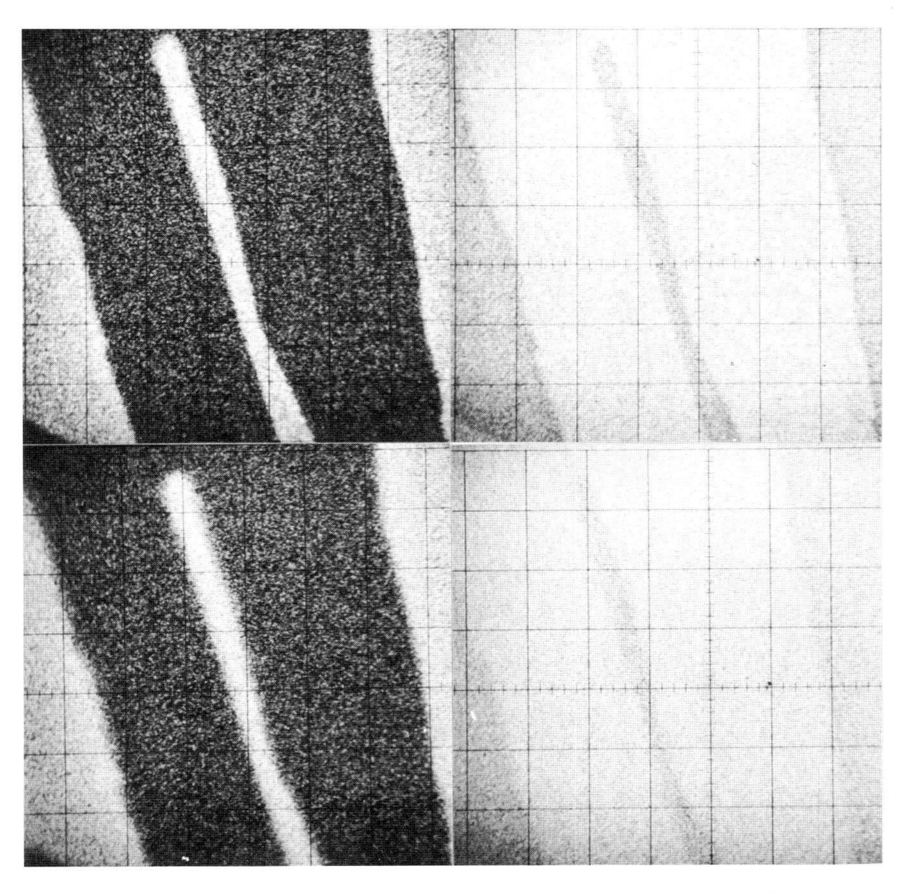

Fig. 15.2. Effect of the electron beam acceleration on spatial resolution. Tazewell meteorite, scanned area: 192 μm × 192 μm. Left: Ni Kα. Right: Fe Kα. Operating voltages: 20 kV (upper images) and 40 kV (lower images) [14.18].

of the scanning electron microscopy signals. If we wish to obtain a sharp X-ray image, on a screen 10 cm wide, an area of at least 200 μm in width must be covered.

However, the maximum useful size of the scanned area is limited, particularly if crystal spectrometers are used. As indicated by Malissa [3.24], the intensity losses due to defocusing of the X-ray optics are worst with crystals which have good focusing properties, and hence good spectral resolution and high line-to-background ratios. The spectrometer characteristics and the permissible drop of X-ray intensity determine how large a beam excursion can be used. Even with crystals of poor resolution such as commercially available lithium fluoride analyzers the loss is significant for scanning excursions larger than 200 μm. Hence, the useful magnification would be limited to a value equal to, or above, 400X.

If the postulated criteria were strictly applied, the analyst would thus be left with a surprisingly narrow range of permissible magnifications. In fact, as long as fully focusing crystal spectrometers with fixed focal circle alignment are used, the conditions for accurate measurement of intensity and for sharpness of image may be mutually exclusive. These limitations are frustrating since they do not equally exist for the scanning electron micrographs which are often made in the initial specimen survey. They render the quantitative evaluation of X-ray scanning images almost impossible, because it is difficult to devise reliable procedures of mathematical correction for the intensity variations due to spectrometer defocusing.

The following possibilities exist for reducing the loss of X-ray intensity by spectrometer defocusing:

a. to use semifocusing X-ray spectrometers [15.3];
b. to generate the scan, or at least one scanning excursion, by mechanical movement of the specimen [15.4];
c. to compensate for the scanning excursion by varying the crystal spectrometer adjustment;
d. to use a Si(Li) spectrometer instead of Bragg spectrometers.

In the semifocusing spectrometer, the crystal is mounted on a fixed shaft, and the detector on an arm which rotates at twice the angular speed of the crystal. The detector has a wide window, and is shielded from the X-ray source by means of a baffle. Although theoretically this spectrometer is focused for one wavelength only, in practice it can be used over a reasonably wide wavelength range. However, the wide window receives a significant amount of scattered radiation from the crystal out of the Bragg condition, and hence the line-to-background ratios are lower than with the fully focusing spectrometer. This device is not used in instruments of recent design but its use in the first scanning instruments has greatly contributed to the diffusion of the scanning electron microprobe technique (Fig. 15.3).

The mechanical scan of the specimen stage was introduced by Rouberol et al. [15.4] in order to combine the advantages of both fully focusing spectrometers of high resolution and area scanning. The defocusing of the spectrometers is fully eliminated since the relative positions of X-ray source and spectrometers remain unchanged during the scan. The big disadvantage of the mechanical scan is its slow speed. A single-frame scan requires 6 to 60 min, and the X-ray scan therefore cannot be observed on the oscilloscope screen. Due to this limitation, the mechanical scan has also fallen into disuse.

The dynamic X-ray spectrometer focusing consists in changing the nominal wavelength setting of the crystal spectrometer as a function of the displacement of the X-ray source in the plan of the spectrometer. As can be seen in Fig. 15.4, the shift of the crystal along the focal circle displaces the position of the focus in such a way that the radiation from the deflected beam can be observed by the crystal at the proper Bragg angle. The change of wavelength setting must be adjusted in proportion to the size of the scanning excursion. Since the detector is displaced at twice the angular speed of the crystal (rather than at the same speed as would be desirable), the technique is of limited usefulness if a narrow slit is positioned in front of the detector, in order to increase the wavelength resolution of the spectrometer. (Figs. 15.5 and 15.6). It should be noted that by the same token, a spectrometer with fixed alignment will admit radiation of a different wavelength if the X-ray source is displaced from the focusing circle. Hence, the scanning displacement affects the effective spectrometer setting as well as the efficiency at the nominal wavelength setting.

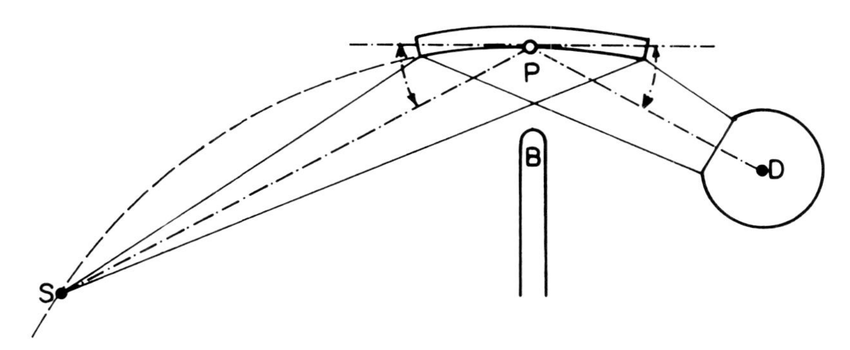

Fig. 15.3. Semifocusing spectrometer, shown in perfect focusing condition. For Bragg angles other than the one shown in the figure, the crystal is rotated around the axis through P, so that the line, SP, is at the correct Bragg angle. The detector, D, is rotated around the same point P. The baffle, B, prevents direct passage of X-rays from the source, S, to the detector.

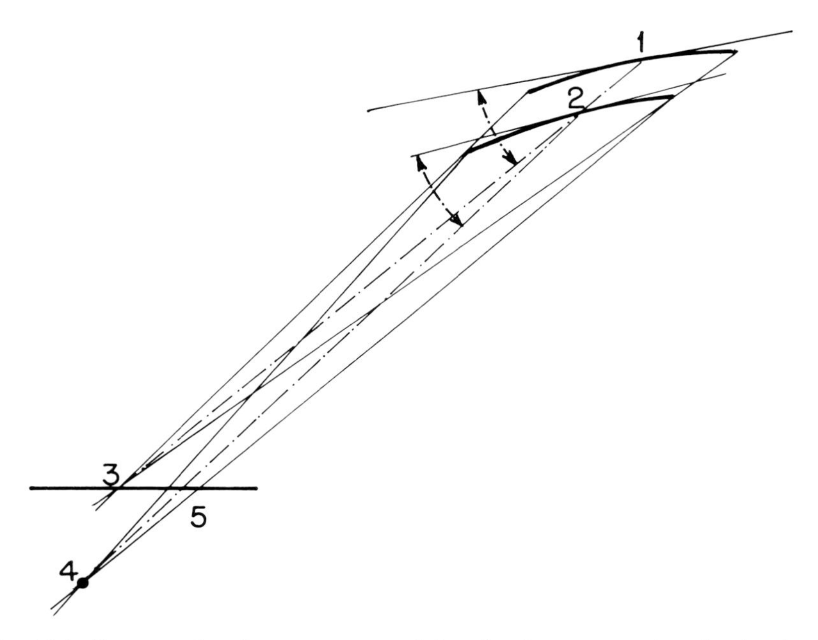

Fig. 15.4. Compensation for spectrometer defocusing in area scanning. As the electron beam is displaced from point 3 to point 5 at the specimen surface, the nominal wavelength setting of the crystal spectrometer is automatically changed, so that the crystal is displaced from position 1 to position 2 (compare with Fig. 5.8). In consequence, the spectrometer focuses now at point 4. X-rays emitted at position 5 are in the path for the Bragg condition, at least for the center of the crystal. Therefore, in spite of the imperfect focusing condition, a high X-ray intensity is detected.

The signal intensity produced by the energy-dispersive detector is much less position dependent than that of the crystal spectrometer. Therefore, signals from the Si(Li) detector produce images of very uniform scanning intensity—even over quite large area scans (Fig. 15.6). The disadvantages of energy-dispersive detection are the high background levels and the possibility of line interferences. As was discussed in Section 12.6, elaborate signal manipulations are required in the quantitative evaluation of energy-dispersive spectra to eliminate background and line interferences. Such procedures cannot be applied simply to every point on an area scan. For these reasons, area scans with Si(Li) and other energy-dispersive detectors must presently be limited to elements present at fairly large concentrations.

The analyst who desires to obtain images of uniform distribution of signal levels is also well advised to check the performance of the camera which is used to take the pictures of the cathode-ray tube screen. In some cases, the brightness of a uniformly illuminated field drops drastically at the corners of the picture,

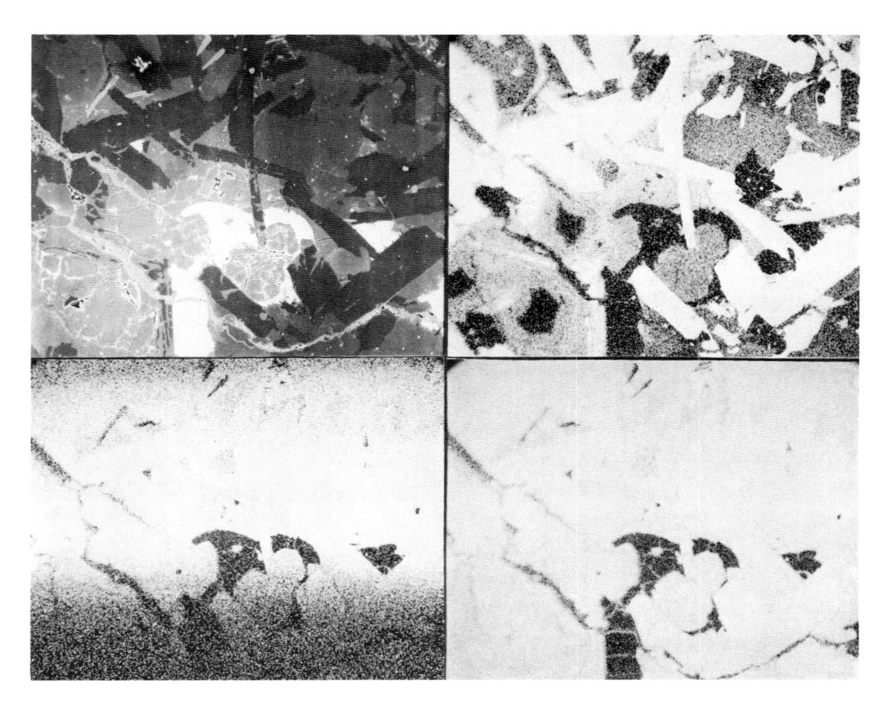

Fig. 15.5. Disco Island basalt, scanned over an area of 1000 μm × 800 μm. E_0 = 20 keV. Upper left: inverted specimen current scan. The white area is ilmenite, $FeO \cdot TiO_2$. Gray areas are iron silicate. Black areas show silicates of metals having low atomic numbers. Upper right: Al Kα, with dynamic spectrometer focusing. Lower left: Si Kα, without dynamic focusing. Lower right: same, but with dynamic spectrometer focusing.

particularly at wide apertures (Figs. 14.8 (upper left), 14.9, 14.28, and 15.14).

The spatial resolution as well as the ratio of line to background depend on the choice of X-ray lines and operating voltages. In general, the principal lines of low photon energy have high line-to-background ratios and are thus useful for low concentration levels. Such lines can be excited at low operating potentials, with the corresponding gain in spatial resolution. This advantage of long-wavelength emission was demonstrated by Andersen [15.5] who compared the spatial resolution obtainable with the Fe Lα-line with that obtained with the Fe Kα-line. The use of very long wavelengths may, however, lead to large X-ray absorption losses. These render the quantitative evaluation of the X-ray intensities more difficult, but are less objectionable in qualitative area scans.

As discussed in Chapter 13, the diffusion of electrons as well as the production of secondary X-rays can be drastically reduced by making the specimen much thinner than the electron depth range. This approach is common in biological

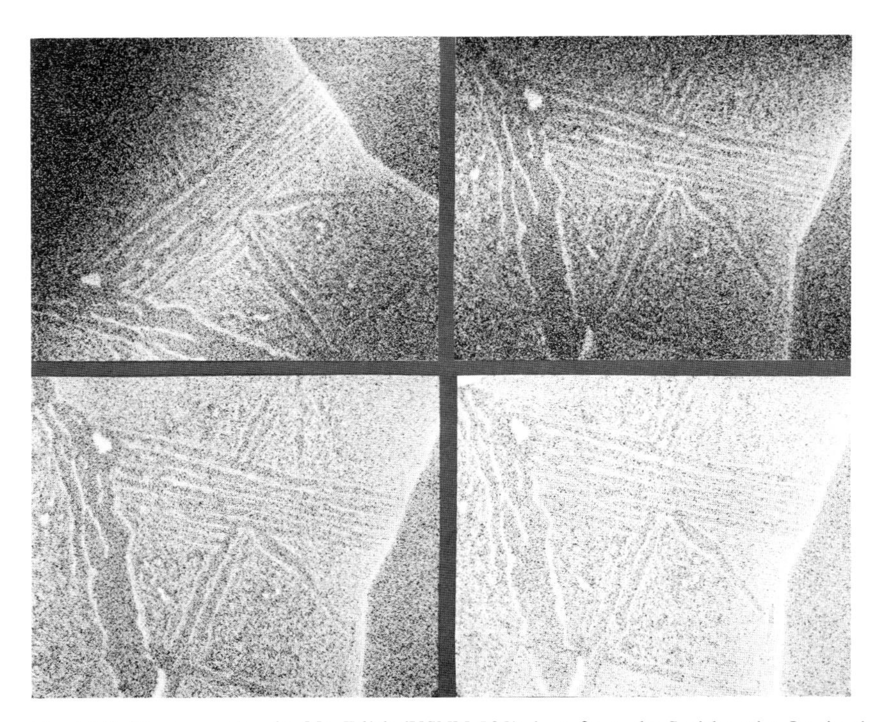

Fig. 15.6. Ferrous meteorite Mt. Edith (USNM 528), loan from the Smithsonian Institution, Washington, D.C. Scanned area: 1000 μm × 800 μm. Upper left: Ni Kα, obtained from a LiF spectrometer. Upper right: the scanning directions are electronically changed to orient the band of focusing parallel to the fast sweep direction. Lower left: same, with dynamic spectrometer focusing adjustment. Lower right: same, with Si(Li) energy-dispersive spectrometer. (Scans by C. Fiori, NBS.)

analyses. However, due to the low density of soft tissues and the low concentrations of most elements of interest in such specimens, the X-ray intensities obtained from such thin sections are quite low. For these reasons, and in view of the statistical limitations which will be discussed next in more detail, it is usually preferable in such thin sections to perform point analyses or line scans rather than area scans.

15.1.1 Scans of Three-Dimensional Objects

Although X-ray scans are mainly obtained from flat polished specimens, there is no good reason why they cannot be used to show the element distributions in specimens of irregular surface. Since such scans are not normally evaluated quantitatively, the resulting uncertainty in the X-ray emergence angle is usually of no practical consequences. The method can thus be applied to crystallized or fracture surfaces as well as to particulate specimens (see Figs. 15.7-15.9 and Fig. 2 on color plate III).

Shading effects can sometimes be observed on the X-ray scans of objects of very irregular surface. For instance, some parts of the silver wire grid in Fig. 14.49 are missing in the X-ray image (lower right). Such effects are due to the blockage of the X-rays emitted from the respective target region, either by the object itself (self-absorption, Fig. 15.9), or by another object, such as the copper grid in Fig. 14.49. Because the shadowing is related to X-ray absorption, rather than to that of electrons, it is prevalently observed when soft X-rays are used (see Fig. 15.9). It also increases with the angle between the electron beam and the detected X-rays, and hence with decreasing X-ray emergence angle. In most cases, such shadows are interpreted in analogy to those observed in visual images, and hence do not pose serious problems. Mistakes may arise from this artifact, however, if a qualitative test is performed of a spot within a strongly entrant part of a specimen.

15.2 STATISTICAL LIMITATIONS TO X-RAY AREA SCANNING

In Section 6.2 we have described how the precision measurement of X-ray intensities is limited by counting statistics. An unbiased estimate of the Poissonian standard deviation can be obtained from the square root of a single count:

$$\sigma_{N} = \sqrt{\overline{N}} \simeq \sqrt{N}. \quad (15.2.1)$$

The Poissonian statistics provide an estimate for the smallest number of counts required to obtain a desired precision if all sources of error other than counting statistics have been eliminated or made insignificant. It is only too obvious that at low count rates excessive statistical requirements will lead to very long counting times and danger of drift. When only a few quantitative measurements are required to perform an analysis, the problem presented by counting statistics is usually manageable. But counting statistics become a severe limitation if demands of precision are set for each point in an area scan with X-ray signals. For a given precision, the duration of signal collection can be reduced by increasing the count rate. This is usually achieved by raising the operating potential and/or the beam current. Loss of spatial resolution or damage, contamination, or destruction of the specimen may set a limit to these recourses. Otherwise, the use of high count rates will ultimately be limited by the detector dead-time. In energy-dispersive systems, high count rates may also produce loss of resolution and peak shifts, as discussed in Chapter 5.

The discussion of counting statistics of the area scan will be simplified if we consider each scanning line of the frame as an array of discrete points or image elements. If these points are closely spaced, the difference between a continuous and a discontinuous scan is not discernible. Hence, the conclusions derived from the counting statistics are equally applicable to both types of scans.

It is reasonable to assume that in an array of points the distances between lines should be equal to those between the points of each line. Therefore, if we have a square frame of 8 cm $\times$ 8 cm which contains 200 lines, this array is in effect a matrix consisting of 4 $\times$ 10^4 points, each 0.4 mm $\times$ 0.4 mm in size.

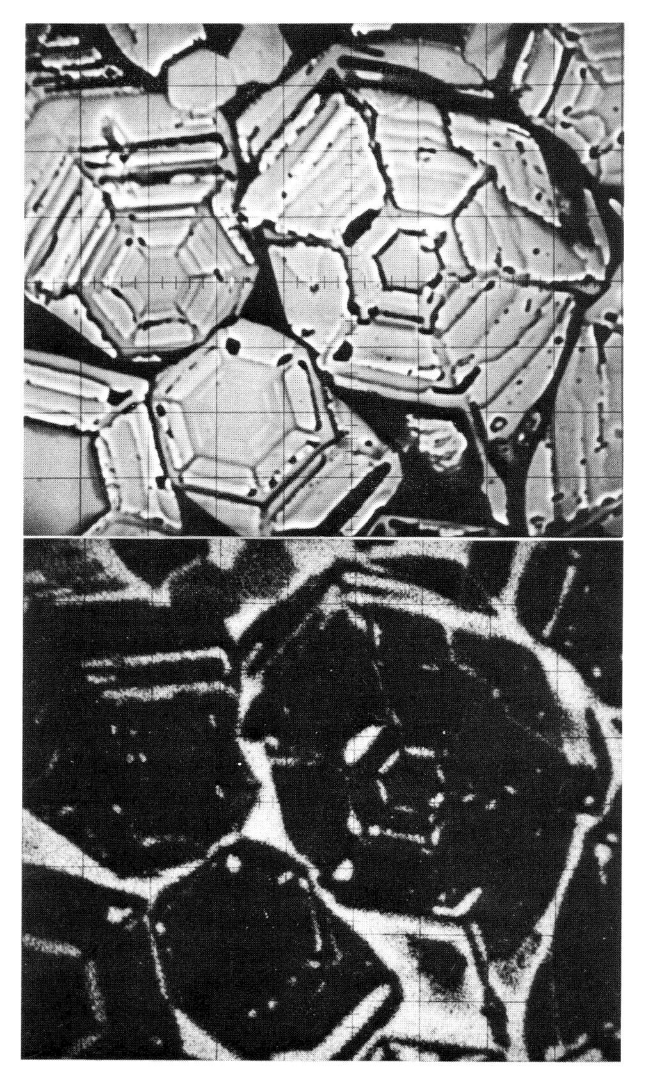

Fig. 15.7. Titanium boride crystals, impurified by iron. Specimen current (upper image) and Fe $K\alpha$ scan (lower image). Width of field: 180 μm. The X-ray emergence angle, ψ, was 52.5°.

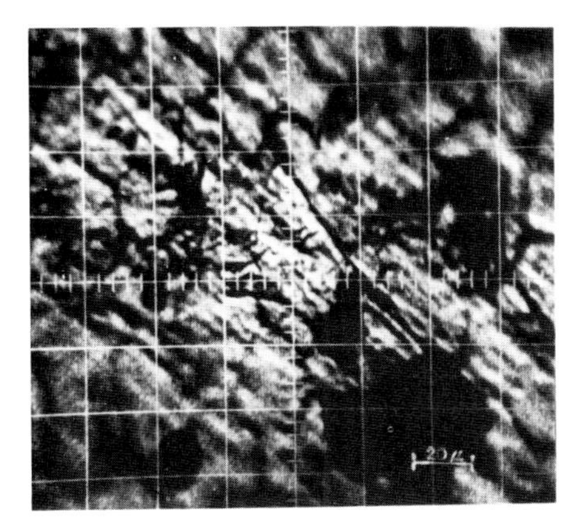

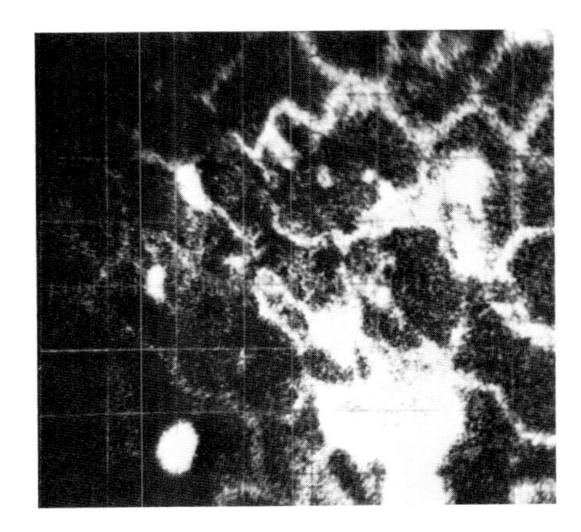

Fig. 15.8. Refractory alloy extruded with glass lubricant. Electron backscatter (left) and Ca Kα area scans (right) of the surface. The X-ray image demonstrates contamination of the alloy sheet in the extrusion process.

Assuming that the X-ray intensities to be displayed at these points are not significantly affected by errors other than counting statistics, we can calculate how many X-ray photons must be collected in order to achieve a predetermined precision. To simplify the calculation, we will assume that these precision criteria are set for the points of highest signal intensity, and that the dwelling time is the same for all points. We will also assume, in our example, that the maximum signal intensity is 2×10^4 counts/sec. Such count rates are common with crystal

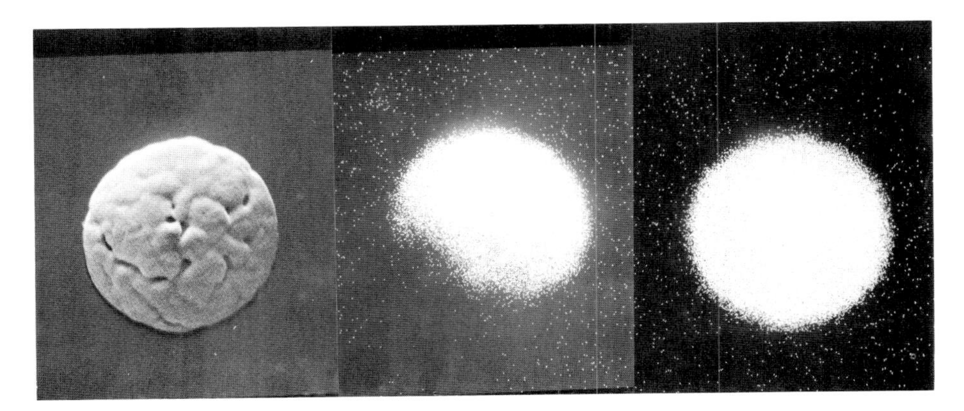

Fig. 15.9. X-ray scanning images of a three-dimensional object. Since the primary beam scan determines the image configuration, all regions in the image are accessible to the primary beam. Hence, shadows are produced by the absorption of X-rays. Sphere of nickel: 12 μm wide. Left: secondary electron image. Center: Ni Lα_1 image, which exhibits a shadow due to its strong absorption. Right: Ni Kα image.

spectrometers, but are very high for a silicon detector. Let the precision criterion be such that the relative standard deviations are not larger than 1%. Therefore, we must accumulate 10^4 counts at the points of highest count rate, and the dwelling time at each point is 0.5 sec. The total counting time is thus 2×10^4 sec, or 5 hr and 33 min. A scan of such duration is possible, but most analysts would consider the time excessive, and under typical conditions the specimen would be badly contaminated.

Let us now assume that we will perform a standard pulse-recording scan. To properly select the time of scanning, we should first consider what concentration differences can be perceived under optimum conditions in such a scan.

The eye can, under good viewing conditions, distinguish brightness differences as small as 1% [15.6], though only between adjacent picture areas. But the number of gray levels which could be identified in a picture and associated with concentration ranges regardless of position, size, and shape of the respective areas is certainly not larger than 10. Hence, even in the absence of background, matrix, and statistical effects, the standard pulse-recording scan is qualitative only or at best coarsely semiquantitative, and not sensitive to small changes in composition. Therefore, the statistical requirements can be relaxed as long as we use this image-forming technique. The number of photons per picture element required to distinguish with a 95% confidence level the highest X-ray intensity from another 10% lower, N, is related to its standard deviation by

$$2\sigma_N = N/10$$

(see Appendix). If the X-ray counts have a Poissonian distribution, we obtain: $2N^{1/2} = N/10$, and therefore, $N = 400$. At a maximum count rate of 2×10^4 counts/sec, a raster of 200^2 points would require an exposure time of 800 sec. This length of time is a quite realistic requirement. If we were to attempt to fill a raster of 400^2 points, the time would quadruple, to 53 min. But even with this long exposure time, some statistical fluctuations will still be perceived as image graininess, since 5% of the points, or 8000 points, can be expected to exceed the 10% limit, and hence ten times the minimum perceptible brightness difference. Obviously, at the signal collection time of 1 to 5 min most operators are willing to invest in the formation of a scanning image, the statistical variation of count rate is a very significant source of uncertainty in the interpretation of brightness levels.

We must therefore accept that in practice all X-ray area scans are grainy. But the most serious effect of statistical intensity variation is a degradation of the spatial resolution of the scanning image (Fig. 15.10). Within a uniform area the eye of the observer tends to disregard the fluctuations; this visual integration breaks down, however, in regions of large concentration changes, such as occur at edges and limits of one-phase regions. The resultant limitation in resolution is

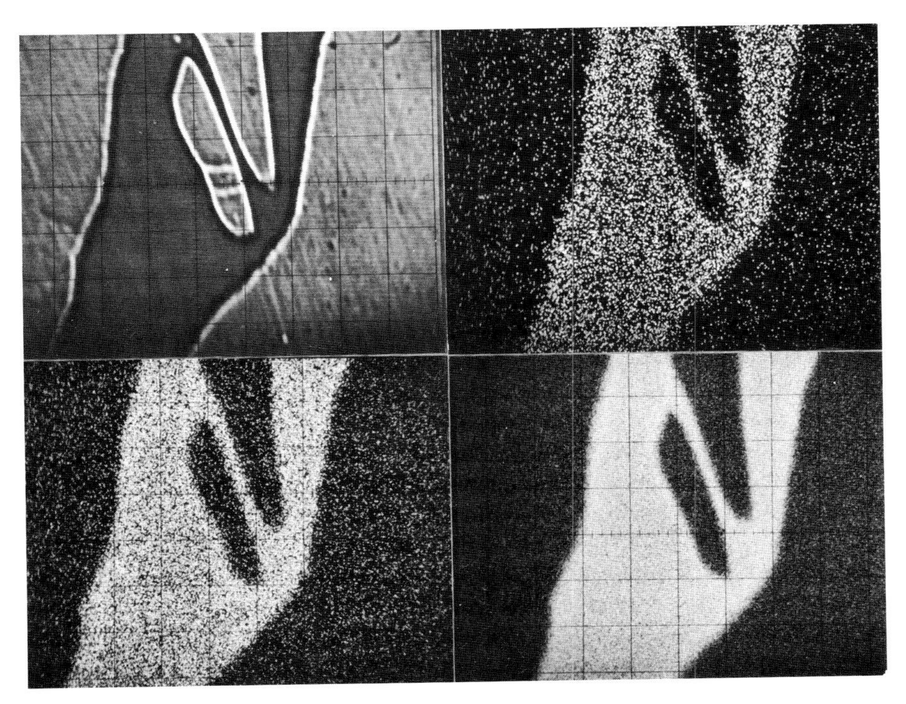

Fig. 15.10. Effect of the number of photons collected in a scanning image on the spatial resolution. The specimen is a ferreous meteorite (Tazewell meteorite). The element and line shown are Ni Kα. $V_0 = 20$ kV. Scanned area: 70 μm × 55 μm. Upper left: target current image. Upper right: 20,000 counts. Lower left: 56,000 counts. Lower right: 560,000 counts [14.18].

usually more serious than those caused by the diffusion of the electrons in the specimen or by the beam size. This limitation is basic, and cannot be removed by signal transforms or other mathematical procedures, unless the number of collected photons is increased, or additional a priori assumptions about the distribution of intensities are made (e.g., in smoothing procedures).

15.3 STANDARD PULSE-RECORDING SCAN

The pulse-recording X-ray area scan has several advantages which account for its popularity. The required equipment is incorporated in most instruments, and the signal transforms are minimal so that the execution and interpretation of the scan are apparently simple. The only variables that must be defined for a specimen region are: the element and line to be observed, the magnification, the number of lines per frame, the total number of photons to be collected, and the level of brightness produced on the oscilloscope screen by a single pulse. The scanning speed is, within limits, immaterial: at high speeds, many frames can be superposed. We recall that in scanning electron microscopy, the speed of scanning may vary over a wide range which includes "television scanning" at speeds at which a virtually continuous visual image is formed. However, if in the pulse-recording scan the speed is raised to the point where the signal dots are stretched to lines, the effect is very unpleasant since the resultant limitation of resolution is put in evidence too obviously. It is therefore advisable to reduce the speed to a level at which the signal pulse—which is usually 10^{-4} sec long—still produces a dot of the desired dimensions. If the length of a signal pulse is 10^{-4} sec, and we wish to obtain 200^2 distinct image elements, one frame requires a minimum time of 4 sec. Scans of much higher speed are not very useful in view of the statistical limitations of X-ray signals, even if the signal dots can be resolved.

The total duration of the scan is determined by the mean count rate of the signal and the number of photons to be accumulated. These can be counted on a scaler during the exposure. The pulse-recording scan by superposition of many frames has the advantage of flexibility. Elements present at high concentration in a few small areas and rough distribution patterns of major components can be identified in a few seconds, and the visual observation of these patterns can help in selecting the areas to be recorded and studied in detail. If, on the contrary, the statistical scatter in one exposure is excessive, the scan can be repeated under essentially identical conditions, but with a larger number of frames collected for the image. Such flexibility is lacking in methods which require a single-frame operation. The pulse-recording scan has, however, two important inherent limitations. The first is the impossibility of manipulating in any way the signal, since every single pulse acts independently upon the photographic image. The second limitation arises from the difficulty or impossibility of quantitative measurement of the X-ray intensities recorded on the image, even if statistical limitations are tolerable. Since every pulse contributes individually to the image formation, it is impossible to separate in this procedure the background contribution, or the effects of atomic number, fluorescence, absorption, dead-time, or line interference. Nor is it possible to subtract a preset intensity level in order to see a limited concentration range with increased contrast (black level adjustment). Such a signal transformation requires the electrical equivalent of the count rate, which must be obtained by digital counting on each picture element or, simpler, by means of a ratemeter.

The impossibility of eliminating the background is the most serious shortcoming of the standard pulse-recording scan since it limits the observation and identification of signals from elements present at low concentrations. A rigorous treatment for background removal such as is employed in the quantitative procedures with a static electron probe is practically impossible in the scanning mode. The operator would, however, frequently like to apply a black level adjustment to remove the background, even if this were to imply a possible loss of weak characteristic X-ray signals. A procedure which approaches black level adjustment consists of inserting in the signal line a gate which determines if a given X-ray pulse is accepted or not. A timer is used to maintain this gate open for an adjustable time after another pulse, or a scaled-down pulse signal (i.e., a signal indicating that a certain number of pulses have arrived) has started the timer. At low count rates, such signals arrive rarely, and so the gate is closed most of the time. The contrast of the image will therefore be enhanced, and low background or signal levels are suppressed [15.7].

We have, up to this point, assumed that the effects of the individual pulses on the image are truly independent, but such an assumption is not quite correct. The effect of a pulse on the image depends of the length and brightness of the dot of light on the oscilloscope, the aperture of the camera, and the sensitivity of the film used in the photographing of the screen. These variables can be arranged in such a fashion that a single signal pulse produces a small white spot on the image. As long as there is no significant superposition of points, the point density in a uniform area is proportional to the X-ray photon detection rate. Let us also make the reasonable assumption that the size of each white spot is equal to that of an image element. Two consequences follow: first, at each spot, the image contains a bit of information in binary code: any one image element indicates either presence (white) or absence (black) of the signal. Secondly, a second pulse falling within the same image element will be of no consequences to the image. Hence, in regions of strong signals, the dots promptly coalesce into a white area, within which no further gradations are possible. With an ideally uniform signal, our set of 200^2 points would be fully saturated by only 4×10^4 pulses. Even a high-resolution screen would be fully saturated by 10^6 pulses. Clearly, this mode of operation, although useful to give approximate distributions of weak signals, does not admit the number of signal pulses which would be necessary for both smooth intermediate gray tones and good contour definition (Fig. 15.11). Furthermore, the background pulses will be clearly visible in this procedure, adding to the confusion of the area scan.

We can reduce the brightness of the single-pulse image to the point where a single pulse produces a barely perceptible dark gray dot, or an invisible latent exposure. In such a case, several gray levels can be obtained through partial or full superposition of individual pulses. This mode of operation is more desirable, since we can use a much larger number of pulses, and hence obtain a better contour definition. It is desirable to accumulate at least 5×10^4 to 10^5 pulses per image, unless the element shown in the scan is concentrated in small regions within the scanned surface. Since a single exposure of an image element produces no visible response, or a very weak one, low pulse rates tend to be suppressed; the effect is similar to black level adjustment: it suppresses a weak background level; unfortunately, it may also eliminate low characteristic signal levels.

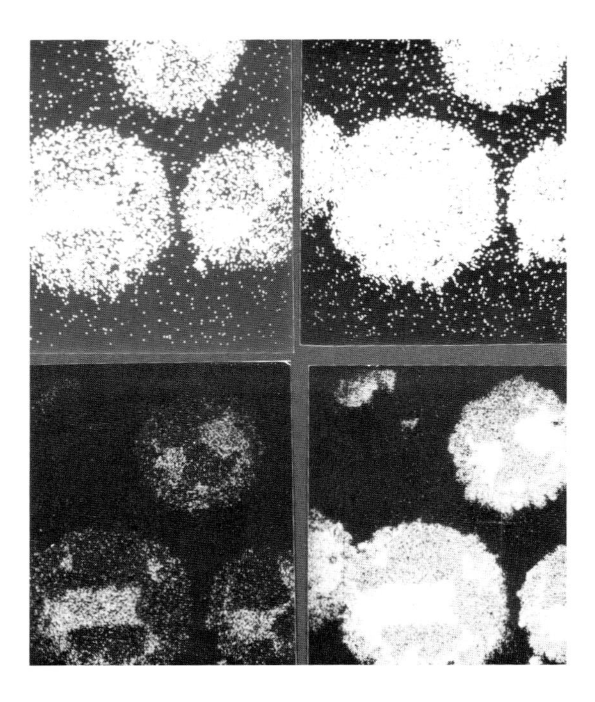

Fig. 15.11. Distribution of gadolinium in an Fe-Co-Gd alloy. E_0 = 20 kV. Scanned area: 80 μm × 90 μm. Upper left: "bright spot" mode, 18 000 Gd Lα pulses. Background noise and poor spatial resolution. Upper right: same brightness, 36 000 pulses. The two gadolinium-bearing phases are difficult to distinguish. Resolution still poor. Lower left: lower CRT brightness, 64 000 pulses. Image is too dark. Lower right: 300 000 pulses, with low CRT brightness: correct X-ray image. (From [15.8].)

The single-pulse technique is further illustrated in Fig. 15.12 which shows three sets of gray scales composed by areas produced with uniform count rates. Adjacent fields differ in count rates by a factor of two. The uppermost set of fields represents the case in which every pulse is clearly depicted. It is obvious from this figure that the observable range of steps in no case exceeds eight, and that in every set there is a clear limit as to high and low count rates beyond which no distinction of count rates is observable. It should also be mentioned that the scales of pulse density are usually not sets calibrated or characterized in a consistent way. Therefore, intuitive interpretations of point-recording area scans in terms of element concentration can be widely misleading.

It has been proposed [15.9] that the graininess of low-density point distributions be eliminated by slightly defocusing the cathode-ray beam which produces them. Such a cosmetic improvement should be used with great care so that the deterioration of resolution which results from it be kept in reasonable bounds.

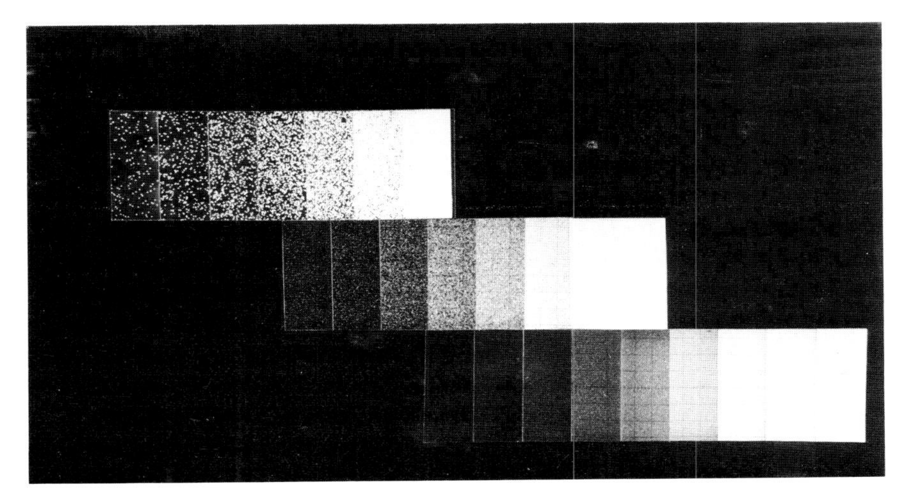

Fig. 15.12. Gray scales obtained from areas of uniform pulse density at three levels of cathode-ray tube brightness. The scale of pulse densities is the same for each series. Adjacent fields in each strip differ in pulse density by a factor of two. (From [15.1].)

15.4 THE USE OF RATEMETER SIGNALS FOR AREA SCANS

If the output of a ratemeter is used to modulate the brightness of the cathode-ray tube scan, the grainy structure of the X-ray image can be reduced or suppressed, depending on the combination of count rate, scanning speed, and of the time constant of the ratemeter. A gray level shift or other transforms of the signal can be applied as well. However, the inherent statistical limitations of the detector signal cannot be eliminated by such manipulations. Unless a black level shift is applied, all the ratemeter does is to smear out the signal along the time axis. If low count rates make the use of a large time constant necessary, the scan must be performed very slowly; otherwise statistical fluctuations will be traded off for loss in spatial resolution. If the duration of the process is of no concern, and if the instrument is stable, images of greatly improved contrast and spatial resolution can be obtained (Fig. 15.13). In the past, ratemeter scan images were performed mainly where it was necessary to subtract background levels [5.22], [5.23]. In most practical cases, the combination of an area scan from an electron signal such as target current, and of a single line scan with the X-ray signal will produce more information in less time. Good scanning images of this type give us, however, a clear indication that the main practical limitation in X-ray area scans resides in the counting statistics.

By means of the "expanded contrast technique" [15.10] or black level shift we can enhance differences in the X-ray signal at high-concentration levels of the element in question (Fig. 15.14). A difference of 5% in the concentration of an

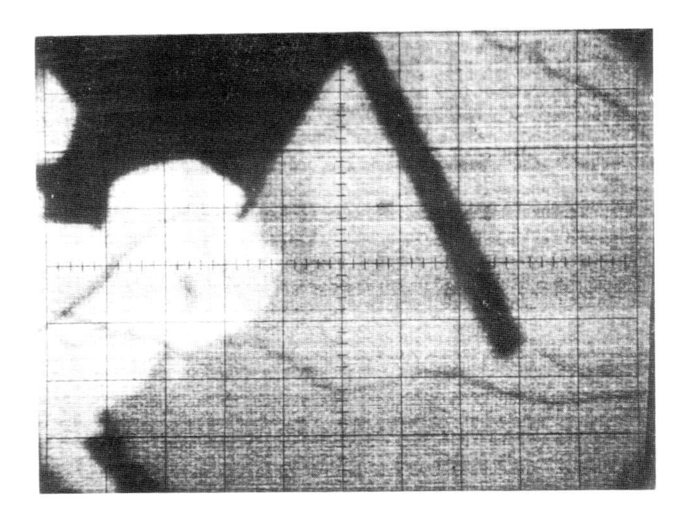

Fig. 15.13. Ratemeter area scan of an area of a basalt; the signal is Mg Kα radiation. Bias (black level shift) was employed. Duration of the scan: 90 min (single-frame scan). Scanned area: 80 μm X 100 μm.

element at the 50% level is difficult to see on a pulse-registration image, but it is clearly observable with expanded contrast, provided that sufficient pulses are collected in the scan. It must be emphasized again that there is no simple way to distinguish a priori a background level from low-signal levels or to compensate for variations in the background levels, especially from a specimen of complex composition or poor flatness. Therefore, the main prerequisite for the imaging of low concentrations of elements is a spectrometer of a high ratio of line to background. For this reason, crystal spectrometers are decidedly an advantage over Si(Li) detectors, as far as X-ray area scans are concerned.

A more quantitative evaluation of the variation of ratemeter signals can be achieved with the *concentration mapping technique* [15.11], [12.49]. By means of optical (Fig. 15.15) or electronic devices, the signal is sorted into a number (two-ten) of preset level ranges. Each range produces a different level of the signal emitted by the concentration mapping device. The output of this device is then applied to modulate the brightness of the oscilloscope screen (Figs. 15.16 and 15.17). The signal transform which produces a concentration map can be generated with the aid of an auxiliary oscilloscope (Fig. 15.15), which is used as an optical signal-level discriminator, or by purely electronic devices. In the first case, the X-ray ratemeter signal is connected to the vertical axis of deflection of a cathode-ray tube. A mask in front of the cathode-ray tube is cut in such a manner that the light on the screen is visible only when the signal intensity from the ratemeter is within a preset range (or ranges). A photomultiplier is placed

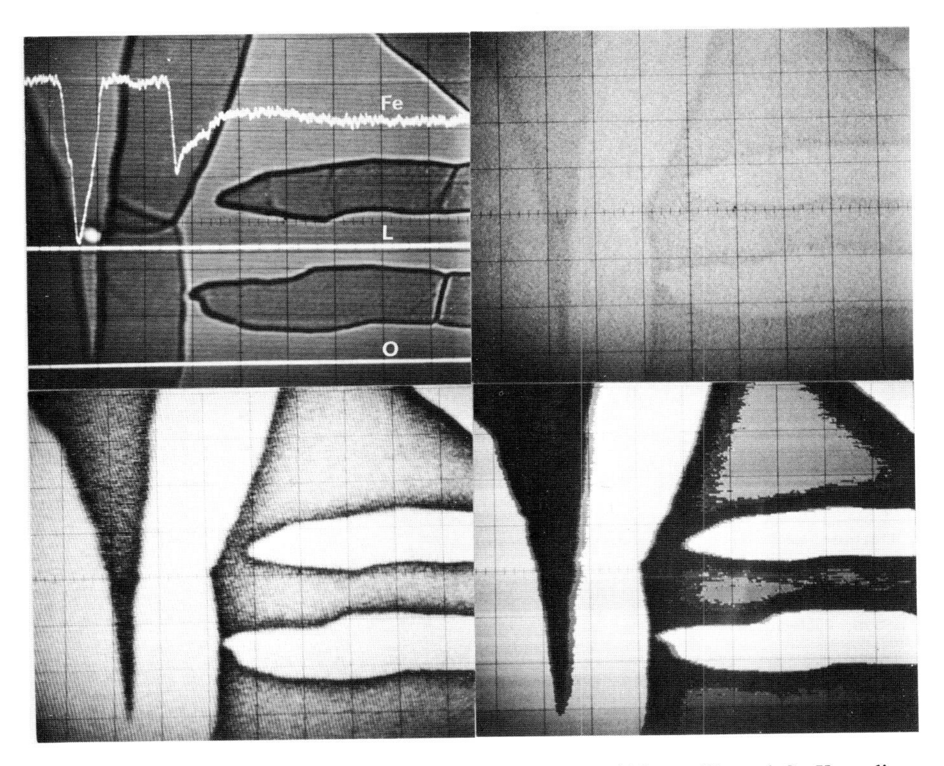

Fig. 15.14. Tazewell meteorite. Scanned area: 138 μm × 110 μm. Upper left: X-ray line scan of Fe Kα (Fe), and zero level of X-ray signal (O), superimposed upon inverted specimen current area scan. The line (L) marks the location of the line scan on the depicted area. Upper right: standard pulse-recording scan. Lower left: X-ray area scan from ratemeter signal, with expanded contrast. Lower right: concentration map [14.18].

in front of the mask, and the signal it produces is used to modulate the brightness of the beam of the image-forming cathode-ray tube (Fig. 15.16). In this configuration, a two-tone image (black and white) is obtained. Gray tones or crosshatched areas can be achieved if the horizontal axis to the auxiliary oscilloscope is connected to a square-wave generator, and if the mask is formed by slots such that in some X-ray signal intensity ranges the beam of the auxiliary oscilloscope is periodically intercepted by the mask (Fig. 15.17). Interesting maps which indicate correlation of element concentrations can be obtained by simultaneously applying the ratemeter signal of one element to one axis of the auxiliary oscilloscope deflection, and that of another element to the other axis.

The characteristic feature of concentration mapping is the discontinuity of the output function with respect to the input signal at preselected input levels,

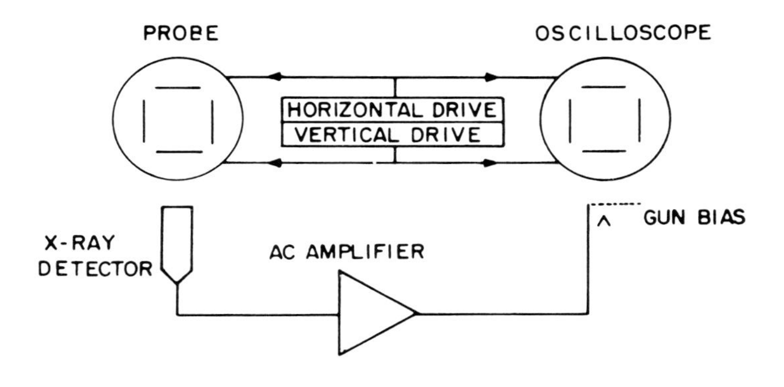

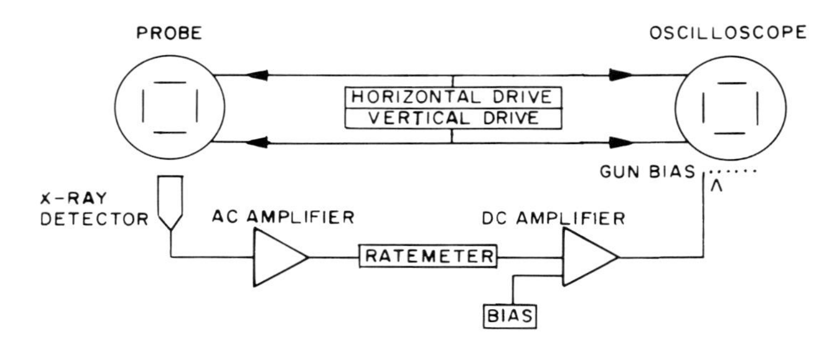

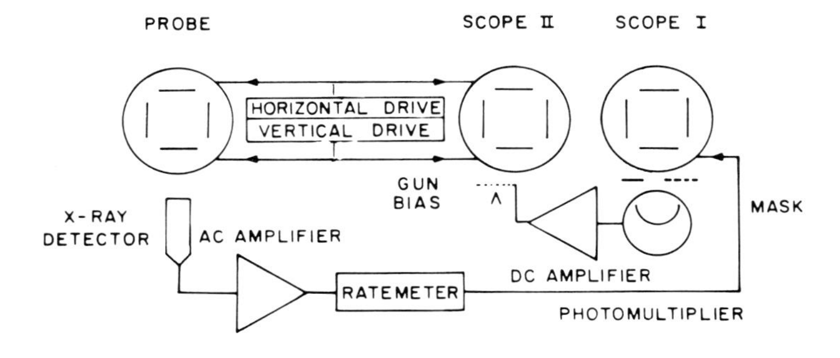

Fig. 15.15. Three procedures for modulating the brightness of X-ray area scans: Top: standard pulse-recording scan. Center: ratemeter signal scan with black level shift. Bottom: concentration mapping with the aid of an auxiliary oscilloscope (scope I), provided with a mask and a photomultiplier [15.11].

and the constancy of the output between these levels. The name of the technique is somewhat optimistic, since variations of the uncorrected signal rather than of concentrations are perceived on the image. This technique can, of course, be applied as well to target current signals, which suffer less from statistical fluctuations (Fig. 14.36).

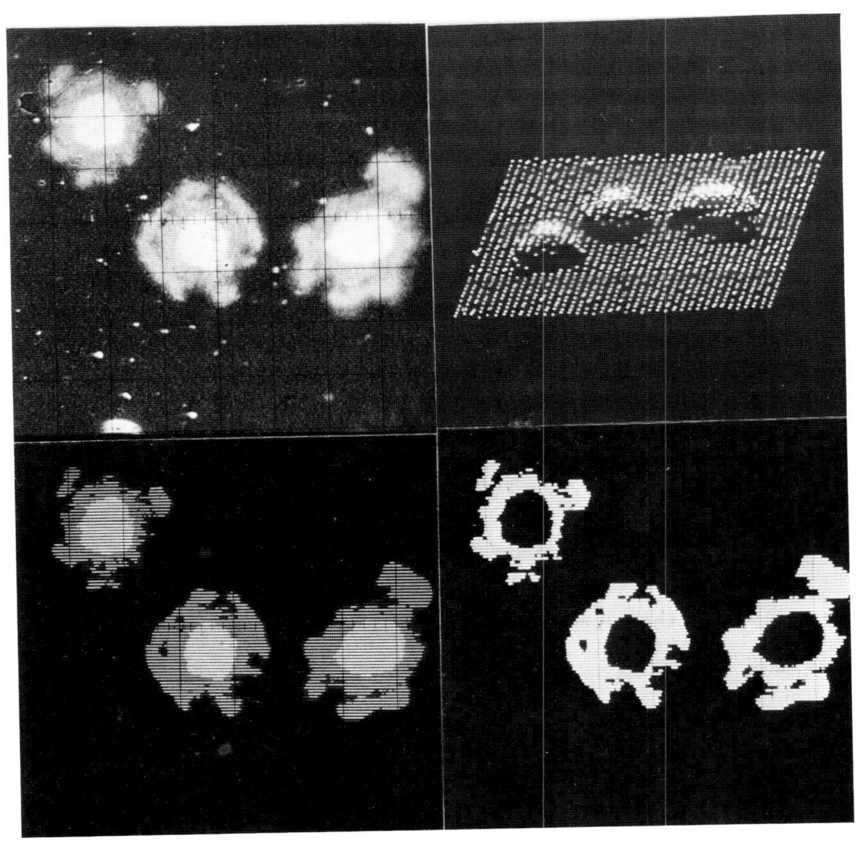

Fig. 15.16. Area images obtained from a specimen in which niobium wires were embedded in tin and heated for diffusion. Upper left: inverted specimen current. Upper right: y deflection and brightness modulation of a 40 × 40 Nb Lα_1 matrix scan. Lower left: three-level concentration map from Nb Lα_1 signal. Lower right: three-level concentration map from Nb Lα_1 signal, in which the intermediate level is white, the low and high levels are black. Phases present are Nb, Sn, and Nb Sn. Width of scanned field: 440 μm × 440 μm.

15.5 DIGITAL MATRIX TECHNIQUES

The accuracy of concentration estimates by the concentration mapping technique depends on all sources of error we have discussed previously: spectrometer defocusing, background variations, statistical fluctuations, line overlapping, and matrix effects. But the main reason for which the application of ratemeter signal scans has found limited use only is that they are much more time consuming and demanding in skill than conventional line or area scans. The black levels and range limits must be carefully established before the scan is begun, and if they, or the camera and cathode-ray tube setting, are in error, or if drift occurs, the defect is usually not observed until the lengthy operation is finished.

The solution to most of the drawbacks of these methods based on analog signals is to gather digitally the counts obtained at every image point, and to

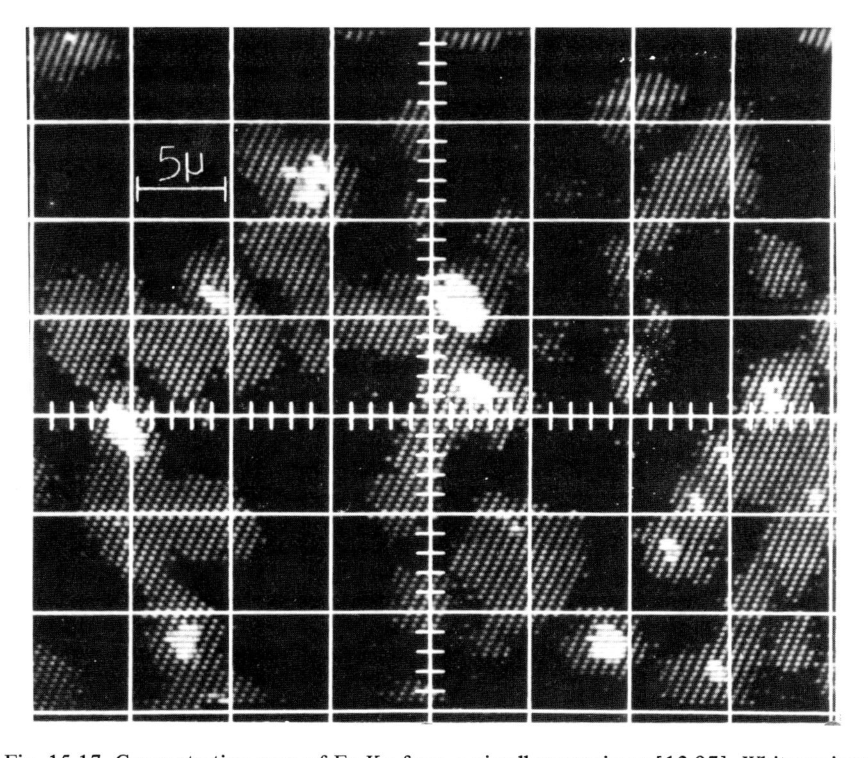

Fig. 15.17. Concentration map of Fe Kα from a zircalloy specimen [12.97]. White regions have a concentration above 1% of iron; in the dotted regions, the concentration is between 0.1% and 1%. E_0 = 20 keV. This map was obtained with the optical concentration mapping described in Fig. 15.15.

apply the transforms and evaluation procedures to the stored information. In such a quantitative image analysis, the proper settings of black levels and other parameters, including smoothing functions where required, can be adjusted without need for repeating the data collection. It will be possible to add background and matrix effect corrections as needed. At present the storage of such an information matrix is not beyond the technical state of art; it is clearly only a matter of time until techniques of image formation based on analog transforms will be displaced by digital techniques. Simple procedures of this type were described by Birks and Batt [5.23], who used, in 1963, a multichannel analyzer as a storage for the signal in a two-dimensional quantitative matrix of X-ray signals. Heinrich et al. [12.3] have developed an electronic device which automatically performs X-ray measurements on points forming a rectangular matrix of up to 100 × 100 points. This device has been used in automatic operations lasting several hours. A careful alignment of the instrument reduced the drift of beam current to 1% or less of the measured intensities. With the lithium-drifted silicon detector, spectrometer defocusing could also be eliminated (Fig. 15.1).

The matrix technique can be further refined if a digital computer is employed to direct the operation and to store the resulting digital information. After the data collection is completed, the computer can evaluate the data in different ways; for instance, it can construct, by means of an oscilloscope or an *x-y* plotter, concentration maps with an accuracy not obtainable by analog techniques. When the matrix technique is extended to matrices of 200 × 200 points or larger, the spatial resolution equals that of the conventional scanning technique. The limits of accuracy are again dictated by counting statistics. In each experiment a compromise must be made between spatial resolution (number of points) and accuracy at each point. The use of the computer will enable the operator to use the available time in the most effective way. Ultimately, however, further progress in quantitative scanning techniques requires the development of faster X-ray detectors and associated electronics.

15.6 MULTIPLE EXPOSURE IMAGES

Each of the images obtained in the investigation of a microscopic region of a complex specimen transmits a partial message about the nature of this region. Thus, a target current scan provides a delicate map of the specimen surface, but without specific information on composition, each X-ray area scan shows the distribution of one element only, usually without clearly marked boundaries, and without an indication of concentrations present, and a line scan provides a cross section along one line only of the intensity of one signal. To interpret the entire information is often tedious, and for this reason it is very useful to combine several pieces of information in a single scan. A simple example is that of a combination of a target current area scan with one or several X-ray signal line scans along a line marked on the line scan (Figs. 15.14 and 15.18). The linear plot of signal intensities permits a semiquantitative or roughly quantitative evaluation of the signal; however, the signal distortions due to statistical fluctuations and limited beam resolution must be taken into account. If the scan is too fast, the dynamic characteristics of signal collection (time constant of ratemeters) can also be significant. This combination is very powerful in practice, and the techniques for obtaining it are not very complicated. Such an image should contain the following:

a. The background image, which must not contain light areas that may distract from the line-scan information.
b. A line marking the location of the line scan. It is desirable that this line be separate from the background level of the X-ray scan, and marked differently. A signal generator producing a dashed line is useful for that purpose.
c. The X-ray intensity profile for one or more elements provided from ratemeters which must have an adequate range of time constants or integration periods. There must be provisions for continuously adjusting the position of the baseline and for the scale of the signals.
d. The baseline which marks the zero level of the X-ray signal, for each X-ray line scan. This line can be obtained by detuning the spectrometer from the

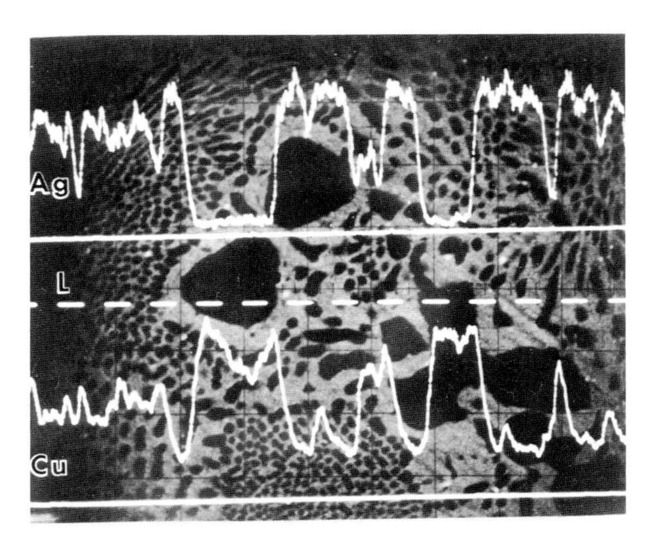

Fig. 15.18. Inverted specimen current area scan and line scans for Ag Lα_1 and Cu Kα emissions. The dashed line marks the location of the line scan on the area image; note zero levels for each X-ray emission. Same specimen as Fig. 14.8.

respective X-ray line. If the background is low, the X-ray scan can be repeated with the electron beam of the detector electronics disconnected.

e. A clear indication of which are the elements, and if necessary, the lines used to produce the X-ray signal.

A somewhat unusual combination of X-ray signals is shown in Fig. 15.19 in which the ratio of signals of two elements (calcium and phosphorus) from small particles in a histological preparation of a human pineal gland was compared with the ratio obtained from small particles. The matrix effects could be neglected with such small particles, and the image indicated that the calcium-phosphorus ratio in all examined particles was lower than that in the standard. The graph was obtained by connecting each ratemeter signal of the two elements to one direction of deflection of the oscilloscope, and performing for each particle an exposure of fixed duration.

It is particularly tempting to combine the pulse-recording scans with area scans from electron signals, since the first provides element specificity, and the second topographic resolution. If black and white exposures are made, the result is almost always disappointing. To distinguish the pulse signal from the continuous background, the individual pulses must be made very coarse, and hence the X-ray part of the image is formed by few photons. The problem is even worse if we try to combine several X-ray area scans. It is possible to use coarse linear raster frames, and then a horizontal raster may represent one element, and a

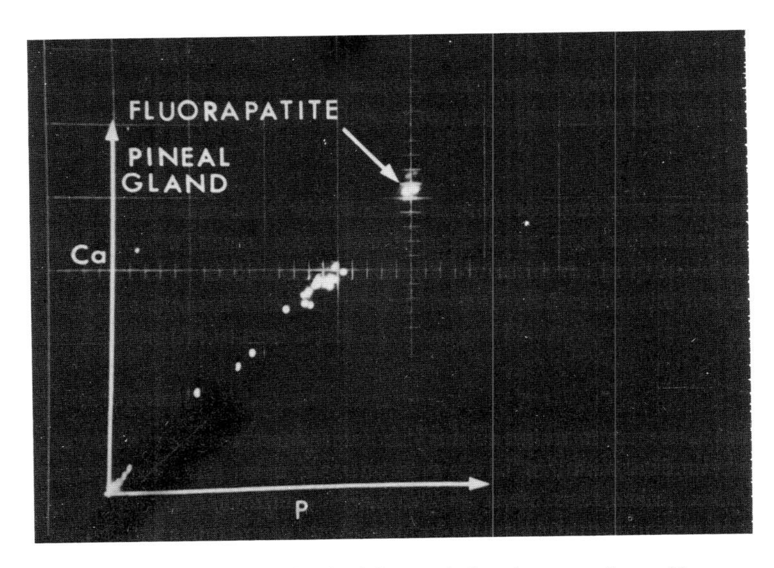

Fig. 15.19. Plot of the ratio of signals of calcium and phosphorus on the oscilloscope screen.

vertical raster the other. But, clearly, such procedures imply a gross loss of spatial resolution of the X-ray signal. Satisfactory combinations of area scans require the use of color, which will be discussed next.

15.6.1 Multiple Area Scans in Colors

Duncumb and Cosslett [15.12] first described a technique based on the additive combination of several area scans in a multicolored image. In such color composites, each element shown is presented in a different color, on a black background. The presence at any location of more than one of these elements produces a mixed color. The perception of pure and mixed colors must therefore be taken into account in the procedure. In most cases, we will wish to interpret every color of the image in terms of the presence of the elements which are depicted. This requires that mixed colors be recognized as combinations of the basic (primary) colors. Since the human eye cannot interpret color mixtures in terms of more than three primary colors, the number of elements shown in a multiple area scan is limited to three. It is best to choose red, green, and blue as primary colors. The binary mixed colors are yellow (red + green), cyan or bluish-green (blue + green), and magenta (red + blue). Absence of all three elements is shown in black. If all three elements are present and give off strong signals, the resulting color is white. Therefore, all binary mixtures can be represented along the sides of a triangle (Fig. 15.20 and Fig. 2 of color plate I.) and ternary mixtures within a double pyramid (Fig. 15.21).

Four methods can be used to obtain a composite color-scan image:

1. The cathode-ray tube displaying the scan is faced by a camera loaded with color film. The exposure for each element is photographed in succession onto the same film, and a different color filter is inserted between the oscilloscope and the camera for each exposure.
2. The partial images are obtained in black and white, by the usual pulse-record-

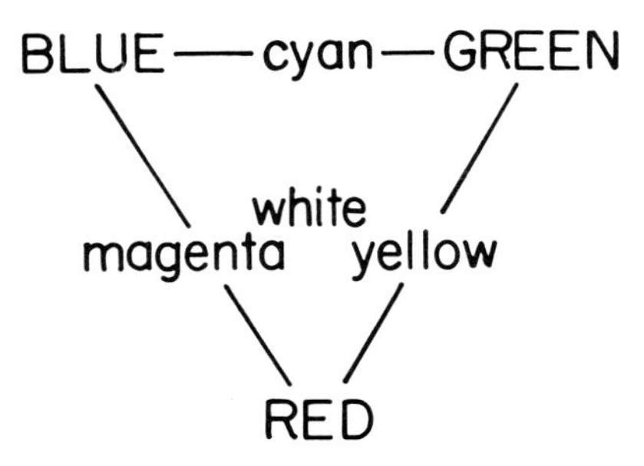

Fig. 15.20. Relation of primary and mixed colors. Addition of colors proceeds along any straight line within the triangle.

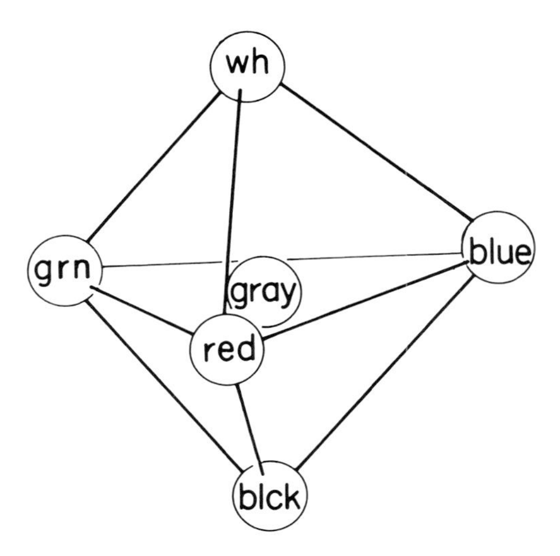

Fig. 15.21. Double pyramid of hues and brightness values. Addition follows straight lines within the double pyramid. Names of secondary colors were omitted in this graph.

ing area scan. The composite picture is obtained by photographing the black and white images by means of a copy camera which is loaded with color film. As each original is copied, a different color filter is inserted in front of the objective of the copy camera (color copy technique) [15.13] (Fig. 3, color plate I; Figs. 1-4, color plate II; Figs. 15.22 and 15.23).

3. The image is produced in a simultaneous process on the screen of a three-color cathode-ray tube, and directly photographed in color (colorscope technique) [15.14].
4. The information from the individual scans is stored in a memory and displayed later in a three-color cathode-ray tube, usually after applying transform operations such as contrast enhancement (black level shift) or concentration mapping [15.15] (Fig. 2, color plate IV).

The direct photography through filters of the black and white picture screen is greatly hampered by the poor spectral distribution of white phosphors used in commercial cathode-ray tubes. Red color is particularly difficult to obtain and requires very intense primary display or long exposures. The effects may be difficult to predict, and in case of an unsatisfactory result the scanning procedure must be repeated. Therefore, this technique is hardly ever employed at present. The color copy technique, which is used frequently in the NBS electron-probe laboratory, requires little in equipment except for the conventional pulse-registration setup and a copy camera such as is frequently available in laboratories. It has the further advantage that in case of an unsatisfactory composite

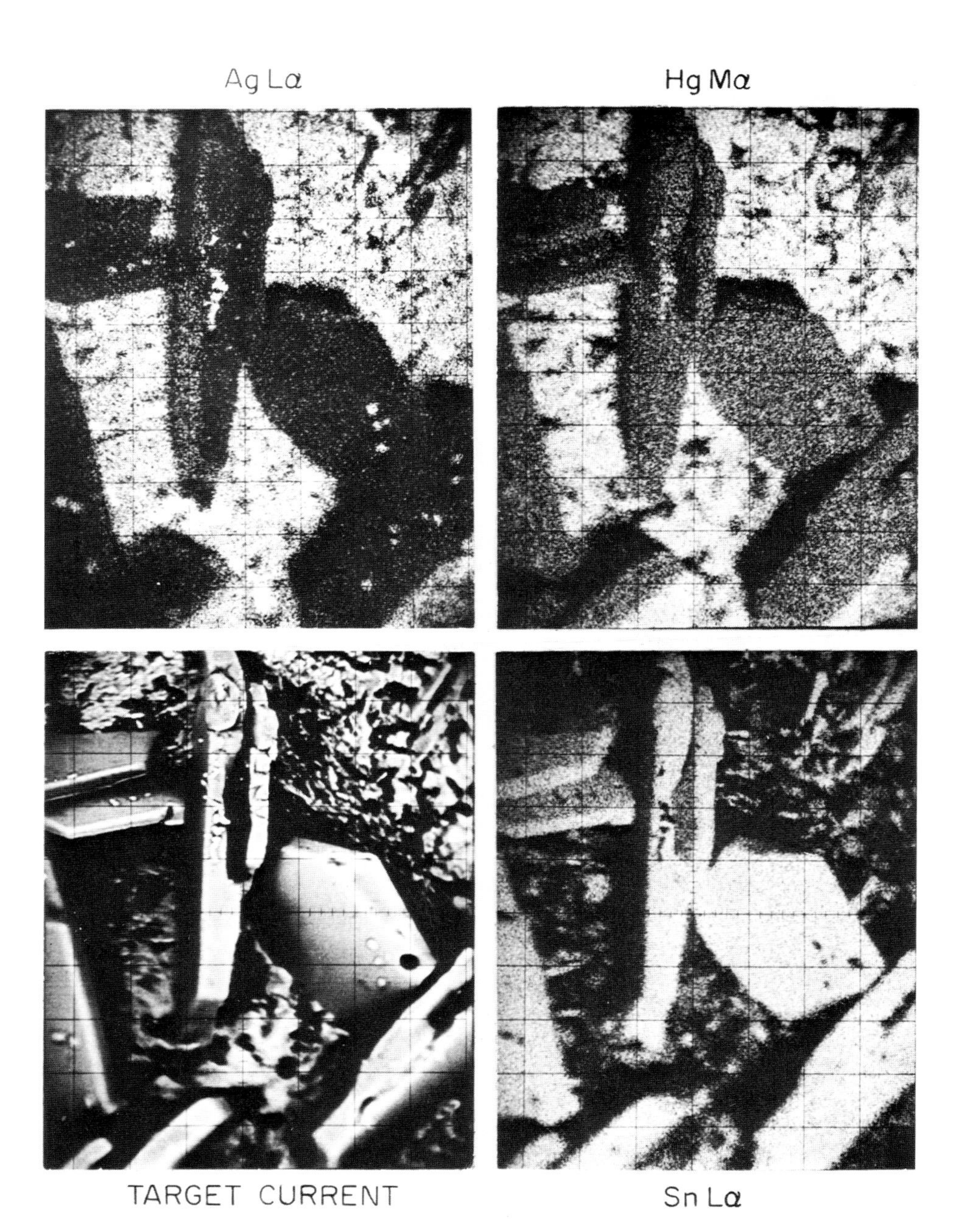

Fig. 15.22. Area scans of the surface of a dental alloy which has crystallized after heating. E_0 = 20 keV. $i_b = 5 \times 10^{-8}$ A. Scanned area: 275 × 220 μm. Upper left: Ag $L\alpha_1$. Upper right: Hg $M\alpha$. Lower left: inverted specimen current image. Lower right: Sn $L\alpha_1$. The X-ray images were used to compose the color image No. 4 on the color plate. (From [14.18].)

picture the electron-probe operation does not have to be repeated. This is an important advantage, since the color composite cannot be observed visually before the photographic image is developed. For this reason it is preferable to use a fast-development film (Polaroid color film or equivalent) so that in case of unsatisfactory results another composite can be prepared. Technique and choice of filters are described in detail in [15.13].

The colorscope technique requires more expensive equipment, but it offers the advantage that the colors can be observed on the screen. The color quality is superior to that of the color copy with fast-development film, but in case of failure, the entire scanning process must be repeated. The use of a memory and of transforms avoids this drawback and is undoubtedly the most efficient, though also the most expensive, way of obtaining color composites. For the analyst who is willing to use color composites occasionally, the color copy technique will be the most likely choice. Its success depends on the proper selection of filters and exposure times. The fact that the composites can be prepared separately from the taking of the X-ray scans is a considerable advantage in a busy laboratory.

In the selection of colors, it should be noted that their effect on the observer (attention value) varies. Red, and the mixed color yellow, attract the eye of the observer much more than blue and green. The perceived brightness also varies; yellow appears as the brightest color, and blue usually as the darkest. Of the three binary colors, yellow is considered by the naive observer as a color in its own right, equivalent to the three primary colors. Therefore, the choice of colors can be used to emphasize or deemphasize certain element signals in the image. If a single element is to be seen prominently, red is the proper choice, and if a binary combination is of particular interest, it should be depicted in yellow (red + green).

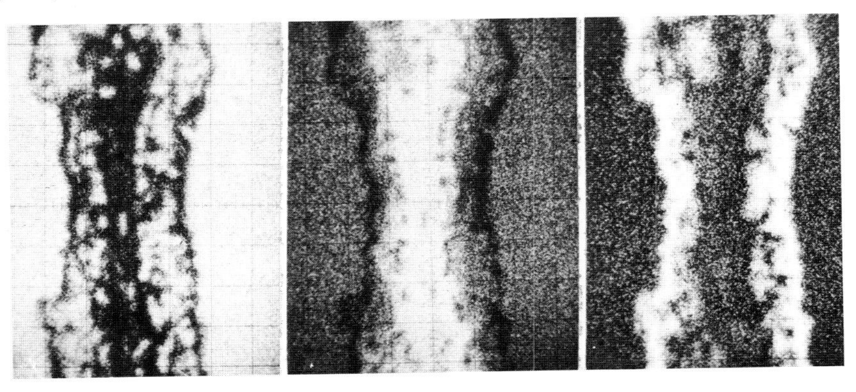

Fig. 15.23. Oxidized crack in a nickel-based superalloy. From left to right: Ni $K\alpha$, Cr $K\alpha$, and Al $K\alpha$. These X-ray area scans were used in composing the color images shown in Plate I, Fig. 3.

In the same manner as the area scans for a single element, the color composites suffer from lack of sharpness of boundaries and absence of the topographic details with which the investigator has become familiarized through preliminary microscopic observation. These deficiencies are removed if a specimen current or secondary electron scan is also superposed upon the color composite (Color plate 1, Fig. 3, plate 2, Figs. 1-4, plate 3, Figs. 1 and 3). In such a procedure, however, we must be careful not to disturb the information conveyed by the X-ray scans. Therefore, the topographic information must be added in a neutral gray hue, and all areas between boundaries should have similar gray levels on the topographic scan. This can be achieved very effectively by partial or total differentiation of the target current or secondary electron image. After some experience, the analyst may increase the emphasis on the topographic information in cases where no confusion is created (Color plate 3, Fig. 3).

In general, color combination techniques work best on images of flat specimens. However, as shown in color plate 2, Fig. 2, this rule admits to exceptions, and satisfactory color composites can be obtained from objects deviating from flatness, such as the crystals high in tin which protrude from the surface in the heat-treated dental alloy shown in Fig. 15.22. It should be noted that this image was obtained without the help of electron signals.

15. REFERENCES

15.1 Heinrich, K. F. J., *Scanning Electron Probe Microanalysis* in *Advanced Optical and Electron Microscopy* 6, 275 (1975), Barer, R. and Cosslett, V. E., Eds., Academic Press, London, 1975.

15.2 Cosslett, V. E. and Duncumb, P., *Nature* 177, 1172 (1956).

15.3 Long, J. V. P. and Cosslett, V. E., in *X-Ray Microscopy and Microradiography*, Cosslett, V. E., Engström, A., and Pattee, H. H., Eds., Academic Press, New York, 1957, p. 435.

15.4 Rouberol, M., Tong, M., Weinryb, E., and Philibert, J., *Mem. Sci. Rev. Metallurg.* 59, 305 (1962).

15.5 Andersen, C. A., Ref. [12.80], p. 58.

15.6 Rose, A., *Adv. in Electronics* 1, 131 (1948).

15.7 Wakabayashi, T., Miyake, T., Date, G., and Soezima, H., Ref. [12.84], p. 287.

15.8 Heinrich, K. F. J., SEM/1977/1, Om Johari, Ed., ITTRI, Chicago, Ill., 1977, p. 605.

15.9 McMillan, W. R., 2nd EPASA Conf., Boston, Mass., Paper 58, 1967.

15.10 Melford, D. A., *J. Inst. Metals* 90, 217, (1962).

15.11 Heinrich, K. F. J., *Rev. Sci. Instr.* 33, 884 (1962).

15.12 Duncumb, P., in *X-Ray Microscopy and Microradiography*, Cosslett, V. E., Engström, A., and Pattee, H. H., Eds., Academic Press, New York, 1957, p. 617.

15.13 Yakowitz, H. and Heinrich, K. F. J., *J. Res. Nat. Bur. Stand., Sect. A* 73, 113 (1969).

15.14 Ficca, J., 3rd EPASA Conf., Chicago, Ill., Paper 15, 1968.

15.15 Schpigler, Private commun., 1976.

PART VI. THE TARGET

16.
Target Characteristics

16. Target Characteristics

In the preceding discussion of analytical theories and procedures we have made several assumptions concerning the specimen and the standard which we must now consider in more detail. Among these assumptions are the following:

a. that the specimen is flat and precisely oriented with respect to the beam, and that the specimen surface is representative of the entire sample, or of the region of the specimen which we wish to investigate;
b. that the irradiated specimen conducts the electrical charge and heat produced in it during analysis, and that it is not altered or destroyed by the interaction with the electron beam;
c. that the analyst can observe the features of interest at the specimen surface and hence select for analysis those parts of the specimen which he wishes to investigate.

As will be seen in the discussion which follows, these requirements often are mutually exclusive. For instance, it may be impossible to find in the optical microscope grain boundaries and other distinctive features on the finely polished surface of an alloy specimen; if these are rendered visible by etching, the surface composition may be altered [16.1], [16.2].

16.1 Specimen Preparation

In most microanalyses, the analyzed specimen region is not of interest by itself, but rather as a representation of a mass of material larger than the emitting region by many orders of magnitude. Hence, we have, in the general case, a sampling problem: the specimen must have a composition and structure equal to that of the material it represents. In the same fashion, the portion of the standard which emits X-rays under the electron bombardment must be identical in composition to that much larger portion which has been chemically analyzed, or whose composition is surmised from stoichiometric relations. If the standard is an element, the emitting region must be free of impurities that can affect the intensity of its X-ray emission. But the analyzed region is at the surface of the specimen, and since natural surfaces can be contaminated or altered, a fresh surface must usually be prepared. Where quantitation is performed, the surface should be flat, and its orientation with respect to the electron beam and X-ray detectors should be accurately known. Furthermore, when the thermal and/or electric conductivity is low, heating or electrostatic charging must be avoided, usually by means of conductive coatings applied to the target surface. However, the application of conductive coating introduces extraneous material in the region of X-ray emission. Complications may also arise in the analysis of thin films or small particles from the presence of mounting materials or supporting films in the region of emission or from decomposition or evaporation of target material under the electron beam.

Many techniques used to grind and polish specimens for observation in the light microscope [16.3] can also be used for the preparation of electron probe targets such as alloys, minerals, ceramics, glasses, bones, and teeth. Brittle structures, such as some oxidation layers, can be reinforced by impregnation with hard plastics, provided that these do not produce interfering X-ray lines [16.4]. A detailed description of methods for the preparation of grains and solid minerals has been given by Taylor and Radtke [16.5].

The importance of flatness and the procedures recommended for polishing were discussed in detail by Yakowitz [16.6]. The preparation of the flat surface is usually accomplished in several steps. If the specimen must be sawed or ground, care must be taken that its structure is not altered by heat. The choice of polishing media and speed depends on the hardness of the specimen. A sequence starting with 600-grit silicon carbide papers and finishing with 1/4-μm diamond powder or an equivalent alumina polish is frequently successful. The presence of minor scratches in the final specimen is of no consequences, but checking the specimen flatness by interferometry is often useful. Sharp steps can often be observed at the interface of phases of greatly differing hardness. If these steps exceed 1 or 2 μm, their presence may cause variations in the observed X-ray intensity due to spectrometer defocusing. Even more harmful is the change in the absorption of emitted radiation of long wavelength, due to relief, produced in polishing and etching where applicable. Electropolishing is particularly dangerous in this respect. Conglomerates of dissimilar phases, such as meteorites containing both stony and metallic grains, are very difficult to prepare. The analysis of inclusions [12.74] is complicated by their tendency to be pulled out of the specimen during specimen preparation. This artifact can be reduced by using as little as possible of lubricant during the polishing, or, in extreme cases, by electropolishing the specimen.

If small-phase regions such as inclusions are embedded in an opaque matrix, the depth to which such bodies extend cannot be determined. Hence, there is the danger that underlying matrix regions will also be excited. The best procedure in such a case is to analyze many inclusions or precipitates, and to reject anomalous data. In critical cases it can be attempted to analyze the region of interest at various operating potentials in order to vary the depth of penetration of the electron beam.

The preparation of soft specimens such as copper, gold, lead, or indium is complicated by the rounding of edges, smearing, and the embedding of abrasive particles such as silicon carbide. In extreme cases, the preparation of surfaces by means of a microtome may be required. This procedure gives excellent results; it is, however, limited to small-surface sectors due to the limitations in size of diamond knifes.

In all polishing procedures, the analyst must be aware of the possibility of polishing material remaining, in spite of rigorous rinsing in holes and crevices of the specimen. For this reason, diamond abrasive is advantageous since in most cases the presence of carbon is not investigated.

Complications in the analysis can also arise from the mounting media. Although occasionally amalgams or other alloys have been used for holding the specimen, the commonly used materials are organic polymers such as diallyl phthalate, phenolic (bakelite), and epoxy resins. Liquid polyesters or cold-setting epoxy resins can be employed where heating and pressures required with phthalate and bakelite are objectionable. Such organic materials are softer than many specimens embedded in them, and are worn away faster in the polishing process. The resulting rounding of the edges of the specimen can be objectionable, particularly if analyses must be performed close to the border of the specimen. To reduce such effects, the analyst may use mounting materials which contain fillers, such as glass fibers or metal particles. The contamination of the specimen by the elements present in the filler must, however, be guarded against. Where edge conservation of the specimen is important, it can be surrounded by a piece of steel tubing so that after embedding the steel ring determines the surface of the polishing action. In critical cases, e.g., in the investigation of a traverse cut through a specimen the surface of which is of interest, a protective piece of steel or another hard material can be glued with epoxy resin against the surface to be investigated, before the assembly is embedded. The degree of preservation of edges also depends, of course, on the polishing media and technique, and general guidelines cannot replace the virtuosity of an expert in specimen preparation.

16.2 ELECTROSTATIC CHARGING

To maintain a stable regime of excitation, the negative charge transferred to the specimen by the electron beam must be removed. Therefore, the specimen surface must be electrically conductive. This condition is naturally present in metallic specimens. For the low currents typically used in electron-probe microanalysis, moderately conductive specimens, including semiconductors, can be handled without special precautions. If conductive specimens are embedded in insulating materials such as bakelite, a conductive path to ground must be created by painting part of the surface with aquadag (a colloidal suspension of graphite) or by a conductive silver paint. For large specimens, a hole can be drilled from the reverse of the mount which reaches the specimen within the mounting material. The hole is filled with Wood's metal which is then connected to the specimen holder. The entire specimen can also be mounted in Wood's metal or in a mercury amalgam. Conductive resins such as bakelite and diallyl phthalate with diverse metal fillers are also commercially available.

In some operations, particularly when area scans are performed, nonconductive regions of the specimen may charge up accidentally if located close to the intended target. Such charged regions may produce distortion and deflection of the beam, or intermittent discharge (see Figs. 14.17 and 14.18). These problems are avoided if the entire specimen surface is coated with a conductive layer. Specimens which should be coated include ceramics, glasses, many oxides, most minerals, and biological tissues. The usual procedure is evaporation of a layer of 100 to 200 Å of carbon or of a metal such as aluminum or copper. The layer must be thin enough to permit the passage of the primary beam electrons. The attenuation of emitted X-rays by the layer is often equally significant. Beryllium, in view of its toxicity, is not a recommendable choice for coating specimens. Carbon is preferred where the transparency of the layer facilitates the orientation of the analyst on a flat specimen surface by microscopic observation. Table 16.1 shows the average energy loss, in kilo electron volts, of an electron which has crossed at a right angle layers of carbon of 200 Å and 1000 Å thickness, and the resultant loss of intensity of generated X-rays, for two elements.

Table 16.1. Mean Energy Loss of Electrons Crossing Carbon Layers, According to Whiddington [Eq. (9.2.31)], and Intensity Losses for Al Kα and Fe Kα Generation, According to Duncumb [Eq. (9.2.33)].

	200 Å			1000 Å		
E_0 (keV)	ΔE(keV)	Al Kα	Fe Kα	ΔE(keV)	Al Kα	Fe Kα
1	0.225	–	–	1.000	–	–
2	0.13	0.000	–	0.82	0.000	–
5	0.07	0.966	–	0.37	0.827	–
10	0.05	0.992	0.971	0.23	0.955	0.871
20	0.03	0.997	0.996	0.15	0.986	0.981
50	0.02	0.999	0.999	0.09	0.997	0.996

Note: Numbers under symbols for X-ray lines are ratios of generated X-ray intensity of coated and uncoated specimens.

Table 16.2. X-Ray Intensity Losses Due to Absorption of X-Rays in Carbon and Aluminum Films (ψ = 45°).

	CARBON			ALUMINUM		
	μ	200 Å	1000 Å	μ	200 Å	1000 Å
Na Kα	1534	0.87%	4.25%	1021	0.78%	3.83%
Al Kα	406	0.28%	1.38%	406	0.31%	1.54%
Ca Kα	42.1	0.017%	0.12%	432	0.33%	1.64%
Cu Kα	4.6	1.8×10^{-5}	1.3×10^{-4}	49.6	3.8×10^{-4}	0.19%

Table 16.2 indicates the losses due to absorption of emitted X-rays on their path through the coating. Errors from both these losses can be minimized if the thickness of coating of standards and specimen is strictly controlled. It is best to mount the standards directly with the specimens where possible, and to coat both simultaneously.

The materials transferred in coating by evaporation in a good vacuum move in a straight line from the source to the target. In specimens which are not flat and may have reentrant parts it is often preferable to use ion-sputtering devices, which produce a more uniform coating in such cases. This is particularly important when particles which do not conduct heat very well are mounted on substrates with which they have poor thermal contact.

16.3 DAMAGE TO THE SPECIMEN

The quantitative analysis of a specimen is impossible if the target changes its composition under the impact of the electron beam because of evaporation of target components, deposition of impurities, chemical changes, or movement of ions within the specimen. The production of heat in the electron-probe target was considered by Castaing [1.1] in his thesis and in later publications [3.22]. The simplest case is that of a massive specimen irradiated by a beam of diameter r_0. The maximum temperature rise for such a target is

$$\theta_m = \frac{W_0}{4\pi J \cdot C_t} (3/r_0 - 2/R) \simeq 3 W_0/(4\pi J \cdot C_t \cdot r_0) \qquad (16.3.1)$$

where $R \gg r_0$ is the radius of the specimen, C_t its thermal conductivity, W_0 the power carried by the beam, and J is the mechanical equivalent of heat. Castaing [3.22] also discusses the conditions of heating in a thin film; a more detailed analysis of the thermal regime of targets including coated targets, was made by Almasi et al. [16.7]. In terms of operating conditions, the above equation translates to

$$\theta_m \simeq 10\ V_0 i C_t^{-1} r_0^{-1} \qquad (16.3.2)$$

where i is the beam current in microamperes, r_0 is the beam radius in micrometers, V_0 is the operating potential in kilovolts, and C_t is the thermal conductivity in W/(cm · °C). Castaing calculated that at a potential of 30 kV and a current of about 0.5 μA, the temperature rise in copper caused by a beam 1 μm in diameter would be 18°C. Although in minerals the temperature rise would be larger, it does not present a problem at typical beam currents and acceleration potentials, since instruments with efficient spectrometers rarely warrant the use of currents higher than 10^{-7} A. Unfortunately, the above calculations are of little use in those cases in which the problem of heat transfer is most serious. Although the energy transferred in the electron-probe microanalyzer is very small, the electron density at the specimen level is much higher than, for instance, in an X-ray tube. If the specimen is a particle loosely affixed to a substrate or a thin slice of soft biological tissue, the thermal contact with the surroundings of the excited specimen region is often poor. Under such conditions catastrophic rises in temperature are not uncommon; silicate particles may be heated to glowing, and biological tissue can be destroyed instantly. Therefore, the analysis of such specimens requires a careful consideration of the thermal regime in the target region.

The resistance to heating of the specimen depends greatly on its composition. Metals are virtually never affected; minerals may lose volatile components such as crystallization water and carbon dioxide; halides and other compounds may react with metal coatings; sulfur and selenium may evaporate. Organic specimens are often seriously affected. If metal coatings are applied, the evaporation of volatile components may produce blistering of the coating, which, in turn causes the heat conductivity to deteriorate locally. Some organic materials used to mount specimens may, under the beam, splatter fragments which can seriously deteriorate the optical microscope objective, or cause beam astigmatism due to local charges, e.g., in apertures. In some glassy materials [12.78], craters may be formed at the point of impact of the electron beam. Materials used as cathodoluminescent screens for beam alignment such as benitoite are often damaged because current densities used in alignment procedures are much higher than in normal usage.

Variation in time of the apparent concentrations of sodium and potassium is frequently observed both in minerals [16.8] and in glasses [16.9]-[16.12], [12.110], upon irradiation with electrons. Usually the signal of these elements diminishes quickly. An explanation for this target behavior was given by Lineweaver [16.11] who argued that the electron beam produces below the conductive coating a negative charge which causes migration of the sodium ions away from the surface. This theory, however, does not explain the increase in the signal from these elements which is sometimes observed subsequent to the initial drop. Borom and Hanneman [16.12] established the hypothesis that the conductive coating is damaged or destroyed in the process, with subsequent diffusion of sodium to the surface. They also showed that the behavior of mobile ions in glasses is strongly affected by minor changes in the composion of the glass. Such instabilities are the major problem in the analysis of glasses containing alkali metals. For this reason, the glasses used as standards should contain little or no sodium or potassium. As in other cases of beam damage, the error caused by these ion migrations can be reduced by using low currents, large beam diameters, and by moving the beam on the surface during quantitation so that continuously new areas are exposed to the beam.

Butt and Vigers [16.13] reported a sixtyfold increase in the sodium signal from the mineral sepiolite after 15 min of exposure to the electron beam, while in other minerals such as nepheline the signal decreased. The increase was shown to be due to movement of the sodium ion in the zeolitic water bound in sepiolite, and was arrested when the mineral was dehydrated prior to analysis. It is probable that at least in some glasses the high temperature produced in the region of the beam impact is required to render the movement of the sodium and potassium ions possible.

16.4 THE PREPARATION OF BIOLOGICAL TISSUE

As mentioned in Chapter 13, the main difficulty in the analysis of soft biological tissue is the high concentration of water in the specimen, the removal of which cannot be achieved without some changes in the spatial distribution of some of the elements present. Such "diffusible" elements include the electrolytes dissolved in the tissue fluids as well as some elements loosely bound to the tissue.

If only elements which stay fixed are of interest, the common histological procedures may be followed. The tissue is first fixed, i.e., treated so as to permanently conserve its structure. A large variety of chemical fixatives can be employed [16.14]. In this process, most of the electrolytes (sodium, potassium, magnesium, calcium, chlorine) are removed [16.15]. Some of the macromolecular compounds may be displaced, and some compounds present in the tissue may be dissolved, depending on the method of fixation. Elements present in the fixing agents, even if present at trace levels, may also contaminate the specimen. Except for tissues fixed and hardened in aldehydes, the specimen must be embedded after fixation in a medium such as paraffine. In the dehydration which must precede the embedding, lipids and other compounds may be dissolved and hence elements associated with them may be removed.

Such chemical preparation procedures can be avoided if the specimen is fixed cryogenically. The water can then be eliminated by freeze-drying [16.16] and the specimen embedded for cutting, or the frozen tissue may be cut without previous dehydration, and the water is then removed from the microtomed tissue. The specimens are mounted on plastic foils or metal grids. If paraffin is present, it must be eliminated before the analysis. The specimen is then coated by evaporation of carbon.

The use of histological stains for easier identification of microstructure must usually be avoided, since elements may be displaced or added in the process. Therefore, it may be difficult to localize in the tissue preparation the microscopic features of interest. It is sometimes advantageous to stain a tissue section adjacent to that which is being analyzed, in order to localize such features.

16.16 REFERENCES

16.1 Blöch, R., Kulmburg, A., and Swoboda, K., *Mikrochim. Acta* (Suppl. I), 232 (1966).
16.2 Dörfler, G., Blöch, R., and Plöckinger, E., *Arch. Hüttenw.* **37**, 375 (1966).
16.3 Kehl, G. L., *The Principles of Metallographic Laboratory Practice*, 3rd ed., McGraw-Hill, New York, 1949.
16.4 Reichard, T. E. and Coakley, W. S., *Anal. Chem.* **37**, 317 (1965).
16.5 Taylor C. M. T. and Radtke, A. S., *Economic Geol.* **60**, 1306 (1965).
16.6 Yakowitz, H., ASTM Special Tech. Publ. 430, American Society for Testing and Analysis, Philadelphia, Pa., 1968, p. 383.
16.7 Almasi, G. S., Blair, J., Ogilvie, R. E, and Schwartz, R. J., *J. Appl. Phys.* **36**, 1848 (1965).
16.8 Ribbe, P. H. and Smith, J. V., *J. Geol.* **74**, 217 (1966).
16.9 Varshneya, A. K., Cooper, A. R., and Cable, M., *J. Appl. Phys.* **37**, 2199 (1966).
16.10 Estour, H., *Verres Réfract.* **25**, 11 (1971).
16.11 Lineweaver, J. L., *J. Appl. Phys.* **34**, 1786 (1963).
16.12 Borom, M. P. and Hanneman, R. E., *J. Appl. Phys.* **38**, 2406 (1967).
16.13 Butt, C. R. M. and Vigers, R. B. W., *X-Ray Spectrometry* **6**, 144 (1977).
16.14 Gray, P., in *Encyclopedia of Microscopy and Microtechnique*, Gray, P., Ed., Van Nostrand Reinhold Co., New York, 1973, p. 165.
16.15 Truchet, M., *J. Microscopie Biol. Cell.* **22**, 465 (1975).
16.16 Middleton, E., in *Encyclopedia of Microscopy and Microtechnique*, Gray, P., Ed., Van Nostrand Reinhold Co., New York, 1973, p. 198.

Appendix • STATISTICS

A complete treatment of statistics would be beyond the scope of this book. The succinct description of statistical terms given in this Appendix has the purpose of defining those which are used in the text. For a more extensive treatment of statistics, the reader is referred to [A.1]-[A.3].

If a measurement is repeated under constant and controlled conditions, variations in the measured values are observable, unless the scale of measurement was too coarse, or the readout system too insensitive. The statistical techniques deal with the limitations of *precision* (repeatability) of the measurement which arise from these statistical (random) fluctuations.

Besides random fluctuations, measurements are usually also affected by *systematic* (recurring) errors, which may severely limit the *accuracy* (correctness) of the result of the measurement, even though the precision may be very high. The statistical devices to be described are used to set limits to precision, rather than the accuracy, of a measurement.

A system of measurement is called *stable* if the measurement can be repeated without systematic change of conditions. A set formed by all possible measurements is called a *population* (or universe). A limited (incomplete) group of measurements drawn from this population is called a *sample*. To be representative of the population, a sample must be drawn at random [A.2].

Populations can be of infinite size (e.g., measurements of temperature at the National Airport, Washington, D.C.), or finite (present ages of the members of the U.S. Senate).

A sample of measurements can be represented by a frequency-distribution diagram. The scale of values of the measured quantity is plotted horizontally, and the observed frequencies vertically. The measured quantity may vary continuously (e.g., the weight of objects), or discontinuously (e.g., number of children per family). Continuously varying quantities must be grouped into classes, or ranges, of equal size (e.g., weight to the nearest gram). Discontinuously varying quantities may also be grouped into classes, by choosing as class interval a multiple of the unit of change. For a comparison of frequency diagrams of samples of different sizes, the frequency axis can be normalized so that the total sample size is equal to unity at the normalized scale. If this procedure is extrapolated for the plotting of infinite populations, then the vertical scale represents probabilities of the occurrence of measurement values within the classes which are represented.

Due to random fluctuations, the frequency distributions of two samples drawn from the same population usually differ. The differences diminish, however, with increasing sample size, and both distributions approach that of the population. This trend is particularly easy to observe in situations such as the throw of a well-balanced dice, in which the outcome of a large number of events can be theoretically predicted.

The characteristics of frequency distributions can be defined by means of certain parameters called statistics. Consider a population constituted of n measurements. Call x_i the value of the measured quantity obtained in the ith measurement. We define the following statistics.

The arithmetic mean (first moment):

$$\mu = \frac{1}{n} \sum_{i=1}^{n} x_i. \tag{A.1}$$

The variance (second central moment):

$$\sigma^2 = \frac{1}{n} \sum_{i=1}^{n} (x_i - \mu)^2. \tag{A.2}$$

The third central moment (in standard units):

$$\alpha = \frac{\mu_3}{\sigma^3} = \frac{1}{\sigma^3} \frac{1}{n} \sum_{i=1}^{n} (x_i - \mu)^3. \tag{A.3}$$

The fourth central moment (in standard units):

$$\beta = \frac{\mu_4}{\sigma^4} = \frac{1}{\sigma^4} \frac{1}{n} \sum_{i=1}^{n} (x_i - \mu)^4. \tag{A.4}$$

The *mean* is a measure of the center of the frequency distribution. It is considered, in the absence of systematic errors, to be the best estimate of the magnitude being measured. For other central parameters (median and mode) see [A.1]. The *central moments* characterize the distribution of frequencies around this mean.

The *variance* is a measure of the spread of values around the mean. It is always positive, and increases with increasing spread of the results of measurement. Positive and negative deviations from the mean have equal effects on the value of the variance. Its dimension is the same as x^2. The positive square root of the variance, σ, is called *standard deviation*. A relative (dimensionless) measure is the relative standard deviation (coefficient of variation), $V = \sigma/\mu$. The relative standard deviation is sometimes expressed as a percentage: $V(\%) = 100\,\sigma/\mu$.

The third central moment is a measure of the skewness (asymmetry) of the frequency distribution. For symmetrical distribution curves, this moment is zero. Positive values of the third central moment indicate a broadening of the distribution toward the high values.

The fourth central moment, which is always positive, is a measure of peakedness (kurtosis). The values of the third and higher moments are strongly affected by the relatively few events located in the tails of the distribution curve. Their values have therefore meaning only for large populations.

Models for Distributions. Several mathematical models have been proposed for frequency distributions of populations. The *Gaussian* or *Normal* distribution reproduces closely many distributions found in populations of real measurements. The equation describing this distribution is

$$dF = y \cdot dx = \frac{1}{\sigma\sqrt{2\pi}} e^{-(x-\mu)^2/2\sigma^2} \cdot dx \qquad \text{(A.5)}$$

where dF is the frequency of the variable from the value x to $x + dx$, and e is the base of the natural logarithms. The equation can be written in a standardized form by defining:

$$u = \frac{x - \mu}{\sigma}$$

so that

$$dF = y \cdot du = (2\pi)^{-\frac{1}{2}} \cdot e^{-u^2/2} du \qquad \text{(A.6)}$$

The Normal curve is defined by two parameters: the arithmetic mean, μ, and the standard deviation, σ. The curve is symmetrical (Fig. A.1), peaks at the point $x = \mu$, and extends from negative infinite to positive infinite. The points of inflexion at both sides from the maximum are at a distance of $\pm\sigma$ from the mean. The total area under the curve represents the probability that a particular value of x will fall between plus and minus infinite, and is thus equal to unity. The area under the curve between the limits $\mu - \sigma$ and $\mu + \sigma$ (i.e., the probability for any value of x to fall between these limits) is close to 0.67, the range between $\mu - 2\sigma$ and $\mu + 2\sigma$ covers 95% of the total area, and that between $\mu - 3\sigma$ and $\mu + 3\sigma$ covers close to 99.25%. Hence, there is a probability of less than 1% that any value of x will fall outside the 3σ range.

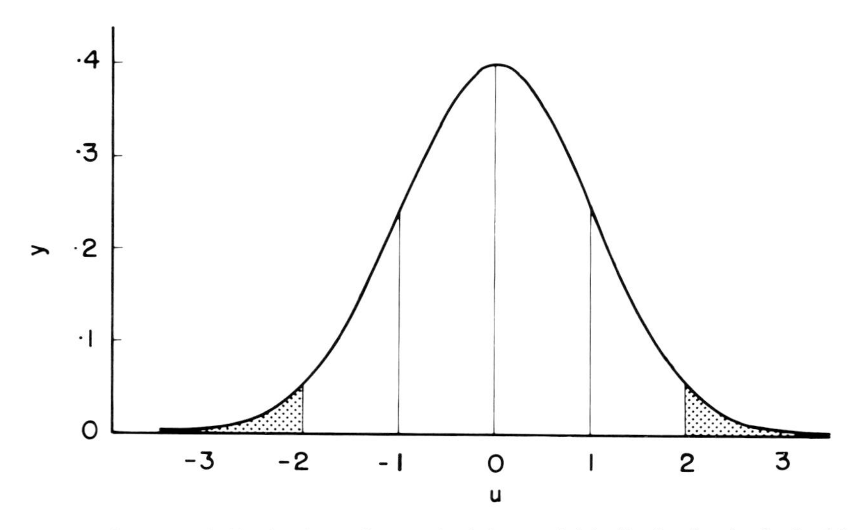

Fig. A.1. The Normal distribution. The standard form of this distribution is obtained by defining a standardized abscissa $u = (x - \mu)/\sigma$. The abscissa y is equal to dF/du. By integrating y over a range of u, a numerical value is obtained that indicates the probability that a value of x falls within the range. For instance,

$$\int_{-2}^{+2} y \cdot du = 0.95$$

therefore, there is a probability of 0.95 that a value of x is between $\mu - 2\sigma$ and $\mu + 2\sigma$.

Since the Normal curve is symmetrical, its third central moment, α, is zero. The value for the fourth central moment, β, is three.

The Normal distribution is frequently a good model for the distribution of measurements affected by random variation. However, not all distributions follow this model, and in general, proof of applicability of this distribution is necessary before its use. A distribution, to approach a Normal model, must be continuous, have a single maximum, and be symmetrical.

The *Poisson distribution* is another model, of great importance in counting statistics. It can be used to predict the frequency of certain events—such as the detection of a photon by a detector—which occur frequently, but at intervals which are, on the average, much longer than the events themselves. The equation of the Poisson distribution is

$$y = \frac{\mu^x}{x!} e^{-\mu} \qquad \text{(A.7)}$$

where the symbols y, x, μ, and e have the same meaning as in the equation for the Normal distribution.

In the application to the probability of events, y is the probability that x events will occur in a predetermined time interval. It follows that x must be a nonnegative integer. This is an important difference between the Poisson distribution and the Normal distribution; in the latter, x could assume noninteger or negative values. However, the mean number of events in the given time period, μ, will not, in general, be an integer. The possible values for x extend from zero to positive infinity. Consequently, the Poisson distribution is skewed. In contrast to the Normal distribution, the Poissonian distribution is defined by a single parameter. It can be shown that the standard deviation of a Poissonian is given by the equality

$$\sigma^2 = \mu. \tag{A.8}$$

In other words, the variance is numerically equal to the mean. It follows that the relative standard deviation is related to the mean by

$$V = \mu^{-1/2}.$$

The higher moments of the Poissonian are as follows:

$$\alpha_3 = \mu^{-1/2}, \quad \alpha_4 = 3 + \mu^{-1}$$

It follows that the skewness diminishes with increasing values of the mean, μ. The shape of the Poisson curve thus becomes increasingly similar to that of the Gaussian curve (Fig. A.2). For values of the mean above 30, little error will be committed in assuming that the curves are identical. We can therefore apply to the Poisson curve the same confidence limits (i.e., ranges within which the probability of occurrence of a particular measurement has a certain value) as previously established for the Normal curve: 0.67 for $\mu \pm \sigma$, 0.95 for $\mu \pm 2\sigma$, and 0.9925 for $\mu \pm 3\sigma$.

The frequency distribution of events will tend to be Poissonian only when the individual events are not causally interrelated. Thus, the primary photons produced in an X-ray target and observed by a detector tend to have a Poissonian distribution, since there is no interaction in the production, propagation, and detection of individual X-ray photons. On the contrary, the number of electrons in a detector avalanche will not be distributed as a Poissonian, since the production of many electrons is related to the same primary ionization event.

Estimates of Statistics of a Population. It follows from the definitions of the statistics of a population [Eqs. (A.1)-(A.4)] that their values cannot be determined unless the entire population is known, or unless they can be predicted by some theoretical assumption not embodied in these equations. However, in practical cases, frequently only a limited sample of a large population is known. In such a case, we must obtain, from the statistics of the sample, an estimate of the statistics of the population, which is the object of the measurement.

The sample mean, of a sample of size n

$$\overline{x} = \frac{1}{n} \sum_{i=1}^{n} x_i \tag{A.9}$$

in an unbiased estimate of the population mean; there is no reason to assume that sample means should, on the average, be either larger or smaller than the population mean. On the contrary, the variance of the sample,

$$\sigma^2_{\text{sample}} = \frac{1}{n} \sum_{i=1}^{n} (x_i - \overline{x})^2 \tag{A.10}$$

is a measure of the dispersion of the sample values around the sample mean, $\overline{x}$, which is smaller than the dispersion of these values around the population mean, μ.

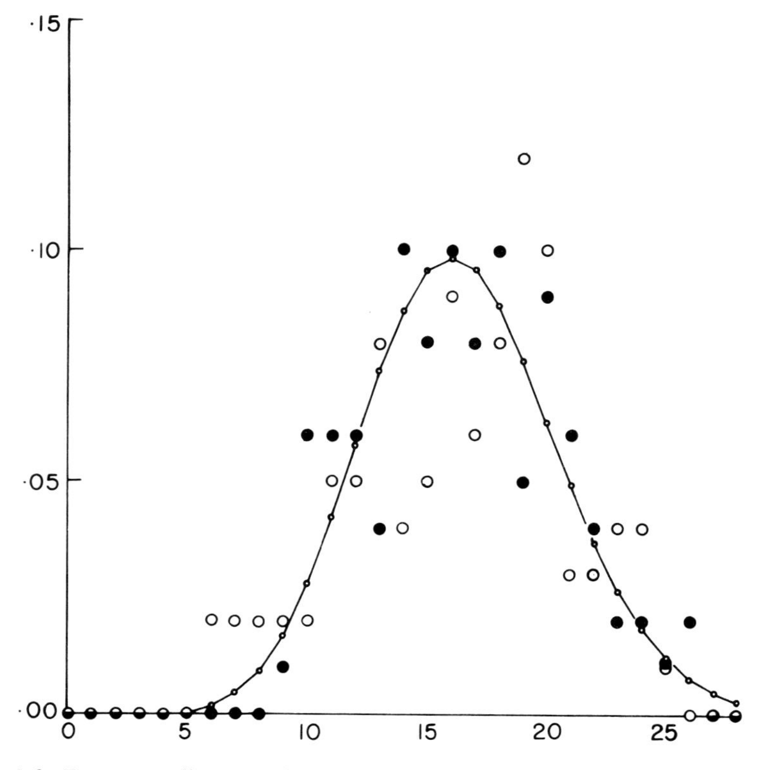

Fig. A.2. Frequency diagrams of two groups of 100 measurements each of a counting experiment in which the average count was μ = 16.5 (black and white points), and the corresponding Poissonian (connected points).

Therefore, the variance of the sample is not an unbiased estimate of the population variance. To obtain such an unbiased estimate, s^2, the sample variance must be multiplied by the factor $n/(n-1)$:

$$s^2 = \frac{1}{n-1} \sum_{i=1}^{n} (x_i - x)^2. \tag{A.11}$$

Because the mean of a sample, $\overline{x}$, differs from the mean of the population, μ, it is of interest to know the standard deviation of the sample means with respect to the population mean. This statistic, commonly called standard error of the mean, is obtained from the standard deviation of the population by the relation

$$\sigma_{\overline{x}}^{\,2} = \sigma_x^{\,2}/n. \tag{A.12}$$

Even in cases where the variable x is not Normally distributed, and even for small samples, the distribution of the sample means, $\overline{x}$, is close to being Normal.

As mentioned previously, levels of probability called confidence levels can be established for ranges defined by a multiple of the standard deviation. Such confidence levels are, however, valid only if the standard deviation is well known. If the standard deviation of the population is estimated from a small sample, the resulting uncertainty produces an uncertainty in the confidence levels, for which the ranges have to be widened. The confidence levels are, in such a case, obtained by means of Student's t distribution [A.1].

It should be noted, however, that the use of the factor $n/(n-1)$ in the estimation of the population variance, and the use of Student's distribution for confidence intervals, are valid only for the estimation of population statistics from a limited sample. They are not applicable when a distribution can be considered Poissonian, and the population variance is thus believed equal to the sample mean. The sample mean of a Poissonian is an unbiased estimate of the population mean, and it is accurate enough to justify the use of the same confidence limits as given for a known Normal population.

A Quick Technique for Estimating the Standard Deviation of a Small Sample. For a sample of two to ten measurements, a quick estimate of the standard deviation is obtained by dividing the range (difference between largest and smallest value) by the square root of the number of values:

$$s = R \cdot n^{-1/2}. \tag{A.13}$$

Although such a test is not meant to replace the more accurate procedure (Eq. (A.11)], it is useful for a quick test of groups of measurements since it is very easy to recall and perform.

Propagation of Errors. If a function depends on various independent parameters, the effects of errors or uncertainties in these parameters on the precision of the value given by the function can be determined as follows. Let f be a function of various independent variables:

$$f = f(x, y, \ldots).$$

If the distributions of random errors of all independent variables can be characterized by the standard deviations σ_x, σ_y etc., the variance of the function f can be determined by

$$\sigma_f^2 = \left(\frac{\partial f}{\partial x}\right)^2 \sigma_x^2 + \left(\frac{\partial f}{\partial y}\right)^2 \sigma_y^2 + \ldots. \tag{A.14}$$

For instance, if $f(x,y) = x/y$, then

$$\sigma_f^2 = \frac{1}{y^2}\sigma_x^2 + \frac{x^2}{y^4}\sigma_y^2 \quad \text{or} \quad \frac{\sigma^2{}_f}{f^2} = \frac{\sigma^2{}_x}{x^2} + \frac{\sigma^2{}_y}{y^2}.$$

For further information, see [12.12].

APPENDIX. REFERENCES

A.1 Snedecor, G. W. and Cochran, W. G., *Statistical Methods*, 6th ed., Iowa State University Press, Ames, Iowa, 1967.

A.2 Natrella, M. G., *Experimental Statistics*, National Bureau of Standards Handbook 91, U.S. Government Printing Office, Washington, D.C., 1966.

A.3 Davies, O. L., *Statistical Methods in Research and Production*, Hafner Publishing Co., New York, 1957.

Name Index

Subject Index